A

CYCLOPÆDIA

OF

QUANTITATIVE CHEMICAL ANALYSIS.

BY

FRANK H. STORER, A. M.,

PROFESSOR OF GENERAL AND ANALYTICAL CHEMISTRY IN THE MASSACHUSETTS INSTITUTE OF TECHNOLOGY.

SEVER, FRANCIS & CO.,

Boston and Cambridge.

1870.

PRESS OF A. A. KINGMAN, BOSTON.

PREFACE TO PART I.

The object of the author in compiling this book has been not only to provide the student and working chemist with a comprehensive dictionary of quantitative processes, but to call the attention of the chemical fraternity to the question of the possibility of presenting this branch of chemical art in a more serviceable and manageable form than has been customary hitherto. The experiment is certainly worth the trying whether a definite system of classifying substances in alphabetical order, and of referring each and every process to the fundamental fact or *principle* upon which it depends, will not greatly facilitate both the study and the practice of analysis.

The difficulty of perfecting the first edition of a work of this kind will be manifest to all. The author has consequently no apology to offer for the manifold short comings of his book. He wishes it to be understood distinctly that he has copied freely from the Handbooks of Rose, Fresenius, Berzelius, and Pfaff; from the Handwœrterbuch der Chemie, and from the Dictionary of Watts. It is to be remarked, in that connection, that the similarity of wording in the descriptions of analytical processes in these various treatises is often so close that it would frequently be very difficult to determine precisely where credit should be accorded, or to judge how much credit should be given to each author. It is evident that in a subject which depends so nearly upon the precise statement of details, no compiler would be likely to depart far from the original description, as set forth by the inventor of a process. A good illustration of this inevitable tendency to repetition may be seen in the similarity of Pfaff's, Rose's and Fresenius's descriptions of Berzelius & Hisinger's process depending upon the insolubility of succinate of iron. References are given in all cases where matter has been taken from the journals and other original sources.

The manuscript of Part I. was completed, as now printed, in June, 1869. Nothing has been added to it, whether as regards matter or arrangement, from either of the treatises which have been published since that date, excepting the description, under Carbonate of Calcium, of Prof. Lawrence Smith's process for estimating alkalies,—copied from Prof. Johnson's edition of Fresenius's Quantitative Analysis, New York, 1870.

It is noteworthy that the tendency of all the works recently published on quantitative analysis is towards condensation and abbreviation, while the aim of the present book is to show that perspicuity can be best gained by amplification, if need be, and *methodical arrangement.* The author believes that the interests of chemists and of chemical students alike demand two kinds of books upon quantitative analysis. The one kind looking to completeness in all directions, while the other is given over either to specific instruction, or to the discussion of special applications of analysis in some one of the various departments of Chemistry.

It is not only important that the analyst should have a dictionary of all known methods, from which to choose the one which seems best to fulfil the conditions and requirements of any new problem which may come before him, but there will always be needed books or tables devoted to schemes for illustrating the various kinds of analyses which are likely to occur in practice. Such, for example, as the works of Wolff and Caldwell, in the department of agricultural chemistry; or, in its day, the Handbuch of Pfaff, in the branch of mineral analysis. But if we were once in possession of a General Encyclopædia, covering the entire field of analysis, it would be easy to draw up long lists of these special schemes in very few words. For example:—

A residual product from some of the chemical works at Stassfurt, in Germany, sent into commerce to be sold as a fertilizer, contains the sulphates of potassium, magnesium and calcium; the chlorides of potassium, sodium and magnesium; and a quantity of oxide of iron, magnesia and sand, insoluble in water. But the commercial value of any given sample of the substance depends upon the proportion of potassium which is contained in it.

Several methods have been employed for estimating the potassium,—after the matters insoluble in water have been got rid of by filtration. They might be briefly described as follows:—

A. Remove the SO_3 with $BaCl_2$, as Sulphate of Barium; the Mg with BaO, H_2O, as Hydrate of Magnesium; the Ca and excess of Ba with $(NH_4)_2O$, CO_2, as Carbonate of Barium, and estimate the K as Chloride and Chloroplatinate of Potassium.

B. Remove SO_3 by means of $BaCl_2$, as Sulphate of Barium, decompose the Chloride of Magnesium with oxalic acid, and separate Mg as Oxide of Magnesium. Determine K as Chloroplatinate of Potassium.

C. Remove the SO_3 with $BaCl_2$, as Sulphate of Barium; the Ba, Ca and Mg with Na_2O, CO_2, as Carbonate of Barium, etc. Acidulate the filtrate with HCl and estimate K as Chloride and Chloroplatinate of Potassium.

D. Remove the SO_3 with $BaCl_2$, as Sulphate of Barium, and estimate K in the filtrate as Chloroplatinate of Potassium (Stohmann's process).

The author of any such tables would of course give reasons why and when either of the processes would be preferred to the others.

In the appendix to the present work, a few examples for students' practice will be given in this sense.

N. B. All temperatures are given in degrees of the Centigrade thermometer,—excepting when otherwise expressly stated.

In the references to authorities, the larger figures indicate the volume, and the smaller figures the page, of the journal or work alluded to; single figures in parentheses ( ) denote the number of the series of the journal.

The names of authors who have labored in concert are connected by the character &,— not by *and.*

To avoid the repeated printing of unnecessary words, cross references are indicated by printing in capitals the initial letters of the names of the substances referred to. Thus, on page 5, column 1, line 25, the capital initials in the words Oxide of Aluminum indicate that the article OXIDE OF ALUMINUM must be consulted, and that further information will be found under that head.

The names of rare elements have been omitted from this edition simply from lack of time to deal with them.

In order that the size and cost of the book might be kept within reasonable bounds, it has been thought best to exclude from it all figures of apparatus. The descriptions of unusual forms of apparatus have, however, been given in minute detail, and, in case of having to use any of this apparatus, the reader will do well to draw rough figures of it for himself with pen or pencil,—following the printed description, step by step. It is believed that if the descriptions be dealt with in this way, no great difficulty will be found in comprehending them, or in choosing that one among several processes which is best suited to the wants of the operator.

All ordinary implements and apparatus will be best understood from the descriptions given in works devoted specially to chemical manipulation.

Boston, July, 1870.

A CYCLOPÆDIA

OF

QUANTITATIVE CHEMICAL ANALYSIS.

Acetic Acid.

Principle I. Power of neutralizing alkaline solutions.

Applications. Estimation of free acetic acid in vinegar, pyroligneous acid and other aqueous solutions. (Method A). Determination of acetic acid in certain acetates from which caustic soda precipitates insoluble hydrates or oxides. (Method B).

Method A. A weighed or measured quantity (10 grms. of vinegar will be enough in most cases) of the solution to be examined is reddened slightly with litmus, and then treated with test-alkali until the whole of the acid is neutralized and the color of the litmus changed to blue. (See Acidimetry).

A solution of caustic soda is usually employed as the test-alkali, though solutions of carbonate of sodium were formerly much used. Greville Williams (*Pharm. Journ. Trans.*, 1854, **13.** 594), has urged that a standard solution of lime in sugar water is to be preferred to soda, and Otto (in his *Lehrbuch der Essigfabrikation*, 1857, p. 77) has shown that dilute ammonia water may sometimes be employed with advantage. Other test-liquors are employed in particular cases, as will be described below.

The test-alkali is added to the acid under examination until the color of the litmus appears distinctly blue, or until a drop of the alkaline liquor is no longer seen to form a blue spot when it falls into the colored liquid.

It has been objected to the use of soda or of potash in estimating acetic acid, that the process must be inaccurate, since the normal acetates of the alkali metals themselves exhibit a slight alkaline reaction with litmus, or, at all events, change the color of litmus to violet. But as Otto (*Annal. Chem. und Pharm.*, 1857, **102.** 71) has shown, the error from this source is of but little significance. It was found not to exceed one-tenth of one per cent in determinations made with solutions containing ten per cent of the monohydrated acid, and is usually even smaller than this, since a slight excess of test-alkali is almost always added to the solution under examination. This source of inaccuracy may practically be eliminated by using a standard soda solution, the value of which has been determined beforehand by means of pure acetic acid. To prepare such a solution, measure off a determined quantity of standard sulphuric acid, mix it with a moderate excess of acetate of sodium, and determine how much of the standard soda solution is required to neutralize the acid mixture. By operating with a soda solution, whose power of changing red litmus to blue in presence of acetate of sodium has been thus determined empirically, the danger of error from the alkaline reaction of the acetate may be in great measure avoided.

According to Merz (*Journ. prakt. Chem.*, **101.** 301), it is well to use a solution of turmeric instead of litmus as the indicator; for acetate of sodium has no action upon the color of turmeric. Enough of the turmeric solution is mixed with the acid to be tested to color it bright yellow; the liquid is heated nearly to boiling, and the soda solution stirred in rapidly until the moment when the color of the yellow liquid changes to brown.

According to Greville Williams, the best way is to dispense with soda altogether, and to employ in its place a solution of lime in sugar water (see Acidimetry); for acetate of calcium has not nearly so much power to disguise the action of acids upon litmus as the acetates of sodium and potassium have, to say nothing of the advantage which the lime solution necessarily possesses, in being at all times free from carbonic acid. For ordinary rapid work in testing vinegars, Williams finds that the process may be fully relied on to one-fourth of one per cent.

In testing crude pyroligneous acid, or highly colored vinegar, the color of the litmus added to the liquid is so much obscured towards the close of the operation by the impurities with which the acid is charged, that the exact point of neutralization cannot be determined by mere inspection of the liquid. It is necessary, as the point of saturation approaches, to test the liquid with litmus paper. To this end, after each fresh addition of the alkali, the point of the glass rod used for stirring is drawn across a narrow sheet of litmus paper, and from the color of the mark thus formed, the progress of the neutralization is inferred. When the paper ceases to be reddened by the liquid, the operation is finished. For some eyes turmeric paper may be advantageously substituted for the litmus paper. A brown spot or ring will form upon the yellow paper as soon as the acid has all been neutralized and the slightest excess of free alkali is present in the liquid.

In the case of vinegars so highly colored that the point of neutralization cannot be satisfactorily determined even with litmus paper, the proportion of acetic acid may be found, according to Carl Mohr, as follows:—Boil a measured sample of the vinegar with a weighed excess of carbonate of barium until no more carbonic acid is given off. Separate the dark colored, soluble acetate of barium by filtration; throw the moist filter, with its contents of carbonate of barium into a beaker, and pour upon it a measured volume of standard nitric acid, more than sufficient to dissolve the whole of the carbonate. Finally determine the free nitric acid with a standard solution of soda, and calculate the proportion of acetic acid from the quantity of carbonate of barium which the nitric acid dissolved. The process is evidently liable to error, inasmuch as acetic acid is readily volatile at the temperature of boiling.

F. Mohr (*Titrirmethode*, 1855, p. 362) claims to have obtained good results in estimating acetic acid by means of the standard solution of ammonio-sulphate of copper, proposed by Kieffer (see Acidimetry), as a substitute for test-alkali. In using this solution, however, the acetic acid must be so highly diluted that a precipitate shall form immediately, from the very first, at the point where the test solution comes in contact with the acid liquor. The close of the operation will be indicated by the fact that the precipitate ceases to redissolve when stirred. Unless the acid be very dilute, considerable quantities of the sub-salt of copper will remain dissolved in the acetate of copper which forms. It may even happen that the solution of acetic acid can be made strongly alkaline, although no precipitate has been seen to form in it. Mohr directs that the beaker which contains the diluted vinegar be placed on a black ground, and that the operator, while stirring the liquid, should look into it from above. — Another method, somewhat related to that of Kieffer, is to use a copper salt as the indicator instead of litmus. A few drops of a solution of pure sulphate of copper are added to the vinegar, and a standard solution of soda is poured into the mixture until the moment when a faint persistent cloudiness, due to precipitated hydrate of copper, appears in the liquid. The point of saturation may be readily hit in this way in vinegars which are transparent and free from suspended matters, even though they be somewhat colored.

Instead of the processes above referred to, which are of comparatively recent invention, vinegar makers have long been accustomed to determine the strength of their products by means of carbonate of sodium or carbonate of potassium. According to Otto, the method of procedure is as follows:—Select a suitable number of clear, transparent, non-effloresced crystals of hydrated carbonate of sodium, rub them to coarse powder in a mortar, and preserve the powder in a tightly stoppered bottle. Weigh out in a capacious flask or beaker fifty grms. (custom prescribes 2 Troy ounces = 960 grains) of the vinegar to be tested, and add to it a couple of drops of a solution of litmus. Set the flask in an inclined position upon a wire gauze support and heat it moderately with a lamp. Put 30 or 40 grms. (or about a Troy ounce) of the powdered carbonate of sodium in a weighing tube and counterpoise the tube with its contents upon a balance. By means of a small spoon or spatula, take portions of the carbonate of sodium from the weighing tube and carefully throw them into the warm vinegar in the flask until the red color of the vinegar is changed to blue. When the point of neutralization has been reached, replace the weighing tube with the unused portion of its contents upon the balance, and by adding weights to make good the loss, determine how much of the carbonate has been required to effect the neutralization. The quantity of acetic acid in the solution tested is readily obtained by the following proportion:—

$$\text{Molecular weight of } Na_2CO_3+10H_2O : \text{Molec. weight of Acetic Acid} = \text{Weight of } Na_2CO_3+10H_2O \text{ used} : x \left(= \text{Wt. of Acetic Acid in the Sample.}\right)$$

The molecular weight of acetic acid in the second term of the proportion may be derived either from the formula C_2H_3O (anhydrous acetic acid), or $C_2H_4O_2$ (monohydrated acetic acid), according as the strength of the sample is to be expressed in terms of anhydrous or of hydrated acid. In practice, the strength of acetic acid is stated sometimes in one way, and sometimes in the other. In case 2 ounces = 960 grains of vinegar, are taken, every 27 grains of crystallized carbonate of sodium required to effect neutralization will represent one per cent of anhydrous acetic acid, or every 22.8 grains one per cent of the monohydrated acid.

To avoid loss of material during the tumultuous ebullition caused by the evolution of carbonic acid, care should be taken that the flask or beaker in which the vinegar is heated is never more than about a quarter full of the liquid. The carbonate of sodium should be taken up by small fractions, and no new portion of it should be added to the liquid until the last portion has ceased to give off gas. Just before each fresh addition of the carbonate the liquid should be shaken in such manner that any drops of it which may have been thrown against the upper side of the inclined vessel shall be washed down into the main body of liquor. The spoon used for transferring the carbonate of sodium should be kept as dry as possible. In case the steam from the flask wet it, it should be rinsed at the mouth of the flask with a wash bottle and afterwards dried.

As long as the color of the solution remains bright red, the carbonate of sodium may be added with freedom and rapidity; but when the point of neutralization is near at hand, only very small portions of the carbonate should be added. In any event the color of the liquid serves merely as a rough indication of the progress of the experiment. Litmus paper must be used as a test of thorough saturation. As soon as the color of the solution becomes obscure, small drops of the liquid must be placed upon litmus paper after each addition of the carbonate, until the paper ceases to be reddened, or is actually turned blue. It is not best to color the vinegar intensely red at the start. Two or three drops of litmus will color it sufficiently. When the vinegar is strongly reddened at first, it is apt to acquire a violet tint towards the close of the operation which tends to obscure the reaction of the solution upon paper. In case the vinegar is strongly heated, it is best not to add any litmus to it until the comparatively sluggish evolution of carbonic acid indicates that the point of neutralization is almost reached. The crystals of carbonate of sodium chosen for this use should be free from any spots of efflorescence. Instead of crystals of sal-soda, some vinegar makers prefer to use the anhydrous, recently ignited carbonate. In any event, carbonate of sodium is to be preferred to carbonate of potassium, although the latter, at one time exclusively employed, is still sometimes used. The use of standard solutions of carbonate of sodium is not to be commended. They are inferior in all respects to solutions of the caustic alkalies. It has been already mentioned that Otto recommends the use of ammonia water four or five times diluted, chiefly because this liquid may be readily obtained of any druggist. For a description of a graduated instrument specially devised for the use of vinegar makers, in which vinegar may be neutralized with standard ammonia water, see Otto's *Lehrbuch der Essigfarbrikation*, p. 77 *et seq.*

An old method of testing vinegar was as follows:—Weigh a lump of pure, dry marble, place it in a measured quantity of vinegar, insufficient to dissolve the whole lump, and warm the liquid slightly, until all the acid is saturated with calcium. Wash the undissolved portion of marble with boiling water, dry it at the same temperature as before, and again weigh it. From the loss of weight calculate the amount of acid in the sample examined. Objections to this process are found in the difficulty of washing out the acetate of calcium from the porous stone as well as in the slow action of acetic acid upon marble, and the impossibility of bringing the marble to absolutely the same state of dryness before and after the action of the acid. (Gr. Williams, *Pharm. Journ. Trans.*, 1854, **13.** 595).

Method B. To determine acetic acid in acetates from which caustic soda precipitates hydrates or oxides, such as ferric acetate or acetate of copper, mix a weighed quantity of the acetate with a measured quantity of a standard solution of caustic soda or carbonate of sodium, more than sufficient to decompose the whole of the acetate. Heat the mixture to boiling, collect the precipitate upon a filter, wash it with hot water, concentrate the filtrate to a convenient bulk and determine the excess of alkali with standard acid. The difference between the amount of soda thus found and the amount of soda taken, gives the quantity of soda which has been neutralized by the acetic acid in the substance analyzed. The amount of acetic acid is found by the proportion:—

$$\begin{matrix}\text{Molec. wt.}\\\text{of NaHO}\end{matrix} : \begin{matrix}\text{Molec. wt.}\\\text{of } C_2H_4O_2\end{matrix} = \begin{matrix}\text{Wt. of Soda}\\\text{neutralized}\end{matrix} : x \left(= \begin{matrix}\text{Wt. of acid}\\\text{in sample.}\end{matrix}\right)$$

Care must be taken that none of the acid is lost by precipitation as a basic acetate in the first instance. Since a good deal of carbonic acid would be absorbed from the air during the evaporation of the filtrate in case caustic soda were employed, it is as well to use a standard solution of carbonate of sodium. (See Acidimetry).

To determine the unused excess of alkali, the solution may be heated to boiling, colored with litmus, and directly saturated with standard acid; or an excess of standard acid may be added at once, the solution boiled to expel carbonic acid, reddened with litmus, and then treated with a standard solution of caustic soda until a blue tint is imparted to it.

Unlike most of the other acids, the strength of a solution of acetic acid cannot be determined by taking the specific gravity of the solution. Not only are the differences in density corresponding to different proportions of acid very small,—ranging only from 1.057, the specific gravity of glacial acetic acid, to 1,000, the specific gravity of water,—but the differences are not regular, or directly proportional to the amounts of acid in the solution. Thus the specific gravity of a solution containing only 37 per cent of anhydrous acetic acid is almost absolutely identical with that of a solution containing 85 per cent. Solutions of vinegar and of crude pyroligneous acid are often met with, moreover, charged with varying proportions of other soluble substances, so that this method of estimation would not admit of general application in any event.

The strength of pure vinegars made from diluted alcohol, whiskey, or other distilled spirit, may, however, be tested with the hydrometer indirectly, as follows:— Add dry slaked lime to a measured quantity of the vinegar, to alkaline reaction, or, better, until the color of the solution suddenly changes to yellow or brown, while a flocculent precipitate forms in the liquid. This change of color, due to the presence of impurities in the vinegar, indicates that an excess of lime has been added. Cool the solution of acetate of calcium to 15° and test it with a hydrometer. By reference to a table of specific gravities and percentage composition of solutions of acetate of calcium, the proportion of this salt and, therefrom, the proportion of acetic acid, may readily be obtained. The process is well adapted for technical determinations. (Ordway, *Amer. Journ. Sci.*, 1861, **31.** 451). It was formerly employed in England by excise officers, but could hardly have afforded accurate results with vinegar made from malt, cider, etc., on account of the various soluble mucilaginous substances with which such vinegar is charged.

Principle II. Volatility.

Application. Estimation of the acid in crude acetate of calcium.

Method. Weigh out about 5 grms. of the acetate of calcium, place it in a small tubulated retort, together with 50 c.c. of water and 50 c.c. of ordinary phosphoric acid of about 1.2 specific gravity, free from nitric acid. Set the retort on a wire gauze support in such a position that its neck may slope slightly upwards. By means of tightly fitting corks and a delivery tube, connect the mouth of the retort with a small Liebig's condenser, and distil almost to dryness. Collect the distillate in a quarter-litre flask, taking care that none of it is lost. Allow the retort to become cold, pour into it 50 c.c. more water, again distil almost to dryness, and afterwards repeat the operation yet a third time. Add water to the distillate by small portions, agitating the mixture after each addition, until the flask is filled up to the quarter-litre mark. Then draw off portions of 50 c.c. or 100 c.c. with a pipette, and determine the acid contained in them by means of a normal solution of caustic soda, as described in A. In case the acetate of calcium is contaminated with a chloride, the salt may be distilled with chlorhydric instead of phosphoric acid. The proportion of chlorhydric acid in the distillate would in that case be determined by titrating one portion of it with a solution of silver (see Chloride of Silver), after the total amount of acetic acid and chlorhydric acids had been determined in another portion with standard soda. The proportion of acetic acid would then be inferred from the difference. Sulphuric acid cannot readily be employed, since sulphurous acid is formed through its action upon organic matters attached to the acetate of calcium. (Fresenius, *Zeitsch. Analyt. Chem.*, 1866, **5.** 315).

Principle III. Feeble solvent action; or rather, power of dissolving certain substances without acting upon others.

Applications. Acetic acid is used to acidulate liquids in many cases where stronger acids are inadmissible. See, for example, Acetate of Aluminum, Acetate of Iron, Chromate of Barium, Phosphate of Iron, Phosphate of Uranium, etc., etc.

The commercial acid of 1.04 specific gravity, containing about 25 per cent of the anhydrous acid, is usually pure enough and strong enough for purposes of analysis. It should be tested with solutions of nitrate of silver, nitrate of barium, and of indigo; with sulphuretted hydrogen, and with ammonia water and sulphydrate of ammonium before use, in cases where the presence of chlorine, sulphuric acid, nitric acid, sulphurous acid, or of either of the heavy metals would be prejudicial. If pure, the acid will neither leave any residue when evaporated on platinum foil, nor emit an empyreumatic odor on being heated after having been saturated with carbonate of sodium.

In order to purify an impure acid, mix it with some acetate of sodium, and slowly distil the liquid almost to dryness in a glass retort. In case sulphurous acid has been detected by the deposition of sulphur on testing with sulphuretted hydrogen, digest the acid for some time with binoxide of lead or precipitated binoxide of manganese, and decant the clear liquid before adding the acetate of sodium.

Acetate of Aluminum.

Principle. Insolubility in water of certain basic acetates of aluminum. These acetates separate when neutral or nearly neutral aqueous solutions of acetate of aluminum, which contain some other salts, are heated. (Compare Dictionary of Solubilities).

Applications. Determination of aluminum in most compounds of aluminum with inorganic or volatile organic acids, which are soluble in water or chlorhydric acid. Separation of Al from Li, K, Na; Ba, Ca, Sr, Mg, Mn, Ur, Fe (see Method B.), Co and Zn; less completely from Ni. The method is employed more particularly when both Al and Fe have to be separated together from the other metals.

Method A. Put the cold, moderately dilute,

somewhat acid solution in a flask or beaker, and add to it carbonate of sodium by small lumps, or ammonia water little by little, until the acid is so nearly saturated that on shaking the liquid the precipitate formed by the last addition of alkali barely dissolves. Add to the clear, cold liquid a few drops of acetic acid, and then pour in as much of a strong aqueous solution of normal acetate of sodium, or acetate of ammonium, as may be needed to convert by double decomposition all the bases present into normal acetates. Heat the mixture to boiling, and continue to boil for a short time. If the operation has been properly conducted, the precipitate will settle readily from the liquid in the form of transparent flocks, when the lamp is removed.

Pour the clear, hot liquor upon a filter, wash the precipitate two or three times by decantation with boiling water, to which some acetate of sodium, or acetate of ammonium, has been added; transfer the precipitate to the filter and finish the washing on the filter with boiling water. — The precipitate is dried, ignited and weighed as Oxide of Aluminum, with the precautions prescribed under that head. Or, in case peculiar accuracy is required, the moist precipitate is dissolved in chlorhydric acid and again thrown down with acetate of sodium, to remove the last traces of the stronger bases.

Method B. In separating iron from aluminum by this process, the two elements are thrown down together as basic acetates. The mixed precipitate is ignited with the precautions enjoined under Acetate of Iron, and weighed as alumina plus ferric oxide. The proportion of iron is then determined by titration. (See *prot* Oxide of Iron). The difference between the weight of ferric oxide thus obtained and that of the mixed precipitate, gives the weight of the alumina.

Precautions. According to Gibbs & Atkinson (*Amer. Journ. Sci.*, 1865, **39.** 60), the metals in the original solution had better be in the form of chlorides. If the original solution contains any considerable excess of acid, this excess should be removed by evaporation, and the solution again diluted with water before adding the carbonate of sodium or the ammonia. The solution should be so dilute that a litre of it would contain no more than two grms. of Al_2O_3, or of Al_2O_3 plus Fe_2O_3, in case both iron and aluminum were present. In case the liquid become turbid through the addition of too much alkali, it should be made clear again by the least possible quantity of chlorhydric acid before proceeding to add the acetic acid and acetate of sodium. After boiling the mixture, it should be filtered hot, for if the supernatant liquid were left to cool in contact with the precipitate, some of the latter would dissolve. For the same reason, it is best to employ a plaited filter and to allow the filter to become empty after each addition of liquid before refilling it. The boiling must not be too long continued, lest the hydrate of aluminum become slimy and stop the pores of the filter. Unless a considerable excess of acetate of sodium be employed, an appreciable quantity of aluminum will escape precipitation; the precipitate, in this case, has a peculiar granular appearance. In case the filtration is slow, some aluminum will pass into the filtrate, and may be separated therefrom on boiling. This subsidiary precipitate should be collected on a separate filter. — The precipitate usually retains insignificant traces of soda, but should not exhibit a strong alkaline reaction when moistened with water after ignition. In case the presence of sodium salts in the filtrate is undesirable, ammonia-water and acetate of ammonium must be employed instead of the carbonate and acetate of sodium. — Though the precipitate cannot be very conveniently filtered and washed, and though traces of aluminum always remain in the filtrate, the process affords satisfactory results when carefully executed. It is less esteemed, however, than the corresponding process for determining iron. (See Acetate of Iron). The analogous process based upon the insolubility of basic Formiate of Aluminum is preferable, inasmuch as the formiate may be washed more readily than the acetate.

Acetate of Barium.

Principle. Power of precipitating sulphates and chromates. Compare Acetic Acid (feeble solvent power of).

Applications. Acetate of barium is used as a reagent to precipitate sulphate of barium or chromate of barium in cases where the presence of chlorhydric or nitric acid would be inadmissible. To prepare the acetate, dissolve pure carbonate of barium in moderately dilute acetic acid, filter, evaporate and set the strong solution aside to crystallize; keep the dry crystals for use.

Acetate of Iron (Ferric Acetate).

Principle I. Insolubility in water of certain basic acetates of iron. These acetates separate when neutral, or nearly neutral, aqueous solutions of ferric acetate, which contain some other salts, are boiled. (Compare Dictionary of Solubilities).

Applications. Determination of iron in salts of iron, with most inorganic or volatile organic acids, which are soluble in water or chlorhydric acid. Separation of Fe from Li, K, Na; Ba, Ca, Sr; Mg, Ur, Al, Mn, Co and Zn; less completely from Ni. Separation of Fe_2O_3 from FeO. (Method B).

Method A. Similar in all essential particulars to that described under Acetate of Aluminum.

Precautions. According to Fresenius, the success of the operation depends on the iron solution being sufficiently dilute, and the free acid sufficiently neutralized, as well as upon the presence of a sufficient excess of acetate of sodium. If these conditions are complied with, all the iron will be thrown down as soon as the liquor boils. It is not necessary in any event to boil the mixture longer than two or three minutes.

Unless it is intended to separate FeO from Fe_2O_3 (as in Method B), the iron must all be in the condition of a ferric salt before the liquid is neutralized. After the addition of the carbonate of sodium, or the ammonia-water, the liquid should assume a deep brownish-red color, but no permanent precipitate should be allowed to form in it before the alkaline acetate is added. The filtration should be rapid, as with the aluminum salt, and the supernatant liquor should not be allowed to cool in contact with the precipitate. If, after boiling, the mixture were left to itself for several hours, a small quantity of iron would go into solution.

Sometimes, though not often, portions of the precipitate pass mechanically through the pores of the filter, and have to be collected upon another filter after standing six or eight hours. The washed precipitate had better be dissolved in chlorhydric acid while still moist, and the iron reprecipitated with ammonia as a hydrate. If the precipitate is small, however, it may be ignited and weighed directly as Oxide of Iron, taking care to oxidize any ferroso-ferric oxide which may form, by a final ignition in oxygen gas, or with addition of nitric acid. The process is convenient, and affords accurate results, though the corresponding process with Formiate of Iron is said to be preferable, inasmuch as formiate of iron is more readily washed than the acetate.

Method B. To separate Fe_2O_3 from FeO, Reichardt (*Zeitsch. analyt. Chem.*, 1866, **5.** 63, 64) mixes the hot chlorhydric acid solution of the two oxides with a tolerably large quantity of chloride of ammonium, dilutes the mixture largely with boiling water, neutralizes the hot solution with ammonia, and immediately redissolves the precipitate with a few drops of chlorhydric acid. The clear liquid is then heated to actual boiling, immediately removed from the fire and treated with an excess, but not too large an excess, of crystals of acetate of sodium. In a very short time after the crystals have been stirred into the liquor, all the iron in the ferric salt will separate as a flocculent precipitate, which may be filtered immediately and washed with boiling water. The province of the chloride of ammonium is to hinder the oxidation of the ferrous salt. It accomplishes this purpose so well that the filtrate from the precipitated ferric acetate might be left standing in the air for hours and still remain clear and colorless. To determine the proportion of ferrous oxide:—Heat the filtrate to boiling, add chlorhydric acid, drop by drop, until the liquid remains permanently clear; throw in small crystals of chlorate of potassium until a drop of the liquid no longer yields a blue color when tested with ferri-cyanide of potassium, remove the boiling liquid from the lamp, and stir into it crystals of acetate of sodium as before, to precipitate the ferric oxide. If the oxidized solution contains no substance, besides ferric oxide, precipitable by ammonia in presence of ammonium salts, the iron may of course be thrown down as Hydrate of Iron in the usual way.

Manganous oxide, or a mixture of manganous and ferrous oxides, may be separated from ferric oxide in a similar way by mixing the original chlorhydric acid solution with chloride of sodium, and hydrate or carbonate of sodium, in place of the chloride of ammonium and ammonia water just described. The FeO and MnO are subsequently oxidized with chlorate of potassium to Fe_2O_3 and Mn_2O_3, and the Fe_2O_3 thrown down by means of a new portion of acetate of sodium.

Reichardt recommends that the acetate of sodium be added in all cases in the form of crystals, as above described, after the boiling liquid has been removed from the lamp, even when iron is to be separated from the alkaline earths, and from phosphates of the alkaline earths. According to him, boiling as directed in Method A renders the precipitate more difficult of filtration, besides being objectionable on other accounts.

Principle II. Colorific Power.

Applications. Estimation of small quantities of iron, as in the analysis of rocks.

Method. A special instrument known as a colorimeter is required. (See Colorimetry). For a description of the process of analysis, which is not yet perfected, see a paper by A. Müller in *Zeitsch. analyt. Chem.*, 1863, **2.** 143.

Acetate of Lead.

Principle. Ready solubility in water, and power of precipitating many acids. Compare, for example, Sulphate of Lead, Chromate of Lead, Phosphate of Lead, etc.

The best commercial acetate of lead is usually pure enough. For use, dissolve one part of the salt in ten parts of water acidulated with a few drops of acetic acid.

Acetate of Sodium (or of Potassium).

Principle. Easy decomposition of the salt by strong acids, and feeble solvent power of the acetic acid set free.

Applications. Substitution of acetic for chlorhydric, sulphuric, or nitric acid in strongly acid solutions. Compare, for example, the Acetates of Aluminum and of Iron, or Phosphate of Iron.

The commercial salt is usually pure enough; it should be free from empyreumatic matter. Dissolve one part of it in ten parts of water.

Acetate of Uranium.

Principle. Power of precipitating Arseniate of Uranium and Phosphate of Uranium.

To prepare the acetate, heat finely powdered pitch blende with dilute nitric acid, treat the filtrate with sulphuretted hydrogen to throw down copper, lead and arsenic; filter again, evaporate the filtrate to dryness, and heat the residue strongly enough to decompose the nitrates of iron, cobalt and manganese. Boil the residue with water and again filter. Evaporate the filtrate to crystallize nitrate of uranium, purify the nitrate by recrystallization, and heat the crystals until a little of the sesquioxide of uranium is reduced. Dissolve the yellowish-red mass in warm acetic acid, filter and let the filtrate crystallize. Acetate of sesquioxide of uranium will crystallize out, while the nitrate of uranium, which escaped decomposition, remains dissolved in the mother liquor. Or, saturate a solution of protochloride of uranium with ammonia-water, and dissolve the precipitate in hot acetic acid.

A solution of acetate of uranium should give no precipitate when tested with sulphuretted hydrogen. But on the addition of carbonate of ammonium, a precipitate, soluble in an excess of the carbonate, should be produced.

Acidimetry.

A term applied to methods of estimating the amount of free acid in acid solutions.

Method A. By Specific Gravity. It has been found by experience that the specific gravity of a mixture of water and an acid is greater than that of pure water; that, as a general rule, the specific gravity of such mixtures increases with the proportion of acid contained in them, and that each particular mixture has a certain definite specific gravity peculiar to itself. Consequently, if the specific gravity corresponding to each and every per cent of any given acid be carefully determined, once for all, by experiments upon mixtures of pure acid and pure water, made expressly for the purpose, it will afterwards be easy to determine the proportion of that acid in any aqueous solution by taking the specific gravity of the latter and comparing the result with those of the standard experiments, provided only that no substance besides the acid in question be dissolved in the solution.

The relations between specific gravity and per cent of pure acid have been determined experimentally for solutions of all the more common acids, and the results of these experiments, expressed in numbers arranged in tabular form, will be found in most of the dictionaries of Chemistry. (See Dictionary of Solubilities.)

In testing the value of an acid mixture in this way, the Hydrometer is sufficiently accurate for most purposes. It is very commonly employed in the scientific laboratory as well as in the work-shop. The precautions to be noticed are, that the test must either be applied at the temperature to which the Standard Table refers, or the observation must be corrected for that temperature, and that the sample of acid tested must be free, or nearly free, from soluble contaminations. Since many of the common acids are volatile, it is easy to detect the presence of nonvolatile impurities by evaporating a portion of the acid to dryness in a suitable dish. Of the acids in common use, acetic acid is the only one which cannot readily be tested with the hydrometer when pure. See Specific Gravity and Hydrometer in this work, or in almost any treatise on Chemical or Physical Manipulation.

Method B. By neutralization with an alkaline solution of determined strength. The principle upon which this method depends is that caustic alkalies and alkaline earths combine readily with free acids to form neutral salts, having no action upon litmus and certain other coloring matters, while the presence of the smallest excess of either acid or alkali is instantly made manifest by its action upon the coloring matter.

After a standard alkaline solution of definite strength has once been prepared, it is easy to determine the amount of acid in any weighed or measured sample of an acid solution by adding to it a few drops of a solution of litmus, and then pouring in the standard alkali until the red color of the litmus just changes to blue. The quantity of alkali required to effect this change being noted in each case, the quantity of acid contained in the sample is readily ascertained by calculation from the known value of the alkali.

The standard alkaline solution. most commonly employed, is prepared from caustic soda; but a solution of caustic potash, or of caustic lime in sugar water, would do as well. Ammonia water is sometimes used, and so are solutions of carbonate of sodium, of borax, and of ammonio-sulphate of copper. — Instead of litmus, various other coloring matters may be employed to indicate the point of saturation. See, in particular, Cochineal, Logwood and Turmeric.

Standard Caustic Soda [or Standard Caustic Potash] is prepared through the intervention of a standard acid (see Alkalimetry) as follows:—Make a solution of caustic soda by boiling carbonate of sodium with hydrate of calcium in the usual way,[1] and dilute the clear caustic liquor until its specific gravity is about 1.06, corresponding to five per cent of hydrate of sodium. Measure off into a porcelain dish, or a beaker placed upon white paper, thirty c.c. of test acid (see Alkalimetry), and add to the liquid as much of a violet solution of litmus as may be necessary to give it a faint red tint. Fill a burette[2] with the solution of caustic soda, and pour the latter into the test acid until the red tint just changes to blue. Repeat the experiment with another portion of

---

[1] Or simply dissolve some caustic potash, such as surgeons use, in water, and boil the solution with enough slaked lime to decompose any carbonate of potassium which may be present.

[2] For descriptions of the various forms of this instrument and of Erdmann's float, used to ensure accuracy in reading, see the works on Chemical Manipulation.

the test acid, and if the two results agree, proceed to dilute the alkaline solution in the manner described under Alkalimetry (test sulphuric acid), so that each volume of it shall be capable of exactly neutralizing a volume of test acid. In case twenty-five c. c. of the soda solution were sufficient to neutralize the thirty c. c. of test acid, five c. c. of water would have to be added to every twenty-five c. c. of that soda solution to reduce it to the proper extent, or for every one thousand c. c. of the soda solution two hundred c. c. of water. When the standard solution is made of such strength that the alkali in one thousand c. c. of it will exactly neutralize an equivalent weight of any acid (expressed in grammes), it is called *normal*.

Normal solution of soda is well adapted for the ordinary operations of acidimetry, but whenever the quantity of acid to be neutralized is small, it is best to use comparatively dilute solutions of standard alkali. To this end the normal solution may be diluted five times, or ten times, with water. To prepare a "one-tenth normal" solution, for example, measure off fifty c. c. of the normal solution with a pipette, allow the liquid to flow into a half-litre flask, fill the latter with water to the mark, and shake the mixture thoroughly.

In order to keep the standard soda solution free from carbonic acid, the bottle which contains it may be provided with a cork carrying a bulb tube open at both ends, filled with soda-lime, so that air free from carbonic acid may pass into or out of the bottle, accordingly as the barometric pressure or temperature of the air changes. Sometimes it will be found convenient to provide a syphon tube through which the soda solution may be drawn off without need of opening the bottle. In this event a bent glass tube reaching nearly to the bottom of the bottle is fitted to the cork beside the bulb tube, and to the end of its outer limb, which reaches to a point below the bottle, is attached a short piece of caoutchouc tubing. After the syphon has once been filled, the caoutchouc tube is kept compressed with a spring clip, excepting at the moments when portions of the solution are allowed to flow out. In most cases, however, this syphon arrangement may be dispensed with.

Instead of taking pains to make the soda solution normal, as above described, it is often more convenient to choose a solution of soda of about the proper degree of concentration, as determined by the hydrometer; to determine its strength accurately by means of a standard acid, and then to use it directly as test alkali without any further preparation. This method of procedure of course necessitates a short calculation in each case, in order to obtain the percentage of acid.

It is convenient sometimes in technical analyses to have the number of cubic centimetres, or half cubic centimetres, of the standard alkali used, express directly the percentage of acid (either anhydrous or hydrated) in the sample tested. This may be accomplished by using normal alkali, and operating upon a weighed quantity of the acid to be tested equal to one-tenth or one twentieth of an equivalent weight of the acid sought for, expressed in grammes. The following table gives the number of grammes of each of the more common acids which must be taken in this event:—

Name of the acid.	No. of grms. equal to one-tenth of an equiv. wt. of the acid.	No. of grms. equal to one-twentieth of an equiv. wt. of the acid.
Anhydrous Acetic Acid,	5.1	2.55
Monohydrated " "	6.0	3.00
Chlorhydric Acid,	3.65	1.83
Anhydrous Nitric Acid,	5.4	2.70
Hydrated " "	6.3	3.15
Anhydrous Oxalic Acid,	3.6	1.80
Crystallized " "	6.3	3.15
Anhydrous Sulphuric Acid,	4.0	2.00
Monohydrated " "	4.9	2.45
Anhydrous Tartaric Acid,	6.6	3.30
Hydrated " "	7.5	3.75

Since the small quantities expressed in the above table could hardly be weighed out directly with sufficient accuracy, it is best to weigh out half equivalent weights (or in other words, five times as many grammes as are indicated in the first column of figures) of the acids, in a five hundred c. c. flask; to fill the flask with water to the mark, and after shaking its contents to measure out portions of the liquid with a pipette, for analysis. One hundred c. c. or fifty c. c. of the liquid must be taken, according as one-tenth or one-twentieth of an equivalent weight of the acid is to be used. In diluting the acid with water, care must be taken if need be, to cool the mixture before adding the last drops of water.

A Standard Solution of Lime in sugar water has the merit of being always free from carbonic acid, for in case the solution absorbs any carbonic acid from the air, insoluble carbonate of calcium is precipitated. To prepare the solution, add slaked lime to a cold, moderately strong, but not too strong solution of white sugar, as long as any of it will dissolve. Filter the solution, determine its strength by means of a standard acid, and dilute with as much water as may be needed. The solution of course becomes weaker in proportion as it absorbs carbonic acid from the air, though the deterioration is found to be very slow when the liquor is kept in tightly stoppered bottles. The strength of the solution must, on this account, be redetermined by means of standard acid at the beginning of any new series of determinations, or according to Gr. Williams (*Pharm. Journ. Trans.*, 1854, **13.** 596), at intervals of four or five weeks. One advantage of the lime solution is found in the fact that acetate of calcium has less action upon the color of litmus than the acetates of sodium and potassium. The lime solution is consequently well fitted for testing Acetic Acid.

Standard Solutions of Lime water and of Baryta water are used with advantage against oxalic acid and sulphuric acid for estimating Carbonic Acid. (After Dalton & Hadfield, and Pettenkofer). — Pasteur also (*Dingler's polytech. Jour.*, **190.** 139 and *Zeitsch. analyt. Chem*, **8.** 86) employs lime water against sulphuric acid for determining the amount of acid in *must* (unfermented grape juice). For this purpose a dilute standard acid is prepared, of such strength that ten c. c. of it shall contain 0.06125 grm. of the monohydrate, and saturate about twenty-seven c. c. of lime water. This quantity of sulphuric. acid is equivalent to 0.0725 grm. of anhydrous malic acid, to 0.09375 grm. of crystallized tartaric acid and to 0.2351 grm. of cream of tartar. Ten cubic centimetres of clear, filtered must are taken for an experiment, and the standardized lime water is poured into it from a burette graduated to tenths of cubic centimeters. Since must always contains substances which are colored by alkalies, it is not necessary to add any litmus to indicate the point of saturation. It is not well, for that matter, even to use litmus paper at the close of the operation, for malate and tartrate of calcium exhibit an acid reaction with litmus, just as the acetate does. (See Acetic Acid). The lime water should merely be added to the must, rapidly and without intermission, until the character of the color of the liquid changes, or in case the must be colorless at first, until the appearance of a tolerably pronounced yellow color. One or two drops should be subtracted from the amount of lime water actually used, in order to reduce the reading from the point of supersaturation to that of saturation.

As a rule, no precipitate of any kind falls at the moment when the color changes, but if the mixture be left at rest for a few minutes, or better, for a half hour or hour, it will become cloudy from deposition of granular crystals of normal tartrate of calcium, or more rarely, of a double compound of tartrate and malate of calcium, which contains one molecule of each salt plus sixteen molecules of water. These precipitates are easily distinguishable under the microscope. No harm is done in case the precipitate should form during the titration.

Whenever there is any difficulty in filtering a sample of must, a quantity of the standard lime water insufficient to effect complete saturation, may be added to a measured quantity of the muddy liquor—enough, for example, to make the liquid give a blue reaction upon sensitive red litmus paper; the solution may then be filtered, and lime water added, drop by drop, to ten c. c. of the clear filtrate, until the color changes. It is then easy to reckon how much lime water in all has been neutralized by acid in the must.

A standard solution of Ammonia-water is to be commended in certain cases, inasmuch as the strong liquor necessary for its preparation may be obtained ready made of every druggist. Were it not for the easy volatility of ammonia, and the consequent difficulty of keeping a standard solution of it unchanged for any great length of time, it would be often employed in acidimetry.

A standard solution of Carbonate of Sodium may readily be prepared by weighing out a convenient quantity of pure, anhydrous carbonate of sodium, and dissolving it in a determined quantity of water;—fifty-three grms. to the litre would be the proportion for a normal solution.

So far as mere preparation is concerned, a standard alkaline solution may evidently be obtained in this way more directly and with less risk of error than by the roundabout methods required in the case of the caustic alkalies. But since the reaction of carbonic acid upon litmus interferes with the reaction of the stronger acids and of the alkalies, a solution of carbonate of sodium is inferior—as a standard alkali—to a solution of caustic soda. It is seldom used in acidimetry, though sometimes employed in the preparation of standard acids, as will be explained under Alkalimetry. (See standard nitric acid). If a solution of cochineal be employed instead of litmus to indicate the point of saturation, the disturbing influence of carbonic acid is far less marked.

Instead of using a coloring matter as the indicator, the point of saturation may be determined by the precipitation of carbonate of barium, as follows:—Add a small quantity of a solution of chloride of barium to the acid to be tested, set the beaker which contains the mixture upon a black ground and pour into it a standard solution of carbonate of sodium, until a persistent cloudiness, due to precipitated carbonate of barium, pervades the liquor. — Acids which, like sulphuric acid, form insoluble compounds with barium, may be treated with a slight excess of chloride of barium, the mixture filtered to separate the insoluble barium salt, and the filtrate, or some definite fraction of it, may then be titrated with the standard carbonate of sodium (Mohr).

Standard solution of Biborate of Sodium. Like carbonate of sodium, borax is an excellent material for preparing a standard alkaline solution, inasmuch as any desired quantity of the dry or crystallized salt may be weighed directly upon the balance. A weighed quantity of borax has only to be dissolved in as much water as may have been determined upon, in order to complete the preparation of the standard solution. — The boracic acid set free when the borax solution is mixed with strong acids has far less action upon the color of litmus than is exerted by carbonic acid. Perfectly satisfactory results can in fact be obtained by titrating acids with the borax solution, even when nothing but litmus is employed to indicate the point of neutralization, especially if the solution to be tested be hot. But if a fresh decoction of Brazil wood be employed as the indicator, instead of litmus, the influence of boracic acid upon the final reaction becomes wholly inappreciable. The solution of Brazil wood should be prepared from clippings taken directly from solid blocks, not from the clippings which are to be found in commerce. The yellow Brazil wood solution is colored purple by alkalies, but becomes of a clearer yellow or reddish tint, when brought into contact with acids. (Stolba, *Journ. pratk. Chem.*, 1864, **90.** 459).

Since borax is rather sparingly soluble, it is impossible to employ a normal solution of it. Even a one-quarter normal solution, which would contain only 47.75 grms. to the litre, sometimes deposits crystals in cold weather. The use of borax solutions is consequently limited to those cases where only a dilute test alkali is needed. (Salzer, *Mohr's Titrirmethode*, 1856, **2.** 102).

Standard Solution of Ammonio-Sulphate of Copper. Instead of using a standard solution of pure alkali to neutralize the acid which is to be estimated, and a solution of litmus or other coloring matter to indicate the point of saturation, it is quite possible to make a single solution serve both purposes. A solution of chloride of silver, for example, in ammonia water, or of oxide of zinc in potash, soda or ammonia, of alumina in potash or soda, or, best of all, of basic sulphate of copper in ammonia water, may be employed both to neutralize the acid, and to show when the neutralization is complete.

The solution of ammonio-sulphate of copper (first proposed by Kieffer, *Annal. der Chem. und Pharm.*, **93.** 386) is prepared by adding ammonia water to a tolerably strong, warm, aqueous solution of sulphate of copper, until the basic sulphate at first thrown down has almost completely dissolved; the liquid is then filtered, and its strength determined by titration with a standard acid, which may be either chlorhydric, nitric or sulphuric. Oxalic acid cannot be used in this case, since an insoluble oxalate of copper would be formed and the liquid obscured. When the ammonio-sulphate of copper is dropped into an acid solution, it neutralizes the free acid just as any other alkaline liquid would, but at the moment when all the acid is saturated the liquid suddenly becomes turbid from the deposition of a quantity of basic sulphate of copper. Though readily soluble in ammonia water and in acids, this sulphate is well nigh insoluble in neutral solutions. Hence that portion of it which was held dissolved in the final drop of the test liquor falls down when the ammonia is taken from it. Moreover, the last drop of the test liquor reacts not upon free acid, but upon sulphate, or some other neutral salt, of copper, which has been formed by the combination of a portion of the acid under examination with some of the copper in the test liquid. At the end, therefore, the ammonia in the test liquid combines with the acid of this neutral copper salt, and the copper contained in it is thrown down in the form of a subsalt at the same time with that from the test liquid, so that the volume of the precipitate is very considerable. The completion of the process is thus made distinctly manifest, for the appearance of the turbidity shows that the point of saturation has been reached.

The solution is brought to the required strength, and is employed for determining acids in precisely the same way as a soda solution, with the single exception that the point of saturation is indicated by the appearance of a precipitate instead of by a change of color, as when litmus or cochineal is employed.

This modification of the ordinary acidimetric process offers no special advantages in so far as relates to pure acids, excepting Acetic Acid, as already described, and is altogether inapplicable in the case of acids which form insoluble compounds with copper, such, for example, as oxalic, tartaric and phosphoric acids. It is peculiarly well adapted, on the other hand, for those cases where a free acid is mixed with any of the so-called acid salts. It may be employed, for example, to determine the free acid in the partially spent liquors—charged with sulphate of zinc—of galvanic batteries, or in the mother liquors resulting from the manufacture of sulphate of copper or sulphate of zinc.

Since the basic sulphate of copper is not absolutely insoluble in solutions of ammonium salts, the method is not to be commended in cases where strict accuracy is required. Carey Lea (*Amer. Journ. Sci.*, 1861, **31.** 190) has shown that the precipitate dissolves with considerable facility in a solution of sulphate of ammonium, that it is likewise soluble, though perhaps to a lesser degree, in solutions of chloride and of nitrate of ammonium, and that a larger or smaller quantity is held dissolved, according as the saline solution is more or less dilute. In practice, therefore, different results will be obtained from solutions containing precisely the same proportion of acid, in case they happen to be more or less highly charged with ammonium salts. Fresenius (*Zeitsch. analyt. Chem.*, **1.** 108), on the other hand, while admitting the justice of Lea's criticism, has shown by experiments that the errors likely to arise from this source are not of sufficient magnitude to condemn the process, in so far as relates to its applications for technical purposes.

It is to be observed that the acid solutions to be tested in this way must be clear; they should be placed in beakers set upon black paper. Since the copper solution is liable to lose ammonia by standing, its strength should be redetermined from time to time, or before each new series of experiments.

The Actual Determination of the proportion of Acid may be performed as follows:—Weigh out in a small beaker or flask as much of the acid to be tested as will probably be sufficient to neutralize from 15 to 30 c. c. of the standard alkali, and add to it litmus enough to color the solution faint red. Fill a burette with the standard soda solution, and drop the latter rather quickly and without intermission into the acid, until the whole of the liquid in the beaker remains distinctly blue for some seconds. The acid must be stirred continually from first to last. After the point of neutralization has once been reached, no attention need be paid to the gradual reversion of the blue color to violet, caused by the action of carbonic acid from the air. — Note the quantity of standard soda solution which has been consumed, and proceed to calculate the proportion of acid. Suppose 4.5 grms. of dilute acetic acid were weighed out, and that 25 c. c. of normal soda solution, containing 40 grms. (= one molecule) of hydrate of sodium to the litre, were required to neutralize the acid, then the proportion

$$1000 : 25 :: 60 \left[= \text{Molecular wt. of } C_2H_4O_2\right] : x \left[= 1.5 = \text{Weight of } C_2H_4O_2 \text{ in sample taken.}\right]$$

will indicate how many grammes of pure acetic acid were contained in the weighed quantity of dilute acid, and the proportion

$$4.5 ; 1.5 = 100 : x \left[= 33.33\right]$$

will give the percentage of pure acid which the sample contained.

If the acid solution be kept hot while the standard soda solution is added to it, the point of saturation can be distinguished without difficulty, even when the soda solution is contaminated with some carbonic acid; but in cases where the operation has to be performed in the cold, as when the acid under examination is volatile, or mixed with ammonium salts, the caustic soda employed should be free, or nearly free from carbonate. No matter how pure the soda, the change of coloration is more readily recognized in hot than in cold solutions. In any event the change of color of the litmus from red to blue is less clearly defined with weak acids, such as most of the organic acids, than with strong acids, such as sulphuric, chlorhydric, nitric, and the like. The blue tint, indicative of the point of saturation, may be obscured also, or modified, by the presence of various substances. If the acid solution contains ammonium salts, for example, the change of color from red to blue is less quick and decided than when they are absent. In all cases of doubt, and particularly when a color peculiar to the liquid obscures the color of the dissolved litmus, it is best to determine the point of neutralization by placing successive drops of the liquid upon litmus paper, in the manner described under Acetic Acid. It is to be observed, however, that in determining the point of saturation with litmus paper, a little more alkali will almost always be used than would be the case if the litmus were in solution. In very delicate experiments it may sometimes be worth while to determine the amount of this excess, and to allow for it in estimating the acid. This may be done by measuring off a volume of water equal to that of the acid tested, and adding to it drops of the soda solution until the reaction of the liquid upon litmus paper is just as strong as that previously exhibited by the liquid, which contained the acid. The quantity of alkali required to effect this result is then subtracted from the quantity actually employed in neutralizing the acid.

See Alkalimetry for the method of using cochineal instead of litmus. See also Cochineal and Logwood.

Estimation of Combined Acids. Though the term acidimetry is commonly understood to apply only to the estimation of free acids in simple solutions, or in solutions charged with substances having no action on alkalies, the process may nevertheless be extended in some instances to the determination of acids combined with metals. — A description of one method of this character, as applied to the estimation of acetic acid when combined with copper, iron, and other metals precipitable as hydrates, oxides or carbonates, by caustic or carbonated soda, has been given already under Acetic Acid.

Another general method of determining the amount of acid in a compound—based upon the insolubility of various metallic sulphides—should here be mentioned. This method is applicable to the estimation of non-volatile acids, which are not acted upon by sulphuretted hydrogen when combined, to form compounds soluble in water, with any metal which can be easily and completely precipitated by sulphuretted hydrogen. The process is as follows:—Weigh out a quantity of the salt to be analyzed, dissolve it in water, boil the solution, and pass a stream of sulphuretted hydrogen through the liquid until the metal is completely precipitated. In order to determine when the precipitation is complete, take out drops of the clear liquid from time to time upon a glass rod, place them upon porcelain, and add a drop of strong sulphuretted hydrogen water, or of any other reagent specially adapted for testing the metal in question. When quantities of salts as large as 5 grammes are operated upon the precipitation is usually complete in half an hour. When all the metal has been thrown down, filter the liquid rapidly, wash the precipitate with hot water until the washings no longer exhibit an acid reaction, and collect the filtrate and wash water in a half litre or a litre flask. Cool the liquid and dilute it with water to the volume of half a litre, or a litre. Agitate the mixture thoroughly, take out with a pipette several portions of 50 or 100 c. c. each, and place them in separate beakers, add a few drops of cochineal or logwood to each of the solutions, and determine the amount of free acid in each with dilute standard alkali in the usual way. As the standard alkali, Gibbs employs one-tenth normal ammonia water. The results of the 1st determination should be regarded as merely approximative, and the mean of the 2d and 3d (or of several successive) determinations taken as the true result. From the quantity of standard alkali used, not only the quantity of acid in the salt examined, but, in many cases, that of the metal also, may be readily calculated.

The precipitation of a metal by sulphuretted hydrogen from boiling solutions is usually comparatively slow, but the boiling is nevertheless necessary in order that the filtered liquid may contain no sulphydric acid. — In operating upon nitrates and chlorides, a quantity of some neutral salt of a fixed organic acid must be added to the solution before passing the sulphuretted hydrogen. A quantity of the organic acid equivalent to that of the mineral acid is then set free, so that none of the mineral acid is lost through volatilization, and none of the nitric acid destroyed by the hot sulphuretted hydrogen. A quantity of Rochelle salt about equal to the quantity of the salt taken to be analyzed, may be used for this purpose.

The process is inapplicable in presence even of very small quantities of iron, aluminum, and various other metals, which, with cochineal and logwood, give reactions not easily to be distinguished from those produced by the caustic alkalies. (Rose, *Pogg. Annal.*, **116.** 125; Gibbs, *Amer. Journ. Sci.*, 1867 [2] **44.** 207).

Principle II. Power of acids to expel carbonic acid from metallic carbonates.

Method. Mix a weighed or measured quantity of the acid to be examined with an excess of bicarbonate of sodium in an appropriate apparatus (see Carbonic Acid), and determine how much carbonic acid is expelled, either by weighing the apparatus before and after the experiment, and calling the loss of weight carbonic acid, or by absorbing the gas in soda-lime and weighing it directly. (See Carbonic Acid). The amount of free acid in the sample tested is then found by the proportion:

$$\text{Molecular wt. of } 2CO_2 : \text{Equivalent wt. of acid tested} = \text{Weight of } CO_2 \text{ found} : \text{Weight of acid in sample.}$$

It will be observed that 2 molecules of CO_2 are set free for every equivalent of acid in the solution:—

$$Na_2O, H_2O, 2CO_2 + H_2N_2O_6 = Na_2O, N_2O_5 + 2H_2O + 2CO_2.$$

Enough acid should be taken to set free one or two grammes of carbonic acid. — The process affords satisfactory results, and may be employed with advantage in testing liquids so highly colored that the ordinary method of acidimetry cannot be applied to them.

Aconitin.

See Iodomercurate of Aconitin.

Albumin.

Principle I. Coagulability by heat.

Applications. Estimation of albumin in alkaline solutions, such as urine and the serum of blood.

Method A. Acidulate the solution slightly with acetic acid and boil for several minutes. Collect the precipitate on a tared filter, wash thoroughly with warm water and dry in a current of warm air at 110° or 115°, or better, in vacuo, over sulphuric acid, until the mass ceases to lose weight. Care must be taken that the precipitate is thoroughly dried, for as soon as the moisture has been driven from its surface the mass acquires the consistence of horn, and forcibly retains the last portions of the water. The addition of acetic acid to the liquid is essential in order that the albumin shall be precipitated completely; but no great excess of the acid should be employed lest some of the albumin be dissolved by it. The precipitate thrown down from solutions thus acidulated is more flocculent and less apt to clog the pores of the filter during the process of washing than that obtained without the addition of an acid. — To determine albumin in urine, place 50 to 100 c. c. of the clear urine, previously filtered, if need be, in a flask large enough to hold twice as much of the liquid. Heat the liquor gradually, with frequent shaking until the albumin begins to coagulate, at about 70°, then throw a couple of drops of acetic acid into the flask from the end of a glass rod, and boil the liquor as above described (Neubauer & Vogel).

Method B. Another method of estimating albumin in urine, devised by Heller (*Heller's Archiv für Chem. und Microsc.*, 1852, p. 266 *et seq.*), is said to afford very accurate results in spite of being indirect:—Evaporate 10 or 15 grms. of the urine to dryness over sulphuric acid, and weigh the residue. Weigh out another quantity of the same urine in a small flask, acidulate it slightly with acetic acid, boil until all the albumin is precipitated, and after the liquid has become cold place the flask upon the balance and add to it, drop by drop, water enough to replace what has evaporated. Filter the contents of the flask and evaporate a weighed portion of the filtrate to dryness over sulphuric acid. The difference between the percentage of residue left by the original urine and that obtained from the urine after the separation of the albumin gives the proportion of the latter ingredient.

Method C. For comparing the quantities of albumin in any two different samples of urine, Dr. John Harley has adopted the following process:—Make 3 small filters from the same sheet of paper, cut down the two heavier to the weight of the lightest, and mark the filters with a pencil, A, B, and O. Take 1000 grain measures of urine A, and having boiled it, pour it while hot upon filter A. Treat urine B in a similar way. Wash the contents of each of the filters with warm water until the last traces of adhering urine have been removed. Then pour upon the albumin an ounce of water containing 2 drops of nitric acid, and subsequently wash out the acid with water. The filter marked O is placed in the first instance between one of the other filters and the funnel, and is thus equally saturated with urine and acid, and equally washed free from both. All 3 filters are dried together, and the empty filter is used as a counterpoise in determining the weight of the albumin upon the others.

Properties. Albumin, as it occurs in the state of solution in the animal economy, is combined not only with water but with minute quantities of certain saline and alkaline ingredients. When a solution of pure albumin is evaporated in vacuo, at temperatures below 50°, there is left a light yellow, translucent mass of soluble albumin, which may be readily rubbed to a fine white powder. When treated with water, this residue swells up to a jelly, without, however, dissolving to any very considerable extent, unless a small quantity of an alkaline salt be present. By the action of most mineral acids, and of many other chemical agents—sometimes by mere contact with atmospheric air, this soluble modification of albumin is changed to the coagulated, insoluble condition.

Insoluble albumin, when recently precipitated and still moist, is a tough, white, opaque, flocculent solid, insoluble in water, alcohol, ether, and most acids when dilute and cold. When left moist in the air it putrefies. It is somewhat soluble in hot acetic, tartaric, phosphoric, and strong chlorhydric acids. When boiled for a long time with water it decomposes and dissolves. On being dried it assumes a yellow color and becomes brittle and translucent like horn. When soaked in water, after drying, it takes up about five times its weight of the liquid and becomes soft and elastic.

Principle II. Opacity.

Applications. Estimation of albumin in aqueous and saline solutions.

Method. Prepare a little trough of sheet iron with glass ends, as follows:—Provide three rectangular sheets of metal, one 7 c. m. square, the others 4 c. m. long by 2.5 c. m. wide. Bend the square sheet into the form of a V-shaped gutter, the upper edges of which are 1 c. m. apart. From each of the smaller sheets cut out wedged-shaped pieces of metal, corresponding to the shape of the trough, so that when the sheets are placed in an upright position, the V-shaped trough may fit into the cuts and be supported as by feet. Place one of these supports near each end of the trough and solder them to the trough. Cut out two V-shaped pieces of window glass, fitted to the trough, and cement ~~them~~ into the ends of the trough with Canada balsam, taking care to place the glasses parallel to one another, and to leave a clear distance of 6.5 c. m. between them. The glasses can be cemented the more readily in case a notch or groove be made upon the iron in the beginning to receive them. The metal of the trough should be painted with asphaltum varnish to protect it from rust.

As applied to the estimation of albumin in urine, the process of testing is as follows:—Filter the urine, in case it is not clear, acidulate it slightly with acetic acid, if it be not already acid, and proceed to determine how much the urine must be diluted to fit it for the test. To this end prepare several dilute solutions by mixing measured portions of the urine with water and boil each of the solutions in regular order, until one is found in which the albumin no longer separates in distinct flocks, but only as a milky cloud. It is a liquid thus clouded by the presence of fine particles of suspended albumin, which admits of being subjected to the test of opacity. — The best way of preparing these dilute solutions is the following:—By means of a little pipette graduated to 0.1 c. c., take up 6 c. c. of the urine and transfer it to a 100 c. c. flask. Fill the flask with water to the mark, shake the mixture thoroughly, pour the liquid into a beaker and leave the flask inverted in order that it may drain. Meanwhile pour 6 or 8 c. c. of the diluted urine into a test tube of 20 or 25 c. c. capacity, heat the liquid to boiling, and afterwards cool it quickly by immersing the tube in cold water. In case the precipitate produced by boiling is so slight that the form of objects placed in strong daylight can be distinguished on looking at them through the liquid, the sample has been too much diluted, and the operator will at once proceed to prepare a more concentrated solution by mixing 12 c. c. of the original urine with water in the 100 c. c. flask. But in case the first solution was not transparent, it may be tested in the trough. — In testing, fill the trough two-thirds full of the cold, boiled liquid, and look through the liquid at the flame of a burning candle in a darkened room. Repeat the experiment with other diluted samples of the urine until a point is reached where the shape of the flame cannot be distinguished, and only diffused light can be seen through the liquid, even when the candle is brought close to the trough. — In case the flame is visible through the first solution, the next trial must be made with a solution containing a few more per cents of urine than the first. But if the shape of the flame cannot be distinguished, the liquor of the subsequent trial must be more dilute, and so on methodically, until a liquid is obtained, through which the reddish yellow cone of flame can only be seen by looking with the strictest attention, as if it were in a thick fog. When this point is reached it is only necessary to add a trace more urine (0.1 or 0.2 per cent) in preparing the next diluted sample, in order that the flame may become completely invisible, and the operation be finished. — To find the percentage of albumin, divide the number 2.3553 by the number of c. c. of urine taken to prepare the dilute solution through which the flame could no longer be seen. This number, 2.3553, is the mean of 35 experiments by Dragendorff, in which the results of the optical test were controlled by precipitating and weighing the albumin.

Precautions. In looking at the flame, the trough should be held before the eye like a spy-glass, and moved forwards and backwards from a distance of 0.5 metre from the candle close up to the flame, while the instrument itself is continually pressed lightly against the eyebrow. Up to a certain point the last glimpses of the cone of flame can be seen more readily, in proportion as the trough is closer to the candle, but if the flame is too near the liquid, the latter is illuminated by a reddish yellow light, through which the flame is seen less readily. The chamber in which the operation is conducted, should always be dark enough that the yellow light of the candle may overpower the daylight.

In case albumin separates from a diluted liquor in small flocks, the liquor may often be made cloudy and fit to be tested by shaking it violently as soon as the flocks appear. Densely clouded liquids, on the other hand, may be used for preliminary, approximative tests, by mixing them with measured quantities of water. But the results of such experiments must always be controlled by testing samples of urine which have been diluted before boiling. Better results can always be obtained by boiling weak solutions of albumin than by boiling comparatively strong solutions and mixing the cloudy liquor with water. — Instead of the 100 c. c. flask above prescribed, a 50 c. c. flask may be used, but rather more accurate results can be obtained when the larger volume of liquid is operated upon. The pipette employed must be graduated to 0.1 c. c., so that quantities of liquid as small as 0.05 can be measured with it.

The chief difficulty of the process is found in endeavoring to properly acidulate the original urine. Many samples of albuminous urine yield no precipitate, or only a comparatively feeble precipitate on boiling, when too strongly acidulated with acetic acid, and, in like manner, less albumin is obtained by the optical test than by the method of precipitation, in case the acid reaction of the urine is indistinct. It is important that the urine should be kept in a cool place in order that it may be as fresh as possible when tested. — The original urine need not be filtered unless it contains a distinct precipitate. Urine that is merely cloudy will usually become clear when mixed with much water. Some samples of albuminous urine, however, become cloudy when treated with a few drops of acetic acid, or with 5 or 10 times their volume of pure water, and it is precisely this kind of urine which is least readily tested by the optical method. Special care must be exercised in adding acid to such urine, since the presence of a trace of acetic acid in excess may present the appearance of any precipitate on boiling. The cloudy solution (paralbumin) produced on mixing the urine with water, need not be filtered. The mixture should be boiled at once, as if it were clear. — The process is easy of execution, and is said to yield very accurate results, excepting perhaps those kinds of urine which become cloudy on the addition of acetic acid. Ordinarily no more than 5 or 6 of the diluted samples of liquid have to be tested in order to hit the point of obscuration, so that the determination will be finished in the course of half an hour. — The several experiments of the series above mentioned agreed with one another in most instances to the second decimal place. Only 3 experiments out of the 35 differed more than 0.1, and only 11 more than 0.05, so that 21 of the trials agreed to 0.05 per cent. (Alfred Vogel, *Zeitsch. analyt. Chem.*, 1868, 7. 152).

Principle III. Power of rotating the plane of vibration of a ray of polarized light.

Applications. Estimation of albumin in aqueous or saline solutions, such as urine and the serum of blood.

Method. When a ray of polarized light is made to pass through a column of albumin solution enclosed in a tube, it is found that the plane of polarization is rotated to the left, and that the angle of deviation is proportional to the length of the column of liquid. In like manner, when a tube of any given length is successively filled with solutions containing different quantities of albumin, the angle of deviation is found to be proportional to the amount of albumin in the liquid.

An apparatus, known as Soleil's Albuminimeter, used for measuring the rotatory power of albumin, resembles the ordinary Saccharimeter (see Sugar), with the exception that in place of the Nichol's prism there used as the analyzer, a double refracting prism is employed, cut in such manner that only a single image shall appear in the field of vision. An intense white light, such, for example, as that of a petroleum lamp, is needed. The lamp is placed in a blackened box provided with a reflector which throws the rays of light upon a moveable lens by which they are concentrated before reaching the apparatus.

After the lamp has been lighted and placed in front of the apparatus, put a piece of red glass in front of the polarizing prism in the path of the luminous rays, and turn the analyzing prism until the luminous image has completely disappeared. The zero point of the apparatus having thus been determined, fill the tube with the solution to be tested, place it in the apparatus and leave the liquid at rest during some minutes. On again looking into the apparatus it will be seen that by virtue of the rotatory power of the liquid in the tube, the luminous ray has again become visible, and it will be found that the index of the apparatus must be turned through a certain number of degrees in order to again extinguish the ray. But by counting the number of degrees and minutes between the two points of extinction, it is easy to determine the amount of the rotation and to estimate therefrom the proportion of albumin in the liquor. — It is important to exclude external light as completely as possible; to make several observations with each sample of liquid, and to read the divisions of the circle and vernier carefully; best with a good lens.

In case serum of blood is to be tested, about 1 grm. of sulphate of sodium should be added to each 100 grms. of the liquid, and the mixture filtered immediately, in order to separate blood globules and other suspended particles. The purpose of the sulphate is merely to facilitate the filtration; in the case of urine none of it need be added. The yellowish orange color exhibited by serum when viewed in thin layers, becomes distinctly red when seen in a long column like that in the tube of the albuminimeter. In general, however, this coloration is not intense enough to do any harm; it simply obviates the need of using the red glass, for, like the latter, it only permits the passage of red rays.

In a series of 50 experiments upon blood serum, Becquerel found that the deviation of the plane of polarization varied between 4° 30′ and 9°, or on the average 7° 30′. In general, the deviation oscillated between 7° and 8°. The corresponding quantities of pure dry albumin were from 4.86 to 9.44 per cent. From these and like observations, it has been calculated that with a column of liquid 20 c. m. long, each minute of deviation corresponds to 0.18 grm. of albumin. (Becquerel, in *Robin & Verdeil's Chimie Anatomique*, 1853, **3.** 316).

Principle IV. Specific Gravity.

Applications. Estimation of albumin in urine.

Method. Take the specific gravity of the urine and note the temperature of the liquid when the observation is made. Acidulate a quantity of the urine with acetic acid, place the liquid in a flask provided with a perforated cork carrying a vertical glass tube, and boil it until all the albumin has separated in the insoluble state. Cool the boiled urine to the temperature at which the specific gravity of the original urine was determined, and take the specific gravity of the clear filtrate. Multiply the difference between the two specific gravities by 210, in order to obtain the percentage of albumin in the sample. In a series of 13 experiments, the least error was 0.005, and the greatest 0.056. (Lang, and Haebler, *Zeitsch. analyt. Chem.*, 1868, **7.** pp. 513, 514).

An article on the estimation of albumin, by gramimetric and volumetric methods, has been published by C. Boedeker, in *Henle & Pfeufer's Zeitschrift für rationelle Medicin, Zurich*, 1859, **5.** 320.

Alcohol.

Principle I. Solvent power.

Applications. Alcohol is used for separating many substances which dissolve in it from others which are insoluble, as when chloride of strontium is removed from chloride of barium; or precipitates, such as the sulphate or malate of calcium, are washed clean by means of it.

Two kinds of alcohol are commonly used in analysis, namely; "Spirit," or ordinary alcohol of 0.83 or 0.84 specific gravity (= 88 or 90 per cent by volume), and "Absolute Alcohol." The latter should be at least as strong as 0.81 specific gravity (= 96 or 97 per cent by volume). To prepare it mix a quantity of ordinary alcohol in a capacious flask or retort, with something more than its own volume of quick-lime, in small pieces. After the lime has slaked, leave the mixture to itself for several hours, and finally distil it slowly upon a water-bath.

Principle II. Reducing power.

Applications. Reduction of chromic acid to sesquioxide of chromium, of binoxide to protoxide of lead, etc.

Method. See, for example, Hydrate of Chromium.

Principle III. Volatility.

Applications. Separation of alcohol from aqueous solutions of non-volatile substances, as a preliminary to the determination of the strength of wines and spirit.

Method. As will appear from the article Alcoholometry, the value of spirits is usually determined by means of the Hydrometer. But since this instrument cannot be employed in case the spirit contain other soluble substances besides water, the alcohol, together with a part of the water, is first separated by distillation before the hydrometer is applied.

A measured volume of the liquid to be tested is placed in a glass flask provided with a cork and delivery tube, connected with a worm, or with a Liebig's condenser. Most of the liquid is then distilled over, the distillate carefully measured, and the proportion of alcohol contained in it is determined with the hydrometer. The amount of alcohol thus found must of course be referred to the volume of liquid originally placed in the flask or still.

Both the liquid to be distilled and the distillate must be brought to some common temperature before measuring, best by immersing the vessels which contain them in flowing water. The joints of the distillatory apparatus must be tightly fitted. In case the substance to be tested contains any free acetic acid or other volatile acid, neutralize it with caustic soda before distilling.

Alcoholometry.[1]

A term applied to methods of estimating the proportion of alcohol in any spirituous liquid.

Method A. By Specific Gravity. See Specific Gravity and Hydrometer. Compare what is said of specific gravity under Acidimetry.

Applications. Determination of the proportion of alcohol in any mixture of alcohol and pure water. This method is convenient, accurate and rapidly executed. It is far more frequently employed than either of the other methods of testing alcohol. It would always be employed for practical purposes, were it not that the presence of sugar, salts, coloring matters, etc., in spirit, sometimes precludes its use.

To estimate alcohol by this method, even in presence of sugar, Zenneck, and after him Aug. Vogel, *Zeitsch. analyt. Chem.*, **6.** 273, proceeds as follows:—Carefully determine, in the first place, the specific gravity of the saccharine spirit at 15° upon a balance. Then weigh out about 150 c. c. of the liquor in a tared flask, and boil to expel alcohol, until about half of the liquid has evaporated. Remove the flask from the lamp, place it upon a balance, and pour in water until the weight of the flask and contents is the same as before boiling. Cool the aqueous solution of sugar to 15°, and determine its specific gravity. Look up in the published tables (Dict. Sols., Art. Sugar) the percentage of Sugar, *S*, which corresponds to the observed specific gravity; 100—S (=A) will then represent the percentage of mixed water and alcohol in the liquor under examination. The specific gravity of this plain spirit may be found by the formula:

$$x = \frac{1.606 \times D \times A}{160.6 - D \times S},$$

in which D = the specific gravity of the original mixture of sugar and spirit. When the specific gravity of the plain spirit is known, it is easy to obtain its percentage composition by referring to the published tables (Dict. Sols., Art. Alcohol), and from the composition of the plain spirit, that of the original saccharine liquor may be easily calculated. — The formula above given is obtained as follows:— The specific gravity of cane sugar being 1.606, and that of the original saccharine liquor D, let x represent the specific gravity of the plain spirit. The volume of the plain spirit will be $\frac{A}{x}$ and that of the percentage of sugar $\frac{S}{1.606}$; hence

$$\text{Sp. gr. of the mixture} : D = 100 : \left(\frac{A}{x} + \frac{S}{1.606}\right)$$

and

$$x = \frac{1.606 \times D \times A}{160.6 - D \times S}.$$

In case the liquor contains no cane sugar, but only grape sugar, the number 1.39 (= specific gravity of grape sugar) must be substituted for 1.606 in the formula.

A prominent advantage of the process is found in the fact that only a comparatively small quantity of liquid is required for an experiment.

Method B. By determining the boiling point of the mixture, and comparing the result with the results of previous experiments made with standard liquids of known composition specially prepared by mixing pure alcohol and pure water. Tables of the relations of boiling points to per cents of alcohol will be found in most dictionaries of general chemistry. (See, for example, Dictionary Solubilities, Art. Alcohol).

Applications. Determination of the proportion of alcohol in mixtures of alcohol and water, and in fermented liquors as well. For mixtures of alcohol and pure water the method is inferior to Method A (by specific gravity). It finds useful application, however, in testing wines and beers, since it has been found that the foreign substances, other than alcohol and water, in fermented liquors, have little influence upon the boiling point of the mixed alcohol and water. Enough sugar or common salt may be added to spirit of 20 per cent, by volume, to reduce the liquor to 0° of Gay-Lussac's alcoholometer without altering the boiling point of the spirit to any appreciable extent. Several forms of Ebullioscopes have been invented for testing wine and beer. For a report upon the various forms of this instrument, see Despretz and others, *Comptes Rendus*, **27.** 374. A recent form of ebullioscope, by Brossard-Vidal, is described in *Zeitsch. analyt. Chem.*, 1864, **3.** 223. Wagner, in his *Jahresbericht chem. Tech.* 1863, **9.** 545, specially commends the ebullioscope of Tabarié. Ebullioscopes are much used for testing wine and beer, and the results obtained by them compare very favorably with those obtained by the method of distillation. (See under Alcohol). — In boiling beer it is well to add to the liquid a trace of tannic acid to prevent frothing.

Method C. By determining the Tension of the Vapor. An instrument called a Vaporimeter has been constructed by Geissler for this purpose. It consists of a vessel provided with a syphon tube, fitted by grinding to an orifice at its top. It is employed as follows:—Metallic mercury is poured into the vessel up to a certain mark, and the space above the mercury filled with the spirit to be tested; the syphon tube is then put in place, and the whole apparatus turned upside down. The quicksilver immediately falls into the bent tube and closes the spirit against contact with the air. The apparatus is then placed in an atmosphere of steam, in the upper part of a vessel in which water is boiling, so that vapor may be evolved from the spirit. The pressure exerted by the

[1] For the details of Alcoholometry, see *Handwoerterbuch der Chemie*, Braunschweig, 1856. **1.** 493. Or almost any of the large Dictionaries of Chemistry.

alcohol vapor, forces a portion of the mercury to rise into the syphon tube, and from the height of this column of mercury, corrected for the atmospheric pressure at the moment of observation, the tension of the vapor is determined. The apparatus is usually provided with an empirical scale, which indicates the proportion of spirit in terms of per cent by volume.

Applications. Determination of the proportion of alcohol in solutions charged with sugar. It has been found that with solutions containing salts, such as chloride of sodium, the tension of the vapor evolved at any given temperature is greater than it would be for a mixture of alcohol and pure water. But sugar exerts no influence. In case the saccharine spirit to be tested contains carbonic acid, or any other volatile acid, the liquid must be saturated with lime before placing it in the vaporimeter. The method is of course one of technical, rather than of scientific, application.

Method D. By Determining the Rate of Expansion of the Liquid by Heat. Instruments have been constructed for this purpose by Silbermann (*Comptes Rendus*, **27.** 418) and by Makins (*Journal of Chemical Soc.*, London, **2.** 224). Silbermann's Dilatometer consists of a thermometer tube, which is filled up to a certain mark with the spirit to be tested, at the temperature of 25°. The air which the liquid holds dissolved is then removed by means of an air pump, and the tube and spirit finally exposed to a temperature of 50°. The amount of expansion is observed, and the proportion of alcohol in the sample found by referring to a scale previously graduated by direct observations upon samples of spirit of known strength.

Applications. The instrument can be used with mixtures containing sugar or salt, since these substances have been found to exert little or no influence on the expansibility of spirit.

Method E. By Capillary Attraction. The instrument known as a Liquometer, in which the determination is made, consists of a capillary glass tube 4 inches long, graduated to 20 degrees, and fitted to a hole in the cover of a glass vessel so that it can be elevated or depressed in the vessel. After the glass vessel has been three quarters filled with the spirit to be tested, the capillary tube is pushed down so that about 0.1 inch of the lower end of the tube shall be immersed in the liquid; a quantity of the spirit is drawn up into the tube by sucking at the upper end to moisten its walls, and the tube again drawn up carefully until its lower end is precisely level with the surface of the spirit. Again suck carefully at the top of the tube, so that the liquid may rise in the tube, and afterwards note the point at which the liquid remains stationary when it is allowed to sink back towards the glass vessel. Compare the degree at which the spirit stands in the tube with the published tables, in order to obtain the percentage of alcohol. (Reynolds, *London Pharm. Journ. and Trans.*, [2.] **9.** 171, and *Zeitsch. analyt. Chem.*, 1868, **7.** 358).

Method F. By Ultimate Analysis. A weighed portion of any mixture of alcohol and pure water could be burned by means of an oxidizing agent (see Carbon), the resulting carbonic acid weighed, and the quantity of alcohol computed from these data. But the method would only be of scientific interest. It has no practical application.

Alkalimetry.

A term applied to the estimation of free alkalies and alkaline carbonates.

Method A. By Specific Gravity. In pure or nearly pure solutions of ammonia, or of the hydrates of sodium and potassium, the proportion of alkali may be inferred with tolerable accuracy from the specific gravity of the solution. (See Acidimetry). But the method is of less general applicability with alkalies than with acids, since, with the exception of ammonia-water, alkaline solutions are rarely found pure.

Method B. By neutralization with an acid solution of determined strength. This method is the precise opposite of the method of Acidimetry by saturation, already described. It consists in determining how much acid of known strength is required to neutralize a definite weight of the sample of alkali under examination.

The chief requisite in this process is of course the *standard* or *test acid;* which may be made either from sulphuric, oxalic, chlorhydric or nitric acid, according to circumstances.

Standard acid may be prepared either by dissolving a definite weight of a crystallized acid, such as oxalic acid, in water, in such manner that each c. c. of the solution shall contain a certain proportion of the dry acid; or by adding the requisite proportion of water to any moderately dilute solution of sulphuric, chlorhydric, or nitric acid, after the proportion of real acid in that particular sample has been determined by analysis or by titration. The details of the several methods will be given below.

Standard acid is often made of such strength that a litre (= 1000 c. c.) of it shall contain exactly one equivalent of the dry acid, expressed in terms of grammes; it is then called "*normal* acid." Thus normal chlorhydric acid should contain 36.5 grms. of HCl to the litre, and normal sulphuric acid 40 grammes of SO_3. Equal volumes of different normal acids have of course the same power of saturating alkalies. Sometimes the test acid is made to contain only a fraction of the quantity of real acid which would be required if it were normal. Thus an acid containing 7.3 grms. HCl to the litre, is called "one-fifth normal," while one that contains 3.65 grms. is "one-tenth normal." Several methods of preparing standard acids will be set forth in the following paragraphs.

Standard Sulphuric Acid.

A. *Normal Sulphuric Acid.* Counterpoise a small flask or beaker with shot or sand upon a rough balance, and weigh out in it 60 grms. of concentrated sulphuric acid. Put 1050 c. c. of water in a large flask and pour the acid into the water while shaking the flask. After the mixture has become cold, pour a part of it into a burette, measure off into beakers two portions, of about 20 c. c. each, of the liquid, and determine the quantity of sulphuric acid in each portion by precipitating it as Sulphate of Barium. If the results of these determinations are concordant, take the mean of the two as representing the amount of acid actually contained in the mixture. Pour the rest of the acid solution into a graduated mixing cylinder, note its volume and pour in as much water as may be required to dilute the liquid to the condition of normal acid (i. e., 40 grms. to the litre). If it were found, for example, that the 20 c. c. of liquid tested contained 0.84 grm. of sulphuric acid, then 1000 c. c. would contain 42 grms., and to every 1000 c. c. of the solution 50 c. c. of water would have to be added, for

$$40 : 1000 = 42 : x \ (= 1050).$$

In default of a mixing cylinder, the dilution may be effected in a litre-flask, as follows:—Fill the flask up to the mark upon its neck with the acid mixture to be diluted, so that the lower edge of the curved depression at the top of the liquor shall coincide with the line upon the glass; empty the liquid carefully into a large stoppered bottle; measure off with a pipette or burette the amount of water required for the dilution; transfer this water to the flask from which the acid has just been poured; shake the flask thoroughly, and add its contents to those of the bottle; shake the bottle thoroughly, pour back half its contents into the litre-flask, agitate the latter, pour back the liquid from the flask into the bottle, again shake the bottle and keep its contents for use.

As is the case with all standard solutions, the bottle should always be shaken just before any portion of its contents are to be poured out for use, unless indeed it be absolutely full of liquid. For when a bottle is but partly filled with a solution, some water is apt to evaporate into the space above the liquid and condense there upon the glass in such manner that if a small portion of liquid was to be poured out of the bottle without shaking, it would wash off the condensed water and become slightly diluted, while the liquid remaining in the bottle would be left a trifle stronger after each pouring.

Instead of determining the sulphuric acid by precipitating it as sulphate of barium, the proportion of acid in the sample chosen may be discovered by noting how much of this acid is required to neutralize a given weight of pure carbonate of sodium. The determination may be made either in the manner described in § α, or, more accurately, by that set forth in § β. It is to be observed, as a general rule, that a standard acid is best prepared by the same method as that for which it is to be subsequently used. If sulphuric acid, for example, is to be used for testing the value of commercial carbonate of sodium, it should be standardized against pure carbonate of sodium by the very method which is afterwards to be applied in testing the commercial carbonate.

α. Weigh out (best from the covered platinum crucible in which it has been ignited) from 4.5 to 5 grms. of pure, anhydrous carbonate of sodium. Place the salt in a capacious flask or porcelain dish, dissolve it in about 200 c. c. of hot water, and color the solution blue with 1 or 2 c. c. of a solution of litmus (or, instead of litmus, color the cold solution with Cochineal, or with Logwood). — Mix about 60 grms. of ordinary, monohydrated sulphuric acid with 500 c. c. of water, allow the solution to cool, and fill a burette with the cold dilute mixture. Stir the solution of carbonate of sodium with a fine glass rod, and gradually pour into it, from the burette, the cold dilute acid, until the color of the solution changes to a wine red; then place the flask or dish over a lamp and heat its contents to boiling. The wine red coloration is caused by the action of carbonic acid which has been set free, and the liquid is heated in order to drive out this volatile acid. As soon as the color of the liquid has become blue again pour more sulphuric acid from the burette into the nearly boiling liquid, until the solution assumes the peculiar deep red color, slightly inclined to orange, which is characteristic of the strong acids.

Since, in liquids thus charged with carbonic acid it is not easy to determine when the operation is finished, if the color of the dissolved litmus be alone relied upon, litmus paper must be used towards the close of the operation in order to detect the precise point of saturation. Hence, after the liquid has become wine red, take care to add the sulphuric acid only by small portions, finally at the rate of only two drops at once, and after each addition of the acid draw the point of the stirring rod across a strip of blue litmus paper, one or two inches wide, so that red streaks may be formed upon the paper in regular order, and each streak correspond to a fresh addition of the acid. Proceed in this manner until the color of the liquid in the dish indicates that the point of saturation has been passed, then note the number of c. c. of acid which have been poured from the burette, dry the strip of litmus paper at a gentle heat, and observe which one of the streaks upon the paper just remains red when dry. The red color of all those streaks which were reddened with carbonic acid will completely disappear as the paper becomes dry, while the reddening due to sulphuric acid will remain visible; it is an easy matter, therefore, to determine how great an excess of the acid has been used, by simply counting the number of red streaks left upon the dry paper. A quarter of a cubic centimetre (or whatever two drops may amount to in terms of the burette employed) is then to be deducted from the total number of c. c. of acid used for every one of the streaks. Besides the sum of these drops used in excess, it is best to subtract also from the total amount of acid used one quarter of one c. c., which has gone to neutralize the alkali in the litmus employed for coloring.

Repeat the experiment with new portions of carbonate of sodium and of the acid, and in case the two results agree, proceed to dilute the acid to the desired degree in the manner already described.

For some eyes, an infusion of *Cochineal* possesses great advantages over the solution of litmus, as a means of recognizing the point of neutralization. To prepare the liquor, put 3 grammes of powdered cochineal in a flask, together with 50 c. c. of strong alcohol and 200 c. c. of distilled water, cork the flask and let the mixture digest for a day or two at the ordinary temperature, shaking it frequently. The clear solution, which may be either decanted or filtered from the residue, has a deep ruby-red color. On gradually diluting it with pure water, free from ammonia, it becomes orange, and finally yellowish-orange. Caustic alkalies and alkaline carbonates change the color to a carmine or violet carmine, and so do the alkaline earths and their carbonates; solutions of strong acids and acid salts make it orange or yellowish-orange, but to carbonic acid it is nearly indifferent. In using cochineal, the solution to be tested should not be heated. It is simply necessary to measure out a given volume, say 20 c. c. of the acid, to dilute it with about 150 c. c. of water, to add 10 drops of the cochineal liquor, and to pour in alkali from a burette until the yellowish liquor in the flask suddenly acquires a violet-carmine tinge through the action of a single drop of the alkali.

In nicer determinations it is important to bring the liquid each time to a given volume, by adding water after the neutralization is nearly finished. To note the level of the proper amount of liquid, say 200 c. c., strips of paper may be pasted upon the beakers or wide-necked flasks employed.

The same amount of coloring matter being thus always diffused in the same volume of the same water, errors due to varying degrees of dilution and varying amounts of ammonia, which is rarely absent from distilled water, are avoided. The contents of one flask, in which the neutralization has been satisfactorily effected, may be kept as a standard of color for the suceeding trials. The tint remains constant for hours; it is unaffected by the carbonic acid of the air. When three or four concordant results have been obtained, the average is taken as expressing the relative strength of the acid and alkali. (Luckow, *Journ. prakt. Chem.*, **84**. 424, through *Amer. Journ. Sci.*, 1863, **35**. 280).

β. Mix concentrated sulphuric acid, of known specific gravity, with enough water that the mixture may contain rather more than one equivalent of the anhydrous acid for every 100 parts of liquid. Weigh out 1 to 1.5 grm. of pure anhydrous carbonate of sodium, place it in a flask or porcelain dish of 300 or 400 c. c. capacity, dissolve it in 100 or 150 c. c. of water, and pour into the liquid a measured quantity a of the cold acid. Take 1 c. c. of the acid for every 0.0053 grm. of the carbonate, so that the acid may be distinctly in excess. Boil the solution to expel the carbonic acid, then color it slightly red with a measured quantity of litmus solution and pour into the hot liquid, from a burette, a dilute solution of caustic soda of undetermined strength, until the last drop of the soda produces a distinct light blue color, unmixed with violet. If the solution is only slightly colored with litmus, and is free from carbonic acid, it is easy to determine the point of saturation with great exactitude, otherwise some difficulty is met with from the blue tint at first formed changing to violet.

Write down the number of c. c., b, of the soda solution required to neutralize the excess of acid. Then measure out a fresh portion, c (= 20 or 30 c. c.), of the dilute acid, and determine how many c. c., b', of the soda solution are required to neutralize it. From this last determination it will appear that 1 c. c. of the soda solution is equivalent to $\frac{c}{b'}$ c. c. of the dilute acid. Hence there was employed in the previous experiment $\frac{b}{b'}c$ cubic centimetres more of this acid than would be needed to neutralize $a \times 0.0053$ grm. of carbonate of sodium; or $a - \frac{cb}{b'}$ cubic centimetres of the acid would have precisely neutralized the carbonate of sodium taken. Hence, if $a - \frac{cb}{b'}$ cubic centimetres of the acid be diluted with water to the bulk of a cubic centimetres, the solution will be normal. Compare Standard Nitric Acid, below.

Standard sulphuric acid may be prepared also by decomposing a known weight of sulphate of copper with sulphydric acid, as proposed by Gibbs (*Amer. Journ. Sci.*, 1867, **44**. 210). Put a quantity of pure powdered sulphate of copper in a porcelain crucible placed within a Hessian crucible, and heat the sulphate for about an hour, taking care to raise the temperature gradually, and that the heat shall at no time exceed low redness. Transfer the hot anhydrous sulphate to a dry weighing tube, close the tube, and after it has become cold weigh out a quantity of the sulphate. Dissolve the weighed sulphate in water, heat the solution to boiling, pass a stream of sulphuretted hydrogen through it to precipitate the copper (see Acidimetry), and dilute the filtrate and wash water to a known volume. The quantity of sulphuric acid in the solution is known from the weight of the anhydrous sulphate taken.

For technical determinations it is convenient to make the standard acid of such strength that 50 c. c. of it will exactly neutralize 5 grms. of pure carbonate of sodium. Such acid may be used for testing the value of either of the caustic or carbonated alkalies; and the number of half c. c. of it required to produce saturation, in any particular case, will correspond directly to the per cent of alkaline carbonate or caustic alkali contained in the sample tested, provided there be weighed out for the analysis a quantity of material equivalent to 5 grms. of carbonate of sodium. The equivalent quantities capable of saturating 50 c. c. of this standard acid are respectively:—

For Carbonate of Sodium	5.000 grms.
" Hydrate " "	3.773 "
" Carbonate of Potassium . . .	6.519 "
" Hydrate " " . . .	5.292 "

If 5 grms. of an impure carbonate of sodium, or 6.519 grms. of common pearlash be weighed out and titrated with the acid in question, the number of half c. c. used will in the one case give the per cent of pure carbonate of sodium, and in the other of pure carbonate of potassium, which the samples contain, without need of any calculation. Where substances poor in alkali are to be tested, some multiple of the numbers above given may be weighed out.

Standard Chlorhydric Acid. Mix 900 c. c. of water with 180 c. c. of chlorhydric acid of 1.12 specific gravity, measure out with a burette two portions of from 10 to 20 c. c., and determine the quantity of chlorhydric acid in each portion by precipitation, as Chloride of Silver. If the two results are concordant, take the mean and calculate therefrom how much water must be added to the acid tested to reduce it to the normal strength. If it were found, for example, that 20 c. c. of the acid contained 0.81 grm. of HCl, then a litre would contain 40.5 grms. and

$$\text{Mol. wt. of HCl} = 36.5 : 1000 :: 40.5 : x\ (= 1111).$$

so that 111 c. c. of water would have to be added to each litre of the acid.

Instead of determining the amount of chlorhydric acid by precipitation, as chloride of silver, it may be estimated with carbonate of sodium in the manner described above, under standard sulphuric acid; but before boiling the liquid to expel carbonic acid, a few grammes of sulphate of sodium must be added to prevent the evolution of chlorhydric acid. Or it may be estimated with carbonate of calcium, as will be described directly under Standard Nitric Acid.

Standard Oxalic Acid. Weigh out 63 grms. (the weight of one equiv.) of crystallized oxalic acid, transfer it to a litre flask, add water enough to nearly fill the flask, shake the mixture until the acid has all dissolved, bring the liquid to the temperature of 16°, and pour in water up to the litre mark; again shake the solution thoroughly, then pour it into a stoppered bottle and keep it protected from sunlight. Since the oxalic acid found in commerce is usually contaminated with potassium, it is best to prepare a pure acid directly by acting upon starch with nitric acid, and recrystallizing the product. (See Oxalic Acid).

Standard Nitric Acid may be readily prepared by the method described above in § β, under the head of standard sulphuric acid. Instead of carbonate of sodium, carbonate of calcium may be employed. It has the advantage of being more readily obtained in a state of purity than carbonate of sodium. An outline of the process may here be re-stated:— Prepare a quantity of dilute nitric acid, in such wise that the strength of this acid, as indicated by the hydrometer, shall be somewhat greater than that of the desired standard acid. Prepare also a solution of caustic soda, about as strong as the acid, and determine by titration how many c. c. of this soda solution are required to neutralize a measured quantity of the acid. Weigh out about 1 grm. of pure carbonate of calcium (either powdered Iceland spar or the precipitated carbonate) which has been dried at 100°, and dissolve it in a measured quantity of the nitric acid. Heat the mixture gently to expel carbonic acid, color it with litmus, and finally pour into the liquid as much of the soda as is needed to neutralize the nitric acid which was used in excess. Note the quantity of soda solution employed, calculate therefrom the number of c. c. of nitric acid which were used in excess, and subtract this amount from the total amount of acid taken to dissolve the carbonate. The remainder will give the number of c. c. of acid which are equivalent to the weighed quantity of carbonate of calcium; in other words, it will indicate what portion of the acid was neutralized by the calcium salt. The weight of dry acid in this portion may now be found by the proportion:

$$\begin{matrix}\text{Equiv.}\\ \text{wt. of}\\ CaCO_3\end{matrix} : \begin{matrix}\text{Equiv.}\\ \text{wt. of}\\ N_2O_5\end{matrix} = \begin{matrix}\text{Wt. of}\\ CaCO_3\\ \text{taken}\end{matrix} : \begin{matrix}\text{Wt. of } N_2O_5 \text{ in the volume}\\ \text{of liquid neutralized}\\ \text{by that } CaCO_3.\end{matrix}$$

The value of the acid having thus been determined, proceed to dilute what remains of it to the required standard, in the manner already described. (Standard Sulphuric Acid).

In some cases it may be found convenient to employ a standard solution of carbonate of sodium instead of the dry salt. The standard solution may be prepared with ease by simply dissolving a weighed quantity of pure carbonate of sodium in the required volume of water. By using measured portions of this solution the operator has it in his power to make several titrations from the product of a single weighing. Reischauer (*Dingler's polytech. Journ.*, **167.** 47) recommends the use of a normal solution of the carbonate, made by dissolving 53 grms. of the salt to the volume of a litre. To prepare a standard acid by means of this solution, measure off 10 c. c. (= 0.53 grm.) of it with a pipette, mix it with an excess of the acid to be standardized, and neutralize this excess with caustic soda, as above described.

Merits of the several Acids. Test acid may be prepared more simply and directly from oxalic acid than from either of the other common acids. Oxalic acid may readily be made pure and dry, in spite of current assertions to the contrary, and were it not for a certain tendency to decomposition which is exhibited by aqueous solutions of this acid when exposed to light, they would doubtless be very generally employed in processes of alkalimetry.

Chlorhydric acid has special merit, in that the fundamental determination of the proportion of acid in any given sample may be made with very great accuracy by precipitation, as chloride of silver.

Sulphuric acid, which is perhaps more frequently employed than either oxalic or chlorhydric acid, has the advantage of being less volatile than the latter, and less liable to change than the former. Dilute solutions of sulphuric acid may be boiled freely without fear of loss. When standard chlorhydric acid, on the contrary, has to be boiled, it is best to mix with it a quantity of sulphate of sodium to hinder the evolution of chlorhydric acid. In any event, care should be taken both with chlorhydric and nitric acids, that only a very slight excess of acid be present when a liquid is to be boiled. Nitric acid has no advantage over chlorhydric acid, excepting that it volatilizes somewhat less readily than the latter when a dilute solution is boiled.

Sulphuric acid is well suited for the estimation of sodium, potassium and magnesium, but cannot be used for determining calcium, barium or strontium; either nitric or chlorhydric acid must be employed when the oxides, hydrates or carbonates of these metals are to be titrated.

Tartaric acid, at one time proposed as a substitute for sulphuric acid in alkalimetry, is not well adapted for use as a standard acid, since aqueous solutions of it are liable to decompose on standing.

The Actual Determination of the proportion of alkali in any sample of unknown value, follows, from what has been said above, without need of further description. A weighed quantity of the material to be tested is dissolved in about 200 c. c. of water, the solution is colored with 1 or 2 c. c. of litmus, and standard acid is poured upon it from a burette until the alkali is saturated. (See standard sulphuric acid α). Or an excess of the standard acid is added in the beginning to the weighed sample of alkali, the mixture boiled to expel carbonic acid, and the excess of the standard acid estimated with a standard solution of caustic soda. (See Acidimetry and standard sulphuric acid β).

The quantity of material to be taken for analysis in any particular case, may be judged of, from what has been said above, under standard sulphuric acid. In order to diminish the errors incidental to weighing, it is often best, in the lack of a delicate balance, to weigh out a quantity of material ten times as large as is really wanted. The weighed substance is then dissolved in half a litre of water, and one or two portions of it, each of 50 c. c., are taken out with a pipette for analysis. In case any portion of the weighed substance refuses to dissolve in water, the liquid should be filtered and the residue washed before proceeding to the acid treatment, or better, the liquid may be allowed to stand until it has become clear, and a definite portion of the clear liquor then taken up with a pipette for the analysis. Care must of course be taken, in the first place, to obtain a fair sample of the material to be tested by taking small portions of it from many parts of the entire mass and rubbing them thoroughly together in a mortar.

According to Baugart & Wildenstein (*Zeitsch. analyt. Chem.*, 1864, **3.** 324), it is well, in technical determinations of the value of alkaline carbonates, to check the frothing, which occurs when acid is added to the liquor, by means of a layer of melted paraffine. A small quantity of paraffine thrown upon the boiling solution of carbonate of sodium, contained, as usual, in a large evaporating dish, will immediately melt and spread over the entire surface of the liquor, in such manner that no permanent froth can form upon the surface of the liquid. The titration may consequently be proceeded with without delay or interruption. A few decigrammes of paraffine are sufficient for a surface 12 c. m. in diameter. In order to reduce the paraffine to convenient shape, dip a not too thin glass rod into a quantity of melted paraffine, bring the rod close to the surface of a quantity of cold water, and then allow the drops of the hot liquid to fall into the water. Dry the solidified drops on filter paper and keep for use; 2 or 3 of them will be found sufficient for a single alkali determination.

Estimation of the proportion of Caustic Alkali and of Alkaline Carbonate in mixtures containing both these substances.

Weigh out from 15 to 20 grms. of the impure carbonate to be tested, and dissolve it in water in a quarter-litre flask. When the soluble portion of the substance has all dissolved, fill the flask with water to the mark and shake its contents thoroughly. Cork the flask in order to exclude the carbonic acid of the air, and allow the liquid to stand until it has become clear. Draw off with a pipette two portions of the solution, each of 100 c. c., and determine the total amount of alkali in one portion by titration with a standard acid. This "total alkali" may be set down either as carbonate of the alkali, or as caustic alkali, according as one or the other of these ingredients preponderates in the sample under examination. Allow the other 100 c. c. portion of the solution to flow into a quarter-litre flask, add to it 100 c. c. of water and a solution of chloride of barium, as long as a precipitate falls. Fill the flask with water to the mark, cork it and leave the mixture at rest, until all the carbonate of barium has been deposited and the liquid has become clear. Draw off 100 c. c. of the clear solution, color it with litmus, add standard chlorhydric acid to distinct acid reaction, and note the quantity of acid used. Neutralize the excess of acid with a standard solution of caustic soda, and subtract this excess from the whole amount of acid taken. The difference will give the amount of acid which has been neutralized by the caustic alkali in the portion of material subjected to analysis, and from the weight of the acid, that of the alkali equivalent to it may readily be found by calculation. Finally subtract the weight of caustic alkali thus obtained from the total weight of alkali as found by titration in the other portion of the solution, in order to obtain the amount of alkali which must be regarded as a carbonate.

According to A. Mueller, the solution to which chloride of barium has been added must not be filtered. It contains caustic baryta as well as caustic alkali and alkaline chloride. It is found that a filter retains some of the baryta, and that a little of the caustic alkali might thus be lost.

Another method of estimating caustic alkali when mixed with an alkaline carbonate, will be given below, under Principle II.

Principle II. Volatility of carbonic acid.

Applications. Determination of the amount of pure alkaline carbonate in salæratus and the other commercial carbonates of sodium and potassium.

Method. A weighed quantity of the alkaline carbonate to be examined is treated with an excess of sulphuric acid in an appropriate apparatus (see Carbonic Acid), and the weight of the carbonic acid expelled from it is determined either by weighing the apparatus before and after the experiment, and calling the loss carbonic acid, or by absorbing the gas in soda lime and weighing it as such. (See Carbonic Acid). From the quantity of carbonic acid found, the proportion of alkaline carbonate in the sample tested is obtained by calculation:—

$$\begin{matrix}\text{Equiv.}\\ \text{wt. of}\\ CO_2\end{matrix} : \begin{matrix}\text{Equiv. wt.}\\ \text{of } Na_2CO_3\\ (\text{or of } K_2CO_3)\end{matrix} :: \begin{matrix}\text{Weight}\\ \text{of } CO_2\\ \text{found}\end{matrix} : x = \begin{matrix}\text{Wt. of } Na_2CO_3\\ (\text{or of } K_2CO_3)\\ \text{in the sample.}\end{matrix}$$

The process yields tolerably good results, but is less accurate and far less convenient and expeditious than the method by neutralization with a standard acid. It is consequently seldom employed in practice. It was at one time somewhat used for determining the proportion of alkaline carbonate in mixtures of caustic and carbonated alkali, after the total alkali value of the sample had been ascertained. The best way of determining the total alkali value is by titration with a standard acid, either directly or indirectly, as has been already described. But in lack of an acid of determined strength, the total alkali value of any sample of mixed caustic and carbonated alkali may readily be determined by the method of expelling carbonic acid. To this end expose a portion of the sample to carbonic acid gas, until all the caustic alkali present has been saturated, ignite to destroy any bicarbonate which may have been formed, and finally determine the carbonic acid in the manner already described. In another portion of the original sample, which has been subjected to no treatment, except drying, determine the amount of alkaline carbonate. The difference between the two determinations will indicate the proportion of caustic alkali in the substance analyzed.

Another method of estimating caustic alkali in presence of an alkaline carbonate, will be found above, under Principle I.

Each of the processes of alkalimetry above described is liable to error when the substance to be tested is contaminated with certain impurities. The special precautions to be taken in order to correct or avoid these errors, will be described further on, under the heads of the several alkalies and alkaline carbonates. For the methods of estimating potassium in presence of sodium, see KCl; 2KCl, $PtCl_4$; $KClO_4$; K_2SO_4.

Aluminate of Ethylamin. See Aluminate of Sodium.

Aluminate of Potassium. See Aluminate of Sodium.

Aluminate of Sodium.

Principle. Solubility in water.

Applications. Separation of Al from Fe, and from small quantities of Mn. Also from Co and Ni (Method B). Method B is specially adapted for the treatment of mixtures of the oxides of iron and aluminum which have been ignited, and so rendered insoluble in caustic lyes.

Method A. Evaporate the chlorhydric acid solution of aluminum, etc., to dryness on a water bath, in order to remove the excess of acid. Take up the residue with water, and in case the solution is cloudy, add a drop of strong chlorhydric acid and warm the mixture upon the water bath to clear it. Pour slowly a quantity of not too dilute soda or potash lye into a porcelain dish, or better, into a large platinum or silver crucible; place the dish or crucible upon a water bath and heat the latter to boiling. Stir the hot lye with a stiff platinum wire, and slowly pour into it the solution of aluminum, etc. Each particle of the aluminum compound is thus brought into intimate contact with a large excess of free alkali, and is converted into a soluble aluminate of the alkali, while the iron is thrown down as a hydrate. When the last portion of the aluminum solution has been washed into the dish which contains the alkali, pour the alkaline mixture upon a filter, wash the ferric hydrate and precipitate the aluminum from the filtrate, as Hydrate of Aluminum, by boiling the liquor with an excess of chloride of ammonium. (Lœwe, *Zeitsch. analyt. Chem.*, 1865, **4**, 357).

After a little experience, the amount of alkali to be employed may be judged of from the quantity of residue left on evaporating the chlorhydric acid solution of aluminum, etc. In most cases, two or three grms. of solid hydrate of sodium will be enough for a single operation. If the mixture to be analyzed contains a large proportion of iron, the ferric hydrate precipitated by the alkali will retain a certain amount of alumina. It is best, therefore, when the amount of ferric hydrate is large, to redissolve the washed precipitate in chlorhydric acid, to evaporate the chlorhydric acid solution as before, and to pour the neutral or nearly neutral solution into a new quantity of hot soda lye. This second alkaline solution must of course be added to the first, after filtering to separate the ferric hydrate, before proceeding to precipitate the aluminum. After the ferric hydrate has been thoroughly washed with hot water to remove the aluminate of sodium, it must be again washed with a hot, but not too strong, solution of chloride of ammonium, to remove a small quantity of alkali which would otherwise be retained by the precipitate. The precipitate is finally washed with hot water until the filtrate no longer gives any reaction when tested with nitrate of silver. Since the aluminum is to be thrown down with chloride of ammonium, the saline wash liquor, above mentioned, need not be kept separate from the remainder of the filtrate.

An older method of procedure is to add the alkali to the solution of aluminum and iron, instead of pouring the aluminum solution into the alkali. In any event, the iron and aluminum are usually precipitated together as hydrates in the first place, and the mixed precipitate collected on a filter and washed. In the old process the subsequent operations are as follows:—Scrape the moist precipitate from the filter with a platinum spatula, and place it in a porcelain, or better, a platinum dish. Set the dish beneath the funnel which holds the filter, and pour drops of hot chlorhydric acid into the latter until all the precipitate which had adhered to it has dissolved. Remove the dish and concentrate its contents, if need be; wash the filter with warm water and collect the washings in a beaker. Add concentrated soda lye to the liquid in the dish until the excess of acid is almost neutralized. Heat the liquid to boiling, remove the lamp and place in the dish a lump of hydrate of sodium or hydrate of potassium, large enough to dissolve all the aluminum which the mixture contains. If the proportion of iron in the mixture be small, the precipitate produced by the soda will, after a short time, contain little or no aluminum, but only ferric or manganic hydrate. Pour the contents of the dish into the beaker which contains the rinsings of the filter, and wash the dish thoroughly with water. Filter off the aluminate of sodium from the insoluble precipitate, wash the latter with boiling water, dissolve it in chlorhydric acid, and precipitate the iron as Hydrate of Iron. Acidulate the filtrate with chlorhydric acid, and precipitate the aluminum as Hydrate of Aluminum.

Since the residual ferric hydrate is liable to retain 1 or 2 per cent of hydrate of aluminum, it must be redissolved in chlorhydric acid, and the solution treated with caustic soda, as before, to ensure the complete removal of the aluminum, unless, indeed, the ferric precipitate is so small that the actual weight of the aluminum contained in it is insignificant.

In the foregoing cases the iron is supposed to be in the form of a ferric salt, but some chemists prefer to reduce the iron to the condition of a ferrous salt, by means of a solution of sulphurous acid, or of sulphite of sodium, before proceeding to separate it from aluminum. Their method is as follows:—Heat the tolerably concentrated, acid solution of aluminum, etc., to boiling, in a flask or capacious dish, best of silver or platinum, remove the lamp and add enough sulphite of sodium to reduce the iron completely to the state of protoxide. Again heat the liquor to boiling, keep it boiling for some time, and neutralize the acid with carbonate of sodium, added cautiously by small pieces. Pour in an excess of caustic soda or potash lye (or of ethylamin), and continue to boil sometime longer. If much iron be present, the voluminous white precipitate of ferrous hydrate thrown down at first will finally be converted into black, granular ferroso-ferric oxide. Remove the lamp and allow the mixture to settle; pour the clear liquid into a filter made of not too porous paper; boil the precipitate with a fresh quantity of soda lye, and wash it thoroughly with hot water, first by decantation, and afterwards upon the filter. The aluminum is determined in the filtrate as Hydrate of Aluminum.

Precautions. If, as is often the case, the mixture of aluminum, etc., to be analyzed, contains magnesium, a portion of the latter will be left combined with the ferric hydrate upon the filter. Some aluminum is also likely to remain undissolved in combination with the magnesium. So, too, in presence of calcium some aluminum is apt to escape solution. If chromium be present, most of it will remain undissolved with the hydrate of iron, but a small quantity sometimes oxidizes and passes into the filtrate as chromate of sodium.

In case the iron is reduced by sulphite of sodium. the liquid is liable to bump violently before it actually boils. To prevent this bumping a spiral coil of platinum wire may be placed in the liquid, or the flask may be shaken continually until its contents boil. When boiling has once begun the bumping ceases.

Special care must be taken that the soda or potash used be free from aluminum and silicon. The use of a porcelain dish should be avoided if possible, since portions of the dish are dissolved by the hot alkali, and impurities thereby added to the substance to be analyzed.

Method B. Fuse the mixed oxides of aluminum, iron, cobalt and nickel, with hydrate of sodium or of potassium, in a silver crucible. Boil the cold mass with water and filter to separate the soluble aluminate from the other oxides, which remain undissolved. The residual oxides, though free from aluminum, hold a certain proportion of sodium or potassium in combination, as well as a small quantity of oxide of silver derived from the crucible. The silver remains as an insoluble powder (chloride of silver) when the ferric oxide is treated with chlorhydric acid.

Anhydrous oxide of aluminum, as it occurs in nature, is not readily attacked by alkalies. Chenevix, for example (cited in *Pfaff's Handbuch analyt. Chem.*, 1824, **1.** 449), found that the mineral corundum could not be decomposed by intense ignition with 6 times its weight of caustic potash. To effect solution the mineral should be fused with bisulphate of sodium. Chenevix fused with 200 to 250 parts of borax glass and treated the product with chlorhydric acid.

Aluminum

Is weighed in the form of anhydrous sesquioxide. It is usually precipitated as a hydrate, though sometimes as a basic acetate or formiate, or as oxide. For the separation of aluminum from the other metals, see the reference list in the Appendix.

Ammonia.

(Compare Nitrogen, Nitrogen compounds, and Hydrate of Ammonium).

Principle I. Comparative lightness of the aqueous solution, in proportion as it contains more ammonia.

Application. Technical estimation of the value of ammonia-water.

Method. Take the Specific Gravity of the liquor with a Hydrometer, and refer to the tables of "specific gravity and per cent ammonia," in any dictionary of chemistry. Compare Acidimetry (Method by specific gravity). The liquid tested must contain no other soluble substance besides water and ammonia. In case the matter under examination be contaminated with any non-volatile impurity, distil off a large fraction of a measured portion of the liquid, as described under Alcohol (volatility of), and take the specific gravity of the distillate. If any portion of the ammonia is combined with an acid, mix a quantity of alkali with the liquid before distilling it, as explained below under the principle Volatility. (Pfaff, *Handbuch analyt. Chem.*, 1825, **2.** 25). The process is far inferior to those which depend upon the neutralization of ammonia by standard acids.

Principle II. Power of neutralizing acids.

Applications. Estimation of ammonia in its aqueous solution. Absorption of ammonia gas by acids. Use of ammonia-water as a neutralizer and precipitant, in a multitude of cases, and as a standard alkali in Acidimetry.

Method A. To estimate ammonia in a solution, weigh or measure out a quantity, say 10 c. c. of the sample to be tested, add 1 or 2 c. c. of litmus solution, and saturate with a standard acid in the manner described under Alkalimetry. Or measure out a definite quantity of the standard acid, color it with litmus, and saturate with the ammonia to be tested. In either case, the point of saturation is hit without difficulty. It is well always to weigh a definite volume of the liquid to be tested, for by dividing the weight of the liquid in grms. by its volume in c. c., we obtain the specific gravity of the solution, and may subsequently dispense with the balance in case any new portion of the liquid has to be taken for analysis. The method of supersaturating the ammonia-water with standard acid and determining the excess of acid with a standard soda solution, is not to be recommended, since ammonium salts, even when neutral, color litmus violet.

Method B. Slightly supersaturate the ammonia with chlorhydric acid, evaporate to dryness at 100° to 200°, and weigh the chloride of ammonium. Or (after Mohr, *Titrirmethode*, 1855, **2.** 58) estimate the chlorine in the dry chloride of ammonium with a dilute standard solution of nitrate of silver (see Chloride of Silver), and calculate how much ammonium would be equivalent to this chlorine.

With regard to the use of ammonia as a reagent, the analyst should remember that the ordinary ammonia-water of commerce is often impure. Besides more or less carbonate of ammonium, it is liable to contain no inconsiderable quantity of a soluble compound of iron and organic matter, which is apt to be dragged down by gelatinous precipitates, and to interfere in many ways with the accuracy of analyses.

For quantitative work, chemically pure ammonia-water should either be procured from a manufacturer of fine chemicals, or prepared expressly by distilling from a glass flask a mixture of 1 part of pure chloride of ammonium, 1.25 parts of slaked lime, and 1 to 1.25 parts of water. The distillate is received in three Woulfe bottles; of which the first, charged with a small quantity of water, or milk of lime, serves to wash the gas. The second bottle should contain about as much water, by weight, as there is chloride of ammonium in the flask, or in case a saturated solution is required, take two-thirds of this weight of water; the bottle should not be more than three-quarters full at first, to allow for expansion. The third bottle should contain but little water. All the bottles should be set in a dish of cold water to facilitate the solution of the gas. Heat the flask upon a sand bath, taking care to avoid foaming, until half the water has distilled from it.

Ammonia-water should be kept in glass-stoppered bottles. It should leave absolutely no residue when evaporated to dryness, should give no precipitate when diluted and tested with lime-water, or with chloride of barium; or when tested with sulphuretted hydrogen or nitrate of silver, after acidulation.

Principle III. Volatility.

Applications. Separation of ammonium from all the elements. Estimation of ammonia in rain and river water (Method A). Estimation of ammonia in copulate ammonio-compounds, in urine, manures, etc. (Methods B and C).

Method A. The mixture is boiled with caustic soda, potash, lime, or baryta.

1. Prepare a distillatory apparatus as follows: Fit to a glass flask a perforated cork or caoutchouc stopper carrying a short delivery tube bent at an obtuse angle; by means of a rubber connector attach this delivery tube to the head of a small worm or of a short Liebig's condenser, the lower end of which passes through a perforated cork into a capacious tubulated receiver; and to the second orifice of the receiver attach a U tube, by means of bent glass connections. According to S. W. Johnson, the worm or condensing tube should be made of block tin, since glass yields a sensible amount of alkali to hot steam.

Pour into the flask as much of a moderately strong solution of caustic soda, caustic potash, or milk of lime, as will fill something more than a third of it. Place the flask in a slanting position upon a wire gauze support and boil its contents until every trace of ammonia, with which the alkali may have been contaminated, is removed. Then cork the flask and leave it until its contents have become thoroughly cold.

Weigh out the substance to be analyzed in a glass tube, 3 or 4 c. m. long by 1. c. m. wide, closed at one end. Place the tube and its contents in the flask after the latter has become cold, and connect the flask with the condensing apparatus.

Measure off a quantity of standard oxalic, chlorhydric or sulphuric acid (see Acidimetry), more than sufficient to absorb all the ammonia which can possibly be expelled from the substance taken, and mix it with 2 or 3 c. c. of a solution of litmus. Pour a little of this acid into the U tube of the apparatus, but no more than will fill the lower part of the tube in such manner that bubbles of air can readily pass through the liquid. Pour the rest of the measured quantity of acid into the tubulated receiver, together with a little water.

After proving that all the joints of the apparatus are tight, heat the flask until its contents boil slowly, and continue the operation until some time after the drops of water falling from the condenser have ceased to give the least tinge of blue at the moment when they strike the liquid in the receiver.

Pour the contents of the receiver and U tube into a beaker, rinse with water and determine the amount of free acid in the liquid by titration with a standard solution of caustic soda (see Acidimetry). By subtracting the amount of acid thus found from the quantity of acid originally taken, we obtain the amount of acid which has been neutralized by the ammonia. The amount of the latter is then calculated (see Alkalimetry).

This method affords accurate results, and can be used in all cases where the substance to be analyzed contains no nitrogenized matter, other than ammonium salts, capable of decomposition by caustic lyes. It is to be observed that the flask in which the decomposition is effected must be placed in a slanting position, so that no particles of the fixed alkaline liquid can be thrown into the delivery tube by the movement of ebullition, and that the end of the condenser must not dip into the liquid in the receiver.

If it be desirable to weigh the ammonium in the form of a solid rather than to estimate it by titration, the receiver and U tube may be charged with chlorhydric acid, and the ammonium weighed as Chloride of Ammonium or as Chloroplatinate of Ammonium.

2. Place a litre of the water to be examined in a retort capable of holding at least four litres, add to the liquid 25 c. c. of baryta water,—or in place of the baryta, lime, potash or soda,— to retain the carbonic acid. Connect the retort with a condensing apparatus, and distil the water slowly until the distillate amounts to a quarter litre. Determine the proportion of ammonia by titrating with normal sulphuric acid (Boussingault), or determine the ammonia in the distillate by Nessler's test. See Iodide of Mercurammonium (Miller).

3. For determining ammonia in urine, F. Mohr (*Titrirmethode*, **2.** 216) has proposed, in place of Method C, the following modification of the process:—Carefully neutralize the urine with a dilute solution of caustic potash, add to the neutral liquor a measured quantity of standard potash, more than sufficient to decom-

pose all the ammonium salts in the solution, boil the mixture as long as ammonia continues to be evolved, and finally, without heeding the ammonia which is set free, determine the amount of caustic alkali left in the urine, by means of a standard acid. The points of neutralization are determined in both cases by means of litmus paper. The presence of urea does not affect the accuracy of the results, although this substance is decomposed by boiling potash, nor does hippuric acid do any harm. In cases where no great degree of accuracy is demanded, the process appears to be applicable for testing human urine, though in that case it indicates rather less ammonia than the processes of Schloesing and Boussingault. But for testing the urine of cattle it cannot be relied on. Rautenberg (*Zeitsch. analyt. Chem.*, 1863, **4.** 500) has shown that with the urine of oxen Mohr's method always indicates a larger proportion of ammonia than can be obtained by the methods of Schloesing and Boussingault, the accuracy of both of which has been well established. This apparent excess seems to be due to the decomposition of various ingredients of the urine which, though originally neutral, become acid when exposed to the action of boiling potash, and so neutralize a portion of the standard alkali.

Method B. The Ammonium Compound is heated in a combustion tube with an excess of Soda-Lime, in the manner described under Nitrogen. The ammonia is collected in acid and determined as in Method A.

Method C. The Ammonia is set free by Hydrate of Calcium, or by potash or soda lye, at a low temperature.

1. *By ebullition in vacuo* (Boussingault's process, *Mémoires de Chimie Agricole*, Paris, 1854, p. 292). To a strong flask of about 1 litre capacity, fit tightly a cork carrying one straight tube provided with a stop-cock and one gas delivery tube, bent at a right angle. The straight tube should reach to within a few m. m. of the bottom of the flask, while the bent tube merely passes through the cork. By means of a short piece of rubber tubing tied tightly to the glass, connect the gas delivery tube with another glass tube of similar bore, bent at a right angle and reaching nearly to the bottom of a narrow cylinder proper to receive a charge of standard acid. The cylinder is fitted with a cork, and is connected by means of a second bent glass tube, provided with a stop-cock, with the receiver of an air pump. The corks of the apparatus must fit tightly enough to support the external pressure which will be exerted upon them by the atmosphere, when the air is pumped out from within the vessels. To strengthen the corks it is well to wind about their heads strips of sheet cork or of sheet lead, so that the head thus thickened may be as wide as the neck of the flask. The whole may then be bound firmly together by means of strips of sheet caoutchouc, wide enough to reach from the top of the cork to a point an inch or more below the rim of the flask. The caoutchouc is finally tied firmly with twine, both to the neck of the flask and to the head of the cork. When everything is ready, place 10 c. c. of standard acid in the cylinder (Boussingault uses sulphuric acid strong enough that 10 c. c. shall saturate 0.2125 grm. of ammonia), and set the cylinder in a beaker of cold water—ice water is best, though the apparatus gives satisfactory results when the temperature of the water is as high as 12° or 15°. Pour into the flask 50 grms. of the urine to be tested, add to it about 5 grms. of slaked lime, cork the flask and set it in a water bath so arranged that it may be kept constantly at a temperature of 35° to 40°. Close both the stop-cocks upon the apparatus, exhaust the receiver of the air pump, and then slowly open the stop-cock on the tube which connects the receiver with the cylinder charged with acid. The liquid in the flask will soon begin to boil. When this happens, close the cock again, and leave the apparatus to itself. Since the cylinder which contains the acid is comparatively cold, the vapors distilled from the flask immediately condense in it. The liquid in the flask consequently continues to boil tranquilly, and after a short time will evaporate completely, so that nothing but a dry residue is left in the flask. In order to sweep forward any ammonia vapor which may be left in the flask, open the stop-cock above it slowly, so that air may enter. Then close the stop-cock, exhaust the receiver, and slowly open the second stop-cock beyond the acid cylinder, so that the air in the flask may be drawn forward through the acid. Finally determine with a standard alkali (Boussingault uses a solution of lime in sugar water) how much of the standard acid originally taken is still left unsaturated in the absorption cylinder. (See Acidimetry). The process requires far less time than No. 2, and is equally accurate. It has the disadvantage of requiring comparatively complex apparatus.

Instead of this method of ebullition in vacuo, Boussingault has attempted to remove the ammonia from mixtures of urine and slaked lime by means of a current of air made to bubble through the liquid continually through 4 or 5 hours, at the temperature of 35° or 40°; but without good results, for a part of the ammonia was always retained in the solution. To ensure the complete and speedy evolution of the ammonia in this way, Boussingault found that the solution must be heated to 90° or 100°. But these high temperatures are inadmissible in the analysis of urine, for ammonia would be formed through the decomposition of the urea which the urine contains.

2. *By Exhalation at the Ordinary Temperature.* (Schloesing's method). Select a shallow, flat-bottomed capsule 10 or 12 c. m. in diameter and weigh it. Measure out something less

than 35 c. c. of the liquid to be tested, and place it in the dish. Weigh the dish and liquid. Then set the dish on a common dinner plate filled with mercury. Bend a thick glass rod into the form of a tripod, place this tripod in the capsule which contains the solution to be tested, set upon it another shallow dish which has been charged with 10 c. c. of standard oxalic or sulphuric acid (see Alkalimetry), and invert a beaker over the whole. Fill a pipette, provided with a wide aperture, with milk of lime, lift up one side of the beaker as far as may be necessary, and allow the contents of the pipette to flow into the solution of the ammonium salt. Immediately replace the beaker and put a weight upon it, so that its lower edge shall be pressed into the mercury. After 48 hours lift the glass at one side and thrust a bit of moistened red litmus paper into the atmosphere within; if the color of the paper remains unchanged, the first stage of the operation is finished; but if the paper become blue the glass must be replaced, and the apparatus left to itself for another term of hours. When all the ammonia has been expelled from the original solution, and has been absorbed by the standard acid, determine how much free acid is left, by titration with a standard solution of caustic soda (see Acidimetry), and calculate the amount of ammonia from that of the acid which it has neutralized.

Instead of the beaker and plate of mercury above described, a bell glass with ground rim may be placed air tight upon a greased ground glass plate. A tubulated bell provided with a ground glass stopper is to be preferred, since, in this case a strip of litmus paper attached to a thread may be introduced into any part of the jar without lifting the latter.

Method C, in both its modifications, is useful in cases where the presence of organic matters decomposable by boiling alkalies precludes the use of Method A. — Reischauer (*Zeitsch. analyt. Chem.*, 1864, **3**. 138) has shown that even after six months' action, cold potash or soda lye has no power to set free ammonia from cyanogen compounds.

According to Schloesing, 48 hours are always sufficient to expel 0.1 to 1 gramme of ammonia from 25 to 35 c. c. of solution; but Fresenius has found that this statement is true only with regard to quantities less than 0.3 grm. When the quantity of ammonia exceeds 0.3 grm., 48 hours is often insufficient for the complete expulsion of the ammonia; it is well, therefore, to operate, if possible, upon quantities of substance which contain no more than this proportion.

Ammonium.

For the separation of ammonium from the several elements, see finding list in Appendix.

Ammonium is usually determined as chloride, chloroplatinate or tartrate; or by titrating ammonia-water with an acid of determined strength. Sometimes the nitrogen of an ammonium compound is collected and measured, and the amount of ammonium calculated from that of the gas. Ammonium may be separated from all metals excepting those of the alkalies by precipitating the metals by means either of H_2S, $(NH_4)HS$, $(NH_4)_2CO_3$, or Na_2HPO_4.

Ammonium Salts.

(Compare Nitrogen and Nitrogen Compounds.)

Principle. Volatility.

Applications. Separation of certain ammonium salts from salts of Li, Na, K; Ba, Ca, Sr; Mg, Zn, Cd; Al, Cr, Ur, Mn, Fe, Co, Ni. The ammonium salt must be wholly volatile, and the mixture to be examined free from other volatile or decomposable matters.

Methods.

1. *Separation of Ammonium Salts from Salts of* Na, Li, K; Ba, Sr, Ca, Mg and Cr. In case the dry mixture to be examined contains but a single acid, snch as chlorhydric or sulphuric acid, heat a weighed portion of it to faint redness in a covered platinum crucible, as long as any fumes are evolved. The difference between the weight of the crucible and contents before and after the ignition gives the weight of the ammonium salt. The crucible must be heated gently at first, but must afterwards be kept for some time at a dull red heat. In case the salts are sulphates, the crucible must be heated with special care, in order to avoid loss of material through decrepitation of the sulphate of ammonium, and the residue must finally be ignited in an atmosphere of carbonate of ammonium (see Sulphate of Potassium) for the purpose of decomposing a quantity of acid sulphate of potassium, which is formed by the decomposition of a part of the sulphate of ammonium. — If the mixture to be analyzed contains more than one acid, it may be moistened with a quantity of free acid, similar to the least volatile of the acids in the mixture, and evaporated to dryness upon a water bath. The operation should be repeated several times, or until the more volatile acids have been completely expelled. A mixture of chloride of ammonium and sulphates of the alkali metals cannot be analyzed in this way, for on igniting the mixture the sulphates would be wholly, or in part, converted into chlorides.

2. *Separation of Ammonium Salts from Salts of* Zn, Cd, Al, Ur, Mn, Fe, Co, Ni.

If there is no chloride of ammonium in the mixture to be analyzed, a weighed quantity of the mixture may be ignited directly, as in No. 1. But if chloride of ammonium be present, the process becomes less accurate. When compounds of aluminum and iron are ignited with chloride of ammonium, a certain amount of chloride of aluminum or of chloride of iron, as the case may be, is lost through volatilization. Under like circumstances manganese compounds are converted into protochloride of manganese mixed with some manganite of manganese; zinc compounds volatilize as chloride of zinc, and compounds of cobalt and nickel are reduced to the metallic state. The safest rule in all these cases is to mix the sample with carbonate of sodium before igniting it, and to determine the metals in the residue by some appropriate process. The ammonia may then be estimated in a separate portion of the mixture by distillation with an alkali. (See Ammonia).

Ammonio-Sesquioxide of Uranium.

See Oxide of Uranium (ammoniated).

Antimonic Acid.

Principle I. Sparing solubility in nitric acid.

Applications. Separation of antimony from Mg, Zn, Cd; Mn, Fe, Co, Ni; Bi, Cu, Hg, Ag and Pb in alloys.

Method. Dissolve the alloy in nitric acid or in aqua regia, as described under *bin*Oxide of Tin. Collect the insoluble residue of antimonic acid upon a filter and ignite to convert it into Antimoniate of Antimony. The results are only approximative, since a small portion of the antimony always remains dissolved in the acid. Alloys of lead and antimony, containing a large proportion of antimony, should be fused with a weighed quantity of pure lead before treating them with nitric acid, lest a part of the alloy escape solution. — According to Pfaff (*Handbuch analyt. Chem.*, 1825, **2.** 416), the antimonic acid is apt to retain a certain quantity of oxide of lead, so firmly combined that it does not dissolve in nitric acid.

This process was formerly much employed for analyzing antimony alloys, and is still used for technical analyses, where no great accuracy is required. It is now recognized, however, that—owing to the solubility of antimonic acid in nitric acid—the process yields far less accurate results than the corresponding method with Oxide of Tin. By repeatedly evaporating the nitric acid solution of antimonic acid to dryness, and treating the residue again and again with fresh portions of nitric acid, it is indeed possible to obtain at last a solution free from antimonic acid, but the operations require so much time that they are not employed in actual practice.

Principle II. Power of changing stannous to stannic chloride, while it is itself reduced to the state of antimonious acid.

Applications. Estimation of antimony in cases where the metal can be converted into antimonic acid.

Method. Mix a weighed quantity of the antimonic acid with a measured volume of a standard solution of protochloride of tin, together with some iodide of potassium solution and starch paste. The quantity of the tin solution must be more than sufficient to reduce the whole of the antimonic acid, and the mixed solution must be heated to 40°. Finally determine how much stannous chloride remains in the solution, by means of a standard solution of *bi*Chromate of Potassium. The process is of limited application, and is said not to yield very accurate results. (See Chloride of Tin).

Antimoniate of Antimony.

(Improperly "Antimonious Acid").

Principle. Fixity of the compound when heated.

Applications. Estimation of antimony in antimonious and antimonic acids, and in compounds of these acids with easily volatile or decomposable oxygenated acids or bases. Determination of antimony in sulphide of antimony.

Method. Antimonic acid may be simply ignited in a platinum crucible until the weight remains constant. Other compounds of antimony should be treated with nitric acid free from chlorhydric acid, and the solution carefully evaporated to dryness before igniting them. Care must be taken to guard the contents of the crucible against the action of reducing gases coming from the filter or the flame.

For converting sulphide of antimony into antimoniate of antimony, Bunsen has devised two methods, as follows:—

1. *With fuming Nitric Acid.* Place the dry sulphide in a weighed porcelain crucible, moisten it with a few drops of nitric acid of 1.42 specific gravity, and cover the crucible loosely with a watch glass or small funnel from which the tube has been cut away. Carefully pour upon the sulphide 8 or 10 times its bulk of red fuming nitric acid, place the crucible on a water bath and allow the acid to evaporate slowly. When the nitric acid is first added to the sulphide, a quantity of sulphur separates in fine powder, but subsequently oxidizes completely during the process of evaporation, so that nothing but a white mixture of antimonic and sulphuric acids is left in the crucible. By igniting this residue it is converted into antimoniate of antimony. — In case the sulphide of antimony under examination happens to be mixed with a large proportion of free sulphur, it must be washed with bisulphide of carbon before adding the nitric acid. The washing may be effected as follows:—By means of a perforated cork, fit the funnel which contains the filter and the dried sulphide of antimony, air tight to the mouth of a test tube; pour enough sulphide of carbon into the filter to cover the dry precipitate, then cover the funnel tightly with a glass plate, and leave the apparatus at rest for several hours. Finally loosen the cork and allow the bisulphide of carbon to flow into the test tube. Repeat the operation, if need be, with a fresh quantity of bisulphide of carbon; 10 or 15 grms. of the bisulphide will, in most cases, be enough for washing a single precipitate; it may always be

easily recovered by distillation and kept for future use.

It is to be observed that *fuming* nitric acid is essential to the success of this process. The ordinary strong nitric acid of 1.42 specific gravity will not answer. The boiling point of the acid of 1.42 specific gravity is almost 10° higher than the melting point of sulphur, while the fuming acid boils at a temperature as low as 86°, far below that at which sulphur melts. The hot fuming acid easily oxidizes the finely divided sulphur with which it is in contact, but when heated with acid of 1.42 specific gravity, the particles of sulphur quickly melt to a solid ball, which obstinately resists oxidation.

No chlorhydric acid should be present lest some of the antimony be lost through volatilization of the terchloride when the dry mass is ignited.

2. *By Ignition with Oxide of Mercury.* Mix the sulphide of antimony with from 30 to 50 times as much precipitated oxide of mercury, and heat the mixture gradually in an open weighed porcelain crucible. Remove the lamp as soon as the appearance of gray fumes of mercury indicates that oxidation has begun. Again heat the mixture as soon as the fumes slacken, and proceed in this way as long as fumes are evolved. Finally ignite the crucible over a blast lamp to remove the last traces of oxide of mercury, and weigh the residual antimoniate of antimony.

Care must be taken that no reducing gases from the lamp gain access to the contents of the crucible. Since oxide of mercury always leaves a small quantity of fixed residue, even after intense ignition, it is well to determine the proportion of this impurity once for all, to weigh roughly the amount of oxide of mercury taken to oxidize any sample of the sulphide, and to subtract the amount of fixed residue contained in the oxide of mercury taken from the weight of the antimoniate of antimony.

If the sulphide of antimony is mixed with free sulphur, this sulphur must be removed by means of bisulphide of carbon, as in No. 1, in order to avoid slight deflagrations and consequent loss of substance, which would otherwise occur when the sulphide came to be heated with oxide of mercury.

The operation may be completed much more quickly in a platinum than in a porcelain crucible. But if a platinum crucible be employed it must be protected from the action of antimony by means of a lining of oxide of mercury. This lining may be made as follows: — Soften the end of a test tube at the blast lamp, place the soft end of the tube in the centre of the platinum crucible, and blow air into the other end of the tube, so that the hot glass may assume the exact form of the crucible. Crack off the bottom of the bulb thus formed, and carefully smooth the sharp edge by fusion. A glass is thus obtained, open at both ends, which exactly fits the crucible. Fill the crucible loosely with oxide of mercury to the brim, and slowly push the glass through the mercury to the bottom of the crucible, occasionally shaking out the oxide from the interior of the glass. A layer of oxide of mercury from half a line to a line thick, may thus be pressed against the crucible so firmly that it will adhere to the platinum after the removal of the glass.

Properties. Antimoniate of Antimony is a white powder when cold, but exhibits a yellowish tint while hot. It neither fuses nor decomposes when ignited in the air. It is scarcely at all soluble in water, though it exhibits an acid reaction when placed upon moist litmus paper. It is not acted upon by sulphydrate of ammonium, and dissolves in chlorhydric acid with very great difficulty. Its composition is: —

$$\begin{array}{lcrcr} Sb & = & 122 & = & 80.26 \\ O_2 & = & 32 & = & 19.74 \\ \hline & & 154 & & 100.00 \end{array}$$

Antimoniate of Mercury.

(Mercurous antimoniate).

Principle. Insolubility in water.

Applications. Separation of antimony from potassium and sodium.

Method. Mix the solution of the antimoniate with an excess of a solution of nitrate of suboxide of mercury. Allow the mixture to stand at rest for many hours, collect the precipitate upon a filter, wash with a solution of mercurous nitrate, dry, ignite strongly, and weigh as Antimoniate of Antimony. Remove the excess of mercury from the filtrate by means of sulphuretted hydrogen or chlorhydric acid, and determine the alkalies in the final filtrate in the usual way. — Antimoniate of mercury, as formed by the mixture of mercurous nitrate and antimoniate of potassium or sodium settles with extreme slowness, but the mixture may nevertheless be filtered without special difficulty, if it be first allowed to stand for a long time.

Antimoniate of Sodium.

Principle. Insolubility in dilute alcohol.

Applications. Separation of antimony from arsenic and tin.

Methods.

1. Place the substance to be analyzed, which may be an alloy of antimony, arsenic and tin, or a mixture of sulphide of antimony and sulphide of tin, in a large beaker, and pour upon it, little by little, nitric acid of 1.4 specific gravity, until the oxidation is completed; an alloy should be reduced to the state of fine powder before weighing. For an alloy, the nitric acid employed must be of the prescribed strength, since acid of 1.52 specific gravity would not attack the alloy; but for the treatment of sulphides the stronger acid is to be preferred. When the reaction has ceased to be violent, transfer the mixture from the beaker to a small porcelain dish, evaporate to dryness

on a water-bath, transfer the residue to a silver crucible, rinse out the porcelain dish with a solution of caustic soda, and again evaporate to dryness. Add to the contents of the crucible as much solid hydrate of sodium as will amount to about eight times the bulk of the residue, and fuse the mixture for some time at a red heat. When the crucible has become cold, cover the fused mass with water, and allow it to soak until it softens, then wash out the contents of the crucible into a beaker, with hot water, and continue to add water until the undissolved residue has assumed the form of a fine powder. Stannate and arseniate of sodium go into solution while most of the antimoniate of sodium remains undissolved. Add to the solution in the beaker as much alcohol of 0.83 specific gravity as will amount to about one-third the volume of the solution. Cover the beaker with a glass plate, and let the mixture stand for 24 hours with frequent stirring. Collect the insoluble matter in a filter, rinse the beaker with spirit made by mixing 1 vol. of alcohol of 0.83 specific gravity with 3 vols. of water, and wash the precipitate upon the filter, first with a mixture of 1 vol. alcohol and 2 vols. water, then with a mixture of equal volumes of alcohol and water, and finally with a mixture of 3 vols. alcohol and 1 vol. water. It is well to mix a few drops of a solution of carbonate of sodium with the dilute alcohol used for washing. The carbonate facilitates the solution of the stannate of sodium, and hinders the antimoniate of sodium from passing through the pores of the filter. — Continue to wash until a portion of the filtrate acidified with chlorhydric acid and mixed with sulphuretted hydrogen water gives no yellowish precipitate of sulphide of tin after long standing.

It is essential that the alcohol used for washing shall be of the prescribed strengths. If strong alcohol were used at first, a quantity of carbonate of sodium formed during the fusion would be left undissolved upon the filter, and would retain stannate of sodium in combination, to such an extent as to occasion losses even as great as 8 or 9 per cent. Weak alcohol, on the other hand, dissolves some antimoniate of sodium, together with the stannate. Water alone cannot be employed for the washing. Not only is antimoniate of sodium somewhat soluble in water, but a portion of the precipitate itself would pass through the pores of the filter as soon as the stannate and carbonate of sodium had been washed away, if nothing but water was employed.

When the antimoniate of sodium has been thoroughly washed, rinse it from the filter into a beaker, leach the paper with a mixture of chlorhydric and tartaric acids, dissolve the precipitate in the same mixture of acids, and precipitate the antimony as Sulphide of Antimony. The antimoniate of sodium cannot be weighed directly, since, though free from stannate of sodium, it always retains some carbonate or sulphate of sodium when washed as above directed. — The process yields excellent results when properly conducted. The antimoniate of sodium settles completely from the alcoholic solution, and the latter is easily filtered. The only real inconvenience arises from the necessity of using a silver crucible, and the consequent liability of contaminating the alkaline filtrate with a small amount of silver.

To determine the tin and arsenic in the filtrate, acidulate the liquid with chlorhydric acid, and without heeding the precipitate of stannic arseniate which forms, pass sulphuretted hydrogen gas for some time through the turbid liquid. Allow the mixture to stand until the odor of sulphuretted hydrogen has well nigh disappeared, collect and weigh the mixture of sulphide of tin, sulphide of arsenic and free sulphur upon a tared filter, and finally heat a portion of the precipitate in a current of hydrogen to expel the sulphur and Sulphide of Arsenic.

If there be no arsenic, but only tin in the filtrate, drive off most of the alcohol by evaporating at a gentle heat, dilute with water and supersaturate with sulphuric acid to precipitate Hydrate of Tin. Or precipitate Sulphide of Tin. The precipitation with sulphuretted hydrogen is safer than the other process, though the sulphide of tin will be contaminated with a trace of sulphide of silver from the crucible, while by using sulphuric acid no compound of silver is thrown down.

In case there be no tin, but only antimony and arsenic is the substance to be analyzed, the arsenic may be determined as Arseniate of Magnesium and Ammonium. To this end, heat the alcoholic filtrate, with the addition of several fresh quantities of water, until the odor of alcohol has almost disappeared, acidulate with chlorhydric acid and proceed in the usual way.

2. In the case of mixtures of the sulphides of arsenic and antimony, together with free sulphur, such as are often obtained in mineral analyses, the process may be modified, as follows:—Place the precipitate in a porcelain crucible, oxidize it with red fuming nitric acid free from chlorine, evaporate nearly to dryness, mix the residue with an excess of carbonate of sodium, together with some nitrate of sodium, and fuse at the lamp. Treat the fused mass as directed in No. 1.

Antimonious Acid, (Sb_2O_3). [For the compound Sb_2O_4—sometimes improperly called antimonious acid, see Antimoniate of Antimony.]

Principle 1. Oxidation by Iodine (A) or by Chlorine (B) in alkaline solution; by Bichromate (C) or Permanganate of Potassium (D), or by Salts of Gold (E).

The Applications of A, C, and D, are limited to the estimation of antimonious acid in pure solutions of this substance—such as the chlorhydric acid solution of sulphide of antimony and, as regards A and D, a solution of tartar-emetic — and to the determination of antimonious acid when mixed with antimonic acid. The applications of B will appear below under the head of Antimony.

Methods. To separate antimonious from antimonic acid, determine the total amount of antimony in one portion of the substance to be analyzed, by precipitation of Sulphide of Antimony. Determine the amount of antimonious acid, in another portion, by one of the processes enumerated below, and calculate the amount of antimonic acid from the difference.

A. Weigh out as much of the substance to be tested as will contain about 0.1 grm. of antimonious acid. Dissolve the weighed substance in 10 or 12 c. c. of a strong aqueous solution of tartaric acid, in case it be not already a tartrate, and add enough carbonate of sodium solution to nearly neutralize the liquid. Mix the solution with 20 c. c. of a cold saturated solution of bicarbonate of sodium, add a few c. c. of thin starch paste, and pour into the mixture, drop by drop, from a burette, a standard solution of Iodine in iodide of potassium until the liquid just remains blue, or better, is of a faint red color. The mixture must of course be stirred continually while the iodine solution is being added to it: —

$$Sb_2O_3 + 2Na_2O + 4I = Sb_2O_5 + 4NaI.$$

So long as there is any antimonious acid present to be oxidized, the blue color formed at the surface of the liquor where the iodine solution first touches the starch, will be destroyed as fast as it forms. But as soon as the last trace of antimonious acid is oxidized, the whole solution will become blue. The operation must be stopped at this moment. The blue color will, in any event, disappear after a few minutes. The value of the iodine solution may be determined beforehand by titrating 0.2 or 0.3 grm. of pure crystallized tartar emetic. (Mohr, *Titrirmethode*, 1855, p. 371).

Though the results obtained by this process are, on the whole, satisfactory, it is, according to Fresenius, essential that the proportion of antimonious acid to that of bicarbonate of sodium be maintained tolerably near that of the quantities above enumerated, in order to ensure accuracy. Bicarbonate of sodium should always be employed as the alkaline liquor, since the monocarbonate has the power of fixing a certain amount of iodine.

Instead of titrating directly with a solution of iodine, as above described, H. Rose directs that the standard solution of iodine be added to the slightly alkaline solution of antimony as long as its color continues to be discharged, and that the excess of iodine be then determined with a standard solution of Hyposulphite of Sodium.

B. *Oxidation by Chlorine.* Same as under Antimony.

C. *Oxidation by Bichromate of Potassium.* To the solution of antimonious acid in diluted chlorhydric acid, add a measured volume of a standard solution of bichromate of potassium, more than sufficient to oxidize the whole of the antimony. Leave the mixture at rest for a short time, and finally determine the amount of unreduced chromate by means of a standard solution of ferrous sulphate, added until a drop of the mixture gives a blue precipitate when touched to a drop of ferricyanide of potassium. (See *bi*Chromate of Potassium). The original chlorhydric acid solution should contain at least one-sixth its volume of chlorhydric acid of 1.12 specific gravity. It will often be found convenient in practice to operate with solutions composed of about equal volumes of water and chlorhydric acid, but in case more acid than this is present, the delicacy and promptitude of the final reaction with ferricyanide of potassium is materially diminished. As with arsenious acid, bichromate of potassium does not act upon antimonious acid in any definite or reliable way when the solution in which the oxidation is to be effected contains less than one-sixth its volume of chlorhydric acid. Since tartaric acid decomposes the bichromate, its presence is inadmissible.

If the substance to be analyzed is free from organic matter, oxides of the heavy metals, and other substances capable of interfering with the titration, it may be dissolved at once in chlorhydric acid; otherwise the antimony must be precipitated, in the first place, as a Sulphide. To prepare the sulphide for titration, place the washed precipitate, together with the filter, in a small flask, cover it with chlorhydric acid, heat the mixture on a water bath until the precipitate has dissolved, add to the liquid as much of a nearly saturated solution of mercuric chloride in chlorhydric acid of 1.12 specific gravity as may be needed to remove the sulphuretted hydrogen, dilute the mixed solution to some definite volume, allow it to settle, and take up a measured volume of the clear liquid for the analysis. (Kessler, *Poggendorff's Annalen*, **95.** 215; **113.** 134; and **118.** 17).

D. *Oxidation by Permanganate of Potassium.* Pour a standard solution of permanganate of potassium from a burette into the chlorhydric acid solution of antimonious acid, until the solution exhibits a permanent red color. The permanganate solution should contain about 1.5 grm. of the crystallized salt to the litre, and the antimony solution at least one-sixth its volume of chlorhydric acid of 1.12 specific gravity. It is not well, however, to have the proportion of acid higher than one-third the volume of the liquid, since the final reaction would be interfered with. The presence of tartaric acid has little or no influence upon the reaction. Hence the process may be employed for analyzing tartar emetic, and the value of the permanganate solution may be determined in the beginning by means of a standard solution of pure tartar emetic.

For the preparation of the antimony solution see above, C. (Kessler, *Poggendorff's Annalen*, **118.** 17).

E. *Oxidation by Salts of Gold.* A process formerly recommended by H. Rose consisted in mixing the antimonicus acid, dissolved in a very large excess of strong chlorhydric acid, with an excess of chloraurate of sodium or chloraurate of ammonium, leaving the mixture at rest during several days at a temperature slightly warmer than that of the air, and collecting and weighing the metallic gold which was deposited, as was supposed, in quantity proportionate to the amount of antimonious acid in the solution. But since Dexter (*Pogg. Ann.*, **100.** 570; compare H. Rose, *ibid.*, **110.** 541) has shown that the process affords neither concordant nor reliable results, it can no longer be commended.

Principle II. Volatility.

Method. To separate antimony from silver, gold, and other noble metals, the alloy may be heated upon a cupel in a muffle. By the action of the hot air the antimony will be converted into antimonious acid, and the latter will go off in the form of vapor, leaving the silver or gold to be weighed. It has been found in practice that a simple alloy of silver and antimony heated upon bone ash in a muffle, until fumes of antimonious acid are no longer visible, still retains about one per cent of antimony, the residual button of silver being dull and gray, and only incompletely soluble in nitric acid. But by again heating the button on a cupel with about 5 times its weight of lead, until the lead has all been oxidized and the melted silver appears bright and lustrous, the antimony may be completely expelled.

Antimony. [Compare Antimony Compounds].

Antimony is estimated as metallic Antimony, as Sulphide of Antimony, Antimoniate of Antimony, Antimoniate of Sodium, or by titration, as has been explained under Antimonious Acid. See also Antimonic Acid, and the finding list in Appendix.

Principle I. Sparing solubility of the metal in chlorhydric acid.

Applications. Estimation of antimony in antimony salts. Separation of antimony from tin.

Method A. Precipitate the antimony by means of metallic zinc from a dilute nitric acid solution. The antimony falls as a black powder, which glistens when burnished. (Pfaff, *Handbuch analyt. Chem.*, 1825, **2.** 411). So long as zinc is present in the liquid the antimony suffers no oxidation or solution. Wash rapidly with hot water and dry at 100°, best in an atmosphere of non-oxidizing gas. — To separate antimony from tin, boil the alloy with chlorhydric acid [taking care to absorb any antimoniuretted hydrogen that is evolved (see below, Method D.)] until all the tin has dissolved, then place a rod of pure tin in the liquor to precipitate any traces of antimony which may have dissolved, and proceed as before. (Pfaff, *loc. cit.*, p. 416).

Method B. To analyze an alloy of tin and antimony, dissolve it completely in chlorhydric acid to which a little nitric acid has been added. Heat the solution nearly to boiling, and throw into it bits of fine iron wire (piano-wire), as long as the latter continues to dissolve. As soon as all the antimony has been precipitated, and the last piece of iron seems to have completely dissolved, add a little more chlorhydric acid; allow the precipitate to settle, decant the clear liquid, and try whether any further precipitate can be produced in it by means of iron. Wash the precipitated antimony at first with hot water acidulated with chlorhydric acid, afterwards with pure hot water, and finally with strong alcohol. To facilitate the operation of drying, and to still further guard against the risk of oxidation, it is well to wash out the alcohol with a few drops of ether. Dry the precipitate quickly at 100°, and weigh. — The process yields good results when the proportion of tin is large, but is less accurate when the solution contains but little tin. (Tookey & Clasen, *Zeitsch. analyt. Chem.*, 1865, **4.** 440).

Method C. Place the alloy or other compound in a small flask, cover it with strong chlorhydric acid, heat the mixture and add small crystals of chlorate of potassium, one by one, until the solution is complete. Dilute the liquid to some definite volume, and divide it into two equal parts. In one part precipitate both the antimony and the tin on a rod of metallic zinc and wash, dry and weigh the powder. Mix the other part with a tolerably large quantity of chlorhydric acid, place a clean strip of metallic tin in the liquid and heat the whole gently for some time. All the antimony will be precipitated while the tin is reduced to the condition of stannous chloride. Wash the precipitate with water acidulated with chlorhydric acid, collect it on a tared filter, dry and weigh. The difference in weight between the first and second precipitates gives the amount of tin. — In case the substance to be examined contains nothing but antimony and tin, the first precipitation with zinc may be omitted, and the difference between the weight of alloy taken and that of antimony found may be regarded as the weight of the tin. Since antimony is not completely precipitated by tin at the ordinary temperature, unless, indeed, after a long time, it is necessary to keep the liquid at a temperature slightly higher than that of the air during the precipitation.

The liquid should contain an excess of acid also from first to last. (Gay-Lussac).

As a modification of Gay-Lussac's process, Levol precipitates the antimony and tin together upon a zinc rod, rinses off the metallic powder which adheres to the zinc when the precipitation is complete, and, without decanting the solution of chloride of zinc, treats the mixed precipitate of tin and antimony with strong chlorhydric acid in order to dissolve the tin. He then weighs the antimony and determines the tin in the filtrate, as Sulphide of Tin. In reply to the criticism of Elsner that this method is inexact, inasmuch as strong chlorhydric acid dissolves some of the antimony as well as the tin, Levol remarks that in presence of chloride of zinc the action of the acid upon antimony is materially lessened. The method can hardly be expected, however, to afford very accurate results in any event.

Method D. A third modification of the process, applicable to cases where the proportion of antimony in the alloy is small, is the following: — Fit to a small flask a cork carrying a thistle tube and two other short tubes, each bent at a right angle. Put the finely divided alloy in the flask, replace the cork, and connect one of the delivery tubes with a source of carbonic acid, and the other with several U-tubes charged with small quantities of red fuming nitric acid free from chlorine. Pour enough strong chlorhydric acid into the flask to seal the thistle tube, and heat the mixture gently. The tin will dissolve completely, and most of the antimony remain in the metallic state, though a part of it goes off in the form of antimoniuretted hydrogen gas. This gas will be oxidized, however, and the antimony retained by the nitric acid in the U-tubes. When the alloy has dissolved, dilute the contents of the flask to some definite volume with recently boiled water, allow the mixture to settle and determine the tin in a measured volume of the clear liquor. Then filter the rest of the liquid, wash the precipitate with acidulated water, place it in a porcelain crucible, add to it the contents of the U-tubes, evaporate to dryness and weigh as Antimoniate of Antimony. If the alloy contain arsenic as well as antimony, the residue obtained by evaporating the contents of the U-tubes would be treated, with the metallic precipitate, as is explained under Antimoniate of Sodium.

Method E. One of the oldest of the processes dependent on the principle now in question, is that of Chaudet (H. Rose, *Handbuch*, 1865, **2.** 301). In this process the antimony is kept in contact with a large proportion of stannous chloride in order that the solvent action of the chlorhydric acid upon the antimony may be hindered. After having determined that the alloy contains nothing but antimony and tin, the first step is to ascertain, approximately, the relative proportions of the two metals. To this end melt one part of the alloy with about 20 parts of metallic tin, roll the product to a sheet, and boil it for a long time with strong chlorhydric acid. The weight of the undissolved matter, after drying, will indicate very nearly the proportion of antimony in the alloy. Next melt very carefully a new portion of the alloy with pure tin, taken in such proportion that there shall be in the melted product 20 parts of tin to 1 of antimony. The weighed metals should be wrapped in paper, placed in a small Hessian crucible, covered with powdered charcoal to prevent oxidation, and ignited for ten minutes in a hot fire. After the crucible has cooled, brush the metallic globule, beat or roll it to a sheet, cut the sheet metal into several pieces, roll the pieces in paper as before, place them in a crucible, cover with powdered charcoal and melt during another space of ten minutes in order to obtain a thoroughly homogeneous alloy. Brush the new globule, roll it to a thin sheet and cut the product into a number of pieces. Weigh out a quantity for analysis, place it in a flask, cover it with strong chlorhydric acid and boil the acid during at least two and a half hours. Dilute the acid with water, collect the finely divided antimony on a weighed filter, dry and weigh. The tin in the original alloy is estimated from the difference. As far as the estimation of antimony is concerned the presence even of a large proportion of lead in the original alloy does no harm.

Properties. Precipitated antimony is a dull black powder, which may be dried at 100° without alteration. It fuses at a moderate red heat. When strongly ignited in hydrogen gas a small part of it volatilizes without chemical change. Nitric acid oxidizes it with formation of antimonious acid, mixed with more or less antimonic acid, according to the strength of the nitric acid. Chlorhydric acid acts upon it but slowly, though an appreciable quantity of it dissolves when left in contact with the acid for several days in open vessels. Dilute acid dissolves more of it than concentrated, and cold acid more than the same acid when boiling. Acid charged with stannous chloride has little or no action upon it.

Principle II. Oxidation by chlorine in alkaline solutions (Method A), by aqua regia, or a mixture of chlorate of potassium and chlorhydric acid (Method B), or by hot air (Method C).

Applications. Separation of Sb from As (B and A); of Sb from Cu and Fe, especially in ores containing sulphur, and from Co and Ni; (Method A). Separation of Sb from Ag, Au and other noble metals. (Method C).

Method A.

1. *To separate Sb from Cu and Fe in sulphuretted ores.* Reduce the mineral to very fine powder, diffuse it through a solution of caustic potash free from sulphuric acid, heat the liquor and pass chlorine gas through it for sev-

eral hours. The sulphur oxidizes rapidly, copper and iron are deposited as oxides and a solution of sulphate and antimoniate of potassium obtained. Antimony may be estimated in the filtrate as Sulphide of Antimony. (Rivot, Beudant & Daguin, *Comptes Rendus*, 1853, **37.** 835).

2. *To separate Sb from As.* Oxidize the mixture with chlorine as in No. 1, and remove the arsenic by precipitation, as Arseniate of Magnesium and Ammonium. Finally determine the antimony as Sulphide.

3. *To separate Sb from Co and Ni*, add to the dilute nitric acid solution of the three metals a large excess of caustic potash, heat the mixture gently and pass chlorine gas into the mixture until the precipitate is black. The precipitate contains the cobalt and nickel as sesquioxides, while the antimony remains in solution as antimoniate of potassium as in No. 1. (Rivot, etc., *loc. cit.*)

Method B. Oxidize the mixed metal or sulphide by boiling with aqua regia, or with chlorhydric acid to which crystals of chlorate of potassium are frequently added, and remove the arsenic acid by precipitation as Arseniate of Magnesium and Ammonium. Determine the antimony in the filtrate as Sulphide.

Method C. See Antimonious Acid, (volatility of).

Antimony Compounds.

Principle I. Reduction of to the metallic state by hydrogen.

Applications. Estimation of antimony in the sulphide or in any other compound of antimony.

Method. Weigh out a small quantity of the dry sulphide, or other compound of antimony to be tested, in a weighed bulb-tube of hard glass; connect the tube with a hydrogen generator and pass a slow stream of dry hydrogen through the tube. Heat the antimony compound, gently at first, until the compound is wholly reduced and the sulphur or other sublimate has been completely expelled from the tube. Remove the lamp and continue the current of hydrogen until the tube is cold, then hold the tube nearly upright for a moment to fill it with air and weigh it together with the antimony which it contains. A minute quantity of antimony is apt to be carried forward either as antimoniuretted hydrogen, or through volatilization of sulphide of antimony, in the current of gas, but most of it may be recovered by heating the narrow part of the tube to redness with a second lamp placed beyond the bulb. A mirror of metallic antimony will be deposited near this second lamp, on the walls of the tube. When the operation is carefully conducted the loss of antimony through volatilization is so small that it hardly amounts to a quarter of one per cent of the total weight of the antimony, even if no second lamp be employed, but it is easy to drive off half a per cent or more by careless heating. The matter in the bulb should never be heated intensely until the reduction is deemed to be well nigh finished.

Principle II. Reduction of to metallic antimony by iron, zinc or tin. See Antimony, sparing solubility of in chlorhydric acid.

Arsenic Acid.

Principle I. Reduction of by sulphur.

Applications. Estimation of arsenic, by loss, in many metallic arseniates. Separation of Fe, Mn, Zn, Pb and Cu from arsenic acid.

Method. Mix the powdered arseniate with pure, powdered sulphur in a porcelain crucible (Rose's reduction-crucible is best) and ignite the mixture in a slow stream of hydrogen gas. The arsenic acid is reduced to metallic arsenic and sulphide of arsenic, which escape in the form of gas, while the metal to be separated from the arsenic remains in the crucible, and is weighed as a sulphide. The apparatus must be so arranged that the arsenic fumes may be carried either into a chimney or into the open air. A simple ignition of the arseniate with sulphur in a covered crucible would be sufficient in most cases to complete the reduction of the arseniate and to expel all the arsenic, but the residual sulphide would then be left in an impure condition. The purpose of the current of hydrogen is to ensure the removal of any excess of sulphur from the residual sulphide.

In case the substance to be analyzed has been thoroughly mixed with sulphur, a single ignition will complete the transformation of the arseniate to a sulphide, but it is always well, after weighing, to mix the residue with a fresh quantity of sulphur and to ignite a second, or if need be, a third time until the results of two consecutive weighings are the same. In order that the amount of arsenic may be determined by the loss, the substance to be analyzed should be made anhydrous by heating it nearly to redness, before weighing, though the reaction with sulphur would take place with an air-dried arseniate as well as with one which had been ignited. The crucible employed must not only be of porcelain but must be provided with a porcelain cover; a platinum cover will not answer since the arsenic fumes would quickly render it brittle and friable. (H. Rose, *Zeitsch. analyt. Chem.*, 1862, **1.** 413).

Principle II. Reduction of by chloride or sulphate of ammonium.

Applications. Estimation of arsenic, by loss. Separation of Na, K, Ba, and other metals from arsenic acid.

Method A. With Chloride of Ammonium. Mix the finely powdered arseniate with from 5 to 8 times its weight of chloride of ammonium and heat the mixture in a covered porcelain crucible until the weight of the residue remains constant. Arseniates of the alkali metals are easily reduced in this way to the condition of chlorides. So, too, is arseniate of barium, though far less easily than the alkaline arseniates; but arseniate of magnesium cannot be completely reduced. The arseniates of iron, cobalt and nickel heated with chloride of ammonium in a stream of hydrogen yield metallic iron, cobalt or nickel, as the case may be, but the product is mixed with so much arsenic that no useful information can be obtained by weighing it. Arseniate of copper is reduced to metallic copper free from arsenic, but during the ignition a quantity of chloride of copper escapes in the form of gas. The process yields really useful results with arseniates of the alkali-metals alone. (H. Rose, *Pogg. Ann.*, **73.** 582; **74.** 562; further *Zeitsch. analyt. Chem.*, 1862, **1.** 422).

Method B. With Sulphate or Acid Sulphate of Ammonium. According to Finkener, acid sulphate of ammonium decomposes many arseniates far more quickly and completely than chloride of ammonium. Though the process, in its present condition, is of no value to the analyst, it is nevertheless worth describing, in the hope that it may be improved. Melt in a porcelain crucible 7 or 8 times as much acid sulphate of ammonium as there is arseniate to be decomposed, and add the latter, little by little, in fine powder to the melted salt. Finally heat the mixture until the excess of ammonium salt has all been driven off and weigh the metallic sulphate which remains in the crucible. It would not be well to heat the mixture of the ammonium salt and arseniate directly, since the mixture would froth violently as it became fluid and some of it would be thrown out of the crucible. Normal sulphate of ammonium can be used instead of the acid salt, though the arsenic is rather more readily expelled by the latter. The process would be valuable, were it not that the crucible is strongly acted upon by the melted ammonium salt in such manner that the metallic sulphate to be weighed is contaminated with other sulphates, formed by the union of sulphuric acid with the ingredients of the crucible, to such an extent that the results are usually much too high. (H. Rose, *Zeitsch. analyt. Chem.*, 1862, **1.** 423).

Principle III. Reduction of by cyanide of potassium. See Arsenic, volatility of.

Principle IV. Solubility in alcohol.

Applications. Separation of arsenic acid from most of the arseniates. The arseniates of iron and aluminum seem to be less easily analyzed in this way than most other arseniates. The magnesium salt, on the other hand, has been thus analyzed with great exactitude.

Method. Mix the arseniate to be analyzed with concentrated sulphuric acid in a platinum dish, and keep the mixture slightly warm until it forms a thick syrup. Add to the syrup a quantity of solid sulphate of ammonium equal to that of the arseniate taken and again heat the mixture until most of the excess of sulphuric acid has been expelled. Allow the mixture to cool and dissolve the viscous mass in the least possible quantity of water, with the aid of gentle heat. Pour upon the solution a large excess of alcohol of 0.83 sp. gr., cover the dish which contains the mixture and leave it at rest for 12 hours. Insoluble double sulphates of ammonium and of the metals to be separated are deposited as fine crystalline powders, while the arsenic acid and the excess of sulphuric acid dissolve in the alcohol. Collect the precipitate upon a filter, wash it with alcohol, dry, ignite to drive off sulphate of ammonium and weigh the residual sulphate of ——, or, in case it cannot be ignited without decomposition, determine the metal in some appropriate way. The arsenic acid may be estimated from the loss or as Arseniate of Magnesium and Ammonium, after diluting with water and expelling the alcohol.

The mixture of sulphate of ammonium and sulphuric and arsenic acids must be dissolved in water before adding the alcohol, or a hard mass will be formed impermeable to alcohol. After the alcoholic mixture has stood 12 hours, ether may be added to it to ensure the precipitation of any traces of sulphates which the alcohol may have dissolved; though as a rule the addition of the ether is not necessary.

Basic Arseniate of Iron (Ferric Arseniate).

Principle. Insolubility in water and fixity when heated.

Applications. Estimation of arsenious and arsenic acids in solutions free from other substances precipitable by ammonia-water, by sesquichloride of iron, on addition of ammonia-water, or by carbonate of barium. Separation of arsenic from Li, Na, K; Ba, Ca, Sr; Zn, Mn, Ni and Co.

Method A. The substance to be analyzed contains no non-volatile metals besides alkali-metals. (Berthier's method). Mix the solution which contains the arsenic acid with a measured quantity of a standard solution of nitrate of sesquioxide of iron and add ammonia-water to the mixture until a decided odor of ammonia persists. In case the substance to be examined contains arsenious acid, it must first be treated with aqua regia, or with nitric acid and chlorate of potassium, to convert the arsenious acid into arsenic acid. If the precipitate produced by ammonia is not of a reddish-brown color another measured portion of the standard iron solution must be added, and afterwards enough ammonia-water to make the solution alkaline. After the mixture has stood for some time at a gentle heat, collect the precipitate upon a filter, wash it with water and allow it to dry. Place the dry precipitate together with the filter ash in a platinum crucible, and heat it very gently for some time in order to expel ammonium compounds at a temperature lower than that at which they can reduce arsenic acid. Increase the heat gradually and at last ignite the crucible intensely. The substance weighed is arsenic acid plus sesquioxide of iron, but the weight of the latter is known from the quantity of the standard iron solution taken. Hence we have only to deduct the weight of ferric oxide taken from the weight of the ignited precipitate to obtain the amount of arsenic acid contained in the substance analyzed.

The best way of making the standard iron solution is to dissolve fine iron wire (piano wire) in hot nitric acid, to dilute the liquid to some definite volume (see Alkalimetry), and then to determine how much sesquioxide of iron is contained in 10 c. c. of the liquid, by precipitation with ammonia-water (see Hydrate of Iron). In this case the presence of a small amount of silica or other precipitable impurity in the iron does no harm, since it is weighed with the oxide of iron both when the strength of the iron solution is determined and in the final arsenic estimation. (Berthier).

Precautions. The process is applicable with sulphuric acid solutions as well as with solutions charged with nitric or with chlorhydric acid, but when sulphuric acid is present it is well to ignite the crucible and precipitate a second time or even a third time after weighing, until the weight remains constant, in order to be sure that all the sulphuric acid has been expelled. This precaution is doubly necessary since it would here be inadmissible to employ carbonate of ammonium (as is done with Sulphate of Potassium), to remove the last traces of sulphuric acid. In case chlorhydric acid is present, the last traces of chloride of ammonium must be washed out of the precipitate lest some of the iron be volatilized as sesquichloride, on ignition. In any event the precipitate must be cautiously heated. If the dry precipitates were strongly ignited at once without any preliminary warming, some of the arsenic acid would be reduced to the condition of arsenious acid or even to metallic arsenic by the action of ammonium compounds contained in the precipitate, and the residue would weigh less than it ought. Even with the most careful heating a small amount of material is usually lost. Besides this liability to loss through reduction of the arsenic acid, an objection to the process is found in the fact that it is sometimes very hard to wash the precipitate completely, and that the last portions of wash water are apt to dissolve some arseniate of iron, in which event the wash water acquires a faint reddish color. This difficulty may be avoided, however, by washing with dilute ammonia-water.

In spite of the greater bulk of the precipitate it is best to employ a large excess of the standard solution of iron, for the highly basic arseniates of iron are less slimy and are far more easily washed than those which contain a larger proportion of arsenic. A good rule is to take one part by weight of metallic iron for every two parts of arsenious acid which the substance to be analyzed is supposed to contain.

The process is of course inapplicable when the arsenical solution contains barium, calcium, strontium or any other metal whose arseniate is insoluble in ammonia-water.

Method B. The substance to be analyzed contains other non-volatile metals besides alkali metals. (V. Kobell's method). Mix the solution of arsenic acid with an excess of a standard solution of nitrate of iron, as in Method A, but instead of ammonia-water add an excess of carbonate of barium to throw down the basic arseniate of iron. If the arsenical solution be strongly acid it should be nearly neutralized with carbonate of sodium before adding the carbonate of barium. The liquid must be kept clear, however, until the moment when the barium carbonate is mixed with it. The mixture must not be heated but should be left to stand for several hours in the cold after the addition of the carbonate of barium. All the iron and all the arsenic will be thrown down in the form of basic arseniate of iron, while an equivalent proportion of the carbonate of barium dissolves. Wash the precipitate, which is of course mixed with the excess of the carbonate of barium, by decantation with cold water, collect it upon a filter, dry, ignite gently for some time (see Carbonate of Barium), and weigh. Dissolve the weighed precipitate in strong chlorhydric acid and determine the amount of barium by precipitation as Sulphate of Barium. From the amount of sulphate of barium found, calculate the equivalent quantity of carbonate of barium, add the weight of this carbonate to the weight of ferric oxide taken, and deduct the sum from the weight of the mixed precipitates of arseniate of iron and carbonate of barium; the difference will give the amount of arsenic acid contained in the substance analyzed.

The method cannot be applied directly in presence of metals precipitable by carbonate of barium in the cold. So, too, if sulphuric acid be present it must be removed by means of chloride of barium before the analysis can be proceeded with.

Properties of the Precipitate. The success of Method A depends materially on the state of aggregation of the precipitate. It must be a highly basic compound, neither slimy nor soluble in ammonia-water. The composition of the precipitate obtained in Berthier's process may be taken as varying from $2Fe_2O_3$, As_2O_5 to $16Fe_2O_3$, As_2O_5. The more basic it is the better. The 16 Fe_2O_3 salt is as insoluble in ammonia-water and as little slimy as ferric hydrate. It is to be observed that normal arseniate of iron ($2Fe_2O_3$, $3H_2O$, $3As_2O_5+9H_2O$), as obtained by mixing a solution of ferric chloride with one of ordinary arseniate of sodium, is a white slimy precipitate soluble in ammonia-water and of no use to the analyst. Pfaff (*Handbuch analyt. Chem.*, Altona, 1824, **1.** 221 and **2.** pp. 335, 344, 459, 476, 478, etc.), who thought highly of the principle now in question, as a means of separating iron from nickel and from manganese, describes the basic arseniate of iron as a white pulverulent precipitate which falls immediately, even in presence of 25,000 parts of water. According to him, it is compact rather than gelatinous, and may be readily collected and washed; it is much more difficultly soluble than the benzoate or succinate of iron,—5 times less soluble than the last named salt. Pfaff obtained his precipitates by adding monoarseniate of potassium to solutions of ferric salts. — It is worth inquiring whether the precipitate thus thrown down might not be weighed directly and then reduced with sulphur by H. Rose's method (see Arsenic Acid), in order to obtain the proportion of iron and of arsenic contained in it.

Arseniate of Lead.

Principle I. Fixity when heated to low redness.

Applications. Estimation of arsenious and arsenic acids in aqueous or nitric acid solutions free from chlorhydric acid and non-volatile substances, as well as ammonium salts and other reducing agents. Valuation of arsenious acid and metallic arsenic.

Method. Weigh a quantity of the solution to be examined in a tolerably large porcelain crucible, and, in case the solution contains no arsenious acid, mix with it directly a weighed quantity of pure, recently ignited, oxide of lead. Take 5 or 6 times as much oxide of lead as there is supposed to be arsenic acid in the substance, and stir it into the solution with a fine glass rod. Evaporate the mixture to dryness on a water bath, heat the residue to low redness and maintain it at that temperature for some time. Finally, weigh the residue and deduct the weight of the oxide of lead taken, in order to obtain the weight of the arsenic acid. The process affords excellent results provided the residue is not heated too strongly. — The oxide of lead is best prepared by igniting pure nitrate of lead.

In determining the value of any sample of metallic arsenic or arsenious acid, or in analyzing a solution of arsenious acid, dissolve the substance in nitric acid, or mix the solution with this acid, and evaporate the liquor to a small bulk before adding the oxide of lead. The dry residue must be ignited with special care in this case in order to avoid loss through decrepitation of the nitrate of lead. The crucible should be placed upon a sand bath at first in order to ensure a gradual elevation of the temperature, and should be kept covered. A similar remark applies of course to all cases in which nitric acid is present.

Principle II. Insolubility in water or acetic acid.

Applications. Separation of lead from many metals—among which may be mentioned Na, K, Li; Zn, Cd and Cu. (Pfaff).

Method. Mix the solution of arsenic acid, which must either be neutral or slightly acidulated with acetic acid, with an excess of a solution of acetate or nitrate of lead. Collect the precipitate upon a filter, wash and dry it, ignite it gently apart from the filter, in a porcelain crucible, weigh and finally determine the amount of lead contained in the precipitate, as Sulphate of Lead. Another method of procedure is to dissolve the substance to be analyzed in nitric acid, to mix the solution with an excess of nitrate of lead, and to carefully evaporate the mixture to dryness in order to expel the free nitric acid. The dry residue, consisting of arseniate of lead, nitrate of lead, and nitrate of the metal previously combined with arsenious acid, is then washed with water and the arseniate of lead collected on a filter. The excess of lead must of course be removed from the filtrate before proceeding to determine the other metal or metals contained in it. Instead of evaporating the nitric acid solution of arsenic it may be carefully neutralized with an alkali and then mixed with acetate of lead. Unlike nitric acid, acetic acid dissolves scarcely any arseniate of lead, hence arsenic acid may be completely precipitated by adding acetate of lead to a solution of free arsenic acid without need of neutralizing or evaporating the liquid. Though inconvenient and almost obsolete, the process may still be employed in some cases. It is inapplicable in presence of chlorhydric acid or a chloride, for in that case insoluble or almost insoluble double compounds of chloride and arseniate of lead would be precipitated.

Properties. Pure arseniate of lead is a white powder, yellowish when hot, which softens at a low red heat. It fuses when heated strongly and is then apt to suffer slight decomposition, a small proportion of arsenic acid being lost in the form of arsenious acid and free oxygen. The substance actually weighed in the first of the two processes above described is not pure arseniate of lead, but a mixture of that substance with free oxide of lead. The product obtained by the second process is a variable mixture of tri- and di-arseniate of lead.

Arseniate of Magnesium and Ammonium.

Principle. Sparing solubility in ammoniated water, and fixity at 100°.

Applications. Estimation of arsenic acid in all solutions free from substances permanently precipitable by ammonia or by a magnesium salt. Separation of arsenic acid from all acids which form soluble magnesium salts. Separation of arsenic from arseniomolybdate of ammonium; from K, Na; Al, Zn, Cd; Mn, Ni, Co, Fe; Cu and Sb. Separation of arsenious from arsenic acid and of Mg from K and Na.

Method A. Absence of substances precipitable by ammonia or by magnesia. Add to the solution which contains arsenic acid ammonia-water in distinct excess, and stir into the liquid a mixture of sulphate of magnesium, chloride of ammonium and ammonia (see Sulphate of Magnesium) as long as a precipitate continues to fall. This "magnesia-mixture" consists of a solution of sulphate of magnesium charged with so much chloride of ammonium that no magnesia can be precipitated from it by ammonia-water. Leave the liquid and precipitate at rest for 12 or 24 hours in the cold. Then collect the precipitate upon a weighed filter which has been dried at 105°–110°. — The liquid should smell strongly of ammonia, after the addition of the magnesia-mixture, but should contain no more chloride of ammonium or other ammonium salt than is absolutely necessary to prevent the precipitation of hydrate of magnesium. The precipitate is in fact considerably more soluble in solutions of ammonium salts than in pure water. The liquid must never be heated, and time enough must be allowed for the precipitate to separate from it completely before filtering. Since the precipitate is by no means absolutely insoluble in ammoniated water, it must be washed with extreme care. In order to avoid using any more water than is absolutely necessary for the washing, transfer the precipitate from the beaker to the filter by means of portions of the filtrate, and afterwards wash it upon the filter with small quantities of a mixture of 1 part strong ammonia-water and 3 parts water until the washings give only a slight opalescence when acidulated with nitric acid and mixed with nitrate of silver. Dry at 105°–110° and weigh.

The process yields satisfactory results, though on account of the solubility of the precipitate they are always somewhat lower than theory requires. According to Fresenius, this error may be partially corrected, as follows:—Measure the filtrate, without the washings, and for every 16 c. c. of the liquid add 1 m. g. to the weight of the precipitate. The wash water must not be measured since it cannot be regarded as a saturated solution of the double arseniate.

Instead of washing with ammoniated water, as above, H. Rose directs that the solution to be examined be concentrated to a small bulk, supersaturated with ammonia-water and mixed with one quarter its volume of strong alcohol before adding the magnesia mixture, and that the precipitate, after standing 48 hours in its liquor, be washed with a mixture of 4 vols. water, 1 vol. absolute alcohol, and a not too small quantity of ammonia-water.

In case any portion of the arsenic in the original solution is not already in the form of arsenic acid or an arseniate, the solution must be gently heated in a flask with nitric or chlorhydric acid, and small portions of chlorate of potassium thrown into the liquid until the latter emits a strong odor of chlorous acid. When all the arsenic acid has thus been oxidized, heat the liquid gently to drive off most of the chlorous acid, and proceed with the addition of ammonia-water and magnesia mixture as before.

Properties. The double arseniate is a white, somewhat transparent, crystalline precipitate, of composition represented by the formula $2MgO, (NH_4)_2O, As_2O_5 + 12H_2O$. When heated to 105°–110° it loses 11 equivalents of water, so that the formula of the compound actually weighed is $2MgO, (NH_4)_2O, As_2O_5 + H_2O$: —

$2MgO$	80	21.05
$(NH_4)_2O$	52	13.68
As_2O_5	230	60.53
H_2O	18	4.74
	380	100.00

The precipitate might be dried at 100° if only time enough were allowed. But at the temperature of a water bath, the precipitate retains 2 or 3 equivalents of water with considerable force.

On ignition the compound loses water and ammonia, and changes to arseniate of magnesium $2MgO, As_2O_5$. It is not impossible to obtain good results by igniting the precipitate and weighing the residue as Levol (*Annales Chim. et Phys.*, (3) **17.** 501), the inventor of the process, originally proposed. Wittstein (*Zeitsch. analyt. Chem.*, 1863, **2.** 20), in reasserting the propriety of this course, directs that the crucible be placed at first on a sand bath and heated with a small spirit lamp until the odor of ammonia is no longer perceptible. Then remove the crucible from the bath and heat it over the free flame more and more strongly, almost to redness. H. Rose admits that by raising the temperature of the precipitate very gradually during several hours, from 100 to 400°, and then as gradually from dull to bright redness, all the water and ammonia may be expelled without losing any arsenic through the reducing action of the ammonia. But in case the dry precipitate is exposed directly to a high temperature, a very considerable portion of the arsenic is lost. According to his experiments, if the precipitate be gently heated for a short time and then brought to redness, no more than 96 per cent of the arsenic contained in the magnesium salt will ever be obtained, while results as low as 95 and 93 per cent are common. If the precipitate be exposed at once to a red heat, only about 88 per cent of the arsenic is left to be weighed. The precipitate might be ignited with bisulphate of ammonium and reduced to the condition of sulphate of magnesium (see Arsenic Acid) before weighing, were it not for the errors introduced by the action of the ammonium salt upon the material of the crucible. The double arseniate cannot be completely reduced, however, by igniting it with sulphur or with carbonate of ammonium, or in a current of hydrogen gas. (H. Rose, *Zeitsch. analyt. Chem.*, 1862, **1.** pp. 418, 424).

Arséniate of magnesium and ammonium is much more soluble in pure water than in dilute ammonia-water or ammoniated spirit. It is more soluble in aqueous solutions of ammonium salts than in water, though the solvent power of these salts is diminished by the presence of free ammonia. At the temperature of 15° according to Fresenius (*Zeitsch. analyt. Chem.*, 1864, **3.** 207), 1 part of the salt (reckoned as dried at 100°) dissolves in 2656 parts of water; in 15038 parts of ammoniated water, made by mixing 1 part of ammonia-water, of 0.96 sp. gr., with 3 parts of water; in 1315 parts of a dilute and 844 parts of a strong solution of chloride of ammonium (containing respectively 70 and 7 parts of water to 1 part of chloride of ammonium); and in 2871 parts of a solution made by mixing 60 parts of water, 10 parts of ammonia-water, of 0.96 sp. gr., and 1 part of chloride of ammonium.

Method B. Separation of Arsenic Acid from Arsenious Acid. Mix the tolerably dilute solution of the two acids with a large quantity of chloride of ammonium and then precipitate the arsenic acid with magnesia mixture, as in A. The arsenious acid may be determined in the filtrate as Sulphide of Arsenic. If the original solution be too concentrated some arsenite of magnesium will go down with the double arseniate. To be sure of the purity of the latter it is well to dissolve it in chlorhydric acid, after weighing, and to test the solution with sulphuretted hydrogen. The immediate formation of a precipitate would indicate the presence of some arsenious acid.

Method C. Separation of Arsenic Acid from Cd, Zn, Al, Ur, Mn, Fe, Co, Ni and Cu. Dissolve the substance to be analyzed in chlorhydric or nitric acid, and in case the arsenic is not all in the condition of arsenic acid, bring it to that state by means of chlorate of potassium as in Method A. Mix the tolerably dilute solution with enough tartaric acid to prevent the formation of a precipitate by ammonia-water, and then with ammonia-water in excess. Finally add the mixture of sulphate of magnesium, chloride of ammonium and ammonia, and proceed as in Method A.

Since the precipitate, in this case, is liable to be contaminated with a difficultly soluble basic tartrate of magnesium, it is best, after washing once with the ammoniated water, to redissolve it in chlorhydric acid, to mix the solution with a very small quantity of tartaric acid, to supersaturate with ammonia and allow the whole to stand during at least 12 hours. The pure precipitate thus obtained may then be collected, washed, dried and weighed. The process is better suited for separating large than small quantities of arsenic acid from the metals in question, for if the proportion of arsenic be very small the amount of arseniate of magnesium and ammonium which is necessarily dissolved in the process of washing may exert a very considerable influence on the accuracy of the result.

With the exception of aluminum, the various metals may be thrown down as Sulphides from the original filtrate, by means of sulphydrate of ammonium. To determine aluminum, evaporate the filtrate to dryness in a platinum dish, roast the residue in a muffle to destroy sulphur and organic matter, fuse with carbonate of sodium, dissolve in chlorhydric acid and precipitate Hydrate of Aluminum.

Method D. Separation of Arsenic from Antimony. Proceed as in Method C. But immediately after adding the tartaric acid mix with the solution a large quantity of chloride of ammonium, before proceeding to saturate with ammonia-water. Any precipitate produced on the addition of the ammonia-water would go to show that the quantity of chloride of ammonium or of tartaric acid previously added was insufficient, and that a further portion of one or the other of these agents must be added before adding the magnesia mixture. — In case the substance to be analyzed is an alloy, or a mixture of sulphides, oxidize it in the manner described under Antimony. Finally determine the antimony in the filtrate as Sulphide of Antimony.

Method E. Determination of Arsenic Acid in a precipitate of Arseniomolybdate of Ammonium. Dissolve the precipitate upon the filter in ammonia-water, wash the paper thoroughly and add enough chlorhydric acid to the solution to neutralize a good part of the ammonia. Take care, however, to leave the solution clear and smelling strongly of ammonia. Finally add the magnesia mixture and proceed as in Method A.

Method F. To separate Magnesium from Potassium or Sodium, mix the solution with a quantity of chloride of ammonium, add ammonia-water in excess, and a solution of arseniate of ammonium as long as any precipitate falls. Allow the mixture to stand and treat the precipitate as in A. To determine the potassium or sodium, evaporate the filtrate from the double arseniate to dryness under a chimney and ignite the dry residue. The excess of arsenic acid goes off, together with the ammonium salts, while the alkali metals are left as chlorides, always contaminated, however, with a little chloride of magnesium. Magnesia may be separated more quickly from the alkalies in this way than by means of the double Phosphate of Magnesium and Ammonium, but the latter process is more accurate, and on the whole, more convenient than the one just described.

Arseniate of dinoxide of Mercury.

Principle. Insolubility in water.

Applications. Separation of arsenic acid from Na, K; Ba, Ca, Sr; Mg, Cd, Zn; Co, Ni, Pb and Cu.

Method. Same as that described under Phosphate of dinoxide of Mercury, excepting that the arsenic acid cannot be determined in the insoluble residue in the way that phosphoric acid is determined.

Arseniate of Potassium (or of Sodium).

Principle. Solubility in water.

Applications. (A). Separation of arsenic acid from Fe, Mn and Cu; less completely from Zn. (B). Separation of arsenic acid from many metals, including those of the alkaline earths.

Method A. Boil the finely powdered, unignited arseniate of iron, or other metal, for some time with a solution of a caustic alkali, or less advantageously with a solution of an alkaline carbonate. Dilute the liquor with water and filter it in order to separate the soluble arseniate of potassium or of sodium from the insoluble oxide or carbonate which remains undissolved. — Since oxide of zinc is soluble in caustic alkalies, arseniate of zinc must always be treated with a carbonated alkali. It is not easily decomposed in any event. (V. Kobell; H. Rose, *Zeitsch. analyt. Chem.*, 1862, **1.** 425). It will not answer to simply precipitate the solution of a metallic arseniate with an excess of a solution of carbonate of potassium or of caustic potash, since more or less arsenic would almost always be retained by the precipitate. — The method is inapplicable for the analysis of native arseniate of iron, or of the artificial product after it has been ignited. Only a part of the arsenic can be removed from these substances by boiling alkali. (Berzelius).

Method B. Mix the powdered substance to be analyzed with 3 parts of dry carbonate of sodium, or better, with 3 parts of a mixture of equal equivalents of the carbonates of sodium and potassium, and melt the mixture at the blast lamp. The mixture of carbonate of sodium and carbonate of potassium melts more easily than either carbonate taken by itself. When the mixture has been thoroughly fused, all the arsenic acid will be found in the condition of arseniate of sodium, and the metals to be separated in the form of insoluble carbonates or oxides. Dissolve the fused mass in water, separate the soluble arseniate by filtration and estimate the arsenic acid as Arseniate of Magnesium and Ammonium.

The process is objectionable, inasmuch as no vessels can be found proper to be used for the fusion. There would be danger of destroying a platinum crucible through the combined action of an alkaline arseniate and reducing gases from the lamp; if a porcelain crucible is used its glazing will be dissolved by the melted alkali, and the solution of arseniate of sodium will thus become contaminated with aluminum and silicon. The risk of destroying a platinum crucible is lessened, and the fusibility of the flux increased, by mixing a quantity of nitrate of potassium with the carbonate; but the process cannot be commended unless the proportion of arsenic in the mixture to be analyzed is very small.

Arseniate of Tin (SnO_2).

Principle. Insolubility of the compound in water and dilute nitric acid, and fixity when ignited.

Applications. Estimation of tin and of arsenic in commercial stannate of sodium.

Method. Mix a weighed quantity of the stannate to be tested with a weighed quantity of arseniate of sodium, more than sufficient to precipitate the whole of the tin. Add an excess of nitric acid to the mixture and boil the liquid; collect the precipitate upon a filter, wash, dry, ignite and weigh. Determine also the arsenic acid in the filtrate by precipitating it as Arseniate of Magnesium and Ammonium. The amount of tin is found from the weight of the ignited arseniate of tin, and that of the arsenic by adding together the quantities of arsenic found in the arseniate of tin and the arseniate of magnesium and ammonium, and subtracting from this sum the quantity of arsenic added in the form of arseniate of sodium.

Properties. The composition of the washed precipitate may be represented by the formula $2SnO_2, As_2O_5 + 10H_2O$; and that of the ignited precipitate by the formula $2SnO_2, As_2O_5$. (E. Hæffely, *Phil. Mag.*, (4.) **10.** 291).

The same principle is involved in a method of estimating arsenic and tin in an alloy of the two metals, proposed by Levol (*Ann. Chim. et Phys.*, (3.) **16.** 493). Levol has found that when an alloy of tin and arsenic, containing less than 5 per cent of arsenic is boiled with nitric acid, the whole of the arsenic will remain in insoluble combination with the oxide of tin, in the form of basic arseniate of tin. Even when the alloy contains 8 per cent of arsenic, little or no arsenic will remain dissolved in the excess of nitric acid, but when the proportion of arsenic exceeds 8 per cent some of it always goes into solution. The process of analysis is as follows:—Treat the finely laminated alloy with tolerably strong nitric acid, first in the cold, in order that stannous oxide may be formed, and afterwards at the temperature of boiling to convert the stannous into stannic oxide. Collect the arseniate of tin upon a filter, dry and weigh it and finally heat it in a current of hydrogen to determine the proportion of Arsenic. In case the filtrate also contains arsenic, estimate the quantity by precipitating as Arseniate of Magnesium and Ammonium or in some other appropriate way.

Levol has applied this method to the estimation of arsenic in metallic copper. He dissolves the metal in nitric acid, mixes the solution with a solution of stannous nitrate, prepared in the cold. and boils the mixture. Stannic arseniate is thrown down from the hot liquor, and the proportion of arsenic contained in it is estimated in the manner already described.

Arseniate of Uranium.

Principle. Insolubility in water and acetic acid.

Applications. Estimation of arsenic in arsenic and arsenious acids and in arseniates and arsenites of the alkali-metals. (Method A). Estimation of arsenic as in A, and also in presence of Ba, Sr, Ca, Mg and Zn; but not in presence of metals, such as Cu, which are precipitated by ferrocyanide of potassium (Method B).

Method A. Gravimetric. Boil the solution of arsenic acid with an excess of potash lye, supersaturate the mixture with acetic acid, and add to the clear solution an excess of acetate of uranium. Wash the precipitate with a very dilute solution of chloride of ammonium and finally remove the chloride of ammonium by washing with a mixture of 1 volume of alcohol and 8 or 9 volumes of water. Dry the precipitate at a gentle heat upon a water bath and ignite it moderately in a porcelain crucible for a long time. The crucible must not be heated to redness.

Precautions. The solution to be analyzed must be free from ammonium salts; if ammonium salts were present ammonia would be thrown down in combination with the precipitate and would reduce some of the arsenic acid when the precipitate came to be heated. No metals of the alkaline earths or other substances precipitable by arsenic acid should be present. The solution must be perfectly clear after the addition of acetic acid. Precipitated arseniate of uranium is apt to be so finely divided that it would pass through the pores of the filter if simply washed with water. This difficulty is avoided by washing with chloride of ammonium instead of water. As a rule the method is less convenient than that depending on the insolubility of Arseniate of Magnesium and Ammonium. (Werther, *Journ. Prakt. Chem.*, **43.** 346).

Properties. The precipitate first thrown down, of composition $2Ur_2O_3$, H_2O, As_2O_5, is insoluble in water, acetic acid and saline solutions, particularly in a solution of chloride of ammonium. The composition of the residue left after ignition may be represented by the formula $2Ur_2O_3$, As_2O_5. The proportion of arsenic can be calculated from it directly.

Method B. Volumetric. Prepare a solution of nitrate of sesquioxide of uranium of such strength that a litre shall contain about 20 grammes of the sesquioxide. This solution should contain as little free acid as possible. To standardize the liquor weigh out a quantity of pure arseniate of sodium, dissolve it in water, supersaturate the solution with ammonia water, and finally make the liquor distinctly acid with acetic acid. Slowly pour the uranium solution from a burette into the acidulated solution of arsenic acid, with constant stirring, until a drop of the liquid placed upon a porcelain plate gives a reddish brown coloration when tested with ferrocyanide of potassium. Note the quantity of uranium solution used; mark the height of the liquid in the beaker by means of a strip of gummed paper; empty and wash the beaker and fill it to the mark with water charged with about as much ammonia and acetic acid as were previously added to the arseniate of sodium. Pour uranium solution from the burette into this acidulated water until a drop of the solution gives the red brown reaction with ferrocyanide of potassium. — A certain excess of the uranium solution has always to be used in order to obtain the reaction with ferrocyanide of potassium, and the purpose of the second titration is to determine the amount of this excess for the particular degree of dilution involved in the given case. The number of c. c. of uranium solution employed in the second titration must be subtracted from the quantity first used, in order to obtain the true value of the solution with reference to arsenic acid. To be sure of the final reaction, spread out a drop of the liquor upon porcelain and place a drop of the ferrocyanide in the centre of the spot; if the titration be finished a distinct reddish brown line will form where the two liquids meet. — Instead of arseniate of sodium pure arsenious acid may be used for standardizing the liquid; after having been weighed it must be converted into arsenic acid by boiling with fuming nitric acid.

In an actual analysis, the substance, after having been brought to the condition of arsenic acid, is treated with ammonia water and acetic acid, and the clear solution titrated with the uranium solution, as above described. The second titration, to determine the correction for dilution, must be made in every analysis precisely as when the test liquor is standardized. In practised hands the process is said to yield good results. A solution of nitrate of uranium is preferable to the acetate, since the latter gradually decomposes when exposed to light. (Bœdeker, *Annal. Chem. und Pharm.*, **117.** 195).

Arsenic. [Compare Arsenious Acid].

Arsenic is weighed either as metallic arsenic, as arseniate of iron (Fe_2O_3), arseniate of lead, arseniate of uranium, arseniate of magnesium and ammonium, or as tersulphide of arsenic. It may be estimated by loss either as As, as $AsCl_3$, or as As_2S_3, by titration with oxidizing agents, as will be explained under arsenious acid, and by indirect methods to be explained below. (See also the finding list in Appendix).

Principle 1. Sparing solubility of the metal in chlorhydric acid.

Applications. Separation of arsenic from tin.

Method. Heat the granulated or laminated alloy gently in a mixture of 1 equivalent of nitric acid and 9 equivalents of chlorhydric acid. The tin alone will dissolve, while metallic arsenic is left in the form of a powder, which may be collected upon a tared filter, washed, first with cold recently boiled water, then with alcohol, and dried at 100°. The acid must not be used in much larger proportion than 1 equivalent nitric acid and 9 equivalents chlorhydric acid to 8 equivalents of the arseniated tin.

When the acids are used in these proportions the solution of the tin is unaccompanied by the evolution of gas, the chlorides of tin and of ammonium being the sole products besides water.

$$4Sn + HNO_3 + 9HCl = 4SnCl_2 + NH_4Cl + 3H_2O.$$

(Gay-Lussac, *Annales Chimie et Phys.*, [3.] **23.** 228).

For the separation of arsenic from an alloy of tin, antimony and arsenic see under Antimony, Principle I., No. 3.

Principle II. Volatility.

Applications. Separation of arsenic from antimony and from tin.

Method A. *To separate arsenic from antimony* weigh out some of the finely divided alloy in a bulb tube, mix with it 2 parts of carbonate of sodium and 2 parts of cyanide of potassium, and connect the tube with a source of dry carbonic acid. After all the air has been expelled, heat the tube gently for a while, afterwards gradually increase the heat to intense ignition, and continue to heat until all the arsenic has been volatilized and driven from the tube. After removing the lamp, continue the current of carbonic acid until the tube has become cold, then lixiviate the contents of the tube, first with a mixture of equal parts of strong alcohol and water and afterwards with water. Finally, dry and weigh the residual antimony. The quantity of arsenic is inferred from the loss. The method is said to yield only approximate results.

In case it is desired to dispense with the carbonate of sodium and cyanide of potassium and to fuse the alloy by itself in the current of carbonic acid, special care must be exercised in heating the alloy or much antimony will be driven off with the arsenic.

2. *To separate arsenic from tin.* Oxidize the alloy with nitric acid in a beaker, wash the product into a porcelain crucible with a solution of carbonate of sodium, evaporate to dryness on a water bath, mix the dry residue with equal parts of carbonate of sodium and cyanide of potassium, and melt the mixture. All the arsenic is driven off in the form of vapor while the tin remains, partly in the form of metal and partly as oxide, and as stannate of sodium. The tin is estimated as Oxide of Tin and the arsenic by the difference.

In case the substance to be analyzed is arseniate of tin, weigh out some of it in a glass boat, introduce the boat into a wide glass tube, pass a current of hydrogen through the tube and heat the tube to dull redness. Metallic arsenic sublimes and is deposited as a coating upon the colder part of the tube. Its weight is determined by cutting off that part of the tube and weighing it first with the arsenic and again after the arsenic has been dissolved. In order to retain any arsenic which might go forward as arseniuretted hydrogen, it is well to lead the escaping gases through a short column of weighed hot metallic copper. Or the whole of the volatilized arsenic may thus be made to combine with copper in one operation.

Since the tin always retains some arsenic, it must be dissolved in chlorhydric acid, and the mixture of hydrogen and arseniuretted hydrogen evolved, after having been freed from chlorhydric acid by washing with soda lye, must be made to pass into a solution of nitrate of silver. The arsenic in the gas is thus converted into arsenious acid while metallic silver is thrown down. In case any solid arseniuretted hydrogen (or insoluble alloy of arsenic) is left undissolved by the chlorhydric acid, it must be dissolved in a few drops of nitric acid. The excess of silver is then thrown down as Chloride of Silver; the filtered arsenious acid is mixed with the nitric acid solution of the solid residue, and the arsenic precipitated as Sulphide of Arsenic. — As a substitute for this method of procedure Berzelius recommends that the arseniuretted hydrogen from the solution of the tin, be decomposed by passing it through a weighed quantity of red-hot copper, and that the solid residue be neglected.

3. The old method of separating arsenic from certain alloys by roasting the mixture in a current of air until the odor of garlic was no longer perceptible, or heating with admixture of charcoal, in case the arsenic were present as arsenic acid, gave only approximative results. For that matter, metals like nickel, cobalt and iron, always retain a certain portion of the arsenic. (Pfaff, *Handbuch analyt. Chem.*, 1825, **2.** 426).

Principle III. Oxidation by chlorine in alkaline solutions (Method A), by aqua regia, by a mixture of Chlorate of Potassium and nitric or chlorhydric acid (Method B), or by Nitrate of Potassium.

Applications. Separation of As from Fe, Co, Ni and Cu (Method A); from Sb (Method B); from Ca, Ba, Sr; Mg, Zn, Cd, Mn, Fe, Co, Ni; Ag, Pb, Hg, Cu, Bi (Method C).

Method A. Oxidation by Chlorine. Same as the methods described under Antimony. A solution of arseniate of potassium is produced.

Method B. Oxidation by Aqua regia or by a mixture of Chlorate of Potassium and Nitric or Chlorhydric Acid. Same as the methods described under Antimony. In order to avoid the volatilization of terchloride of arsenic, care must be taken never to evaporate a solution which contains arsenious acid and chlorhydric acid unless free chlorine or chlorous or nitric acid be also present.

Method C. Oxidation by Nitrate of Potassium. Fuse the alloy with a mixture of 3 parts of carbonate of sodium and 3 parts of nitrate of potassium. Boil the fused mass with water, separate the soluble alkaline arseniates from the insoluble oxides and carbonates, by filtration and washing with hot water, and determine the arsenic as Arseniate of Magnesium and Ammonium. — In case the proportion of arsenic in the alloy is not large, the fusion may be effected in a platinum crucible, in the bottom of which a layer of carbonate of potassium has been placed before introducing the mixture. When a porcelain crucible is used, part of it is dissolved by the melted alkali and the fused mass is consequently contaminated with silicon and aluminum. (Wœhler, *Pogg. Ann.*, **25.** 302).

In order to avoid loss of arsenic by volatilization, alloys rich in arsenic had better be oxidized with nitric acid in the first place and the dry residue subsequently fused with 3 parts of carbonate of sodium and 1 part of nitrate of potassium.

Arsenious Acid. [Compare Arsenic Acid].

Principle. Oxidation by Iodine (A), by Chlorine (B), by Bichromate (C), Permanganate (D), Nitrate (E), or Chlorate (F), of Potassium; by aqua regia or by Salts of Gold (G).

Applications. Methods A B C and G may be employed for determining arsenious acid in arsenites and in the commercial acid; for separating arsenious from arsenic acid and for determining arsenic in boiled clorhydric acid solutions of sulphide of arsenic. Method A is well suited for determining the amount of arsenious acid in commercial realgar and orpiment. It may be used for estimating iodine, and also for standardizing iodine solutions,—as will be explained under Iodine, — by which to estimate Chlorine, Bromine, Ozone; Hypochlorous, Chloric, Chromic, Sulphurous and Sulphydric Acids; the higher Oxides of Manganese, Cobalt and Nickel; in short, all oxides which evolve chlorine when heated with chlorhydric acid; and tin in Stannous salts.

Method B is largely used for estimating the value of bleaching salts. — Method E serves to separate arsenic from Ca, Ba, Sr; Mg, Cd, Zn, Mn, Fe, Co, Ni; Ag, Pb, Hg, Cu and Bi. Method F may be used for determining nitric acid, as well as for converting arsenious into arsenic acid in the ordinary course of analysis.

Method A. Oxidation by Iodine in Alkaline Solutions.

$$As_2O_3 + 2Na_2O + 4I = As_2O_5 + 4NaI.$$

Weigh out as much of the substance to be analyzed as will contain about 0.1 grm. of arsenious acid, and dissolve it by boiling with 20 c. c. of a saturated aqueous solution of pure bicarbonate of sodium. Stir some starch paste into the solution, and afterwards pour in a standard solution of Iodine from a burette, until the whole liquid just becomes blue. If the substance to be analyzed is an acid solution it must be neutralized with pure carbonate of sodium, and if alkaline with pure chlorhydric acid, before mixing it with the bicarbonate. Results accurate.

To determine arsenious acid when mixed with sulphide of arsenic, boil the mixture with carbonate of sodium, dilute with water and acidulate with acetic acid; filter, to remove the sulphide of arsenic which is thrown down, supersaturate the filtrate with bicarbonate of sodium, and proceed with the titration. The substance to be analyzed and the carbonate of sodium must be wholly free from sulphurous acid, hyposulphites and other substances capable of acting upon iodine. (F. Mohr, *Titrirmethode*, 1855, p. 295).

To separate arsenious from arsenic acid determine the amount of arsenious acid in one portion of the substance, by titration with iodine as above, then oxidize the arsenious acid in another portion by means of nitric acid and chlorate of potassium (Method F), or by chlorine in alkaline solution, and determine the total amount of arsenic as Arseniate of Magnesium and Ammonium. Calculate the amount of arsenic acid from the difference.

Method B. Oxidation by Chlorine.

$$As_2O_3 + 2H_2O + 4Cl = As_2O_5 + 4HCl.$$

1. *By chlorine acting in an acid solution.*

The old method of chlorimetry devised by Gay-Lussac (*Annal. der Chem. und Pharm.*, **18.** 18), which was formerly much used, depends upon the action of chlorine upon arsenious acid in chlorhydric acid solution.

A weighed quantity of pure arsenious acid was dissolved in hot concentrated chlorhydric acid, and the solution colored with two or three drops of a solution of indigo in sulphuric acid. A weighed quantity of the bleaching powder to be tested was mixed with water to a definite volume, and portions of the milky liquor were poured from a burette into the solution of arsenious acid, until the blue color of the indigo disappeared. — The process is distinctly inferior in several respects to that in which the chlorinated liquor is made to act upon arsenious acid in alkaline solution. Notably, because the oxidizing power of chlorine is less decided in presence of an acid than in that of an alkali, and because some chlorine is always set free by the action of the acid and lost, at the moment when the bleaching powder solution comes in contact with the arsenical liquor. As an indicator, moreover, indigo is inferior to the mixture of iodide of potassium and starch employed in the alkaline process. In practice, the color of the indigo does not disappear all at once; it is destroyed gradually by the local action of chlorine, wherever that agent happens to be in excess in any part of the liquor, and is not reproduced by the action of unoxidized arsenious acid in another part. Hence it is often necessary to add a fresh drop of indigo towards the close of the titration in order to be sure of the point of saturation.

2. *By Chlorine acting in an alkaline solution.* [Compare the description of this method under Antimony].

As applied to the estimation of the value of bleaching powder (after Penot, *Bulletin de la Société Industrielle de Mulhouse*, 1852, and *Dingler's polytech. Journ.*, 1853, **127.** 134), the process may be conducted as follows:—Heat a mixture of 4.4341 grms. of pure arsenious acid and about 13 grms. of crystallized carbonate of sodium in 600 or 700 c. c. of water, until the arsenic has all dissolved. Cool the solution and afterwards dilute it to 1 litre. Each cubic centimetre of this solution contains 0.004434 grm. of As_2O_3, corresponding to 1 c. c. of chlorine gas at 0^0 and 76 m. m. pressure.

Weigh out 10 grammes of the bleaching powder, rub it in a tolerably large mortar with a little water, to a smooth paste; gradually add more water, until the mass becomes fluid, and transfer the liquid little by little to a litre flask, taking care to rub the residues in the mortar with fresh quantities of water as long as any heavy lumps remain. Fill the flask with water to the litre mark, and mix the milky liquor thoroughly by shaking. Each c. c. of this solution will contain 0.01 grm. of the bleaching powder. — By means of a marked pipette, measure into a beaker 50 c. c. of the freshly shaken, milky solution of bleaching powder, and pour into it the arsenical solution from a 50 c. c. burette, until a drop of liquid taken from the beaker fails to produce a blue spot on iodo-starch paper. The contents of the beaker must be stirred continually during the addition of the arsenical liquid, and the latter must be poured in slowly,—at last drop by drop. It is easy to hit the point of saturation exactly, since the successive blue spots produced on the paper gradually become fainter and fainter, and warn the operator to proceed more and more slowly. — If the operations were reversed, and the solution of bleaching powder were poured into the arsenical solution until a drop of the mixture gave a blue spot, this premonition would be lost. Another objection to this inverse method of working would be found in the tendency of bleaching powder solutions to froth,—whereby it becomes difficult to read the indications of the burette,—and to soil the burette. But, as Penot has suggested, it is sometimes well to resort to the inverse method for the sake of controlling the accuracy of an experiment made in the ordinary way.

The number of half cubic centimetres used of the standard arsenious solution, indicates directly the number of French chlorometrical degrees; *i. e.*, the number of litres of chlorine gas, at 0° and 76 m. m. pressure contained in 1 kilogramme of the bleaching powder. For example: In case 35 c. c. of the arsenical solution were consumed, then the quantity of bleaching powder used would contain 35 c. c. of chlorine gas. But the portion of bleaching powder solution taken (50 c. c.) contained 0.5 grm. of the powder; hence 0.5 grm. of the bleaching powder contains 35 c. c. of chlorine. and 1000 grms. contain 70,000 c. c. = 70 litres. It is for the sake of these degrees that the standard solution of arsenious acid is made of the given strength, as will appear from the following proportion: —

$$\underset{142}{\text{Wt. of 4 Atoms of chlorine}} : \underset{198}{\text{Wt. of 1 Molec. of } As_2O_3} = \underset{3.18002}{\text{Wt. of 1 litre of Cl gas}} : x$$

where x = 4·4341, or the quantity of arsenious acid that 1 litre of chlorine can convert into arsenic acid. — To obtain directly the chlorometrical degrees employed in England and America, take 6.9718 grms. of arsenious acid for the standard solution instead of the quantity stated above, and dissolve to the volume of 1 litre. 1 c. c. of this liquid will be equivalent to 1 c. c. = 0.005 grm. of chlorine. The number of c. c. used will consequently indicate the per cent of chlorine, if 0.5 grm. (50 c. c.) of the bleaching powder be taken as before. — The method of converting French or "Gay-Lussac" degrees into per cents, will appear from the following examples:—Bleaching powder of 90° Gay-Lussac contains 90 × 3.18002 = 286.2018 grms. chlorine in 1000 grms., or 28.62 in 100. A sample containing 34.2 per cent chlorine would mark 107.5°, for if 100 grms. contain 34.2 grms. Cl, and 1000 grms. contain 342 grm. Cl, then 342 ÷ 3.18002 = 107.5, *i. e.*, 1000 grms. of the powder contain 107.5 litres of chlorine.

For a complete table of per cents corresponding to Gay-Lussac degrees, see L. Müller in *Dingler's polytech. Journ.*, 1853, **129.** 286.

To prepare the *iodo-starch paper*, grind up 3 grms. of white starch in 250 c. c. of water, boil the mixture, with constant stirring; add an aqueous solution of 1 grm. iodide of potassium and 1 grm. crystallized carbonate of sodium, dilute with water to 500 c. c.; soak strips of printing paper in the liquor, dry them quickly in pure air, and keep in tightly stoppered bottles for use.

The process is accurate and of easy application. The standard arsenical solution undergoes no change on keeping, provided the arsenious acid and the carbonate of sodium used for preparing it are both absolutely free from easily oxidizable matters, such as sulphides or sulphites. (Mohr, *Titrirmethode*, 2d Edition, **1.** 290). When prepared from ordinary materials the alkaline solution of arsenious acid is liable to oxidize slowly in the air. But even in this case, all difficulty may be avoided by keeping the solution in a number of small glass-stoppered bottles, each completely filled with the liquid, and using a fresh bottle for every new series of experiments. (Fresenius).

According to Fresenius (*System Quant. Anal.*, Art. Chlorimetry), much more constant and accurate results are obtained by operating with the turbid solution, or rather emulsion, of bleaching powder above described, than can be got by filtering the liquor, or allowing it to settle, and operating upon the clear portion. The emulsion should always be shaken just before any portion of it is taken for analysis.

Attempts of Mohr (*Titrirmethode*, 1855, p. 287) to do away with the iodo-starch paper, by mixing a definite quantity of the standard arsenical solution with starch paste and iodide of potassium, and pouring bleaching powder solution into the mixture, from a burette, resulted in failure. So much of the starch was destroyed by the hypochlorite, wherever it happened to be in excess for a moment, that no useful results could be obtained.

Another modification of the process, however, suggested by Mohr (*Ibid.*, pp. 323, 290), gives accurate results. In this case a definite quantity of the bleaching powder solution is measured into a beaker, and rapidly titrated with a standard solution of arsenious acid,—prepared by dissolving 4.9 grms. of pure arsenious acid in an aqueous solution of 20–25 grms. of pure crystallized carbonate of sodium, with the aid of heat, and subsequently diluting to 1 litre,—until a drop of the mixture ceases to turn iodo-starch paper blue. The excess of arsenious acid which has been used is then accurately determined by adding some starch paste to the mixture in the beaker, and pouring in a standard solution of Iodine, until the starch becomes blue. The amount of arsenious acid which corresponds to the quantity of iodine solution used, is then subtracted from the amount of arsenious acid which was poured into the beaker, in order to obtain the precise amount consumed by the chlorine of the bleaching powder.

Other applications. Both processes are of course applicable to the valuation of "chloride of soda," and other hypochlorites, as well as to that of bleaching powder. They may be used also for analyzing compounds of the other chlor-oxygen acids; for determining chlorine, and for analyzing many oxygen acids and peroxides which evolve chlorine when heated with chlorhydric acid.

An aqueous solution of chlorine may be mixed immediately with a measured quantity of the standard solution of arsenious acid; while gaseous chlorine may either be received directly in a measured quantity of the standard arsenic solution, or better, in an unmeasured excess of bicarbonate of sodium solution. In the first case, some starch solution is added to the mixture after the chlorine has been absorbed, and the excess of arsenious acid is estimated by means of a standard solution of Iodine. In the second case, the standard solution of arsenious acid is poured into the alkaline mixture, until a drop of the latter placed upon iodo-starch paper ceases to color it; some drops of starch solution are then added to the liquor, and the slight excess of arsenious acid which has been added is determined by means of the standard iodine. — In neither case should any starch be added to the solution, until after the chlorine has all been absorbed and combined by the action of the arsenious acid.

In case the value of binoxide of manganese, chlorate of potassium, or bichromate of potassium is to be estimated by this process, or any analogous compound or peroxide is to be analyzed, boil the substance with an excess of concentrated chlorhydric acid, absorb the chlorine in arsenite or bicarbonate of sodium, and estimate it as above. Compare the next Method (by nascent chlorine).

3. *By nascent chlorine.* See Chromic Acid. (Reduction of by chlorhydric acid). If a weighed quantity of a chromate, such as bichromate of potassium, for example, is boiled with concentrated chlorhydric acid, chromic acid is reduced and chlorine set free in equivalent proportion, in accordance with the equation:—

$$K_2O, 2CrO_3 + 14HCl = Cr_2Cl_6 + 2KCl + 7H_2O + 6Cl.$$

The precise amount of Chlorine set free may readily be determined in a variety of ways. (Compare the last paragraph of the preceding Method). One method is to conduct the gas into a solution of iodide of potassium and to estimate the iodine, which the chlorine liberates, by means of a standard solution of hyposulphite of sodium, or by sulphurous acid. (See Iodine). But if arsenious acid, or an arsenite be added to the mixture of bichromate of potassium and chlorhydric acid before this mixture is heated, a definite quantity of the chlorine will be consumed in converting arsenious acid into arsenic acid, in accordance with the equation:—

$$As_2O_3 + 4Cl + 2H_2O = As_2O_5 + 4HCl,$$

and just so much less Cl will pass forward to be absorbed by the solution of iodide of potassium.

By operating with weighed quantities of bichromate of potassium it is easy to determine in one sample how much chlorine is set free in the absence of arsenious acid, and in a second sample how much less chlorine is given off after the addition of a weighed quantity of arsenious acid. The quantity of arsenious acid must of course always be less than sufficient to consume all the chlorine which the quantity of bichromate taken can evolve. Besides arsenious acid, the process may be applied for determining ferrous oxide, either by itself or in presence of ferric oxide, and other substances which are easily and completely oxidized by nascent chlorine, as will appear under the several headings. (Bunsen, *Annal. der Chem. und Pharm.*, 1853, **86.** 290). — It may here be remarked that the process, though rapid and accurate, requires care and a sensitive balance. If no arsenious acid were present, 0.2 grm. of a chromate containing 50 per cent of chromic acid would be enough for an analysis. If more were taken, an inconveniently large quantity of hyposulphite of sodium would have to be used. A gramme of bichromate of potassium would require 500 c. c. of hyposulphite of the ordinary strength. (Hinman). The cost of iodide of potassium is an objection, moreover, when any considerable number of analyses are to be made.

Method C. Oxidation by Bichromate of Potassium.

1. *Volumetric.* Method and precautions described under Antimonious Acid. See also biChromate of Potassium (Kessler, *loc. cit.*). See also the preceding paragraph,—oxidation of As_2O_3 by nascent chlorine.

2. *Gravimetric.* Mix the substance to be examined with a weighed quantity of pure, dry bichromate of potassium, place it in a suitable vessel and add strong sulphuric acid. A quantity of chromic acid will be reduced by the arsenious acid, in accordance with the equation:—

$$3As_2O_3 + 4CrO_3 = 3As_2O_5 + 2Cr_2O_3.$$

Determine how much chromic acid is left, by means of oxalic acid (see Chromic Acid, reduction of by oxalic acid), and calculate how much arsenious acid must have been present to reduce the remainder of the chromic acid taken. (Vohl, *Annalen Chem. und Pharm.*, **94.** 219).

Method D. Oxidation by Permanganate of Potassium. The process is attended with difficulties which have not yet been fully overcome. (Kessler, *Poggendorff's Annalen*, **118.** 17; Lenssen, *Journ. prakt. Chemie*, **78.** 197).

Method E. Oxidation by Nitrate of Potassium. Same Method as that described under Arsenic, with the exception that a smaller proportion of nitrate of potassium is needed. A mixture of 3 parts carbonate of sodium and 1 part nitrate of potassium, is sufficient for the fusion of an arsenite. The fusion may be effected in a platinum crucible.

Method F. Oxidation by Aqua regia, or by Chlorate of Potassium. Method described under Antimony and Arsenic.

According to Stein, this principle may be used for estimating nitric acid. To this end, mix the nitrate to be analyzed with 3 parts of arsenious acid, dissolve the mixture in concentrated chlorhydric acid, and evaporate to dryness. Take up the residue with water, and determine the arsenic acid by precipitating it as Arseniate of Magnesium and Ammonium.

Method G. Oxidation by Salts of Gold. Mix the solution which contains arsenious acid with an excess of a solution of chloraurate of sodium, or chloraurate of ammonium, and allow the mixture to stand in a covered beaker for several days at the ordinary temperature. Collect and weigh the precipitated Gold, and from its weight calculate that of the arsenic in the solution, in accordance with the reaction:—

$$3As_2O_3 + 2Au_2O_3 = 3As_2O_5 + 4Au.$$

In case the original solution is very dilute, it had better be left to stand in a place slightly warmer than the ordinary air. It is well to keep the filtrate from the metallic gold, and to collect any further portion of gold which may be deposited from it. Even from the original solution the gold is deposited very slowly upon the walls of the beaker. The beaker should be new and smooth, for it is difficult to wash out the precipitated gold from a beaker which has become roughened by use. The solution must be free from nitric acid and other oxidizing agents, and must be protected from dust also. A considerable excess of chlorhydric acid, however, appears to do no harm. In case the substance to be analyzed is solid, it should be dissolved in chlorhydric acid.

Though inferior to Method A, the gold process is reputed to yield accurate results, and to be free from the sources of error which have led to the abandonment of the process as applied to the determination of Antimonious Acid. The inventor of the process, H. Rose, remarks that its results would probably be less accurate in case too large a proportion of chloride of sodium, or of chloride of ammonium, were mixed with the chloride of gold in preparing the chloraurate.

In order to be fit for use as a reagent, arsenious acid should volatilize without residue, and should yield no brownish coloration when 10 or 12 grms. of it are dissolved in a solution of pure carbonate of sodium, and tested with a drop or two of a solution of acetate of lead. The brown color thus produced is due to the presence of sulphide of arsenic; and in case a residue is left on heating, which turns black when ignited in a current of hydrogen, the presence of antimony is indicated.

The solid cakes of arsenious acid to be obtained of importers of pure chemicals, are generally quite pure.

Arseniomolybdate of Ammonium.

Principle. Insolubility in water, and in acid and saline solutions.

Applications. Separation of arsenic acid from Na, K, NH_4, Li; Ba, Sr, Ca, Mg; Al, Cr, Fe, Mn, Co, Ni, Zn, Cd, Pb, Bi, Ag, Hg and Cu. Estimation of arsenic acid in solutions free from phosphoric and silicic acids, and from substances capable of decomposing molybdic acid.

Method. Prepare a solution of molybdate of ammonium by heating 10 grms. of that salt with 40 c. c. of ammonia water of 0.96 specific gravity, pour the solution into a mixture of 120 c. c. strong nitric acid, and 40 c. c. of water, and let the whole digest for 8 or 10 hours, at 40°. Allow the mixture to settle, and decant the clear liquid.

Mix the solution which contains the arsenic acid with a large proportion of the aforesaid solution of molybdate of ammonium, add enough nitric or chlorhydric acid to redissolve the precipitate of molybdic acid which forms at first, and boil the solution for a long time. If molybdic acid has been added in excess, arseniomolybdate of ammonium will be thrown down in the form of a yellow precipitate. The subsequent operations are similar to those described under Phosphomolybdate of Ammonium. Since it is not easy to separate molybdic acid from the filtrate, a new portion of the original substance had better be taken for the determination of the metals with which the arsenic acid was originally combined.

Properties. Arseniomolybdate of ammonium is well nigh insoluble in water, saline solutions and acids, especially nitric acid, provided an excess of molybdate of ammonium be present, together with a moderate excess of nitric or chlorhydric acid. The composition of a sample of the precipitate analyzed by Seligsohn (*Journ. prakt. Chem.*, **67.** 481), was molybdic acid 87.67 per cent; arsenic acid 6.31 per cent; ammonia 4.26 per cent, and water 1.77 per cent. Since its composition is variable, the precipitate is never weighed directly. It is dissolved in ammonia-water, and Arseniate of Magnesium and Ammonium is precipitated from the ammoniacal solution.

Atropin. See Iodo Mercurate of Atropin.

Barium is usually determined as Sulphate or Carbonate, but sometimes as Fluosilicate of Barium. For its alkalimetrical estimation, see Oxide and Carbonate of Barium. See also the finding list for Barium, in the Appendix.

Baryta. See Hydrate of Barium.

Basic **Benzoate of Iron** (Ferric benzoate).

Principle. Insolubility in water.

Applications. Separation of Fe from Mn, Ni and Zn.

Method. Same as that described under Succinate of Iron.

When benzoic acid, or rather, when the alkaline benzoates can be obtained at less cost than succinic acid and the succinates, it is better to precipitate iron as a benzoate (Hisinger & Berzelius). Basic benzoate of iron is indeed slightly less soluble than the corresponding succinate, but this advantage is balanced by the fact that the precipitated benzoate is more voluminous than the succinate. Moreover, it contains much more carbon than the latter; but this difficulty can be overcome, as with the succinate, by leaching the precipitate with dilute ammonia-water, until most of the benzoic acid is removed.

Bismuth may be determined as metallic Bismuth, Carbonate, Chromate, Oxide or Sulphide. For a list of methods for separating it from the other metals, see Appendix.

Principle I. Insolubility of the metal in dilute nitric or acetic acids in presence of metallic lead or zinc.

Applications. Separation of bismuth from certain other metals,—notably from lead.

Method A. Precipitate the bismuth by metallic zinc, from a dilute nitric acid solution. (*Pfaff, Handbuch analyt. Chem.*, 1825, **2.** 401).

Method B. Dissolve the alloy in nitric acid, dilute the solution with a large quantity of water, and place in it a strip of pure metallic lead. The bismuth will be precipitated rapidly and completely, in the form of a black powder. As soon as the precipitation is finished, decant the clear acidulated liquor, wash off adhering particles of bismuth from the strip of lead, and wash the precipitate rapidly by decantation, first with water, and then with alcohol. Collect the precipitate upon a small weighed filter, dry and weigh. (Patera, *Zeitsch. analyt. Chem.*, 1866, **5.** 226).

Method C. After having brought the mixture of bismuth and lead into solution, throw down both the metals as carbonates, by means of carbonate of ammonium, dissolve the precipitate in acetic acid, and place in the solution a weighed piece of pure sheet lead. Cover the beaker, and leave it at rest for several hours, taking care that no part of the lead projects above the surface of the liquid. When all the bismuth has been precipitated, wash the strip of lead, and dry and weigh it. The difference between the first and last weight of the lead indicates how much of this metal has been added to the solution. Collect the precipitated bismuth upon a filter, and wash it with water which has been boiled to expel air, and afterwards cooled. Dissolve in nitric acid, and precipitate Carbonate of Bismuth; or evaporate the nitric acid solution to dryness, and ignite the dry residue to obtain Oxide of Bismuth. Lead may be determined in the filtrate by precipitating it as Carbonate of Lead. (Ullgren, *Berzelius's Jahresbericht*, **21.** 148).

Principle II. Fixity, when heated.

Applications. Estimation of bismuth in the oxide, sulphide, oxychloride and salts of this metal.

Method. Fuse the compound to be analyzed in a capacious porcelain crucible, with 5 times its weight of ordinary cyanide of potassium. Soak the fused mass in water; wash the kernels of metallic bismuth rapidly, first with water, then with weak alcohol and finally with strong alcohol; collect upon a small, tared filter, dry and weigh. In case oxide or oxychloride of bismuth be operated upon, the reduction is soon completed, at a low heat, but the sulphide requires more time and a higher temperature. When, on treating the fused mass with water, only metallic grains remain undissolved, the reduction may be deemed complete; but if in reducing the sulphide, there is left a quantity of the latter, in the form of black powder, mixed with the metal, this powder should be collected and again fused with a fresh quantity of cyanide of potassium. The crucible should be weighed before the experiment and after it; also in connection with the tared filter; for it sometimes happens that particles of porcelain torn off from the crucible during the process of reduction, remain mixed with the metallic bismuth. The method yields good results. (H. Rose, *Poggendorff's Annalen*, **91.** 104, and **110.** pp. 136, 425).

Boiling Point.

To determine the boiling point of any substance, place a quantity of the liquid in a small, tubulated, long-necked glass retort; throw into the liquid two or three fragments of coke, each as large as a small pea, a globule of mercury, a few small scraps of platinum foil, or better, a piece of metallic sodium, in case the liquid under examination is a hydrocarbon. As much as 100 or 200 c. c. of the liquid should be taken, if it can be had. The presence of the coke or metal, to facilitate ebullition, is essential. In case no fragments of solid matter are placed in the retort, the liquid is not only liable to "bump" when heated, but even to boil at a higher temperature,—in the case of some liquids, 5° or 6° higher,—than its real boiling point.

Close the tubulure of the retort with a cork, to a perforation in which a thermometer has been fitted, in such manner that when the cork is pushed tightly into the tubulure the bulb of the thermometer shall extend into the liquid, and reach almost, but not quite, to the bottom of the retort. (Warren, *Amer. Journ. Sci.*, 1865, **40.** 222). To that part of the stem of the thermometer which projects above the cork, tie a second thermometer, in such manner that it can be slipped up or down without difficulty, and place a broad sheet of paper or pasteboard across the top of the cork to screen the upper thermometer from any heat which might radiate from the retort. Set the retort upon a wire gauze support, above a lamp, in such position that its neck shall incline slightly upwards; place a screen below the retort to protect its sides from being overheated, twist moistened cloths around the neck of the retort to keep it cool, and in case the liquid is easily volatile, place bits of ice upon the cloth. Finally heat the liquid until it boils steadily, and the mercury in the thermometer ceases to rise. Some little time will often be required in order that that portion of the thermometer which projects above the cork may acquire the highest temperature which the boiling liquid can communicate to it. — In case the liquid boils at a lower temperature than that of the air, place the retort in a bath of ice-water, and gradually raise the temperature of the bath by means of a small lamp-flame. — Arrange the upper, moveable thermometer so that its bulb shall be in contact with the stem of the fixed thermometer, at a point midway between the upper end of the mercury column in that thermometer and the centre of the cork. Then note the height of the barometer and of both thermometers. As soon as the mercury has become stationary in the thermometer, remove the lamp for a moment from beneath the retort, turn the neck of the retort downward, and connect it with a Liebig's condenser. Again place the lamp beneath the retort, and continue to boil the liquid until most of it has been distilled over out of the retort. In the course of a few seconds after the lamp has been replaced, the mercury in the thermometer will again become stationary. At that moment note the time and the height of the mercury in the thermometers, and repeat these observations at frequent intervals, until the distillation is finished. If the liquid under examination be pure, the height of the thermometer will scarcely vary from first to last. In case there are slight differences between the earlier and later observations, either the average of the several observations, or better, of those corresponding to the longest intervals of time, may be chosen. Any material differences between the several observations would of course indicate that the liquid under examination was a mixture of two or more substances of unlike boiling points.

In case the boiling point of a mixture, such as ordinary spirit, is to be determined, only the first observation, made when the neck of the retort is inclined upwards, is of value; for the strength, and consequently the boiling point of the spirit, would change if the liquid were subjected to distillation. The process of distillation above described, is to be omitted, therefore, when a mixture is tested.

The purpose of the upper or side thermometer is merely to determine approximately the mean temperature of that portion of the mercury column in the fixed thermometer, which is above the cork, in order to ascertain what amount of correction should be made to compensate for the cooling action of the air. The upper thermometer would be useless in case no portion of the mercury column were to project above the cork, for in that event the whole of the metal, being surrounded with the vapor of the boiling liquid, would have one and the same temperature. But when, as is usually the case, a part of the mercury rises above the cork, and is thus exposed to a lower temperature than that of the boiling liquid, the upper extremity of the column will stand at a slightly lower point than it would if no portion of the mercury had been cooled. The correction to be applied to the observed temperature T°, is found by means of the following formula:—

$$N\,(T-t)\,\delta,$$

in which t is the observed height of the upper, moveable thermometer, N the difference between T° and that degree of the thermometer scale which is situated at the middle of the cork; or, in other words, the length of the column of mercury which has been cooled to t° by contact with the air, and δ the coefficient of apparent expansion of mercury in the glass of which the thermometer is constructed. The value of δ may be taken as equal to 0.0001545. (H. Kopp, *Poggendorff's Annal.*, 1847, **72.**

38). For example,—if a liquid is observed to boil at 171.3° (= T) and the 25th degree of the thermometer be at the middle of the cork, then N will equal 171.3°—25° = 146.3°. And if the mean temperature of the external column of mercury were 43° (= t), as observed by the upper thermometer, then T—t would equal

$$171.3 - 43 = 128.3.$$

The correction to be applied will consequently be

$$146.3 \times 128.3 \times 0.0001545 = 2.9,$$

and the boiling point of the liquid "corrected for external column of mercury," will be

$$171.3 + 2.9 = 174.2.$$

Since the boiling point of a liquid depends on the pressure to which the liquid is subjected, another correction must be applied in order to reduce observations made under the varying pressure of the atmosphere, to the values which would obtain if the atmospheric pressure were normal,—that is to say, equal to 760 millim. of mercury at 0°. It is customary, on this account, to apply a correction of ± 0.1° C., for every 2.7 millim. that the mercury in the barometer stands above or below 760. Strictly speaking, the correction to be applied to the boiling point of a liquid observed under an atmospheric pressure other than the normal pressure, varies with the nature of the liquid; for equal alterations of pressure do not cause precisely equal changes in the boiling points of different liquids. The value above given, of 0.1° for a variation of pressure of 2.7 millim., has been deduced from direct determinations of the boiling point of water under different pressures, and is absolutely accurate for that liquid alone. But since the greatest variations which ever occur in the pressure of the atmosphere are relatively very small, they may, without any appreciable error, be regarded as affecting the boiling points of all liquids equally.

Boracic Acid is determined either indirectly; as Fluoborate of Potassium; or as Borate of Magnesium. [See finding list for Boron, in the Appendix].

Principle I. Power of expelling carbonic acid from Carbonate of Sodium.

Applications. Estimation of boracic acid in aqueous solutions free from all other substances, except ammonium. In case a heavy metal has been separated from boracic acid by precipitating with sulphuretted hydrogen, the method now in question may be employed for determining boracic acid in the filtrate, after the excess of sulphuretted hydrogen has been expelled, by passing a current of carbonic acid through the liquor. The filtrate must, however, contain no acid besides boracic acid.

Method. Weigh out 1 or 2 times as much pure, fused carbonate of sodium, as there is supposed to be boracic acid in the solution to be analyzed. Dissolve the carbonate in the weighed solution, evaporate the mixture to dryness at a gentle heat, and carefully heat the residue until it fuses to a thin liquid. A quantity of carbonic acid, equivalent to the quantity of boracic acid present, will be expelled, and there will be left a mixture containing a known quantity of sodium, and unknown quantities of carbonic and boracic acids. Weigh the residue, and determine the amount of Carbonic Acid contained in it. From the difference between the quantity of carbonic acid in the carbonate of sodium taken, and that found in the residue, calculate the amount of boracic acid (H. Rose).

In case the proportion of boracic acid in the solution to be analyzed is known approximately, between 1 and 2 equivs. of carbonate of sodium should be taken for two equivalents of boracic acid. If this be done, all the carbonic acid will be expelled on heating, and we have only to deduct from the weight of the residue the weight of oxide of sodium taken, in order to find the weight of the boracic acid. Since the tumultuous escape of carbonic acid might occasion loss, it is best not to ignite the residue left by evaporation all at once, but after thorough drying, to project it by small portions into a red hot crucible. (Schaffgotsch, *Pogg. Ann.*, **107.** 427).

Properties. It may be remarked, in this connection, that boracic acid cannot be determined by evaporating its aqueous or alcoholic solution to dryness, and weighing the residue; for though little volatile of itself, considerable portions of the acid are carried away by the aqueous or alcoholic vapor. The same remark applies to solutions to which an excess of oxide of lead, of ammonia-water, or of triphosphate of sodium, has been added.

In the cold, no carbonic acid is expelled by boracic from carbonate of sodium, and only a small quantity escapes during the evaporation of the aqueous solution. It is only when the dry mass is ignited, that the evolution of carbonic acid really begins.

Principle II. Power of expelling water from certain refractory hydrates; *i. e.*, fixity and power of combining with bases to form fixed products.

Applications. Estimation of water in the hydrates of Na, K, Ba and Sr.

Method. Mix a weighed quantity of the substance to be examined, with a weighed quantity of powdered anhydrous boracic acid, more than sufficient to saturate the whole of the alkali. Fuse the mixture in a platinum crucible until all the water has been expelled, and weigh the cold product. The difference between the final weight and the sum of the weights of boracic acid and substance taken, will represent the weight of the water. (Pfaff, *Handbuch analyt. Chem.*, 1824, **1.** 117). The substance analyzed must either be free from carbonic and other volatile acids, or the proportion of these substances must be determined in separate portions of the material. The process can hardly be capable of yielding absolutely accurate results, since, as is well known, boracic acid is somewhat volatile in an atmosphere of aqueous vapor. Compare *bi*Borate of Sodium.

Principle III. Power of decomposing silicious minerals.

An old method employed by Davy, Gehlen and others, for the decomposition of refractory minerals in which alkalies were to be determined, deserves mention, though it has long since been superseded by the methods of fusing with alkaline earths, or treating with fluorhydric acid.

Davy fused the powdered mineral with twice its weight of boracic acid glass, for half an hour, in an intense fire; decomposed the fused mass with dilute nitric acid; evaporated to separate silica; precipitated with carbonate of ammonium; supersaturated with nitric acid and separated the boracic acid by crystallization from the nitrates of the alkalies. (Pfaff, *Handbuch analyt. Chem.*, 1824, **1.** 467). Pfaff proposed to separate the alkalies from the boracic acid by converting them into sulphates, and then washing out the last portions of boracic acid with alcohol.

Basic Borate of Magnesium.

Principle. Insolubility in water, and fixity when ignited.

Applications. Estimation of boracic acid in presence of alkalies.

Method. Neutralize the solution with chlorhydric acid, add enough double chloride of magnesium and ammonium, that at least .2 parts of magnesia may be present in the solution for 1 part of boracic acid, pour in some ammonia-water, and evaporate the mixture to dryness in a weighed platinum dish. If, on the addition of the ammonia-water, a precipitate falls which does not dissolve readily on warming, a new quantity of chloride of ammonium must be added to the mixture. From time to time during the evaporation, add a few drops of ammonia-water. Ignite the dry residue, and afterwards treat it with boiling water. Collect the insoluble mixture of borate and oxide of magnesium upon a filter, and wash it with boiling water, until the wash water ceases to give any precipitate when mixed with nitrate of silver. Mix the filtrate and washings with ammonia-water, again evaporate to dryness, ignite and wash with boiling water, as before. Place both the insoluble residues in the platinum dish, and ignite them for a long time as strongly as possible, in order to decompose the small quantity of chloride of magnesium which may still be mixed with the borate. Weigh the cold product, dissolve it in chlorhydric acid, and determine how much magnesia is contained in it by precipitating as Phosphate of Magnesium and Ammonium, or estimate the magnesia at once by the process of Alkalimetry. Subtract the weight of the magnesia found from the weight of the borate of magnesium, in order to obtain the weight of the boracic acid. The process yields satisfactory results. (Marignac, *Zeitsch. analyt. Chem.*, **1.** 406).

Precautions. The platinum dish must be weighed in the first place, since small portions of platinum are apt to appear on dissolving the residual borate in chlorhydric acid. If the weight of the dish were unknown, it would be necessary to weigh this abraded platinum, and subtract its weight from that of the borate of magnesium. When the volume of liquid to be operated upon is considerable, it had better be concentrated to a small bulk in a porcelain dish, and afterwards evaporated to dryness in the platinum vessel. — In case boracic acid is to be separated from a heavy metal precipitable by sulphuretted hydrogen, the filtrate from the metallic sulphide should be mixed with hydrate and nitrate of potassium, evaporated to dryness and ignited, before adding the chloride of magnesium, as above described.

Borate of Potassium. See Borate of Sodium.

Borate of Sodium.

Principle. Solubility in water.

Applications. Separation of boracic acid from most of the metals, other than alkali metals.

Method. Boil or fuse the borate to be analyzed with carbonate or hydrate of potassium, or of sodium, filter off the precipitated carbonate or hydrate of the metal, and determine boracic acid in the filtrate as Borate of Magnesium, or as Fluoborate of Potassium.

*Bi*Borate of Sodium. (Borax).

Principle. Power of expelling Carbonic, Nitric and Silicic acids, when fused with salts of these acids. See the respective acids and Boracic Acid.

It is to be observed, that though "borax-glass" may be kept fused at a red heat for quarter or half an hour without loss, through volatilization, an appreciable quantity of the compound exhales at a white heat. A decided loss of weight occurs, for example, when a platinum crucible containing dry borax-glass is ignited even for a few minutes at the blast lamp. (Fresenius, *Zeitsch. analyt. Chem.*, **1.** 65).

Borax was formerly somewhat used for decomposing refractory minerals. Thus Berzelius & Hisinger (*Gehlen's neues Journ. der Chemie*, **6.** 304) fused finely pulverized spinel with twice its weight of borate of potassium, in a platinum crucible, and dissolved the fused mass in chlorhydric acid. (Compare Aluminate of Sodium). — Borax-glass is still used, in conjunction with carbonate of sodium, in the excellent method of Plattner (*Probirkunst mit dem Löthrohre*, 1853, p. 689), for analyzing, with the aid of the mouth blowpipe, those iron ores which cannot be decomposed by acids. It is to be observed, however, that the process depends upon the fusibility of borate of iron, and the solubility of this substance in acids rather than upon any superior affinity of boracic acid for iron. The details of the method are as follows:—Grind a considerable quantity of the ore to fine powder, best in an agate mortar, mix the powder thoroughly, spread it to a thin layer, and draw a fair sample of it by taking a number of small quantities from different parts of the layer. Dry the sample at 100°, weigh out 0.1 grm. of the powder on a delicate assay balance, transfer it to a large test tube, and pour upon it, little by little, a quantity of strong chlorhydric acid. As soon as the acid has ceased to act in the cold, heat it over a spirit lamp as long as anything dissolves; throw the mixture upon a small filter, and wash the residue with boiling water. Dry the residue, which will exhibit a yellowish, red, or gray color, over a spirit lamp. If the ore were of such character as to be decomposed completely by chlorhydric acid, the residue upon the filter would be colorless, and free from iron; so that the process now in question, *i. e.*, the fusion with borax, would not be called for. But the residue does, in fact, contain silicates of iron and the earths, as well as quartz, and rarely some sulphate of barium, also. As soon as it is thoroughly dry, mix it in the filter with as much powdered borax-glass, and 3 times its bulk of dry carbonate of sodium. Roll the paper into the form of a small ball or cylinder, place the ball in a hole bored across the grain of a good piece of charcoal, and melt the mixture, with the oxidizing blowpipe flame, to a clear, transparent bead. After the bead has become cold crush it in a steel mortar, place the powder in a light porcelain cup covered with a watch glass, and treat with dilute chlorhydric acid; then evaporate the solution to dryness, together with the previous solution obtained by acting upon the ore with acid, to separate Silicic Acid. The filtrate from silicic acid is boiled with nitric acid to oxidize the iron; the latter is precipitated, together with alumina, as Hydrate of Iron, and the alumina dissolved out as Aluminate of Sodium, in the usual way. — The process has great merit from its simplicity. It may be put in practice almost anywhere, since it neither requires a well appointed laboratory nor expensive apparatus. It is evident that besides iron, silicic acid, and many other constituents of refractory silicates, may readily be determined in this way. Analyses of red ochre, made under my direction by this method, and by a modification of it, in which either pure silicic acid or silicate of sodium (water glass) was substituted for the borax, showed clearly the importance of the borax, and the impropriety of trying to do without it.

To prepare borax-glass, heat ordinary crystallized borax in a platinum or porcelain dish until it ceases to swell up. Grind the porous residue to powder, and again heat it in a platinum crucible, until it fuses to a transparent mass. Pour the semi-fluid, viscid substance upon a piece of porcelain, and keep the cold solid in a well-stoppered bottle. The borax-glass must, in most cases, be reheated just before use to ensure its absolute freedom from moisture.

The purity of commercial borax may be tested by adding carbonate of sodium to the aqueous solution, and by adding solutions of nitrate of barium and nitrate of silver to other portions of the solution after acidulating it with nitric acid. Neither of these reagents should occasion any alteration in the appearance of the solution. In case the borax is impure, it may be purified by recrystallization.

Bromates may be converted to Bromides by ignition, and weighed as such; or they may be mixed with an excess of chlorhydric acid, the mixture evaporated to dryness, and the chlorine estimated as Chloride of Silver, by titration. From the weight of chlorine found, the equivalent weight of bromic acid may then be calculated. (Mohr).

Bromides of volatile elements, such as the bromides of arsenic, antimony, phosphorus and sulphur, are completely decomposed by water, with formation of bromhydric acid, which may be determined by precipitation, as Bromide of Silver.

Bromide of Mercury.

Principle. Power of Oxide of Mercury to absorb bromine at the ordinary temperature, and fixity of the compound formed.

Application. Estimation of bromine in organic substances.

Method. Heat the substance to be analyzed in a stream of hydrogen, and burn the hydrogen thus charged with the vapor or products of distillation of the substance, in a stream of oxygen. Pass the products of the combustion over a layer of strong sulphuric acid, to free them from water, absorb the bromine in a tube filled with fragments of precipitated oxide of mercury, and the carbonic acid in potash lye. The oxide of mercury should be in the form of pellets, or coarse masses, and should be followed by a short column of chloride of calcium, placed at the end of the tube farthest from the burning hydrogen. For the details of the process see A. Mitscherlich's paper in *Zeitsch. analyt. Chem.*, 1867, **6**. pp. 151, 153, 156, 141. Compare Carbon, oxidation of with oxygen gas.

Bromide of Silver.

Principle I. Insolubility in water and nitric acid.

Applications. Separation of bromine from most of the metals, and from organic substances. Estimation of bromhydric acid and of free bromine. Separation of bromhydric acid from As_2O_3, As_2O_5, P_2O_5, CrO_3, SO_3, B_2O_3, $C_2H_2O_4$, HFl, SiO_2 and CO_2.

Since bromide of silver is precipitated before

chloride of silver when nitrate of silver is added to a cold solution containing both chlorides and bromides, the insolubility of the bromide may even be used for separating bromine from chlorine as will be described below.

The Methods and Precautions are similar to those described under Chloride of Silver. The process yields excellent results.

Properties. As prepared in the wet way, bromide of silver is a yellowish-white precipitate, insoluble in water and nitric acid, tolerably soluble in ammonia-water and readily soluble in solutions of the hyposulphites and cyanides of alkali-metals. Though insoluble in very dilute solutions of alkaline chlorides and bromides, some of it dissolves in concentrated solutions of these salts. Traces of it dissolve also in solutions of the alkaline nitrates. A solution of iodide of potassium converts it into iodide of silver. — When heated, it melts to a reddish liquid which solidifies to a yellow, horn-like mass on cooling. It may be weighed, however, as well after having been dried at 100° as after melting. When heated in chlorine gas it is converted into chloride of silver and when heated in hydrogen to metallic silver. The precipitated bromide gradually becomes gray and finally black when exposed to light. Both the precipitated and the fused bromide are decomposed by metallic zinc with formation of metallic silver. The composition of the bromide is as follows: —

Ag	=	108	—	57.44
Br	=	80	—	42.56
		188		100.00

Determination of Bromine in bromides of Organic Substances soluble in water. [Compare Chloride of Silver and Iodide of Silver]. Weigh out about 1 grm. of nitrate of silver, dissolve it in water and add chlorhydric acid until all the silver is precipitated as chloride. Wash the chloride of silver with hot water, by decantation, and pass the washings through a small weighed filter to retain any suspended particles of the chloride. Pour upon the washed mass of chloride of silver an aqueous solution of about 0.5 grm. of the substance to be analyzed. Warm the mixture and let it stand for a couple of hours, then collect the mixed bromide and chloride of silver upon the weighed filter, above mentioned, wash it thoroughly, dry, and weigh. From the weight of the mixed precipitate and the known weight of silver taken, calculate the weight of the bromine in the precipitate as explained below, under Principle III.

One merit of the process, at least as applied to rare and costly substances, consists in the fact that the substance is not destroyed. It is found again in the filtrate from the bromide of silver in the form of a chloride. The process is not applicable to all organic substances containing bromine. Only that portion of bromine (as in a bromide) which is precipitable by nitrate of silver can be thus determined, but on this very account the process is valuable for determining the rational constitution of certain bromine compounds. (Maly, *Zeitsch. analyt. Chem.*, 1866, **5.** 68; Kraut, *Ibid.*, 1865, **4.** 167).

Separation of Bromine from Chlorine. [Compare Chloride of Silver and Iodide of Silver]. Several methods of separating bromine from chlorine depend upon the principle now in question.

A. *Fehling's method.* Mix the cold solution to be analyzed with a quantity of nitrate of silver solution somewhat more than sufficient to precipitate all the bromine, but not nearly sufficient to throw down all the chlorine. Shake the mixture strongly and leave it standing for some time, with occasional agitation. Wash the mixed precipitate of bromide and chloride of silver with especial care and thoroughness, dry, ignite, weigh, and treat a weighed portion of the residue with chlorine, as directed below, under Principle III, or with zinc (see Silver). To find the quantity of chlorine, precipitate another weighed portion of the original solution completely with nitrate of silver and from the weight of this precipitate deduct that of the bromide of silver found. — The following rule indicates in a general way how much nitrate of silver should be used. If the solution to be analyzed contains 0.1 per cent of bromine, add $\frac{1}{5}$ or $\frac{1}{6}$ as much of the solution of nitrate of silver as would be required to effect complete precipitation; — if 0.01 per cent, $\frac{1}{10}$; — if 0.002, $\frac{1}{30}$; — if 0.001, $\frac{1}{60}$. (Fehling, *Journ. prakt. Chem.*, **45.** 269).

B. *Mohr's method.* Precipitate all the bromine and part of the chlorine by means of a known weight of silver, added in the form of nitrate of silver. Wash and weigh the mixed precipitate. Calculate the amount of chloride of silver equivalent to the metallic silver taken, subtract this weight from that of the mixed precipitate and finally calculate how much bromide of silver was present, by the method given below, under Principle III. — This method has the merit of being more convenient and expeditious than the old method with chlorine described below (Principle III), but is probably somewhat less accurate than the latter, especially when the proportion of bromine in the mixture is small. In criticising the process, Fresenius urges that the supposition that a weighed quantity of silver will yield a precisely equivalent quantity of chloride of silver is practically incorrect, errors to the extent of some milligrammes being scarcely avoidable; it might thus happen that bromine could be calculated from a supposed difference even in cases where no trace of this element was present. (Compare Principle III).

The silver used as the precipitant may either be weighed in the metallic state and then be dissolved in nitric acid, or it may be added in the form of a standard solution of nitrate of silver. (F. Mohr, *Ann. Chem. und Pharm.*, **93.** 76).

C. *Wittstein's method.* See under Chloride of Silver.

D. *Pisani's method.* Add to the mixed solution of bromides and chlorides a known quantity of a standard solution of nitrate of silver in slight excess, filter and determine the silver in the filtrate by titration with iodine. (See Iodide of Silver). The precipitate is weighed and the proportion of bromine calculated as in B. The principle of the method is slightly different from that of the preceding methods, inasmuch as it precludes the partial precipitation. (Pisani, *Comptes Rendus*, **44.** 352).

Principle II. Reduction of to metallic silver by hot hydrogen.

Applications. Estimation of bromine in presence of chlorine.

Method. Weigh out a portion of the mixed precipitate of bromide and chloride of silver in a light bulb-tube : pass a slow current of hydrogen through the tube, heat the substance in the bulb until it fuses, and occasionally shake the fused mass so that new surfaces may be exposed to the action of the hydrogen. When the reduction seems to be complete, allow the tube to cool, hold it in an inclined position for a moment so that the hydrogen within it may be replaced by air, weigh it, again heat its contents in a current of hydrogen and continue these operations until the two last weighings agree. Calculate how much chloride of silver is equivalent to the metallic silver found, subtract this chloride of silver from the weight of the mixed chloride and bromide of silver taken, and calculate how much bromine was present, by the method given below, under Principle III. This method of reducing by hydrogen offers an excellent control to the method by chlorine described below, under Principle III. — Compare also the details of manipulation there described.

Principle III. Conversion into chloride of silver by hot chlorine. [Compare Chloride of Silver, decomposition of by bromide of potassium].

Applications. Estimation of bromine in presence of chlorine.

Method. After the mixed precipitate of bromide and Chloride of Silver has been collected and weighed in the usual way, fuse it in the crucible and pour out upon a piece of porcelain as much of the fused mass as will be sufficient for an analysis. Place the fragments in a weighed bulb tube, and again weigh the tube with its contents. Pass a slow stream of pure, dry chlorine gas through the tube, heat the silver salts to fusion and shake the bulb occasionally, in order to expose fresh surfaces of the fused mass to the chlorine. After the lapse of 20 or 30 minutes, allow the tube to cool, detach it from the chlorine generator, hold it in an oblique position until the chlorine has been replaced by air, and weigh. Again heat the contents of the tube in a current of chlorine during 10 or 12 minutes and weigh as before. Repeat the operations of heating and weighing until two consecutive weighings give the same result. The decrease of weight is simply the difference between the weight of the bromine originally present in the mixed salts and that of the chlorine which has replaced it. The weight of bromine originally present is therefore deduced from the proportion.

$$\begin{array}{c}\text{Difference between}\\ \text{At. Wts. of Cl and Br}\\ 44.5\end{array} : \begin{array}{c}\text{At. Wt.}\\ \text{Br}\\ 80\end{array} = \begin{array}{c}\text{Decrease of}\\ \text{Weight}\end{array} : \begin{array}{c}\text{Amt. of}\\ \text{Br present.}\end{array}$$

It results from this proportion that the weight of bromine may be obtained by multiplying the observed decrease by 1.797. If the amount of chlorine originally present is also desired, calculate the quantity of silver in the pure chloride of silver last weighed, and subtract this weight of silver and the weight of bromine deduced as above described, from the original weight of the mixed chloride and bromide. The remainder is the required amount of chlorine. — To determine the quantity of chlorine, precipitate completely with nitrate of silver a new portion of the substance to be analyzed and deduct from the weight of this precipitate that of the bromide of silver found as above.

Precautions. Though the method is tedious and inconvenient, very accurate results can be obtained by it if the proportion of bromine in the mixture is not too small. Care must be taken to expel the last traces of bromine and to weigh with scrupulous accuracy in each instance. The method is less liable to error than Mohr's process of precipitating with a known weight of silver, above described, (under Principle I). According to Fresenius, a simple experiment will show that pure chloride of silver heated cautiously in a current of chlorine in a light bulb tube suffers no alteration of weight. An error of 0.5 milligramme in this operation would be more surprising than an error of 2 mgrs. in the conversion of 2 or 3 grms. of silver into chloride of silver, more especially if a filter were used in the process; and the filter can hardly be dispensed with, since a precipitate almost always subsides less readily and completely in case of partial precipitation than when the precipitation is complete.

The process is not directly applicable for determining small traces of bromine in presence of large quantities of chlorine, as in the waters of saline springs, for example. In such cases the bromine must be concentrated either by fractional precipitation with nitrate of silver, as has been described above (Principle I), by fractional solution (see Bromide of Sodium), or by fractional distillation (see Bromine).

It is easy to control the results obtained by this method by treating the residual chloride of silver with hydrogen, as described under Principle II. The difference between the weight of the mixed bromide and chloride of silver, and the amount of chloride of silver equivalent to it found in the direct way by the method now in question, and by calculation in the method by hydrogen, should be the same.

Bromide of Potassium.

Principle. Power of decomposing freshly precipitated chloride of silver.

Application. Separation of Bromine, Iodine and Chlorine.

Method. See Iodide of Potassium.

Bromide of Sodium.

Principle. Solubility in strong alcohol.

Applications. Separation of bromine from chlorine, or rather concentration of bromine when mixed with chlorides.

Method. Add an excess of carbonate of sodium to the mixed solution of bromides and chlorides, filter if need be, evaporate nearly to dryness, and extract the residue with hot absolute alcohol. The whole of the bromide of sodium [or of potassium] will dissolve together with a small portion of chloride of sodium. Add to the solution a drop or two of a solution of carbonate of sodium, evaporate to dryness, dissolve the residue in water, acidulate with nitric acid, precipitate with nitrate of silver and analyze the mixture of bromide and chloride of silver, as described under Bromide of Silver. (Compare Marchand, *Journ. prakt. Chem.*, **47.** 363).

Bromine is commonly determined as Bromide of Silver, or as Bromine (by volumetric and colorimetric methods, or by loss). [See the finding list in Appendix].

Principle I. Volatility.

Applications. Estimation of bromine in metallic bromides. Separation of bromine from chlorine, or rather concentration of bromine when mixed with chlorides.

Method A. Place the dry weighed bromide in a porcelain crucible, mix it with pure sulphuric acid in slight excess, and evaporate until no more fumes are seen to arise from the dry residue. Weigh the residual metallic sulphate, calculate the amount of metal in this sulphate and subtract it from the weight of the bromide taken; the difference gives the weight of bromine in the sample analyzed. The method is inapplicable for the analysis of the bromides of silver, lead, mercury and tin, since these compounds are not readily decomposed by sulphuric acid. — Platinum crucibles cannot be employed; the metal would be attacked by the escaping bromine.

Method B. Supersaturate the bromide to be analyzed with chlorhydric acid, evaporate to dryness to expel the excess of acid, and estimate the chlorine in the residue as Chloride of Silver, by titration. From the weight of chlorine found calculate the equivalent weight of bromine. (Mohr).

To concentrate bromine when mixed with a large proportion of chlorides, mix the solution in a flask with chlorhydric acid and binoxide of manganese, connect the flask with two wide Woulfe bottles, by means of bent delivery tubes. Charge the bottles with strong ammonia-water and slowly heat the mixture in the flask. The whole of the bromine will pass over before much, if any, chlorine is evolved. The bottles must be large enough and the distillation slow enough that no vapors shall escape. When all the bromine has been evolved, as may be seen by the color of the gas in the tubes and in the flask above the liquid, loosen the cork of the flask to prevent the return of bromide of ammonium fumes, let the bottles cool and unite their contents. The liquid, which contains all the bromine in the substance analyzed, together with a relatively small proportion of chlorine, may then be analyzed by one of the methods described under Bromide of Silver.

Principle II. Power of decomposing iodide of potassium and other metallic iodides.

Applications. Estimation of free bromine (Method A); estimation of iodine in presence of bromine and chlorine (Method B).

Method A. Bring the gaseous bromine or the bromine water to be analyzed into contact with a solution of iodide of potassium and determine how much iodine is set free, by titrating with a standard solution of hyposulphite of sodium, sulphurous acid, arsenite of sodium or some other reducing agent (see Iodine). One equivalent of iodine is set free by each equivalent of bromine in the substance analyzed.

If bromine water is to be tested, the portion to be analyzed may be measured in a pipette provided with a tube charged with moist hydrate of potassium, to protect the lungs of the operator, and the liquid may simply be stirred into a solution of 1 part iodide of potassium in 10 parts of water. The formation of a black precipitate of iodine in the liquid would indicate that an insufficient quantity of iodide of potassium had been taken. — If the bromine is evolved in the form of gas, it may be collected in a series of two or three small flasks, or miniature Woulfe bottles charged with the solution of iodide of potassium and tightly connected with one another and with the generating flask. At the mouth of the flask in which the bromine is generated, it is well to have a wide bulb tube, slanting back towards the flask, in order to condense aqueous vapor and return it to the flask. — The method yields accurate results and is easily executed.

If the solution to be analyzed contains bromhydric acid or a metallic bromide, as well as free bromine, determine the free bromine in a weighed portion of it with iodide of potassium as above described. Mix another portion with an excess of sulphurous acid in aqueous solution, allow the mixture to stand for some time, acidulate with nitric acid and determine the whole of the bromine as Bromide of Silver. The difference between the two determinations gives the weight of the combined bromine.

If the bromine under examination contains chlorine, the proportion of the two elements may be determined as follows: — Let A represent the weight of the impure bromine taken, i the weight of the iodine found, y the weight of the chlorine in A, and x that of the bromine in A, then

$$y = \frac{i - 1.5866\,A}{1.991}; \text{ and } x = A - y,$$

(Bunsen, *Annal. Chem. und Pharm.*, **86.** 276).

Method B. Prepare a standard solution of bromine by dissolving 1 grm. of bromine in 4 litres of water. At the moment of use place 40 c. c. of this solution (= 0.01 grm. of bromine) in a litre flask and dilute with water to the mark (see Alkalimetry) in order to obtain a solution containing $\frac{1}{100}$ milligramme of bromine in each cubic centimetre. If the iodide solution to be tested is alkaline, neutralize it with dilute nitric acid, then add to it, by means of a graduated pipette, a definite number of drops (say one c. c.) of bisulphide of carbon or of chloroform and from another pipette or burette pour upon the mixture a stated number of drops of the weak bromine water. On shaking the mixture the bisulphide of carbon or the chloroform will dissolve the iodine which has been set free and will become violet-colored. As soon as the color ceases to become deeper on shaking the mixture after the addition of a drop of the bromine water, remove the colored bisulphide with a pipette and replace it with a fresh quantity, equal to that first taken, add a further portion of the bromine water, and continue to repeat these operations until the last portion of bisulphide or of chloroform fails to become violet-colored after the addition of the bromine. The quantity of standard bromine water used, minus the last portion which failed to impart any violet color to the test liquid, is equivalent to the amount of iodine in the substance tested. — Both bisulphide of carbon and chloroform are capable of indicating the presence of exceedingly minute traces of free iodine. Bromine alone colors bisulphide of carbon yellow. An excess of bromine in presence of iodine yields bromide of iodine which imparts no violet color to the bisulphide. In order to judge of the intensity of the coloration, similar amounts of the bisulphide should be added throughout the experiment. (De Luca, *Comptes Rendus*, **37.** 866). According to Fresenius this method is neither so convenient nor so practical as the analagous method of Dupré, in which Chlorine is used in place of bromine.

According to Casaseca (*Ann. Ch. et Phys.*, (3), **45.** 482) the method above described is faulty, inasmuch as bromine water decomposes rapidly. A standard solution of it cannot be kept during 24 hours. He finds, moreover, that the amount of iodine set free is not always equivalent to the quantity of bromine used; in one experiment as much as three equivalents of bromine were required to liberate the iodine. Before proceeding to the actual analysis he therefore determines empirically the value of his bromine water by titrating with it a known weight of iodide of potassium. In case iodine is to be determined in solutions containing but little more than 1 milligramme of the element in 10 c. c., a standard solution of iodide of potassium is first prepared of such strength that each c. c. of it shall contain 1 milligramme of the iodide. 1 c. c. of this solution is placed in a graduated tube, diluted with water to 10 c. c. and tested with chloroform and bromine water, according to De Luca's directions. The bromine water should contain about 1 milligrm. of bromine to the c. c. and had better be poured from a burette marking tenths of cubic centimetres. The quantity of bromine water which represents 1 milligrm. of KI, or the 0.000763 grm. I, therein contained, having thus been determined, the actual analysis of any unknown iodide solution of the given strength may be undertaken. A similar method of procedure is applicable to solutions containing less than 0.5 milligramme of iodine in 10 c. c.; but in that case two standard solutions of bromine water are needed, the one containing 1 milligrm. of bromine and the other 10 milligrms. to the c. c. The strength of any unknown solution must be determined approximately by a preliminary trial.

In case bromine, chlorine and iodine are all to be determined, De Luca precipitates one portion of the solution completely with a standard solution of nitrate of silver (see Chloride of Silver); determines the iodine as above described in another portion, and the bromine and iodine together in a third portion by means of a standard solution of Chlorine.

Principle III. Power of oxidizing the lower oxides of iron, manganese, and some other metals, ferrous salts, sulphurous acid, hyposulphites, etc., etc.

Applications. Precipitation of manganese as binoxide. Estimation of free bromine, either in aqueous solution or in the gaseous form. Estimation of free bromine in presence of chlorine.

Methods. Similar to those described under Chlorine. For precipitating *bin*Oxide of Manganese, in particular, from acetic acid solutions, and for oxidizing manganous salts before precipitating the metal as Hydrate *of sesquioxide* of Manganese, bromine will be found (after Henry) to be a much more convenient agent than chlorine. — In case free bromine contaminated with chlorine is to be estimated, weigh out some of the substance in a small glass bulb, treat it with a cold aqueous solution of sulphurous acid and precipitate with nitrate of silver. Digest the mixed precipitate of bromide and chloride of silver with nitric acid to dissolve any sulphite of silver which may have gone down, wash, dry and weigh. — If A equal the weight of substance taken, B that of the AgCl plus AgBr obtained, x the weight of bromine in the substance taken, and y the weight of the chlorine, then

$$y = \frac{B - 2.35A}{1.695};\ \text{and}\ x = A - y.$$

(Bunsen, *Annal. Chem. und Pharm.*, **86**. 276).

Principle IV. Decoloration of by oil of turpentine (substitution of Br for H in the oil).

Application. Estimation of free bromine.

Method. Dissolve 20 grammes of perfectly pure oil of turpentine in enough absolute alcohol that the mixture may be equal to 200 c. c. 34 c. c. of this liquid (= 1 equivalent of oil of turpentine) correspond to 8 grms., or 1 equivalent, of bromine. Place the solution of bromine to be analyzed in a stoppered bottle, add the oil of turpentine solution drop by drop, shaking the bottle after each addition, until the mixture has become perfectly colorless. One equivalent of oil of turpentine takes up and decolorizes 1 equivalent of bromine. The process yields satisfactory results, but is less convenient than the method with iodide of potassium (see Principle II). In case very small quantities of bromine are to be estimated, a more dilute solution of oil of turpentine should be used than that given above.

Principle V. Power of decomposing ammonia with evolution of nitrogen. (See Nitrogen Compounds).

Principle VI. Power of coloring ether, water, and chloroform.

Applications. Estimation of bromhydric acid or combined bromine, even in presence of chlorides. Especially useful for determining small quantities of bromine.

Methods. See Chlorine, power of decomposing metallic bromides.

Principle VII. Affinity for metals of the alkalies and alkaline earths at high temperatures.

Application. Estimation of bromine in organic substances.

Methods. Ignite the substance with quicklime, or the like, dissolve the resulting bromide of calcium in water and precipitate Bromide of Silver. For the details of the ignition see Chlorine.

Brucin. [Compare Iodo-mercurate of Brucin.]

Principle. Solubility in benzole.

Application. Separation of brucin from strychnin.

Method. See Strychnin.

Cadmium may be determined as Oxide, Sulphide or Carbonate, or as metallic cadmium. For the separation of cadmium from other elements see finding list in the Appendix.

Principle. Insolubility in dilute acids when in presence of metallic zinc.

Method. Place a rod of pure zinc in the dilute sulphuric, chlorhydric, or even nitric acid solution of the cadmium compound, wash the precipitated cadmium with hot water, dry (best in an atmosphere of some nonoxidizing gas), and weigh. (*Pfaff, Handbuch analyt. Chem.*, 1825, **2**. 391; Meissner, *Gilbert's Annalen*, **49**. 99). According to Wollaston (*Schweigger's Jahrb.*, **4**. 371), it is best, after any heavy metals which may be present have been thrown down by iron and separated by filtration, to put the chlorhydric acid solution of cadmium, etc., in a platinum dish with a piece of zinc. The metallic cadmium will in this case adhere firmly to the platinum and may consequently be washed with peculiar ease. It may either be weighed as such, after drying, or dissolved in chlorhydric acid and reprecipitated as Carbonate of Cadmium.

Calcium may be determined as Carbonate, Oxide, Oxalate, or Sulphate. For its estimation by alkalimetric methods see the Carbonate, Oxalate and Oxide. For the separation of calcium from each of the other elements see finding list in the Appendix.

Cantharidin.

Principle. Sparing solubility in bisulphide of carbon, and in alcohol.

Applications. Separation of cantharidin from fats and oils. Estimation of cantharidin in Spanish flies.

Method. Plug the throat of a funnel or percolation cylinder with cotton wool, pour fine sand upon the cotton to the depth of 10 or 15 m. m., and place about 40 grms. of finely powdered cantharides above the sand. Exhaust the powder thoroughly and methodically with ether or chloroform. Evaporate the solution to dryness and heat the residue, at a temperature no higher than 40°, until it ceases to smell of the ether or chloroform. Allow the residue to cool and pour upon it 50 or 60 c. c. of bisulphide of carbon. When the mass becomes pulverulent, transfer it to a weighed filter and wash the crystals of cantharidin with bisulphide of carbon until the last traces of oil have been removed. Finally dry and weigh the filter with its contents. (Mortreux, *Journ. Pharm. et Chim.*, **46**. 33).

Dragendorff & Blum (*Zeitsch. analyt. Chem.*, 1867, **6**. 126) mix 25 or 30 grms. of powdered cantharides with 8 or 10 grms. of calcined magnesia, moisten the mixture with water and rub it to a paste. The paste is then dried on a water bath, the dry product rubbed to powder and enough dilute sulphuric acid added to slightly supersaturate the magnesia. Immediately after adding the acid the mixture is shaken with small successive quantities of ether as long as any cantharidin continues to be dissolved. The several ethereal solutions are mixed, the mixture is shaken with water and the ether recovered by distillation. The residue from the distillation, — consisting of crystals of cantharidin, fat and a yellow substance, — is transferred to a tared filter and washed first with bisulphide of carbon to remove fat, and afterwards with alcohol to dissolve the yellow substance. The cantharidin is then dried at 100° and weighed. Corrections should be applied to compensate for the solubility of cantharidin in bisulphide of carbon and in alcohol, as follows: — For every 10 c. c. of bisulphide of carbon used add 0.0085 grm. to the amount of cantharidin actually weighed, and for every 10 c. c. of alcohol add 0.0024 grm. — Instead of magnesia, oxide of zinc may be used; and, instead of extracting at once with ether, the slightly acid mixture of sulphate of magnesium, sulphuric acid and cantharides may be evaporated to dryness and the dry mass extracted with ether or with chloroform. No cantharidin is lost by volatilization during the distillation of the ethereal solution.

Carbon. [Compare Graphite; see also Nitrogen and Oxygen].

Principle I. Insolubility in dilute chlorhydric acid,[1] or in solutions of sulphate of copper, chloride of copper, chloride of iron, or acidulated chromate of potassium.

Application. Separation of carbon from iron; — as in the analysis of cast-iron and steel.

Method A. Solution of the iron in dilute chlorhydric acid, with the aid of a weak galvanic current.

Cast-Iron. In case the material to be operated upon is white or gray cast-iron, weigh out a lump (10 or 15 grms.) of the iron and suspend it in dilute chlorhydric acid by means of platinum-pointed pincers, in such manner that that portion of the mass which is in contact with the platinum shall not be moistened by the acid. Or, in default of pincers, lay the iron upon a small sieve of platinum and sink the sieve in the dilute acid. The dilute acid may be prepared by mixing about 12 vols. of water with 1 vol. of chlorhydric acid of 1.12 sp. gr. Connect the pincers with the positive pole of a single Bunsen element, by means of a copper wire. Immerse a strip of platinum foil in the dish of chlorhydric acid and by means of another wire attach it to the negative pole of the element. Regulate the strength of the galvanic current in such wise that no sesquichloride of iron shall be formed. That is to say, increase or diminish the distance between the electrodes. To do this, move the foil away from or towards the lump of iron. The presence of any ferric chloride is, for that matter, immediately made manifest by the yellowish tinge which it imparts to the stream of concentrated ferrous chloride flowing downward from the piece of iron. During the process of solution the external appearance of the lump of iron undergoes but little change, for the carbon and other insoluble matter retains the original form of the lump.

After about 12 hours, when all the iron which was immersed in the acid has dissolved, break off from the mass of carbon the compact piece of iron which has been nipped by the pincers, wash, dry and weigh it, and subtract its weight from that of the original lump of iron, in order to obtain the weight of iron which has really dissolved. Collect the carbon and other insoluble matter upon a tared filter of asbestos, wash with hot water, dry at 120°–130° in a current of air, and weigh. Transfer the dry residue to a weighed porcelain boat, rinse out the filtering tube with dry oxide of mercury and pour the rinsings upon the carbon in the boat. Place the boat in a glass tube behind a column of oxide of copper, heat the latter to redness, ignite the contents of the boat in a current of oxygen gas, in the manner explained below (Principle II), and collect the Carbonic Acid in a weighed quantity of potash lye or soda-lime. An unweighed chloride of calcium tube may be employed to catch the vapor of mercury. Finally weigh the incombustible residue of slag and silica in the boat. — The process yields good results as compared with most other processes, and has the very great advantage of dispensing with the necessity of reducing the iron to powder.

Precautions. Care must be taken to maintain a weak galvanic current. The iron should dissolve as protochloride and all the hydrogen should escape from the surface of the platinum foil. Little or no gas should be evolved from the surface of the lump of metal. A strong current of galvanism would not only be apt to render the iron passive, but might occasion an evolution of chlorine from the surface of the iron, whereby some of the carbon would be oxidized and lost as carbonic oxide or carbonic acid. Another portion of the carbon would unite with chlorine, and the compound being subsequently decomposed by the galvanic current, carbon would be set free at the negative pole, together with hydro-

[1] For an account of the liquid and gaseous hydrocarbons which may be formed by the action of dilute acids upon white and gray cast-iron, see Hahn, *Annalen Chem. und Pharm.*, **129.** 57. See also Rinman, *Zeitsch. analyt. Chem.*, 1865, **4.** 159, who finds that pieces of thoroughly hardened steel leave no carbonaceous residue when treated with chlorhydric acid of 1.12 sp. gr., and that the carbon which separates from soft steel can all be destroyed by dissolving the steel in boiling chlorhydric acid of 1.12, and continuing to boil the mixed acid and residue for half an hour after the metal has dissolved. On the other hand, a considerable residue of carbon insoluble in hot acid is left when soft steel is slowly dissolved in cold chlorhydric acid. Sulphuric acid diluted with 5 parts of water yields results similar to those obtained with chlorhydric acid, only the sulphuric acid must be boiled rather more vigorously than the chlorhydric in order that all the carbon shall be expelled as a hydrocarbon gas.

gen, in such manner that some of it might be lost in the form of a hydrocarbon gas.

The points of contact between the platinum pincers and the lump of metal must not be allowed to become moist lest the current be impeded by the separation of a film of carbon between the platinum and the iron (Weyl, *Pogg. Ann.*, **114.** 507).

The process as above described succeeds well with specular iron, according to Weyl and Fresenius (*Zeitsch. analyt. Chem.*, 1864, **3.** 337, note).

Steel. In the case of steel a new precaution is required, since the carbon, which separates when steel is dissolved in this way, is so finely divided that the particles do not cohere and remain as a permanent mass at the positive pole of the battery. If the operation were conducted as above described most of the carbon would be transported to the negative pole of the battery, there to be deposited in part and in part to be lost through union with the nascent hydrogen and evolution as carburetted hydrogen gas. To prevent this waste of carbon, a porous membrane must be interposed between the fragment of steel and the negative pole of the battery. The apparatus may be prepared as follows: — Tie a piece of bladder over one end of a wide glass tube or broken cylinder, support the cylinder in a beaker half full of dilute chlorhydric acid, so that the bladder end shall be immersed in the liquid near the bottom of the beaker, and pour dilute chlorhydric acid into the cylinder until the acid stands at sensibly the same level within and without. Hang the weighed lump of steel in the acid in the cylinder, immerse the platinum foil which serves as the negative pole of the battery in the acid in the beaker, and proceed as above described. After several hours a black deposit will sometimes form at the negative pole of the battery, but it will be found to be readily soluble in chlorhydric acid, and to consist of nothing but iron resulting from the electrolysis of chloride of iron which has passed through the membrane. (Weyl, *Zeitsch. analyt. Chem.*, 1865, **4.** 157).

The Asbestos Filter may be prepared as follows: — Select a not too narrow glass tube, heat a portion of it, near one end, cautiously in the lamp, and slowly draw out the softened and thickened glass in such manner that this part of the tube shall be made much narrower than the rest. Push a loose plug of asbestos down the longer portion of the original tube until it rests upon the beginning of the narrowed portion. Fix the tube in an upright position, and pour into it, upon the asbestos, the substance to be filtered. The asbestos should be thoroughly boiled beforehand with chlorhydric acid to remove soluble impurities, and then ignited in a current of moist air to expel any contamination of fluorine.

Method B. Solution of the iron in sulphate of copper.

Weigh out about 2 grms. of the cast-iron, in the form of borings if it be gray, or of coarse powder if white, place it in a small beaker, pour upon it a solution of 10 grms. of sulphate of copper in 50 c. c. of water, heat the liquid gently and stir it frequently until the whole of the iron has dissolved. Allow the mixture to settle, decant the clear solution and oxidize the moist residue with chromic acid in the manner explained below (Principle II, Method 10). (Ullgren, *Annal. Chem. und Pharm.*, **124.** 59). Or, collect the carbon upon a filter of asbestos, as already explained (A).

Sulphate of copper is preferable to chloride of copper (Method C) since it may be heated without detriment to the analysis. The solution of the iron is consequently far more rapid in this case than when chloride of copper is employed.

Method C. Solution of the iron in chloride of copper. Weigh out about 5 grms. of the finely divided metal, pour upon it a quantity of a concentrated, neutral, or almost neutral, solution of cupric chloride, and let the mixture stand at the ordinary temperature, with occasional stirring, until the whole of the iron has dissolved. The solution of chloride of copper must be as free from acid as possible, and rather more of it must be taken than would be sufficient to convert all the iron into ferrous chloride. The mixture must not be warmed, lest subchloride of copper be formed and a small amount of carburetted hydrogen evolved. (Karsten; Hahn, *Annal. Chem. und Pharm.*, **129.** 76). By pressing the undissolved matter with a glass rod it is easy to determine if any hard lumps of iron are still mixed with it. When the last fragments of iron have dissolved and nothing is left but a crumbly mass of metallic copper, free carbon and the insoluble impurities of the iron, mix the solution with a quantity of chlorhydric acid, together with some more chloride of copper, if this be needed, and wait until the whole of the copper has dissolved to cupreous chloride. Collect the carbon on a tared filter of asbestos (see A) wash it first with water, then with chlorhydric acid to remove subchloride of copper, and afterwards with water to remove the acid. Dry at 120°–130°, weigh and burn to carbonic acid, as in A. The carbonic acid should be made to pass through a small chloride of calcium tube on its way to the soda lime, since the carbon is always contaminated with a small proportion of some hydrogen compound. This method, originally proposed by Berzelius (*Pogg. Ann.*, **46.** 42), is still esteemed. (Compare M. Buchner, *Berg. und Hüttenm. Zeitung*, **24.** 84). According to Ullgren (*Annalen Chem. und Pharm.*, **124.** 59) it yields results somewhat too high, since no allowance is made for the nitrogen in the iron.

Method D. Solution of the iron in chloride of iron. Add to a solution of ferric chloride enough carbonate of calcium to neutralize the free acid, and pour the filtered liquid upon the iron to be analyzed. The metallic iron will dissolve to ferrous chloride, and a mixture of carbon and other insoluble impurities, together with a quantity of muddy oxide of iron will be left. Dissolve this oxide in chlorhydric acid, and collect and determine the carbon as before. Both C and D succeed best when the iron is finely divided.

In order to pulverize hard cast-iron, beat the metal to moderately small fragments upon an anvil, crush the fragments in a steel mortar and sift the powder in a leaden sieve provided with small apertures. The softer kinds of iron cannot be broken in this way. They must be filed to powder with a sharp file, after the external crust of dirt or oxide has been filed off. — In powdering highly graphitic iron a very considerable loss of carbon may occur. (Booth & Morfit, *Chemical Gazette*, Vol. 11).

Method E. Solution of the iron in acidulated bichromate of potassium.

Dilute a saturated aqueous solution of bichromate of potassium with an equal volume of water and add as much sulphuric acid as will be sufficient to saturate both the potash and the chromic and ferric oxides which are to be formed. By means of platinum pincers or wire, hang a lump of iron (10 or 15 grms.) just beneath the surface of the liquid and leave it at rest. The iron will dissolve rapidly, without evolution of gas, while the carbon remains intact. The hydrogen which may be supposed to result from the action of the acid on the iron is immediately oxidized by the chromic acid before it can become free or combine with any of the carbon. The ferric salt sinks to the bottom of the beaker as fast as it is formed and soon renders the entire liquid opaque. The carbonaceous residue contains a large proportion of iron and appears to be a chemical compound of the two elements. That obtained from steel dissolves completely in chlorhydric acid with violent evolution of hydrogen and carburetted hydrogen. For the method of estimating the carbon in the residue, see A, B and C.

Like Method A, this process dispenses with the trouble of pulverizing the iron. It appears to be specially adapted for the analysis of steel and white iron, since with specular iron a small quantity of hydrogen and carburetted hydrogen is evolved during the process of solution, unless the solution of chromic acid be very concentrated. (Weyl, *Zeitsch. analyt. Chem.*, 1865, **4.** 158).

Principle II. Oxidation to carbonic acid, by oxide of copper, chromate of lead, chromic acid, oxygen or air.

Applications. Estimation of carbon in any mixture or compound.

Method 1. Combustion with Oxide of copper. (Liebig's method).

The process is well suited for the analysis of *easily combustible*, non-volatile, solid or liquid organic compounds. The apparatus required consists of a *combustion tube* of hard Bohemian glass, drawn up to a point at one end, as described in works relating to chemical manipulation; a *combustion furnace;* a light tube full of Chloride of Calcium (*q. v.*) for absorbing water; a set of potash bulbs (see Hydrate of Potassium), and a soda lime tube (see Soda Lime) for carbonic acid; an *air pump* or exhausting syringe, and a small copper dish or porcelain mortar for mixing the substance with oxide of copper; besides corks, and connections of narrow caoutchouc tubing.

The combustion tube may be some 40 or 45 c. m. long and should have an internal diameter of 12 or 14 c. m.; the glass of the tube should be about 2 m. m. thick, and of the most infusible quality. Clean the tube from dust by wiping it out with a rag or piece of paper attached to a wooden ramrod or to a blunt wire. The tube will be dried at a later period, as will be explained. Fit to the tube three excellent *corks*, and carefully perforate two of them, with a fine round file, in such manner that the narrow portion of the chloride of calcium tube, described below, may fit the hole, air tight. Each of the corks should fit the combustion tube so tightly that by pressing strongly with the fingers no more than one third of the cork can be screwed into the tube. The corks should be smooth, soft, and as free as possible from visible pores; they should be dried for a long time at 100°. The purpose of the second perforated cork is to replace the first in case of accident. Instead of ordinary corks, perforated caoutchouc stoppers of good quality may be employed in most cases.

Charge the *potash bulbs* with a clear solution of caustic potash of about 1.27 sp. gr., tolerably free from carbonate. — Caustic soda will not answer so well, since solutions of it are liable to froth. — Wipe the outside of the bulbs with a dry cloth and the insides of the ends of the tubes with slips of twisted filter paper. Close the two ends of the apparatus with little caps formed by plugging short pieces of rubber tubing with bits of glass rod. Hang the bulbs in a room not liable to wide or sudden fluctuations of temperature, and leave them there half an hour or more before weighing.

Fill a bulbed *chloride of calcium tube* with small lumps of the porous chloride, wipe out the narrow portions of the tube, close the two ends of the tube with plugged caps as before, and place it in the room of constant temperature with the potash bulbs.

Fill three quarters of another bulbed tube with granulated *soda lime* and the other quarter with small lumps of porous chloride of calcium, plug the ends of the tube and place it with the others.

Meanwhile put from 0.33 to 0.6 grm. of the *substance to be analyzed* in a small dry glass tube, 4 or 5 c. m. long by about 1 c. m. wide, slip the tube into another somewhat larger tube in such manner that it may be closed almost air tight, and weigh the tube and substance. It is well to weigh the empty tubes roughly beforehand in order that a quantity of the substance proper for an analysis may be taken with certainty. The material to be analyzed should be perfectly dry and in the state of fine powder; more or less of it should be taken according as it is supposed to be rich or poor in oxygen.

Weigh the potash bulbs and soda lime tube, and finally the chloride of calcium tube. In case the balance is sensitive when heavily laden, the potash bulbs and soda lime tube may be weighed together. Take care to remove the stoppers of the tubes before weighing and to replace them afterwards.

Fill a Hessian crucible of about 100 c. c. capacity nearly full with soft black oxide of copper, — prepared by igniting nitrate of copper until no more nitrous fumes escape from the mass, — cover the crucible carefully with a tightly fitting cover, heat the crucible with its contents to dull redness in a small fire of charcoal, and set it aside to cool. In case the substance to be analyzed is solid, a porcelain mortar will be needed in which to mix the substance with the oxide of copper. — The *mortar* should be wide rather then deep, and should have a lip. It should not be glazed inside but must be free from cracks and flaws. Before use, the mortar should be washed clean, dried thoroughly and kept in a warm place until needed. A shallow, oblong copper dish or saucer, provided with a lip at one end, may be employed with advantage for mixing the substance with oxide of copper.

The clean *combustion tube* should now be thoroughly dried by moving it quickly to and fro over a lamp — so that the entire length of the tube may be heated, — and repeatedly sucking out the hot air from within the tube, through a small, long glass tube which reaches to its bottom.

As soon as the *oxide of copper* has cooled to such an extent that the crucible which contains it can just be held in the hand, pour a little of it into the porcelain mortar and another small portion into the combustion tube, rinse out both mortar and tube with the warm oxide, and then throw it aside.

Spread a large sheet of glazed paper, such as bookbinders use, upon a clean table, and place the mortar upon it in order that nothing may be lost in case of spilling. Fill nearly two-thirds the length of the combustion tube with the warm oxide of copper, either by repeatedly thrusting the mouth of the tube into the oxide in the crucible held somewhat inclined, and then inverting the tube, or by dipping up the oxide with a tea spoon of German silver and pouring it into the tube through a small, warm copper funnel with a short neck. — Pour some of the oxide of copper from the combustion tube into the mortar, throw out upon this oxide of copper the substance to be analyzed, from the tube in which it was weighed, taking care to shake the tube so that little or none of the substance shall remain adhering to it. Preserve the empty tube carefully since it must be reweighed. Pour more oxide of copper from the combustion tube to cover the substance in the mortar and mix the substance with the oxide by carefully rubbing the two together with the pestle, taking care not to press upon the latter too strongly. Add to the mixture in the mortar all the oxide of copper in the tube excepting a layer of 3 or 4 c. m. at the very end and incorporate the matter thoroughly. Take out the pestle from the mortar, shake it to remove adhering particles of the mixture, and lay it upon the glazed paper. Carefully transfer the contents of the mortar to the combustion tube by repeatedly thrusting the mouth of the tube into the mixture and then inverting the tube: to transfer the last remnants of the mixture pour them upon a smooth card and thence into the tube. Pour a small quantity of oxide of copper from the crucible into the mortar, rub it about with the pestle so that any particles of the mixture still adhering to the porcelain may be rinsed off, and transfer it to the tube. Repeat this operation of rinsing until the combustion tube is full to within 3 or 4 c. m. of its mouth, then push a not too tight plug of asbestos or of copper turnings against the oxide of copper to keep it in place and close the tube temporarily with a dry cork.

The asbestos should be ignited beforehand in a stream of moist air to remove fluorine, and the copper turnings first in air and then in hydrogen to free them from dirt.

Rap the tube gently against the table until its contents settle together to such an extent that a narrow air-channel is left open above the oxide of copper from one end of the tube to the other, for the passage of the gases which are to be evolved, and that the upturned posterior point of the tube is left clear. Carry the tube to an air pump or simple exhausting syringe to which is attached a long tube full of dry chloride of calcium, and connect the combustion tube with the latter, by means of a perforated cork. Place the combustion tube in a narrow wooden box or trough, fill the box with hot sand so that the entire length of the tube shall be covered with the sand and then slowly pump out the air which is contained in the tube. The sand must not be hot enough to singe paper. The pumping must be careful and deliberate or some of the contents of the combustion tube will be carried out with the escaping air. After a moment slowly open the stop-cock of the pump so that fresh air may enter the combustion tube. All the moisture contained in this fresh air will be stopped by the chloride of calcium, as well as that brought out of the tube in the exhausted air. Again carefully pump out air from the tube, then admit more air and continue to exhaust and to admit air alternately some 10 or 12 times in order to remove the last traces of moisture which may have been absorbed by the oxide of copper during the operation of mixing.

Instead of operating as above described, a better way is to place the combustion tube upright in a retort holder, to mix the substance with oxide of copper in a small copper dish, and to pour the mixture into the tube through a smooth, warm copper funnel. The anterior portion of the tube may then be filled with a tightly packed layer, 20 c. m. long, of hard, gray, granulated oxide of copper, which, unlike the soft oxide above described, has but little power of absorbing moisture from the air. (Mulder). By proceeding in this way the necessity of pumping may be done away with.

Bunsen's Modification. Another way of avoiding the moisture of the air and of dispensing with the need of an air-pump has been indicated by Bunsen. This chemist directs that the substance to be analyzed be thrown directly into the combustion tube and there mixed with oxide of copper by means of a twisted wire, instead of being rubbed with the oxide in a mortar or other dish. His process is particularly well adapted for the analysis of highly hygroscopic bodies and of substances which would be decomposed by warm oxide of copper.

The hot oxide of copper is transferred from the crucible in which it was ignited to a warm, dry glass flask, or wide tube closed at one end, which is then corked tightly and left to cool. The substance to be analyzed is meanwhile weighed in a long tube of thin glass provided with a cap cover, as described above. This weighing tube should be about 20 c. m. long and 6 or 7 m. m. wide. — As soon as the oxide of copper has become cold, uncork the flask or tube which contains it, thrust the end of the dry combustion tube through the neck of the flask into the oxide of copper, in such manner that a small quantity of the oxide may enter the tube; rinse the combustion tube with this oxide and throw the rinsings aside. Again thrust the combustion tube through the neck of the flask into the oxide of copper and take up enough of the latter to form a column or layer about 10 c. m. deep at the posterior end of the tube. — The transfer of oxide from the flask to the tube is readily effected by holding the tube in a slightly inclined position and gently tapping the tube. — Open the weighing tube, thrust it as far as possible into the combustion tube, held slightly inclined, and pour out the substance to be analyzed. In doing this, turn the weighing tube about so that its contents may fall out more readily, and at the same time press its rim against the upper side of the combustion tube in order to keep it from touching the substance after it has once fallen out. As soon as the substance to be analyzed has been poured from the weighing tube, bring the combustion tube into a horizontal position so that the weighing tube, still pressed against its upper side, shall be slightly inclined, with the closed end downwards. Then slowly withdraw the weighing tube, taking care to turn it so that any portions of the substance which may have remained attached to the rim of the tube may fall back into it. Close the empty tube and put it aside to be weighed. — Transfer from the filling flask to the combustion tube another quantity of oxide of copper equal to the first, so that there shall be a column of oxide of copper 20 c. m. long at the end of the tube, with the substance in the middle of it. To mix the substance with the oxide of copper, provide a long, bright, stiff iron wire, bent to a ring or loop at one end, for the handle, and at the other end pointed and twisted like a cork screw, with a single twist. Push the screw end of the wire deep into the oxide of copper and move it about rapidly for a few minutes in all directions, so that the substance and oxide may be intimately mixed. Withdraw the wire, transfer a new quantity of oxide of copper from the filling flask to the tube, wipe the wire in this oxide, and finally fill the tube with oxide of copper. — So little water is absorbed by the oxide of copper in this process that a single charge of the filling flask may be made to serve for several analyses. If the flask be provided with a tight cork the oxide of copper will remain several days fit for use, even though the flask be repeatedly opened and portions of its contents withdrawn. (Bunsen, *Handwœrterbuch der Chemie*, Supplement, p. 186).

As soon as the combustion tube has been properly filled, and freed from hygroscopic moisture, thrust the narrow end of the weighed chloride of calcium tube through one of the dry perforated corks, twist the cork tightly into the mouth of the combustion tube, and place the latter in a "combustion furnace." The combustion furnace may be fed with charcoal, alcohol, or far better, with gas, in accordance with almost any one of the numerous plans described by works on chemical manipulation. Compare, for example, Baumhauer (*Annal. Chem. und Pharm.*, **90.** 21), J. Lehmann (*ibid*, **102.** 180), Heintz (*Pogg. Ann.*, **103,** 142), Hofmann (*Journal Chem. Soc.*, **11.** 30).

Copper foil for wrapping soft tubes. In case the combustion tube has to be made of glass which is not infusible enough to withstand the heat of the furnace, the tube should be wrapped in thin copper foil or gauze, and wound around with iron wire, before it is placed in the furnace. Or the tube may be laid in a shallow trough of sheet iron.

The combustion tube should incline forward slightly, and its mouth should project at least an inch beyond the edge of the furnace, and a screen of sheet iron should be placed at the edge of the furnace so as to protect the mouth of the tube from excessive heat. Throughout the experiment the projecting part of the tube should be kept so hot that the fingers can hardly bear the shortest contact with it; but no hotter than this, lest a portion of the cork be burned, and the analysis thereby vitiated.

By means of a short rubber connector, attach the potash bulbs to the free end of the chloride of calcium tube, taking care to place the largest bulb next to the chloride of calcium tube, and tie the ends of the connectors firmly to the glass with fine cords. During the operation of tightening and tying the cords, the ends of the two thumbs should be pressed firmly together, so that no part of the apparatus need be broken in case a cord happens to give way. It is well, also, to rest the fragile bulbs upon a folded cloth, or some other soft substance. To the free end of the potash bulbs attach, with another connector, the supplementary soda lime tube, and support it in a horizontal position by means of a ring-stand, or any other suitable prop.

In order to determine whether the fittings of the apparatus are air-tight, hold a tolerably large piece of glowing charcoal near the largest potash bulb, so that the air within the bulb may be expanded, and in part driven out of the apparatus. After a certain amount of air has been expelled in this way, take away the coal. Note the height to which the potash solution rises in the large bulb to replace the lost air, and observe whether the liquid remains at this height for the space of three or four minutes. If the apparatus be tight, the liquid will remain at the highest level to which it rose as the bulb cooled; hence, if it gradually recedes from the large bulb, and comes to a common level in both limbs of the apparatus, either one of the corks or connectors must be leaky, or the apparatus somewhere cracked.

The weight of the empty weighing tube may be conveniently taken while the tightness of the tube is being tested.

Heating the tube. After the apparatus has been proved to be tight, heat carefully two or three inches of the anterior portion of the combustion tube until it is red hot, then slowly work backwards inch by inch, taking care to bring each section of the tube to redness before proceeding to heat the next section. If charcoal be used as the fuel, the unheated portion of the tube must be protected from the radiant heat of the fire by means of a sheet-iron screen, which may be moved backwards at will with a pair of tongs or pincers.

When heat is first applied to the tube, some bubbles of air will be driven through the potash bulbs by virtue of simple expansion; afterwards, when the heat reaches that portion of the oxide of copper which was used to rinse the mixing mortar, or to wipe the mixing wire, and carbonic acid begins to be evolved, the rest of the air in the apparatus will escape in large bubbles. But when the actual mixture is reached, the bubbles which pass into the potash are nearly pure carbonic acid, so that only now and then a solitary air bubble will escape through the liquid. — The heating of the tube should be so regulated that the gas bubbles may follow one another at intervals of from one-half to one second.

After about three quarters of an hour, when no more gas is evolved, although the tube is red hot from end to end, take away the fire from the posterior end of the tube, so that the upturned point may be free, and place a screen between the point and the fire. The cooling of the end of the tube thus caused, taken in connection with the absorption of carbonic acid by the potash lye, will cause the latter to be forced back into the large bulb. The liquid will rise slowly at first, but with increased rapidity after it has once entered the large bulb, but there is no danger of its flowing into the chloride of calcium tube if the bulbs were properly filled in the first place, and are now set level. It may here be said, that during the combustion it is well enough to place a cork, or a piece of wood as thick as a man's finger, beneath that end of the potash bulbs (Liebig's bulbs) which is farthest from the large bulb, so that the liquid shall tend to flow into the large bulb. But as soon as gas bubbles cease to come forward, this prop must be removed, and the bulbs brought to a level position. — At the moment when the large bulb has become about half full of the potash lye, crush the end of the upturned point of the combustion tube with a pair of stout pincers, and push over the stump a dry glass tube about 60 c. m. long, and open at both ends. Support this tube in an upright position by means of a ring-stand. Attach a long caoutchouc tube to the supplementary soda-lime tube beyond the potash bulbs, bring the potash bulbs to their original oblique position, and with the mouth, or better with a small aspirator, suck air through the combustion tube until the bubbles which pass through the potash bulbs cease to diminish in size. An aspirator has an advantage in that it affords ocular evidence of the volume of air which is drawn through the apparatus.

Take the apparatus to pieces, stop the ends of the chloride of calcium tubes and of the potash bulbs with rubber connectors plugged with glass, set the larger chloride of calcium tube in a vertical position, with its bulb upward, and leave the several pieces in the room of constant temperature for half an hour before weighing. The increased weight of the larger chloride of calcium tube gives the amount of water produced in the combustion, and from this weight that of the hydrogen in the substance analyzed is obtained by the following proportion:—

$$\underset{18}{\text{Molecular wt. of } H_2O} : \underset{1}{\text{Wt. of an atom of H}} :: \text{Wt. of water found} : x \left(= \text{Wt of H in substance taken.}\right)$$

The increased weight of the potash bulbs and supplementary soda lime tube gives the weight of the carbonic acid, whence the weight of carbon is derived by the proportion:—

$$\underset{44}{\text{Molecular wt. of } CO_2} : \underset{12}{\text{Weight of an atom of C}} :: \text{Wt. of } CO_2 \text{ found} : x \left(= \text{Weight of C in the sample.}\right)$$

In practised hands, this comparatively old process gives excellent results. It is, however, far less convenient and trustworthy than the process of combustion with oxide of copper in a current of oxygen gas (see below, Method 2), and is therefore seldom employed. — In operating with easily combustible substances (and the process is really only suitable for the analysis of such), it is easy to determine in this way the proportion of carbon with great accuracy, but as regards hydrogen, the results obtained are usually about 0.1 or 0.15 per cent higher than the truth. This excess of hydrogen comes in part from moisture absorbed by the oxide of copper, but mainly from the moisture of the air, which is drawn through the apparatus at the close of the combustion to remove the carbonic acid. The error can be corrected in good part by attaching, with a perforated cork or rubber connector, a tube full of solid hydrate of potassium to the posterior point of the combustion tube,—in place of the upright open tube above described,—before beginning to suck air through the tube.

Volatile substances, and those liable to lose water, or to undergo other alteration at 100°, may be analyzed in this way, though less readily than by Method 2, by mixing them with cold oxide of copper, in the manner proposed by Bunsen, see above, p. 61.

Volatile liquids, such as alcohol, ether, essential oils, and the like, had better be analyzed by Method 2 or 6. In case they are analyzed by Method 1, the combustion tube should be 50 or 60 c. m. long, the oxide of copper should be cooled in a flask (p. 61), and the liquid weighed in two or three small bulbs similar to those described under Method 2 only smaller. — Pour a layer of the cold oxide of copper, 6 c. m. deep, into the combustion tube, scratch the stem of one of the bulbs with a file or steel glass-knife, break off the point quickly with the thumb and finger, and drop both bulb and point into the combustion tube. Pour another layer of oxide of copper, 6 or 8 c. m. deep, into the tube to cover the bulb, throw in the second bulb in the same way as the first, and add another layer of oxide of copper. If the substance to be analyzed contains a large proportion of carbon, and is rather difficultly volatile, the quantity (about 0.4 grm.) taken for analysis, had better be weighed in three than in two bulbs, in order that no particles of carbon may be left unconsumed with the mass of reduced copper. Rap the tube gently against the table to clear a passage for the gases which are to be evolved, and finally fill the anterior half of the tube with small lumps of hard oxide of copper, or with copper turnings which have been superficially oxidized, so that there may be free passage for gases and vapors, although little or no visible channel is left open above the mass. — After the chloride of calcium tube, potash bulbs, etc., have been attached to the combustion tube, place a screen at the middle of the tube and heat the anterior column of oxide of copper to redness. Heat the upturned, posterior point of the tube so that no vapor shall condense in it, and place a hot coal, or an exceedingly small flame, near that part of the tube which contains one of the bulbs. The contents of the bulb should be driven out very slowly, and the whole operation conducted with extreme care. It is an easy matter to lose an analysis by distilling the substance rapidly, so that considerable quantities of partially burned material can escape through the potash bulbs. — After the contents of the first weighed bulb have been driven out and consumed, that part of the tube which contains the second bulb may be heated. The combustion tube is finally heated from end to end, and the analysis finished in the usual way. — In the case of liquids which are but slightly volatile, it is well to empty the bulbs before the combustion begins, instead of heating them, as above described. To this end attach the filled combustion tube to an air pump, and give a single slow stroke with the pump handle. The bubbles of air within the weighing bulbs will expand, and the liquid be forced out to be absorbed by the oxide of copper. Bodies rich in carbon should never be analyzed in this way; a supply of oxygen gas is needed in order that their carbon may be completely consumed. See Methods 2 and 6.

Nonvolatile Liquids are analyzed by Methods 2, 3, 5 and 6.

Nitrogenous Compounds. When substances containing nitrogen are ignited with oxide of copper, most of the nitrogen goes forward in the gaseous form, together with the carbonic acid and water, and escapes into the air as free nitrogen. A small quantity of the nitrogen, however, is converted into nitric oxide, and the latter, on coming in contact with the potash lye, is partially decomposed to nitrous acid, which combines with the potash. A part of the nitric oxide changes into hyponitric acid also, by coming in contact with the air in the apparatus, and the acid thus formed is absorbed, as well as the carbonic acid, in the potash bulbs. Though the amount of nitrous and nitric acids thus generated is never very large, there is still enough formed in many cases, especially in the combustion of substances rich in oxygen, to vitiate the determination of carbon. The difficulty may be obviated in some cases by mixing the substance very intimately with the oxide of copper, and conducting the combustion very slowly. But as a general rule, the oxides of nitrogen must be decomposed by bringing them in contact with red hot metallic copper. The metal is used either in the form of turnings, or of rolls or spirals, made of wire or of strips of sheet copper. The rolls or spirals may be 8 or 10 c. m. long, and just thick enough to be admitted to the combustion tube. If turnings are used, they may be compressed to a cylindrical form, by forcing them while hot into a short tube, a little narrower than that in which the combustion is to be made. According to Schroetter and Lautemann (*Jour. prakt. Chem.*, **77.** 316), the copper rolls or turnings cannot be replaced by the metallic powder obtained by reducing oxide of copper with hydrogen, since the powder obstinately retains hydrogen, which, by reacting upon carbonic acid in the process of the combustion, causes an appreciable quantity of carbon to be lost as carbonic oxide. — The rolls or plugs of copper are first heated to redness in the air in a Hessian crucible, until the surface of the metal is oxidized, and the last trace of dust and oil has been burned off; they are then heated in a tube in a stream of hydrogen, until the oxide has all been reduced. Since recently reduced copper retains hydrogen gas, and on exposure to the air absorbs aqueous vapor, it should always be heated to 100° in the air for some time before it is used, and should be as nearly as possible at this temperature when introduced into the combustion tube. — The combustion tube, which should be 12 or 15 c. m. longer than if it were to be employed for analyzing a body free from nitrogen, is filled in the usual way, with this exception, that enough metallic copper is placed at the anterior end, to form a column 10 or 12 c. m. long. As soon as the tube is laid in the furnace, the metallic copper is heated to bright redness before proceeding to the ordinary steps of the analysis. Hot copper decomposes all the oxides of nitrogen, fixing the oxygen while nitrogen goes free; but since this action occurs only when the metal is intensely ignited, care must be taken to keep the metal hot throughout the entire combustion. Compare the heading, Nitrogenous Compounds, under Method 2.

Sulphur Compounds. In determining the carbon of compounds which contain sulphur as well as carbon, hydrogen, oxygen or nitrogen, it has hitherto been customary to proceed, as above described, as if nothing but carbon, hydrogen and oxygen (or nitrogen) were present, but to place between the chloride of calcium tube and the potash bulbs a narrow tube, 10 or 12 c. m. long, filled with dry binoxide of lead, in order to absorb the sulphurous acid formed by the oxidation of the sulphur, which would otherwise be absorbed in the potash bulbs. (Liebig & Wœhler). But according to Carius (*Annal. Chem. und Pharm.*, **116.** 28), all the sulphurous acid cannot be retained in this way when the substance analyzed contains a large proportion of sulphur, and on the other hand, Bunsen has shown that binoxide of lead is capable of absorbing no inconsiderable quantity of carbonic acid. Carius urges that sulphur compounds had better be burned with chromate of lead. (See Method 8). See also Sulphur.

Chlorine, Bromine and Iodine Compounds. When organic substances containing chlorine, bromine or iodine are burned with oxide of copper, subchloride, (bromide or iodide) of copper is formed, some of which is liable to condense in the chloride of calcium tube and vitiate the determination of the hydrogen. Hence the compounds in question are usually burned not with oxide of copper, but with chromate of lead (Method 8). Compare Method 2.

In Analyzing Compounds which contain Inorganic Constituents, other than those allowed for in the preceding paragraphs, the proportion of inorganic matter must be determined in a special portion of the material. The analysis of such compounds presents no particular difficulty unless a volatile metal like mercury or a metal capable of retaining more or less carbonic acid, — such as potassium, sodium, calcium, strontium or barium, be present. In the combustion of substances which contain mercury, a layer of copper turnings (see Nitrogenous substances, above) may be placed in the anterior part of the combustion tube in order to retain the metal within the tube. Care must be taken not to heat this copper too strongly. If the substance to be analyzed contains metals capable of retaining carbonic acid, a quantity of some substance capable of decomposing the carbonates in question at high temperature may be added to the oxide of copper with which the substance is mixed; either teroxide of antimony, phosphate of copper or boracic acid, will answer the purpose. See also Method 9, and the remarks on same subject under Method 2.

Method 2. Combustion with oxide of copper, in conjunction with oxygen gas forced from a gas-holder.

In this process the substance to be analyzed, or at least the volatile products given off from it by distillation, is heated in contact with oxide of copper in the midst of a slow current of oxygen gas which is made to flow in continually upon the mixture from a gas-holder. The method, besides being suitable for the analysis of difficultly combustible substances, is of general applicability for the estimation of carbon and hydrogen in all organic substances. It will be found particularly convenient when several analyses are to be made in succession, and in cases where from our inability to pulverize the substance to be analyzed it cannot be intimately mixed with oxide of copper or any other solid oxidizing agent.

Besides the apparatus described under Method 1, the process now in question requires a couple of gas-holders and a permanent set of drying tubes. For a simple and inexpensive form of gas-holder, see Eliot & Storer's Manual of Inorganic Chemistry, Appendix § 11, Fig. xvii.

Provide a straight combustion tube open at both ends, about 60 c. m. long, and of any width to which corks can be conveniently fitted. Fit sound corks to both ends of the tube. Perforate the corks so that one of them may fit the weighed chloride of calcium tube, and the other a short, straight tube of diameter proper to be connected with the permanent drying tubes; then dry the corks at 100°.

Place a tolerable compact plug of clean copper turnings or a loose plug of asbestos (previously ignited in a current of moist air in order to remove fluorine) in the combustion tube at a distance of 4 or 5 c. m. from its anterior end, pour enough oxide of copper into the tube to fill two-thirds of it, and push down upon the oxide of copper another loose plug of asbestos to keep the column in place. A space about 20 c. m. long should be left open at the posterior end of the tube. The oxide of copper had better be in the form of coarse granules free from dust; it need not be dried or ignited before being placed in the tube. Or, instead of mere oxide of copper, the tube may be charged with a mixture of asbestos and fine oxide of copper. When mixed with asbestos, the oxide of copper in the tube will be light as well as porous, and the tube will better preserve its shape when heated.

Lay the tube in a shallow trough or gutter of sheet iron, upon a thin layer of calcined magnesia or of asbestos, and place the gutter and tube in the combustion furnace.

Unless the glass of the combustion tube is of the most refractory character the tube had better be coated with clay or with asbestos, and then covered with copper foil bound round with copper or iron wire, before it is placed in the furnace. The furnace may be heated with alcohol or with charcoal, in default of gas.

The posterior end of the combustion tube is in the next place connected with the *Permanent drying (or rather, cleaning) tubes.*

These tubes, which serve to purify air and oxygen for the analysis, stand between the combustion tube and the gas-holders which supply air and oxygen; they may consist of a set of large potash bulbs filled with oil of vitriol, a large U-tube full of soda-lime, and another U-tube filled with chloride of calcium. The sulphuric acid bulbs are attached to the gas-holder, while the chloride of calcium tube is connected with the posterior end of the combustion tube. Where many analyses are to be made, it is well to have two or three soda-lime tubes and as many more charged with chloride of calcium. In order that a single set of the tubes may serve for both gas-holders without inconvenience, a supplementary flask may be attached to the set, as follows: — Fit to a small, wide mouthed flask a caoutchouc stopper, with three perforations. Provide three glass tubes suited to the stopper and each bent at a right angle. One of the tubes is a simple abduction tube only long enough to pass through the stopper, while the other two must be long enough to reach almost to the bottom of the flask. Fill the flask one-third full of strong potash lye, place it between the gas-holders and the sulphuric acid bulbs and attach it to the latter by means of a rubber connector, tied to the short abduction tube. To the other end of each of the longer tubes tie a piece of thick caoutchouc tubing about 3 inches long, close each of these rubber connectors with a spring clip, and by means of glass tubes attach one of them to the opening of the gas-holder which contains air and the other to the oxygen-holder. By opening one or the other of the clips, air or oxygen may be made to flow into the combustion tube at will, and by renewing from time to time the potash lye and the sulphuric acid in the flask and bulbs the efficiency of the apparatus may be kept up for a long time. (Piria, *Kopp & Will's Jahresbericht*, 1857, p. 573).

As soon as the combustion tube has been laid in the furnace, start a slow current of air through the tube and heat it throughout its entire length, at first very gently but afterwards to low redness, in order to dry the oxide of copper. During this preliminary ignition leave the anterior end of the tube open, but as soon as the tube has been thoroughly heated close it with a dry cork carrying an unweighed chloride of calcium tube, extinguish the fire, and without interrupting the slow current of air, allow the tube to cool.

Weigh out the substance to be analyzed, in a small boat of platinum, copper, porcelain or glass, inclosed in a glass weighing tube which has been weighed together with the boat before the introduction of the substance. Push the loaded boat into the posterior end of the combustion tube until it almost touches the asbestos plug, replace the cork at the end of the combustion tube and for the time being shut off the current of air. It is well to lay two or three fibres of asbestos beneath the boat to prevent it from fusing to the glass. From 0.3 to 0.5 grm. of substance should be taken as a general rule, though less of the substance will be needed in proportion as it contains more carbon; 0.2 of a grm. of material is sufficient for the analysis of many hydrocarbons.

After the boat has been introduced, remove the unweighed chloride of calcium tube from the anterior end of the combustion tube, replace it with the weighed chloride of calcium tube and attach to the latter the weighed potash bulbs and the soda-lime tube as described under Method 1 (see p. 62).

In order to determine whether the apparatus is tight, open the cock of the oxygen gas-holder slightly so that a very slow current of the gas may pass through the tube, and after a moment suddenly close it again. Then watch the level of the liquid in the potash bulbs. If the liquid does not sink back from the outermost bulb in the course of a few minutes the apparatus is tight enough for use. In applying this test the tube must of course be as cold as the surrounding air. — Place a sheet iron screen across the end of the tube to protect the cork from the fire and proceed to heat the oxide of copper in the tube, with the exception of a couple of inches next the substance to be analyzed, and as soon as it has become red hot start a slow current of oxygen through the tube. Then slowly heat the rest of the oxide of copper, and finally the substance itself, with extreme care. The substance must neither be heated too quickly nor the stream of oxygen made too strong. It is to be remembered, however, that an amply supply of oxygen will be most needed at the moment when the substance is distilling most rapidly. There is no harm in using so much oxygen that an excess of it shall slowly bubble through the potash bulbs from first to last. In order that the current of oxygen may be readily controlled and nicely adjusted the cock of the oxygen gas-holder should be provided with a long lever.

When the substance to be analyzed has been completely burned, and the character of the bubbles in the potash apparatus indicates that no more carbonic acid is coming forward, shut off the oxygen, turn on air, and allow the apparatus to cool in a slow stream of air. Since oxygen is heavier than air it is important to expel it thoroughly from the potash bulbs. Enough air will have been passed through the tube when a glowing splinter of wood ceases either to burst into flame or to glow vividly when held in the air which issues from the weighed soda-lime tube. — As soon as all the oxygen has been forced out of the apparatus, remove the chloride of calcium tube, the soda-lime tube and the potash bulbs to the room of constant temperature, and after half an hour weigh them as directed in Method 1 (p. 62). Withdraw the boat from the combustion tube and weigh the ashes contained in it, if any there be. Unless some accident occur, the combustion tube and oxide of copper are left in perfect order for a new analysis. The operator has only to push another boat charged with a new quantity of substance into the tube and to attach another set of weighed chloride of calcium tubes, etc., to the anterior end of the tube.

To guard against the possibility of the escape of any of the products of distillation backwards to the cork and drying tubes, Piria, (*Kopp & Will's Jahresbericht*, 1857, p. 573) interposes hot oxide of copper between the substance to be analyzed and the posterior end of the tube, as follows: — The anterior portion of a tube 80–85 c. m. long is filled with granulated oxide of copper, as above described, the boat loaded with the substance is pushed in nearly to the oxide of copper and a couple of coils of superficially oxidized copper foil are thrust in behind the boat. The tube and oxide of copper are dried in a current of air in the usual way, but in making the combustion the posterior portion of the tube is heated to redness, as well as the anterior portion, before the substance to be analyzed is warmed. Screens are employed to protect that portion of the tube which contains the substance, while the rest of the tube is being heated.

Instead of placing the substance to be analyzed in a boat, as above described, it may be mixed directly with a part of the oxide of copper in the tube. To this end, weigh the substance in a long, narrow, weighing tube (see p. 61), and pour it upon the posterior end of the column of oxide of copper (see p. 65) after the latter has been dried in a current of air in the usual way, and allowed to become cold. Mix the substance with the oxide of copper by means of a twisted iron wire, such as has been described on p. 61, and then fill the tube to within about 12 c. m. of its end with coarse oxide of copper, which has previously been ignited, and cooled in a corked flask (see p. 61). Tap the tube gently against the table to shake down the oxide of copper, so that a narrow passage may be left above it. Replace the tube in the furnace, and connect its posterior end with the permanent drying tubes, and attach to it the weighed chloride of calcium tube, potash bulbs, etc. Start a very slow current of oxygen through the tube and proceed to heat first the anterior column of oxide of copper, then the mixture, and finally the posterior column of oxide,—working backwards always from the front end of the tube. Take care to protect both the corks with sheet-iron screens. A moveable screen will be found useful, also, for regulating the heat at that part of the tube which contains the mixture. Throughout the analysis cause a slow current of oxygen to pass through the tube. The stream should be so slow, however, that no oxygen shall escape through the potash bulbs while the substance is actually burning. After the evolution of carbonic acid has ceased, force a more rapid current of oxygen into the tube until the reduced oxide of copper has been completely revivified, and finally throw in a current of air to sweep out the excess of oxygen. As a general rule, however, it will be found more convenient and satisfactory to make use of a boat instead of mixing the substance with oxide of copper. When boats are employed, the same combustion tube may not only be used over and over again for many analyses, but needs absolutely no preparation after one analysis, to fit it for the next.

Liquid Substances. — In case the substance to

be analyzed is liquid, but not readily volatile, such, for example, as a fatty oil, it may be placed in the combustion tube in a boat, as described. Wax, and other easily fusible matters, may be melted and allowed to cool in the boat before weighing. The combustion tube had better be rather long, and provided with a posterior column of oxide of copper (compare Piria, above). When the oxide of copper has been heated to redness at both ends of the tube, place a piece of red hot charcoal near the boat, so that the substance may be slowly distilled. Increase or diminish the heat at the boat, and at the same time regulate the stream of oxygen in such manner that the copper reduced by the products of distillation may be oxidized as fast as it is formed. At the close of the combustion take care to burn off all the carbon from the boat.

In analyzing substances, such as some of the heavier components of petroleum, which volatize only at temperatures so high that their vapor would ignite as soon as formed, and explode in the atmosphere of oxygen which fills the tube, a cap or cover of asbestos fibres, roughly woven, attached to a stiff wire, may be placed over the boat after it has been pushed into the tube. The asbestos cloth acts as a safety screen to prevent explosions, like the wire gauze of Davy's lamp. (Peckham).

Volatile Liquids. For the analysis of liquids which volatilize at comparatively low temperatures, select a stick of combustion tubing 50 or 60 c. m. long, and bend it slightly, at a distance of about 12 c. m. from one end, so that the posterior end of the tube may point upwards, and the lower edge of the extremity of the tube reach to a height of 8 or 9 c. m. above the level of the sheet-iron trough in which the horizontal part of the tube reposes. Pack the horizontal portion of the tube either with a mixture of oxide of copper and asbestos, or with coarse oxide of copper, secured with asbestos plugs, and dry the tube in the furnace with a current of air, in the usual way.

In order to introduce the substance into the tube, blow a couple of light weighing bulbs, about a centimetre in diameter (not too large to slip readily into the combustion tube), with capillary stems 5 or 6 c. m. long. Weigh the empty bulbs one at a time, and place them in paper trays bearing descriptive marks or numbers. Pour some of the liquid to be analyzed into a porcelain crucible or small dish—kept cool with ice, if need be—warm each bulb in succession at a lamp, and as soon as the glass is hot thrust its stem into the liquid in the dish. As the glass cools, the liquid will rise up into the bulb and fill it more or less completely. In case the liquid is highly volatile, a portion of that which first enters the partially cooled bulb will be converted into vapor, so that, for the moment, the rest of the liquid will be driven out; but as soon as the vapor in the bulb condenses, a new portion of the liquid will rise into the bulb and fill it more completely than before. With less volatile liquids, only a small quantity of the fluid will enter the bulb at first, but it is easy to fill the bulb completely by heating the liquid which first enters until part of it is converted into vapor, and again thrusting the stem into the liquid. To throw out any excess of liquid which may remain in the stem, suddenly jerk the bulb; then hold the point of the stem in a fine blowpipe jet until the glass fuses, and the bulb is closed. Weigh each of the bulbs with its contents. Each bulb should contain from 0.3 to 0.4 grm.—enough for a single analysis. The duplicate bulb will replace the first in case of accident.

After the combustion tube has been dried, cooled, connected with the weighed chloride of calcium tube, potash bulbs, etc., and proved to be tight, take out the posterior cork of the combustion tube and make a slight scratch upon the stem of one of the bulbs full of substance, by means of a sharp file, or a knife of hard steel, proper for scratching glass. This file-mark should be near the point of the stem. Place the stem of the bulb within the combustion tube, and press its point firmly against the side of the tube until it breaks at the file mark. The moment the stem breaks, drop the bulb, point downwards, into the tube, replace the cork, put a screen in front of the substance, 9 or 10 c. m. from the bend in the tube, heat the anterior column of oxide of copper, and start a slow current of oxygen through the tube. When the copper has become hot remove the screen, or set it backwards towards the substance, accordingly as the latter is more or less volatile, and finally heat the substance itself carefully with a hot coal, or better, with a thick rod of copper. The copper bar may be laid across the ring of a lamp stand, in such manner that while one end of the bar can be brought close to the combustion tube, above the substance to be analyzed, the other end can be heated to redness by means of a Bunsen's lamp. By moving the lamp-stand to and fro, so that more or less of the hot copper is brought near the bulb, the distillation of the substance may, in most instances, be easily controlled. (Warren).

Great care must always be exercised in heating the bulb in order that its contents may not distil too rapidly. Unless due attention be paid to this particular the combustion of the substance is liable to be incomplete, and gaseous carbon compounds will pass off unabsorbed through the potash bulbs. Sudden heating of the substance would, in many cases, occasion a rush of gas strong enough to throw some of the potash lye out of its bulbs, and to project vapors backwards into the permanent drying tubes. — In order to avoid explosions, it is essential that the empty, inclined portion of the combustion tube shall never be heated to a temperature high enough to ignite the vapor of the substance, until after the last portions of this vapor have been swept forward by the current of oxygen. Throughout the analysis the supply of oxygen gas must be sufficient to re-oxidize the copper almost or quite as fast as it is reduced.

To prevent any portion of the vapor of the substance from being lost at the posterior end of the tube, it is well to admit the oxygen through a special tube of hard glass loosely packed with asbestos and kept hot during the combustion by means of a Bunsen's lamp. This tube is placed between the combustion tube and the permanent drying apparatus and is attached to the latter by means of a perforated cork; at the anterior end, which enters the combustion tube, it is drawn out to a short, blunt point, having an opening no larger than will admit a small needle. The oxygen is thus made to enter the combustion tube in a rapid stream, against which little or no vapor can pass back. The hot oxygen, moreover, prevents the condensation of any vapor near or upon the cork. (Warren).

Hygroscopic Substances. Compounds which absorb water so rapidly from the air that they cannot be readily dried and weighed in the ordinary way may be dried in the combustion tube, as follows: — Pack the combustion tube with oxide of copper, as above described, lay the tube in the furnace upon a sheet-iron gutter so short that it does not reach behind the column of oxide of copper, dry the tube and oxide in the usual way and attach to it the weighed chloride of calcium tube. Weigh out the air-dried substance in a boat and push the boat into the combustion tube as far as the oxide of copper. Heat the empty, posterior part of the tube, at a distance of 3 or 4 inches from the boat, and pass a slow current of dry air through the tube. At the same time heat the column of oxide of copper gently so that no water can be deposited in the anterior part of the tube. Keep up the stream of hot air as long as any water is seen to be deposited in the neck of the chloride of calcium tube, then allow the combustion tube to cool, without checking the current of air, and re-weigh the chloride of calcium tube. Subtract the weight of the water thus found from the weight of the substance taken, in order to obtain the weight of really dry material to be analyzed. Replace the chloride of calcium tube, attach the potash bulbs and the soda lime tube and proceed with the analysis.

In a similar way, the water of crystallization of many substances may be determined: — To this end suspend a sheet of copper-foil beneath the combustion tube, between the substance and the source of heat, place the bulb of a thermometer above the foil and light the fuel beneath the foil; heat the tube to the temperture necessary to expel the water from the substance. In case the substance to be dried is liable to decomposition, the weighed chloride of calcium tube should be connected with a set of weighed potash bulbs. By re-weighing the latter after the substance has been dried, it is easy to determine whether any carbonic acid has been produced. (W. Stein, *Journ. prakt. Chem.*, **100.** 55).

Nitrogenous Compounds. In case the substance to be analyzed contains nitrogen, choose a combustion tube about 80 c. m. long, pack it at the anterior end with clean copper turnings to a depth of 15 or 18 c. m. (Compare the heading Nitrogenous Compounds under Method 1), fill in with oxide of copper in the usual way and proceed with the analysis. Take care to regulate the streams of air and oxygen so that the anterior half at least of the column of copper turnings shall not be oxidized either in the process of drying or during the actual combustion. The stream of oxygen must be very slow. When the combustion of the substance is complete and it is seen that the copper turnings are rapidly oxidizing, shut off the oxygen and let the tube cool in a slow current of air.

According to Stein & Calberla (*Journ. prakt. Chem.*, **104.** 232) silver turnings may be used with advantage, instead of copper, for reducing oxides of nitrogen, — as well as for retaining chlorine, as described below. Red-hot silver decomposes nitric oxide completely but has no action upon carbonic acid.

Sulphur Compounds. See this heading, under Method 4.

Chlorine, Bromine and Iodine Compounds. When chlorine, bromine or iodine compounds are burned with oxide of copper in conjunction with oxygen gas, there is not only danger of some dichloride (bromide or iodide) of copper being carried forward into the chloride of calcium tube, as has been already explained (Method 1, p. 64), but some of the dichloride is always decomposed by the oxygen gas into oxide of copper and free chlorine. Part of the chlorine thus evolved is retained in the chloride of calcium tube and part of it is absorbed by the potash lye. Though the error from this source is usually small, it must always be carefully guarded against by placing a column of metallic copper in the front part of the tube (see Nitrogenous compounds, above) and keeping the metal red hot throughout the combustion. At the close of the analysis the current of oxygen should be arrested as soon as the copper begins to oxidize lest the chloride of copper which has formed upon it be again decomposed. (Staedeler, *Annalen Chem. und Pharm.*, **69.** 334).

According to Kraut (*Zeitsch. analyt. Chem.*, 1863, **2.** 242) it is best to push back the column of metallic copper about 5 inches from the mouth of the tube, and to place a roll of silver foil in front of the copper. If this be done the stream of oxygen may be kept up as long as may seem fit, at the close of the operation, without any risk of chlorine being carried forward into the potash bulbs. The silver is of use also, inasmuch as, unlike the copper, it prevents any dichloride (bromide or iodide) of copper from passing into the chloride of calcium tube. The same roll of silver may be used over and over again in many analyses. Only after repeated use will it be necessary to ignite it in a stream of hydrogen. In the analysis of iodine compounds the column of copper turnings may be dispensed with altogether and the silver foil employed by itself.

Kekulé places several pieces of fused chromate of lead in the front part of the combustion tube, to stop chlorine and bromine; and to the same end Vœlcker (*Chemical Gazette*, 1849, **7.** 245) recommends that the oxide of copper be mixed with one-fifth its weight of oxide of lead. Compare Method 4.

Substances which contain Fixed Inorganic Constituents may usually be analyzed more readily by the process now in question than by Method 1 (see p. 64), for the ash which remains in the boat after the completion of the combustion can in most cases be weighed. In the analysis of compounds containing K, Na, Ca, Sr, or Ba the amount of carbonic acid retained by the ashes in the boat can sometimes be ascertained by simple calculation and added to the quantity found in the potash bulbs. In case the composition of the ashes cannot be inferred *a priori*, the amount of carbonic acid contained in them had better be determined by fusing the residue with borax glass (see under Carbonic Acid).

Method 3. Combustion with Oxide of Copper, in conjunction with Oxygen gas obtained by heating Chlorate or Perchlorate of Potassium, at or within the combustion tube.

The operation of filling the combustion tube is similar to that described under Method 1, with the exception that the column of oxide of copper at the posterior end of the tube is made about 5 c. m. long, instead of 3 or 4, and is mixed, by shaking, with 3 or 4 grms. of fused chlorate of potassium, in the form of coarse powder. The fused chlorate should be still warm when thrown into the tube. A layer of oxide of copper about 2 c. m. thick should be placed above the chlorate mixture to separate it from that which contains the substance to be analyzed. — After the anterior column of oxide of copper and the mixture of oxide of copper and substance to be analyzed have been heated, as in Method 1, the mixture of chlorate of potassium and oxide of copper is slowly heated, so that the combustion tube may be filled with oxygen gas. Any particles of carbon which the oxide of copper has failed to consume, will now be completely oxidized, as well as the metallic copper which has been reduced in the previous steps of the analysis. The mixture of chlorate of potassium and oxide of copper must be heated with extreme care in order that no great quantity of the chlorate may be decomposed at once. A tumultuous current of oxygen would be lia le to throw some of the potash lye out of the bulbs, and ruin the analysis. After all the reduced copper has been oxidized, the oxygen set free from the chlorate will of course sweep forward the carbonic acid which was contained in the tube, and will fill the potash bulbs and chloride of calcium and soda-lime tubes. To remove this oxygen, connect the bulbs and tubes with an aspirator, after disconnecting them from the combustion tube, and draw through them a quantity of air free from moisture and carbonic acid, before placing them in the room of constant temperature to be weighed.

Instead of placing chlorate of potassium in the combustion tube, Laurent (*Annales de Chim. et Phys.*, (3.) **19.** 360; *Gerhardt's Traité de Chim. Organ.*, Paris, 1853, **1.** 35) proceeds as follows:—After having dried the combustion tube and thrown a small quantity of warm, coarse oxide of copper into its posterior end, introduce the substance to be analyzed, mix it roughly with a small quantity of warm oxide of copper, and fill the tube with oxide of copper, taken directly from the crucible in which it has been ignited, and still as hot as 200° or 250°. Place the tube in the combustion furnace, and by means of a bent glass tube and rubber connector attach the closed upturned point of the tube with a U-tube, one arm of which is filled with chloride of calcium, and the other with fragments of caustic potash. That arm of the U-tube which is farthest from the combustion tube contains the potash, and is provided with a cork pierced with two holes. One of the holes carries a tube bent at an angle somewhat less than a right angle, and the other carries a straight, upright tube, drawn to a fine point and closed at its upper extremity. The outer end of the bent tube is attached, by means of a perforated cork, to the mouth of an ignition tube of hard glass, 30 or 40 c. m. long, and charged with 3 or 4 grammes of fused chlorate of potassium. The ignition tube rests upon a wire grate in such manner that the chlorate can be heated by placing bits of hot charcoal around the tube. Detach the rubber connector from the posterior point of the combustion tube, and proceed to heat the latter in the usual way. — After the combustion in the tube has been pushed to such an extent that little or no more carbonic acid goes forward, suck a small quantity of air from the open end of the soda-lime tube, in order to establish a partial vacuum within the combustion tube, and crush the end of the upturned point of the combustion tube with a pair of pincers. Slip on the rubber tube which connects the apparatus with the U-tube and the tube charged with chlorate of potassium, and place live coals about the latter so that oxygen may be slowly evolved from it. The tube must be heated very slowly, and care must be taken that the upper part of the tube is kept hot enough to prevent the solidification there of the melted chlorate. When the oxygen ceases to be absorbed by the reduced copper, and begins to pass rapidly through the potash bulbs, break off the point of the upright tube attached to the U-tube, and draw air through the latter and the rest of the apparatus until the whole of the oxygen has been expelled.

Since chlorate of potassium decomposes with a certain degree of violence, Bunsen recommends that it be replaced by perchlorate of potassium prepared, in the ordinary way, by heating the chlorate. A few grammes of the fused and still hot perchlorate are thrown into the posterior end of the combustion tube, a loose plug of recently ignited asbestos is pushed down to keep the potassium salt in place, and the rest of the tube then filled in the ordinary way. In case the substance to be analyzed is mixed with oxide of copper by means of a twisted wire, as explained on p. 61, perchlorate of potassium should always be used instead of the chlorate.

Non-volatile Liquid Substances. Weigh the substance (a fatty oil for example) in a small glass tube, kept upright by a wire support. Place a quantity of oxide of copper and chlorate of potassium in the end of the rather long combustion tube, as above described, drop in the loaded weighing tube and cause the oil to run out into the combustion tube and spread about upon the surface of the glass, excepting the anterior third or fourth of the tube and the upper side of the tube where the open channel is to be left. Fill the tube with oxide of copper which has been cooled in a tight flask, as described on p. 61, and take care that the weighing tube is filled with the oxide. Place the tube in hot sand, so that the oil may become more fluid and so be completely absorbed by the oxide of copper, and proceed with the analysis as above described. — Solid fats or waxes may be fused in a small weighed boat of glass, made from a tube divided lengthwise, and then cooled and weighed. The boat and contents may then be dropped into a combustion tube, into the end of which oxide of copper and chlorate of potassium have been previously thrown, and heat applied so that the substance may be spread about the tube and subsequently absorbed by the oxide of copper as just described.

Method 4. Combustion with Oxide of copper in conjunction with Atmospheric Air. (Method of Cloëz).

This method of analyzing organic compounds is characterized by the simplicity and cheapness of the apparatus required, as well as by its very general applicability. It may be employed not only for the analysis of compounds of carbon, hydrogen and oxygen, whether solid or liquid, volatile or non-volatile, but for the determination of carbon and hydrogen in substances contaminated with nitrogen, sulphur, chlorine, bromine, iodine, or fixed inorganic materials.

In some respects the process resembles Method 2, but differs from it inasmuch as a wrought iron pipe is employed as the combustion tube, in place of the glass tube above described, and that atmospheric air is used as an oxidizing agent instead of oxygen gas. Since the iron tube is well nigh indestructible, the apparatus, once mounted, becomes a permanent fixture, and may be used for an almost indefinite number of analyses.

Select a piece of wrought iron gas pipe, about 115 c. m. long and 20 or 22 m. m. in diameter. Provide a combustion furnace about 75 c. m. long, so that each end of the tube shall project 20 c. m. beyond the fire. Heat the tube to redness and pass in a current of steam to oxidize the inner surface of the tube. Fill the middle part of the tube with a long column of coarse, hard oxide of copper, and to keep the oxide of copper in place, plug it with spiral rolls of superficially oxidized copper foil. Provide two boats of stout sheet iron, one 20 the other 30 c. m. long, of such size that they can be readily pushed into or drawn out from the combustion tube; for this purpose attach a short iron wire to one end of each of the boats. The longer boat belongs at the posterior end of the combustion tube, the shorter at the front. If the substance to be burned contains nothing but carbon, hydrogen and oxygen, the anterior boat is to be filled with coarse oxide of copper, — or the boat may be left out altogether in case the substance to be analyzed is readily combustible. If, on the contrary, the substance to be analyzed is nitrogenous, fill the anterior boat with copper turnings (compare p. 64); if it contain sulphur, chlorine, iodine, or bromine, fill the boat either with red lead or with chromate of lead.

Solids. In case the substance to be burned is a solid consisting only of carbon, hydrogen and oxygen, both the boats are filled with oxide of copper and placed in the combustion tube, which is then heated as far as the furnace reaches during 10 or 15 minutes, while a slow stream of air is forced through the tube. Allow the posterior portion of the tube to cool somewhat, then take hold of the tube with a pair of roughened pincers (gas-fitters' pliers), remove the stopper, draw out the boat and cork it up in a special iron tube, kept for the purpose, until it has become well nigh cold. Meanwhile attach the weighed chloride of calcium tubes, etc., to the anterior end of the tube, which has hitherto remained open.

As soon as the oxide of copper is cold enough, place the boat on a sheet of thin copper foil, and with a polished iron hook transfer a portion of the oxide to a small, shallow rectangular box or shovel of brass, open at one end. Quickly scatter the substance to be analyzed over the oxide of copper left in the boat, cover it with the oxide in the shovel, push back the boat into the combustion tube, replace the stopper and pass a slow current of air through the apparatus. The combustion is conducted in the usual way in so far that the anterior end of the boat is first heated, while the oxide of copper in the middle and front part of the tube is kept red hot. By noting the rate at which the air bubbles pass through the potash lye in the permanent set of purifying tubes behind the furnace, and the weighed potash bulbs in front, it is easy to follow the process of the combustion, and determine when the operation is finished. — Instead of transferring the boat full of oxide of copper to a special cooling tube, it may be left to cool in the combustion tube if the operator prefer.

After the combustion is finished, and the weighed absorption tubes have been removed to the room of constant temperature, force a strong current of air through the tube in order to reoxidize the reduced copper, and proceed to the next analysis.

Liquids. Non-volatile liquids are weighed in a tube drawn out to a fine point, and are then transferred to the oxide of copper in the boat; the weight of the substance taken is determined by weighing the empty tube. Volatile liquids may be weighed in a tube provided with a glass stopper at one end, and drawn out to a point at the other. This tube is laid on the oxide of copper in the boat, the stopper is removed, the boat immediately pushed into the combustion tube, and a slow stream of air forced through the tube. In this case the oxide of copper in the posterior boat is not heated until the stream of cold air has ceased to carry forward vapors of the substance into the column of hot oxide at the middle of the tube.

Nitrogenous Substances. In analyzing nitrogenous substances, a copper boat filled with copper turnings (see Nitrogenous substances under Methods 1 and 2) is placed in the fore part of the tube, and only a very slow stream of air is kept up during the first part of the combustion. The current of air may be made more rapid towards the close of the analysis, though the anterior portion of the layer of metallic copper should remain unoxidized to the last.

Sulphur, Chlorine, Bromine or Iodine. In the analysis of substances containing sulphur, chlorine, bromine or iodine, the anterior boat is filled with dry red lead or chromate of lead, and the substance is mixed with fused and powdered chromate of lead in the posterior boat. The anterior boat should be heated only to incipient redness, in order that its contents may not fuse.

Ash Determinations. In case the proportion of ashes in the substance analyzed is to be determined as well as the carbon and hydrogen, weigh the substance in a porcelain boat, place the latter on a piece of platinum foil turned up at the edges and provided with a wire for withdrawing it from the tube, push the boat up to the permanent column of oxide of copper in the middle of the tube, and proceed with the combustion in the ordinary way. After the products of the dry distillation of the substance have been consumed, the residual carbon is burnt at the expense of the oxygen in the stream of air. A somewhat longer time is required in this case than when the combustion is finished in oxygen, but the results obtained are said to be equally accurate.

To dry and purify the air required for this process, Cloëz employs a set of permanent tubes and flasks, consisting: 1st, of a small bottle containing dilute potash lye,—the tube which brings air to this flask barely dips beneath the surface of the liquid; 2d, of a tall, drying cylinder, filled with bits of pumice stone soaked in sulphuric acid; 3d, of a long horizontal tube, with turned-up ends, filled with porous chloride of calcium; and 4th, of a similar tube filled with fragments of caustic potash.

Instead of a weighed chloride of calcium tube to absorb the water produced by the combustion, he employs a U-tube filled with fragments of pumice stone moistened with oil of vitriol. (Cloëz, *Annales Chimie et Phys.*, (3.) **68.** 394.)

Method 5. Combustion in Atmospheric Air, in conjunction with Oxide of Copper. (Method of Fresenius).

This method is applied to the estimation of the chemically combined carbon in cast or wrought irons, particularly in those kinds of iron which contain much graphite and but little combined carbon.

Method. Reduce a quantity of the iron to fine powder. Weigh out from 1 to 1.5 grms. of the powder, and place it in a flask of about 150 c. c. capacity. Fit a two-holed caoutchouc stopper to the flask, and to one of the holes fit a glass tube rather more than twice as long as the flask is high. Bend this tube twice near the middle into the form of an S, in such manner that the bent portion may be wholly outside the cork, while enough of the tube is left straight at either end to reach almost to the bottom of the flask; blow a small bulb at the middle of the bent portion, *i. e.*, at the centre of the S. After the S-tube has been fitted to the stopper, pour a few drops of quicksilver into the lower bend of the S. To the other hole in the stopper fit a delivery tube. This delivery tube should pass straight upwards through a distance about as great as the height of the flask, and then bend at a right angle to meet a horizontal combustion tube. Before fixing this delivery tube to the stopper, slip a short piece of rather wide glass tube over its upright part, and by means of a perforated cork fasten the lower end of this outer sleeve tightly to the delivery tube. During the experiment, this sleeve is kept full of water to cool the mixture of air and hydrogen passing forward from the flask.

The combustion tube may be about 30 c. m. long. Half its length—the end next the flask,—is filled with loosely packed asbestos and the other half with coarse oxide of copper, the latter being kept in place by a final plug of asbestos. To the end of the combustion tube which is farthest from the flask attach a large unweighed U chloride of calcium tube, to the latter attach a small, weighed soda-lime tube, and beyond the soda-lime place another unweighed U-tube, filled half with soda-lime and half with chloride of calcium, to protect the weighed tube. To this last U-tube attach an aspirator.

Ignite the contents of the combustion tube in a current of air free from carbonic acid; connect the combustion tube with the delivery tube of the flask, test the tightness of the apparatus, and heat the oxide of copper half of the combustion tube to low redness. Place a small thistle tube in the top of the S-tube of the flask, fill the thistle tube with dilute sulphuric acid, made by mixing 1 part of strong acid with 5 parts of water, and by means of the aspirator draw over into the flask, past the mercury, a quantity of the acid sufficient to dissolve the iron. Remove the thistle tube, replace it with an unweighed soda-lime tube, attach the latter to the S-tube with a caoutchouc connector and, by means of the aspirator, draw a slow current of air through the apparatus while the iron is dissolving. — Heat the flask carefully in a water bath, or upon a metallic plate, so that the iron may dissolve with tolerable rapidity.

The mixture of air, hydrogen and carburetted hydrogen burns to water and carbonic acid, at the edge of the asbestos in the combustion tube, so completely that but little of the oxide of copper in the combustion tube is ever reduced. In case any of the oxide of copper happens to be reduced for a moment the metal is immediately reoxidized by the excess of air. The column of asbestos prevents the possibility of the flame's passing back into the flask. The carbonic acid formed during the combustion is absorbed in the weighed soda-lime and its amount determined by reweighing the tube at the close of the experiment.

As soon as the evolution of hydrogen ceases, heat the column of asbestos in the combustion tube in order to oxidize any traces of hydrocarbons which may have condensed on the asbestos, and afterwards allow the apparatus to cool in a slow current of air. Weigh the soda-lime tube and from the weight of carbonic acid thus found calculate that of the carbon which has been evolved from the iron as a gaseous compound. The traces of hydrocarbons retained by the iron solution in the flask are practically insignificant, though sufficient to impart an odor to the liquid.

The iron in the flask of course dissolves more or less rapidly, according to the kind of metal, the fineness of the powder and the degree of heat, but as a rule the experiment requires 1 or 2 hours for its completion. — In operating upon samples of iron containing but little graphite it may be well to throw into the flask a small quantity of platinum sponge to accelerate the solution. — The process is inapplicable for the analysis of certain kinds of iron which deposit a portion of the combined carbon in the solid form (together with the graphite) when treated with dilute sulphuric acid. (Buchner, *Journ. prakt. Chem.*, **72.** 364). It is easy to determine whether carbon has been thus deposited in any given case by examining the residual graphite. To this end notice, in determining the Graphite, whether any coloration is imparted to the liquids, — boiling water, potash lye, alcohol, and ether, — with which the graphite is washed. Observe also whether the alcohol or ether, with which the graphite has been washed, after removal of the potash lye by water, leave any residue of organic matter when evaporated. (Fresenius, *Zeitsch. analyt. Chem.*, 1865, **4.** 73). The process yields accurate results. (Compare Tosh, *Chem. News*, 1867, **16.** 67).

Method 6. Combustion with Oxygen gas. (Warren's method).

Specially adapted for the analysis of volatile liquids, and of gases.

Bend a combustion tube in the manner described under Method 2, in the paragraph relating to volatile liquids; pack the horizontal part of the tube with small, fibrous pieces of asbestos, or other inert substance, so loosely that gases may pass freely through all parts of the mass, and at the same time so closely that no mixture of oxygen and hydrogen or other inflammable gas can explode when heated in its midst. In packing the tube, the asbestos should be added little by little and each new portion pressed gently against the preceding by means of a stiff wire. Little attention need be paid to the arrangement of the asbestos in the middle of the tube, since that portion of the column will come right of itself if the material is properly packed against the glass. A good way is to hold the tube in the left hand and turn it continually, while the wire in the other hand is made to pass around against the sides of the tube, gently tapping the asbestos and compressing it so that only very small open spaces can be seen in the finished column. It is best to make the horizontal part of the tube long enough to hold a column of asbestos 10 or 12 inches long, and to place a short column (2 or 3 inches) of coarse oxide of copper in front of the asbestos as a safeguard against accidents.

The substance to be analyzed is introduced

at the posterior part of the tube in the manner described in the paragraph just cited, and the analysis is conducted as there directed. Special care must, however, be exercised in regulating the stream of oxygen so that an excess of this gas may always be present in the combustion tube. To this end, the permanent drying apparatus should contain a set of potash bulbs of the same size as the weighed potash bulbs at the front of the tube, so that the number of bubbles, that is to say volumes, of oxygen entering the tube may be readily compared with the bubbles of carbonic acid which pass out. About 2 vols. of oxygen should be made to pass through the posterior potash bulbs for every bubble of mixed carbonic acid and oxygen which appears at the anterior bulbs. Or, better, the posterior set of bulbs may be made of such size that each of their bubbles shall be twice as large as those of the anterior bulbs. In case the supply of oxygen were at any time decidedly deficient the fact would be indicated by the reduction of some of the oxide of copper at the front of the tube. For a description and figure of an attachment for saving the excess of oxygen, and at the same time preventing the possibility of losing an analysis through the escape of unconsumed carbonaceous gases see Warren's original memoir, as cited below.

In spite of the fact that by far the larger part of the substance is consumed in a very small portion of the column of asbestos, it has not been found advisable to make the combustion tube any shorter than has been recommended above. With a short tube it is far more difficult to control the distillation of the substance and to regulate the supply of oxygen. The method of preventing any of the substance from escaping backwards from the combustion tube has been already described. See Method 2, paragraph on Volatile liquids. (Warren, *Amer. Journ. Sci.*, 1864, **38.** 387).

Method 7. Combustion with Oxygen after volatilization in a stream of Hydrogen. (A. Mitscherlich's method).

Applications. Simultaneous estimation of C, Cl, Br, I, S, and N in solid, liquid, or gaseous, substances without need of any combustion furnace.

Method. Heat the substance to be analyzed in a stream of hydrogen, and burn the hydrogen, thus charged with the vapor or products of distillation of the substance, in a stream of oxygen. If the substance contains no sulphur, pass the products of the combustion over a sheet of strong sulphuric acid to remove the water, then through a strong solution of nitrate of lead to absorb the chlorhydric acid, next through a tube filled with oxide of mercury to absorb bromine, and then through potash lye to absorb the carbonic acid. Finally pass the residual nitrogen and the excess of oxygen through a tube containing phosphorus, in order that the oxygen may be absorbed, and collect and measure the nitrogen in a graduated cylinder.

The amounts of Cl, Br and C are determined by weighing the several absorption tubes before and after the experiment.

If the substance contains sulphur, replace the solution of nitrate of lead with a concentrated solution of bichromate of potassium to absorb the sulphurous acid. In order to get rid of a small quantity of sulphuric acid which is formed during the combustion of the mixed hydrogen and sulphur compound, place some chloride of calcium and a little sulphite of calcium near the hydrogen flame. The moist sulphuric acid is absorbed by the sulphite of calcium as fast as it is formed, and at the same time a quantity of sulphurous acid, equivalent to that of the sulphuric acid, is set free from the sulphite to be absorbed by the chromate of potassium in due course. — If the substance contains iodine, the latter is allowed to deposit itself near the hydrogen flame at first and is subsequently sublimed into a clean glass tube, of known weight, and weighed as such.

In case any portion of the substance remains unvolatilized after having been thoroughly heated in the stream of hydrogen, the residue is weighed directly and estimated either as carbon or ash, as the case may be. For the details of this interesting process, see A. Mitscherlich's memoir in *Zeitsch. analyt. Chem.*, 1867. **6.** pp. 151–166, and 141.

Method 8. Combustion with Chromate of Lead.

This method was formerly much employed as a substitute for Method 1, in analyzing difficultly combustible substances, such as resins, coal, fats, etc.; it is, however, less convenient than Methods 2, 4 and 6.

Anderson (*Annalen Chem. und Pharm.*, **122.** 300) has observed but cannot explain the fact that in analyzing hydrocarbons with pure chromate of lead, the amount of hydrogen found is often considerably larger than it should be.

The details of the process, which may be used for analyzing solids and liquids, whether volatile or non-volatile, are similar to those of Method 1, with the exception that fused, powdered chromate of lead is substituted for the oxide of copper. The combustion tube need not be so large as in Method 1, since chromate of lead contains far more available oxygen than an equal volume of oxide of copper. (See Chromate of Lead). The chromate is simply heated in a porcelain or platinum dish, over a lamp, until it begins to turn brown, and then allowed to cool at 100°, or to a lower temperature, before it is transferred to the combustion tube or mixing mortar. Since chromate of lead is as hygroscopic as oxide of copper, the tube must finally be dried with an exhausting syringe, as described under Method 1.

The combustion is conducted in the same

way as with oxide of copper, but towards the close that part of the tube which contains the substance must be heated intensely, in order to fuse the chromate of lead so that no particles of carbon may escape combustion. This power of fusing at a moderate heat is one of the chief advantages of the chromate as compared with oxide of copper. Care must be taken, however, not to fuse the chromate at the anterior part of the tube, nor to heat it so hot that it shall lose its porosity, lest it become powerless to act upon gaseous products which have been only incompletely oxidized at the other end of the tube. It is well, for that matter, to fill the anterior part of the tube with coarse oxide of copper instead of chromate of lead, or with copper turnings which have been superficially oxidized by heating them in the air.

For the use of bichromate of potassium in conjunction with chromate of lead, see below, Method 9.

Sulphur Compounds may be burned with chromate of lead in a tube 60 or 80 c. m. long, if care be taken that 10 or 20 c. m. of the fore part of the column are never heated above low redness. (Carius, *Annalen Chem. und Pharm.*, **116.** 28). For the use of chromate of lead in conjunction with oxide of copper and air, see above, Method 4.

Chlorine and Iodine Compounds may be analyzed with chromate of lead without the risk of errors such as are liable to occur when oxide of copper is used. The chlorine or iodine is converted into chloride or iodide of lead and so retained in the combustion tube.

Some Bromine Compounds, however, are not readily analyzed in this way, since the metallic bromide formed during the combustion is liable to fuse and enclose some particles of carbon so completely that they cannot be burned. To avoid this difficulty Gorup-Besanez, *Zeitsch. analyt. Chem.*, 1862, **1.** 439) recommends the following method: — Choose a combustion tube with a rather long, upturned point, place in the tube a three-inch layer of oxide of copper, then a plug of asbestos, then a porcelain boat containing the substance to be analyzed in fine powder, mixed with about an equal weight of dry oxide of lead; then another asbestos plug, a column of granular oxide of copper, and finally chromate of lead or copper turnings. First heat to redness the anterior and posterior portions of the tube, and then warm the tube at the boat very carefully and gradually. The combustible portions of the substance will distil over in the form of vapor and be burnt by the hot oxide of copper, and nothing but bromide and oxide of lead will be left in the boat. Complete the combustion in a stream of oxygen, taking care that no more of the gas is passed than is necessary and that the contents of the boat are not too strongly heated. To prevent bromide of copper from subliming into the chloride of calcium tube, see Method 2, p. 68.

Method 9. Use of Bichromate of Potassium.

A. *Bichromate of Potassium with Chromate of Lead.* In analyzing some very difficultly combustible substances, such as graphite, for example, it has been found advantageous to mix the chromate of lead with about one-tenth its weight of fused and powdered bichromate of potassium. This mixture melts readily and gives off towards the close of the process a little more oxygen than would be evolved from mere chromate of lead. (Liebig; Mayer, *Annal. Chem. und Pharm.*, **95.** 204). The mixture of chromate of lead and bichromate of potassium is well fitted also for the analysis of organic compounds containing potassium, sodium, barium, calcium or strontium. No trace of carbonic acid is retained by either of these metals when their compounds are ignited in presence of the mixed chromates.

B. *Bichromate of Potassium with Oxide of Copper.* It has been suggested by Rochleder that a fused mixture of oxide of copper and bichromate of potassium might be used instead of chromate of lead for burning difficultly combustible substances, and, considered merely as an oxidizing agent, the mixture is said to yield excellent results. But the fused mass is so hard to pulverize that it is practically less convenient than chromate of lead.

Schulze has employed these ingredients in a somewhat different way for determining the proportion of humus in soils. His process is as follows: — Rub 10 grms. of the finely powdered and sifted soil together with 20 grms. of a mixture of equal parts of oxide of copper and bichromate of potassium. Place the mixture in the closed end of a combustion tube, put a layer of oxide of copper in front of the mixture, a layer of finely divided metallic copper in front of the oxide of copper, and leave a space of several inches empty at the front of the tube. Connect the anterior end of the tube with the collecting bottle of Schulze's apparatus for estimating carbonic acid (see Carbonic Acid, estimation of by measuring the gas), turn down the water delivery tube of the connecting bottle and proceed to heat the mixture in the combustion tube precisely as in an ordinary combustion; *i. e.*, from the front backwards, taking care not to heat the tube so strongly as to distort it. Measure the water expelled from the collecting jar, as directed in the paragraph above referred to, in order to determine the volume of carbonic acid. — The metallic copper at the front of of the tube will retain any oxygen which may be given off from the bichromate. A special experiment must be made, as directed under Carbonic Acid, in order to ascertain how much of the carbonic acid obtained by the combustion is derived from carbonate of calcium in the soil. The difference between the two determinations will give the amount of carbonic acid derived from the carbon burnt. In case the carbonic acid is mixed with nitrogen or ther foreign gas, it may be passed into baryta water and precipitated as Carbonate of Barium. Schulze, *Zeitsch. analyt. Chem*, 1863, **2.** 98).

Another method of employing the two ingredients, is to mix them in the combustion tube, as has been proposed by Gintl (*Zeitsch. analyt. Chem.*, 1868, **7.** 302). — Into an ordinary combustion tube, drawn up to a point and closed at one end, pour a layer of about two inches of coarse oxide of copper, then a layer of several inches of fused and powdered bichromate of potassium, then the organic substance to be analyzed and then a layer of several inches of oxide of copper. Both the powdered bichromate and the oxide of copper are kept free from moisture in tightly closed tubes or flasks, as explained on p. 61. By means of a twisted wire (Compare p. 61), mix the substance, the oxide and the bichromate intimately, then fill the anterior part of the tube with oxide of copper and proceed with the combustion in the usual way. (See Method 1). Since some free oxygen resulting from the partial decomposition of the bichromate is apt to escape from the tube, care must be taken at the close of the combustion to draw air through the apparatus until the last trace of oxygen gas has been displaced from the potash bulbs.

The process yields excellent results, the combustion is easily completed at a comparatively low temperature and the materials employed are much cheaper than chromate of lead.

C. *Bichromate of Potassium with dilute Sulphuric Acid.* (Method of analysis by gentle or "Limited" oxidation). Compare Brunner (below, under Method 10).

Many organic substances, such, for example, as the fatty acids, undergo no change when treated with a dilute solution of bichromate of potassium mixed with weak sulphuric acid; while other substances, such as compounds of lactic acid and diethoxalic acid are readily decomposed by the mixture, with evolution of a definite quantity of carbonic acid. Hence it is possible to analyze certain mixtures of organic compounds, — after the mode of decomposition of the easily oxidizable constituents has once been determined, — by heating the mixture with the acidulated solution of bichromate and collecting and weighing the carbonic acid which is formed.

For decomposing one class of substances, such as the lactates, Chapman & M. Smith (*Journal London Chem. Soc.*, 1867, **20.** 173) employ the following process:—Place a weighed quantity of the substance to be analyzed in a wide-mouthed flask, and fit to the flask a caoutchouc stopper with two perforations. In one hole of the stopper place a short delivery tube bent at a right angle, and in the other hole a long tube reaching almost to the bottom of the flask, provided with a bulb above the stopper, and closed at the upper end with a bit of caoutchouc tubing and glass plug. To the delivery tube attach a Will & Varrentrap's nitrogen bulb filled with concentrated sulphuric acid and set in a basin of cold water; to the nitrogen bulb attach a set of Liebig's potash bulbs, and to the latter attach a small tube filled with fragments of solid hydrate of potassium. Since the purpose of the sulphuric acid is merely to purify the carbonic acid, it need not be weighed, but both the potash bulbs and tube must be weighed before and after each experiment. — Prepare a quantity of the oxidizing solution by placing 100 grammes of bichromate of potassium and 125 grms. of oil of vitriol in a litre flask and filling the flask with water to the mark. Pour about 150 c. c. of this solution through the bulb-tube into the flask, close the tube with the plug, and heat the contents of the flask upon a water bath. The carbonic acid set free is dried and otherwise purified by the sulphuric acid in the nitrogen bulbs, and is absorbed by the potash in the weighed apparatus beyond. In the course of 20 minutes bubbles of carbonic acid cease to pass through the sulphuric acid, and a quantity of the potash lye rises into the large bulb of the Liebig apparatus. When this occurs, remove the plug from the top of the bulb-tube at the flask, attach a soda-lime tube to the bulb-tube to purify the air, and by means of an aspirator draw air through the flask until all the carbonic acid has been swept forward. The increase in weight of the potash bulbs and tube gives the amount of carbonic acid which has been evolved from the substance. — It is important to keep the sulphuric acid in the nitrogen bulbs cool, lest it act upon some of the aldelyde or other volatile products of the decomposition, which pass forward with the carbonic acid, and so generate sulphurous acid. Since the province of the sulphuric acid is to absorb aldelyde as well as water, chloride of calcium cannot be used in its place.

Another form of apparatus employed by Chapman & Smith in certain cases for estimating the carbonic acid by loss, is figured on p. 180 of their memoir.

Method 10. Combustion with Chromic Acid.

This process is used for estimating carbon in cast-iron (see above, Principle 1, D), and for determining carbon in soils, vegetables and other substances whose ashes would be likely to interfere with the ordinary methods of analysis. Like Method 7 it differs essentially from Methods 1 to 6 and 8, inasmuch as it is applicable to the determination of carbon without regard to the hydrogen which usually accompanies this element.

Provide a two-necked flask of about 150 c. c. capacity. To one neck of the flask fit a perforated cork carrying a tube bent at a right angle, and to the other opening attach a wide tube provided with a bulb of 70 or 80 c. c. capacity. The purpose of this bulb is to condense aqueous vapor generated in the flask; it must be supported in a vertical position above the flask. The inner limb of the right-angled tube above mentioned must be long enough to reach to the bottom of the flask; the other limb extends a few inches outwards from the cork and is plugged at the end with a bit of glass rod fitted to a short piece of rubber tubing. Set the flask in a basket of wire gauze on a ring of a lamp-stand. Connect the vertical bulb-tube with a permanent drying apparatus consisting of a cylinder of about one-quarter litre capacity, filled with pumice stone, which has been soaked in strong sulphuric acid and afterwards heated to expel any traces of chlorhydric or fluorhydric acid which may have been contained in the stone, and a U-tube about 60 c. m. long full of chloride of calcium. To absorb the carbonic acid, attach to the chloride of calcium tube a small, weighed U-tube, full of soda-lime or potash-pumice, with a short layer of chloride of calcium at the outer end.

In case the substance to be analyzed consists of the rèsidue from 2 grms. of cast iron, as explained on p. 58, the details of the process will be as follows: — Remove from the flask the cork which carries the right-angled tube, and draw up the inner limb of this tube through the cork so that it shall not reach far into the flask. By means of a wash bottle with fine jet, wash the carbonaceous matter into the flask, taking care to use as little water as possible. If the quantity of liquid in the flask does not exceed 25 c. c. add to it 40 c. c. of concentrated sulphuric acid, or a proportionally larger quantity of acid if the volume of water is larger, and allow the mixture to cool. Attach the flask to the system of drying and absorption tubes, throw into it about 8 grammes of solid chromic acid, free from potassium, and immediately replace the cork. — Heat the flask gradually until the evolution of gas becomes so violent that the mixture threatens to boil over, and maintain the temperature at this point as long as there is a lively evolution of gas. When the current of gas slackens, heat the flask more strongly until white fumes appear in the vertical bulb-tube, and maintain the mixture at this temperature until the evolution of gas becomes very weak. Attach an aspirator to the weighed soda-lime tube and open its cock slightly; push down the right-angled tube in the flask until it dips beneath the surface of the liquid, remove the plug from the outer limb of the right-angled tube, and attach to the latter a tube filled with fragments of caustic potash. Then open the cock of the aspirator somewhat wider, so that bubbles of air may pass through the liquid in the flask at the rate of about 2 a second. When 5 or 6 litres of water have flowed from the aspirator the whole of the carbonic acid will have been swept out of the apparatus and absorbed in the weighed soda-lime tube. Allow the latter to cool before weighing it.

In using this method for determining carbon in an organic substance, about 16.6 grms. of chromic acid and 25 c. c. of concentrated sulphuric acid should be taken for every gramme of the substance. About 2 vols. of water should be placed in the flask for every 3 or 4 vols. of strong sulphuric acid used. If the organic substance contains chlorine, bromine or iodine, a tube filled with pulverulent iron, copper or silver must be placed between the vertical bulb-tube and the permanent drying cylinder to absorb the foreign gas. (Ullgren, *Annal. Chem. und Pharm.*, **124.** 59). — Ullgren prefers chromic acid to the bichromate of potassium recommended by the brothers Rogers, and by Brunner (*Pogg. Ann.*, **95.** 379), since it is possible in this way to avoid the formation of an annoying precipitate of anhydrous chrome alum which interferes with the reaction, when bichromate of potassium is employed. The chrome alum thus thrown down as a green, slimy powder, almost insoluble in water, acids, or alkalies, not only delays the oxidation, but tends to conceal its completion. — According to Brunner, it is possible, by his process, to estimate the ordinary modification of carbon in presence of graphite, by oxidizing first with a mixture of bichromate of potassium and dilute sulphuric acid and afterwards using strong acid; but the separation did not succeed in Ullgren's hands. For the application of this process to organic analysis, compare W. Knop, *Chem. Centralblatt*, 1861, p. 17.

Method 11. Combustion with a mixture of Chromate of Lead and Chlorate of Potassium.

Applications. Estimation of carbon in cast-iron.

Method. Close a combustion tube of hard glass at one end and place at the closed end a two-inch layer of a mixture of equal parts of chromate of lead and chlorate of potassium. Mix about 3 grms. of the finely powdered iron with 50 grms. of a mixture of 40 parts chromate of lead and 6 parts of fused chlorate of potassium, and place the mixture in the combustion tube. Finally, place a layer of chromate of lead in front of the mixture, and connect the tube with a chloride of calcium tube and a weighed soda-lime tube or set of potash bulbs. Heat the contents of the combustion tube carefully, beginning at the anterior end. When the mixture of iron and the oxidizing agents is heated to dull redness the metal burns with incandescence, and the carbon contained in it is converted into carbonic acid. At the close of the operation the free oxygen evolved from the mixture of chlorate and chromate at the end of the tube sweeps forward the carbonic acid and, if need be, completes the oxidation of the iron. — The process usually yields good results, though the percentage of carbon found is sometimes rather lower than the truth. The addition of chlorate of potassium is essential to success in determining carbon in cast-iron; with chromate of lead alone the amount of carbon obtained from the cast-iron is always too small. (Regnault, *Annal. Chem. und Pharm.*, **30.** 352; Bromeis, *Ibid.*, **43.** 241; Tosh, *Chem. News*, 1867, **16.** 67). Take care to remove the oxygen from the absorption tube before weighing. Or, instead of that, the soda-lime tube may be filled with oxygen in the first place, and always be weighed full of oxygen.

Method 12. Combustion with Chlorate of Potassium.

The original process of Gay-Lussac & Thénard was as follows: — An intimate mixture of known weights of chlorate of potassium and of the substance to be analyzed was made into small pellets and the latter were thrown, one by one, with proper precautions, into a vertical tube kept at a red heat. The carbonic acid, oxygen and nitrogen (if present) were conducted through a lateral delivery tube to a mercury trough and there collected in a graduated cylinder. After the total volume of gas had been measured, the amount of carbonic acid was determined by absorption with caustic potash and that of the oxygen by exploding the residue with hydrogen in a eudiometer.

The process was faulty, as regards nitrogen compounds, since a portion of the nitrogen was converted into nitrous acid. It was inapplicable, moreover, to the analysis of liquids or of volatile substances. Berzelius improved it by heating the mixture of chlorate and substance in a horizontal tube and collecting and weighing the water generated by the combustion. There was consequently no longer any need of weighing the chlorate of potassium. He mixed the chlorate, moreover, with a large proportion of chloride of sodium, in order that the combustion might be tranquil.

The use of chlorate of potassium in this way was, however, soon given up, and the process superseded by that in which oxide of copper is employed. Quite recently, suggestions have again been made in favor of employing the chlorate in certain cases. Mène (*Comptes Rendus*, **56.** 446) recommends the following method. Close one end of a glass combustion tube about 1 m. m. thick, 50 c. m. long and 1 c. m. in diameter, throw in enough fused and powdered chlorate of potassium to fill 2 c. m. of the length of the tube, fill the rest of the tube almost completely with an intimate mixture of the substance to be analyzed and of fused and powdered chlorate of potassium, and secure the whole by placing a plug of asbestos at the front of the column. In making the mixture proceed as follows: — Measure out in the tube itself as much of the chlorate of potassium as the tube will hold, pour out all but the last 2 c. m. of the chlorate, mix it with the weighed and finely powdered substance and pour back the mixture into the tube through a funnel. Since, according to Mène, the fused chlorate is scarcely at all hygroscopic, the mixing and transfer of the substance may be made at the ordinary temperature without special precautions.

Hang the combustion tube, or support it upon wires, in a horizontal position, attach the weighed absorption apparatus (a large chloride of calcium tube and a couple of soda-lime tubes will be best) and proceed to heat the mixture of chlorate and substance with a single small spirit lamp or Bunsen's burner. The tube is first heated at the asbestos plug and then backwards step by step, until each portion of the chlorate has been melted and made to react upon the organic matter. It often happens that the reaction is attended with evolution of light or even deflagration, particularly when the substance contains a large proportion of carbon, but the deflagration, according to Mène, does no harm even when potash bulbs are used to collect the carbonic acid. At the close of the combustion, the last traces of carbonic acid are swept forward out of the tube by the oxygen set free from the pure chlorate of potassium at the end of the tube. Care must be taken that the absorption apparatus is either weighed full of oxygen at the start, or that air is drawn into it after the combustion. The entire operation lasts only 20 minutes. It is to be observed that the process requires no special furnace and that by means of it a comparatively large weight of the substance may be taken for each analysis.

Another method, proposed by Schulze, is adapted to the analysis of very small quantities of carbon compounds. A mixture of weighed quantities of the substance to be analyzed and of fused, powdered chlorate of potassium is sealed up and ignited in a combustion tube of known capacity, after the air has been pumped out of the tube; and the mixture of gases resulting from the combustion is measured and analyzed by the methods of absorption and eudiometry ordinarily employed in gas analysis. — The combustion tube is made of glass of the quality and strength usually employed for organic analyses, but the capacity of the tube to be selected should be determined approximately by reference to the supposed composition of the substance to be analyzed. In the case of sugar, for example, 16 milligrm. of oxygen would be half as much again as is absolutely necessary for the combustion of 10 milligrm. of sugar. The gas given off from the quantity of chlorate of potassium, about 41 milligrm., necessary to produce 16 milligrm. of oxygen would occupy a space of about 12 c. c., measured at the ordinary temperature and pressure. But if this amount of gas were enclosed in a tube of 25 c. c. capacity and heated to 500° or 600°, only a comparatively feeble pressure could be exerted by it upon the walls of the tube. It may be admitted that 100 milligrm. of chlorate of potassium yield 20.805 c. c. of oxygen, and that 1000 c. c. of oxygen at 0° and under the pressure of 1 metre of mercury, weigh 1.8819 gramme.

The combustion tube is drawn out to a point at one end and the point closed; the mixture of substance and chlorate is poured into the tube in such manner that no portion of the mixture shall remain adhering to the tube near the open end. The tube is then drawn out at the lamp near the open end, to a conical point fit to receive a caoutchouc connector. Behind this point the tube is narrowed to such an extent that it can be readily closed at any moment by touching it with a blowpipe flame. Neither the point nor the depression in the tube must be made too thin, lest the glass give way when the tube is exhausted of air.

To exhaust the tube, fit to its aperture a flexible pipe which has been rendered impermeable to air by soaking in hot linseed oil, and connect the other end of the pipe with an air pump, the precise degree of efficiency of which has been determined beforehand by experiments on empty tubes and flasks. Schulze found that the pump employed in his experiments could remove all but 0.001715 of the original volume of air, so that for a combustion tube of 50 c. c. capacity a residuum of 0.08575 c. c. of air had to be allowed for, in calculating the results of an analysis.

When the tube has been exhausted as completely as possible, close it by throwing a blowpipe flame against the narrow place behind the rubber connector, place the tube in an iron gas-pipe a little larger than itself, and heat the iron almost to redness during 20 minutes, over a row of Bunsen's burners. In case several analyses are to be made, a couple of combustion tubes may be placed end to end in the same iron pipe, and heated together. After the lamps have been extinguished, the apparatus may be cooled by pouring water upon the iron. As a general rule the glass tubes may be withdrawn from the gas-pipe half an hour after the lamps were lighted.

To collect the gaseous products of the ignition, provide a rather tall, tubulated bell-glass about 3 c. m. in diameter, to the neck of which has been cemented a fine steel stop-cock connected with a straight projecting tube about 3 m. m. long by 1 m. m. in diameter. Immerse the posterior end of the combustion tube in mercury contained in a tall cylinder, break off the lower point of the tube in the mercury, push the tube down into the mercury and place the bell-glass full of mercury over the top of the tube. Then break the upper, anterior point of the combustion tube so that the contents of the tube may rise into the bell. From the bell the gas may be readily transferred to a measuring tube by simply sinking the bell beneath the tube and opening the stop-cock. — After the total volume of gas has been measured, with the proper precautions, the carbonic acid is determined by absorption with hydrate of potassium, in the usual way.

Since the fundamental idea of the method is to operate upon very small quantities of material, a very accurate balance is required. According to Schulze, the balance should be delicate enough to indicate, at the least, quantities as small as $\frac{1}{200}$ of the substance weighed. Balances on the principle of that of Ritchie (*Phil. Trans.*, 1830, p. 402; see also *Karsten's allgemein. Encyclop. der Physik*, Lief. **16.** p. 606) may be employed with advantage.

Hygroscopic substances may be weighed, and analyzed in the moist condition, it being only necessary to determine the proportion of moisture by drying a special portion of the material. (F. Schulze, *Zeitsch. analyt. Chem.*, 1866, **5.** 269.)

Carbonates (various).

Principle. Decomposition of by acids. See Carbonic Acid (Volatility of); and Alkalimetry.

Carbonic Acid. Compare the various Carbonates, below, notably the carbonates of barium, calcium, and mercury. — It may be here remarked that for estimating the amount of carbonic acid in normal carbonates of the alkalies and alkaline earths, it is well to determine the base by means of a standard acid (see Alkalimetry) and to calculate upon the amount of base an equivalent quantity of carbonic acid.

Principle I. Affinity for alkalies.

Applications. Estimation of carbonic acid. Separation of carbonic acid from other gases. Estimation of carbon in organic compounds.

Methods. See Hydrate of Potassium; Soda-Lime, and Carbon. Compare also, below, Carbonic Acid (Volatility of, absorption of the gas in alkali), and Hydrate of Calcium.

Principle II. Volatility.

Applications. Estimation of carbonic and boracic acids. Separation of carbonic acid from all the bases and from all other acids. Alkalimetric determination of the commercial value of the carbonates of sodium and potassium. Estimation of carbonate of calcium and similar carbonates. Acidimetric determination of the value of commercial acids. Indirect separation of strontium from calcium.

Method A. By simple ignition. This method is applicable to the estimation of carbonic acid in all carbonates, — such as those of Mg, Cd, Zn, Pb, Cu, Ni, etc, — which give off the acid readily and completely on being heated. Even with carbonate of calcium carbonic acid can be determined in this way, if only a small quantity of substance be operated upon and the heat be made intense.

If the carbonate to be analyzed is anhydrous, and contain no metal liable to form a mixture of several oxides, when heated, it will be enough to ignite a weighed quantity of it in a crucible until the weight of the crucible and contents remains constant. The loss of weight will represent the carbonic acid. In case carbonate of lead, carbonate of cadmium, or any other easily reducible carbonate is to be analyzed, the crucible must be of porcelain.

If the residual oxide is prone to absorb oxygen when heated in the air, replace the crucible with a bulb-tube of hard glass, and pass a stream of carbonic acid through the tube during the process of ignition. This method yields very accurate results.

If the substance to be analyzed contains water, it is sometimes possible to estimate the water before the carbonic acid, by first heating a weighed quantity of the material gently until the water has been expelled, and afterwards intensely to drive out the carbonic acid. In case the water cannot be determined in this way it may be estimated as follows: — Weigh out a quantity of the substance in a small boat, and place the latter in a tolerably wide glass tube. By means of dry, tight corks, attach a weighed Chloride of Calcium tube to one end of the wide tube, and an unweighed chloride of calcium tube to the other end. Connect the other end of the weighed chloride of calcium tube with an aspirator, and draw a slow stream of air through the apparatus. Heat the boat and, from time to time, the whole of the wide tube, until all the carbonic acid and water have been expelled from the carbonate and carried forward into the chloride of calcium tube. Allow the apparatus to cool and weigh the boat and the chloride of calcium tube. The difference between the first and second weights of the contents of the boat gives the total amount of carbonic acid and water; the increase in weight of the chloride of calcium tube gives the amount of water, and the difference between the two, the carbonic acid. — Care must be taken neither to leave any moisture in the wide tube nor to burn its corks. To prevent the possibility of any moisture passing back from the aspirator into the weighed chloride of calcium tube, it is well to interpose an unweighed chloride of calcium tube between the two.

Instead of weighing the substance in a boat, it may just as well be placed in a bulb-tube. In case the substance to be analyzed is likely to absorb oxygen when heated in air it should be ignited in a current of dry carbonic acid.

Indirect separation of Calcium from Strontium. Precipitate the two metals as carbonates, — see Carbonate of Calcium, — weigh the mixed precipitate and call its weight W. Afterwards ignite the precipitate at a moderate white heat, until all the carbonic acid has been driven out from both the carbonates, and weigh the residual oxides. The difference between the two weights will give the weight of the carbonic acid. Calculate this carbonic acid as if the mixed carbonates were really nothing but carbonate of strontium, thus,

$$\underset{44}{\text{Molec. wt. of } CO_2} : \underset{147.5}{\text{Molec. wt. of } SrO, CO_2} = \text{Weight of } CO_2 \text{ found} : x \left(= \text{Weight of supposed } SrO, CO_2\right)$$

The difference (x–W) between the weight of the supposed carbonate of strontium, and that of the mixed carbonates actually found, is proportional to the quantity of carbonate of calcium in the mixture, so that the weight of the last named salt may be obtained by the proportion:

$$\underset{47.5}{\text{Difference between the molec. weights } SrO, CO_2 \text{ and } CaO, CO_2} : \underset{100}{\text{Molec. wt. } CaO, CO_2} = (x - W) : y \left(= \text{Wt. of } CaO, CO_2 \text{ in the mixed carb's}\right)$$

In short, multiply the carbonic acid found by 147.5 ÷ 44 (= 3.3534), deduct from the product the weight of the carbonates, and multiply the difference by 100 ÷ 47.5 (= 2.10526). The product will express the weight of the carbonate of calcium. The difference between this weight and that of the mixed carbonates will give the weight of the carbonate of strontium. The process gives good results unless the quantity of one or the other of the metals in the mixture is too minute.

Precautions. The mixed precipitate should be thrown down from hot solutions. During the process of ignition the agglomerated mass should be turned over now and then in the platinum crucible, and pressed down carefully against the hot metal until, after repeated ignitions, its weight remains constant. If the operator prefer, the carbonic acid may be estimated by fusing with borax glass (see below), but a moderate white heat such as may readily be obtained by a blast lamp is sufficient for the purpose. (Schaffgotsch, *Pogg. Ann.*, **113.** 615).

Method B. By ignition with a solid, non-volatile acid, or acid salt. By this method, carbonic acid can be separated from all the bases which form anhydrous carbonates, and its amount determined with great accuracy. The principle is applied also in organic analysis, for expelling carbonic acid from the carbonates of K, Na, Ca, Ba and Sr, in cases where either of these compounds is formed during the combustion of the organic substances. The same principle is involved, moreover, in the method of estimating boracic acid indirectly by igniting a carbonate with borax glass, as has been explained under Boracic Acid. It is applied also to the indirect separation of barium from strontium or calcium, and of strontium from calcium, as will be explained immediately.

1. *Fusion with Borax-glass.* Fuse a quantity of borax-glass (see *bi*Borate of Sodium) in a weighed platinum crucible, place the crucible with its contents in a dessicator to cool, and weigh when cold. Place in the crucible a quantity of the carbonate to be analyzed, and again weigh so that both the weight of the carbonate and that of the borax may be known. The carbonate should be perfectly dry, and should weigh about a quarter as much as the borax glass. Gradually heat the mixture to redness, and keep it at that temperature until the contents of the crucible are in a state of tranquil

fusion and no more bubbles of carbonic acid escape from the molten liquid. No heed need be taken of a few persistent bubbles of gas which are liable to remain in the liquid. Again place the crucible in the dessicator and weigh it, with its contents, when cold. The loss of weight represents the amount of carbonic acid in the substance analyzed. The mixture must not be ignited over a blast lamp, lest a certain amount of borax glass be lost through volatilization. (Schaffgotsch).

To separate barium from strontium or from calcium, by the indirect method, throw the two metals down together as carbonates, see Carbonate of Barium (insolubility of), and note the weight of the mixed precipitate. Then expel the carbonic acid by fusing with borax-glass, and note the weight of the acid. The proportion of the different metals in the mixed precipitate may be calculated as follows. Compare the method by simple ignition, above.

To separate Ba from Sr: — Calculate the carbonic acid as if it belonged to strontium, thus:—

$$\underset{44}{\text{Molec. wt. of } CO_2} : \underset{147.5}{\text{Molec. wt. of } SrO, CO_2} = \text{Wt. of } CO_2 \text{ found} : x \left(= \text{Weight of supposed } SrO, CO_2\right)$$

and subtract from this supposititious weight the weight of the mixed carbonates (W) found. The number (x–W) thus obtained, is proportional to the quantity of carbonate of barium in the mixture; hence the weight of the barium salt may be obtained by the proportion.

$$\underset{49.5}{\text{Diff. between molec. wts. of } SrO, CO_2 \text{ and } BaO, CO_2} : \underset{197}{\text{Molec. wt. of } BaO, CO_2} = (x - W) : y \left(= \text{Wt. of } BaO, CO_2 \text{ in the mixed carbonates}\right)$$

In short, multiply the carbonic acid found by 147.5 ÷ 44 (= 3.3534), deduct from the product the weight of the mixed carbonates, and multiply the difference with 197 ÷ 49.5 (= 4.4246). The product will express the weight of the carbonate of barium. The difference between this weight and that of the mixed carbonates will give the weight of the carbonate of strontium.

To separate Ba from Ca, proceed in a similar way, calculating the carbonic acid, at first, as if it were all carbonate of calcium.

2. *Fusion with Bichromate of Potassium.* Select a combustion tube of hard glass 50 or 60 c. m. long, and bend it in such manner that it shall be very slightly U-shaped at the middle, while the two ends remain horizontal. Fasten to one end of this tube a pipe closed with a copper stop-cock and connected with a permanent set of potash and chloride of calcium tubes, suitable for removing carbonic acid and water from the air. To the other end of the combustion tube attach a weighed chloride of calcium tube and a set of potash bulbs, or a soda-lime tube, precisely as in the estimation of Carbon, and to the potash bulbs attach an aspirator. It is well to attach a small supplementary weighed chloride of calcium tube to the potash bulbs, and also to interpose an unweighed chloride of calcium tube between the absorption tubes and the aspirator, to guard against the aqueous vapor which may arise from the latter. — Place in the combustion tube from 30 to 60 grms. of bichromate of potassium, which has been fused just before the experiment, together with from 1 to 3 grms. of the carbonate to be analyzed, and place a loose plug of ignited asbestos in front of the mixture. Carbonates of the alkali metals may be introduced in the form of fragments, but most of the other insoluble carbonates must be finely powdered and mixed with the bichromate. After the mixture has been placed in the combustion tube, and the latter connected with the absorption tubes, heat the mixture of bichromate and substance, and allow a slow stream of water to flow from the aspirator, so that a current of air shall be drawn through the apparatus. The evolution of carbonic acid begins as soon as the bichromate melts, and it admits of being easily regulated. The operation is finished when the whole of the mixture is seen to be in a state of tranquil fusion. The increase in weight of the potash bulbs or soda-lime tube gives the weight of the carbonic acid, and that of the chloride of calcium tube the weight of the water in the substance analyzed, in case any water was present. The presence of sulphites or hyposulphites, or of sulphides or oxysulphides of the alkali or alkaline-earthy metals, does no harm, but the substance analyzed must of course be free from carbonaceous matters.

The process may be applied not only to the estimation of carbonic acid for its own sake, but for determining the alkalimetrical value of the carbonates of potassium and sodium. (See below, under "Expulsion of Carbonic Acid by Acids"). (Persoz, *Comptes Rendus*, **53.** 239).

3. In the analysis of *organic compounds* which contain an alkali-metal or a metal of one of the alkaline earths, mix the substance with a quantity of boracic or antimonious acid, or with phosphate of copper, before placing it in the combustion tube. (Compare Carbon).

Method C. Expulsion by acids at the ordinary temperature. By this method carbonic acid may be separated from all bases, without exception. The evolved gas may then be measured as such, or determined as loss, or estimated indirectly from the weight or by the analysis of the residue, or, better, it may be determined by absorption in an alkali in accordance with Principle I.

1. *By Loss.* The earlier chemists were content to throw a weighed quantity of the carbonate to be analyzed into 3 or 4 times as much chlorhydric, nitric, or sulphuric acid contained in a tall cylinder, and counterpoised upon a balance. When the effervescence ceased, the increase of weight on the part of the cylinder was noted, and the difference between that quantity and the weight of carbonate taken

was reckoned as carbonic acid. Care was taken to avoid loss by spirting, and that the operation should proceed rather rapidly, so that as little acid as might be should be lost by evaporation. The acid usually employed was either nitric acid of 1.15 sp. gr., or a mixture of strong chlorhydric acid with an equal bulk of water. (Pfaff, *Handbuch analyt. Chem.*, 1824, **1.** 505; **2.** 18). Of late years, however, the method has been greatly perfected by the invention of apparatus specially arranged so as to avoid the risk of loss by evaporation or transportation of the solvent acid.

A great many different forms of the apparatus have been devised by chemists, see, for example, *Mohr's Titrirmethode*, 1855, p. 122, and *Fresenius's Quantitative Analysis*, 1865, p. 300. The essential features of the apparatus are, 1st, a small glass flask, in which to place the substance to be analyzed, and in which, at the proper moment, to mix the substance and the acid by which it is to be decomposed; 2d, a tube or flask to hold this acid and keep it away from the carbonate until the time when the operator sees fit to mix the materials; 3d, a chloride of calcium tube, or other apparatus, for drying the gas evolved by the action of the acid on the carbonate, so that nothing but absolutely dry carbonic acid can escape from the flask.

A simple form of the apparatus may be made as follows: Choose a small, wide-mouthed flask, and fit to it a soft cork or rubber stopper with two perforations. Pass the stem of a chloride of calcium tube through one hole of the cork, and insert in the other hole a narrow glass tube bent at a right angle and reaching nearly to the bottom of the flask. Make a small bottle, like a homœopathic phial, from a piece of tolerably wide glass tubing, by closing one end of the tube at the lamp, and pressing the softened glass flat, so that the short tube, or bottle, may be made to stand upright. This little bottle must be capable of passing readily through the neck of the flask. To the top of the bottle tie a thread, or better, a piece of fine platinum wire, a little longer than the flask is high.

The actual analysis is performed as follows: Place in the flask a weighed quantity of the carbonate to be analyzed, and fill the little bottle two-thirds full of acid. By means of the thread or wire lower the bottle carefully into the flask, leaving the upper end of the thread projecting from the flask. Press the cork into the neck of the flask so that the thread or wire may be held firmly between the cork and the glass, taking care not to spill any of the acid out of the bottle. Plug the outer end of the plain glass tube with a small ball of wax, or better, with a bit of rubber tubing closed with a short piece of glass rod, and leave the apparatus at rest during 15 or 20 minutes in a room of tolerably constant temperature before weighing or counterpoising it. Then hold the flask in an inclined position, so that the acid in the bottle may run out, little by little, upon the carbonate. The carbonic acid set free is dried more or less completely as it passes out through the chloride of calcium tube. — After the evolution of gas has ceased, remove the stopper from the narrow glass tube, attach to the latter an unweighed chloride of calcium tube, connect the permanent chloride of calcium tube with an aspirator, and draw a quantity of air through the flask. After all the gaseous contents of the flask have been swept out by the current of air, there still remains a certain quantity of carbonic acid dissolved in the liquid at the bottom of the flask. In order to remove this dissolved gas, disconnect the flask from the temporary chloride of calcium tube and aspirator, plug the narrow glass tube, and slowly heat the contents of the flask to incipient boiling. Then remove the plug, re-attach the chloride of calcium tube and the aspirator, and again draw a quantity of air through the flask. When all the carbonic acid has been swept out, disconnect the flask from the aspirator and temporary drying tube, plug the narrow glass tube, allow the flask to cool for half an hour in the room of constant temperature, replace it upon the balance and determine how much less it weighs than at first. The loss of weight represents the carbonic acid in the substance analyzed. — Good results are sometimes obtained in this way, but it is nevertheless a matter of common experience that the loss of weight of the apparatus cannot be counted as carbonic acid with certainty. If the dilute acid liquor in the flask be heated long enough and strongly enough to expel all the carbonic acid which is dissolved by it, some aqueous vapor will be driven out of the apparatus also, and lost. In most cases, where good results are obtained, the agreement is really due to the compensation of opposite errors, as may be seen by trying the experiment of repeatedly heating the flask of the apparatus and sucking air through it. If the suction be continued just long enough, the diminished weight of the apparatus will exactly correspond to the carbonic acid that was contained in the substance, but further exhaustion of the air will diminish the weight of the apparatus, not by complete removal of the carbonic acid, but by loss of aqueous vapor, which easily escapes through the dessicating material. By continued working on a carbonate of known composition, one may soon learn how long to exhaust in order to bring about the proper loss, but where the analyst is out of practice an error of 1 or 2 per cent is not unlikely to happen, and the process itself furnishes no means of judging when it will give a correct result. (S. W. Johnson, *Amer. Journ. Sci.*, 1869, **48.** 111). A far better method is to charge the apparatus with carbonic acid gas in the first place, and to weigh it full of carbonic acid as soon as the disengagement of gas has ceased. See below, *Johnson's modification.*

Choice of the decomposing acid. Usually the acid employed for decomposing the carbonate is either sulphuric, nitric, chlorhydric or oxalic. When the substance analyzed contains a fluoride, however, some weak, non-volatile acid, such as tartaric or citric acid must be employed to set free the carbonic acid, for if either of the mineral acids were used, some fluorhydric acid would be evolved. Sulphuric acid is to be preferred for the treatment of the carbonates of all those metals which form soluble sulphates, but is unfit for decomposing the carbonates of metals which form insoluble compounds with sulphuric acid, for the insoluble layer of sulphate formed by the action of the first portions of the acid would cover over a part of the carbonate, and prevent the acid from coming in contact with it. Nitric, or in some instances chlorhydric, acid must therefore often be substituted for the sulphuric acid.

When chlorhydric acid is used to decompose a carbonate, or in case the latter is contaminated with a chloride, the escaping carbonic acid should be made to pass through a tube filled with some substance, like anhydrous sulphate of copper, which has power to retain both water and chlorhydric acid. To prepare the sulphate of copper, boil fragments of pumice stone in a concentrated aqueous solution of blue-vitriol, and dry and heat the stone until the copper salt is completely dehydrated and has become white. A U-tube about 8 c. m. high and 1 c. m. bore filled with the sulphate will usually be found sufficiently large; it may be used as long as a third of its contents remain uncolored. (Stolba, *Dingler's polytech. Journ.*, **164.** 128). In case a carbonate fit to be treated with sulphuric acid happens to be contaminated with a chloride, the risk of evolving chlorhydric acid may be avoided by mixing the weighed carbonate with a quantity of a solution of sulphate of silver before adding the acid. So too, if the carbonate contains a sulphite or a sulphide, these salts may be oxidized by treating the weighed carbonate with a solution of yellow chromate of potassium before adding the acid.

In some cases it will be found advantageous to determine the metallic base of the carbonate, by Alkalimetry, in the solution left in the flask. In this event, mix the weighed carbonate with a weighed quantity of crystallized oxalic acid more than sufficient to decompose the whole of the carbonate, charge the tube with water, and at the proper time pour the water upon the mixture of acid and carbonate. After all the carbonic acid has been expelled from the flask, determine how much of the oxalic acid has been left free by titrating with an alkaline solution of known strength (see Alkalimetry). Instead of using solid oxalic acid, the tube might be charged with tolerably strong standard sulphuric, nitric, or chlorhydric acid. As thus modified, the process is peculiarly useful for determining the relative proportions of carbonic acid and base in unweighed quantities of moist, recently precipitated carbonates, and in carbonates which cannot be dried without suffering decomposition. (Stolba, *Journ. prakt. Chem.*, **97.** 312).

Application to Alkalimetry. The method now in question is sometimes employed as an alkalimetric process for determining the value of commercial carbonates of sodium or potassium in cases where the carbonates are strongly colored. For this alkalimetric determination a form of apparatus devised by Fresenius & Will is usually employed. Though similar in principle to the apparatus above described, the alkalimeter of these chemists is peculiar in some of its details. It may be readily made as follows:—Select two small, light Berlin flasks. Fit to each flask a cork or caoutchouc stopper, and bore two holes in each of the corks. Place the flasks upon a table, side by side, so that they shall almost, but not quite, touch one another. Call the flask at the left hand L, and the other R. Bend a glass tube at two right angles, in such manner that one leg of the tube may pass into the flask L, and the other into the flask R. Push one leg of the bent tube through one of the corks and the other through the other. Cut off one leg of the tube just below the cork of the flask L, but leave the other leg long enough to reach almost to the bottom of the flask R. Through the second hole of the cork in L pass a straight tube, open at both ends, long enough to reach almost to the bottom of the flask, and through the corresponding hole of the cork in R push a similar, but shorter, tube long enough to project a few m. m. below the cork. — The size of the flasks must be determined by the capacity and delicacy of the balance at the disposal of the operator. If the apparatus is to be weighed upon a delicate chemical balance, it may be made so light that when fully charged with acid it need not weigh more than 70 or 80 grammes; but as a general rule it will be found more convenient to use somewhat larger flasks and to operate upon a coarser balance.

To use the apparatus, plug the upright tube in the flask L, as directed in a previous paragraph; pour the weighed carbonate into L, and add to it enough water to fill about a third of the flask. Pour enough strong sulphuric acid into the other flask, R, to half fill it; re-place the corks and push them tightly into the mouths of the flasks. Let the apparatus stand 15 or 20 minutes in a room of tolerably constant temperature, and then counterpoise it upon a balance. Attach a piece of caoutchouc tubing to the upright tube in R, and proceed to test the tightness of the apparatus. To this end, suck out a small quantity of air from the flask R, so that a few bubbles of air may escape from L into R, through the sulphuric acid, to supply the partial vacuum which has been created in R. Then remove the lips from the caoutchouc tube, and suffer the external air to enter R. If the apparatus is tight, some sulphuric acid will be pressed up towards L, in the bent glass tube, and will remain stationary in the tube. Watch the column of sulphuric acid for a few minutes to see that it does not sink back into R, and then suck a quantity of air from R, somewhat larger than before. On removing the lips from the caoutchouc tube, a portion of the sulphuric acid in R is forced over into L, and decomposes a corresponding quantity of the carbonate therein contained. The carbonic acid set free passes through the sulphuric acid in R, and is thereby dried before escaping into the air through the open, upright tube. As soon as the effervescence in L slackens, suck out more air from R, so that a new quantity of sulphuric acid may be forced over into L, and repeat the operation at intervals until the whole of the carbonate is decomposed. When the effervescence has entirely ceased, force over a comparatively large quantity of the sulphuric acid into L, so that the contents of that flask may be considerably heated. After a few moments, remove the plug from the upright tube in L, attach to the tube a chloride of calcium tube, and by means of the caoutchouc tube at the other flask, suck air through the apparatus until this air no longer tastes of carbonic acid. Let the apparatus stand at rest in the room of constant temperature for 2 or 3 hours, replace it upon the balance, and bring the latter into equilibrium by means of weights. The sum of the weights added indicates the amount of carbonic acid which has been expelled from the substance analyzed.

$$\frac{\text{Molec. wt. of}}{CO_2} : \frac{\text{Molec. wt. of}}{Na_2CO_3} = \frac{\text{Wt. of } CO_2}{\text{found}} : \frac{\text{Wt. of } Na_2CO_3}{\text{in sample}}$$

Tolerably accurate results may be obtained by this method, unless the proportion of carbonic acid in the substance analyzed is very small.

To fit the apparatus just described for the analysis of lime-stone and the carbonates of other metals forming insoluble sulphates, replace the straight, upright tube in L by a tube which has been blown to a somewhat capacious bulb near the top, and drawn to a fine point at the lower end. The apparatus is charged as before with the exception that the bulb-tube is filled with dilute nitric acid, and that at the start the lower part of the tube is not pushed so deep into the flask as in the other case. Care must be taken that the point of the bulb-tube does not touch the water in L until after the apparatus has been counterpoised on the balance.

Like the straight tube of the apparatus previously described, the bulb-tube must be closed at the top with a moveable plug so that none of the nitric acid can flow out until the plug is loosened. The quantity of nitric acid taken must be more than sufficient to decompose the whole of the carbonate.

In case the metallic base of the carbonate is to be determined as well as the carbonic acid (see above, p. 82), a measured quantity of standard acid must be introduced into the bulb-tube. To this end, hold the tube point upwards, warm the point over a lamp, and touch it with a bit of tallow in such manner that a little of the tallow may solidify in the tube, and close it water tight. Then invert the tube, fill it through the upper opening with the acid, measured from a Mohr's burette with fine point, plug the upper opening in the usual way, and heat the point of the tube gently until the tallow is melted.

After the apparatus has been counterpoised carefully push down the bulb-tube until it touches the water, and at intervals loosen the plug so that small quantities of the nitric acid may flow out upon the carbonate. As soon as the carbonate has been completely decomposed, suck air through the apparatus as before, and finally heat the contents of the flask L to incipient boiling, in order to set free the last traces of the gas.

Application to Acidimetry. The volatility of carbonic acid has been made the basis, not only of a process of alkalimetry, as just described, but of a method of acidimetry as well. It is in fact easy to estimate the strength of any sample of acid, by weighing the carbonic acid set free from bicarbonate of sodium by the acid in question. The operation is as follows:— In the flask L of the apparatus just described, weigh out a quantity of the acid to be examined, and if it be concentrated, dilute it with water. In either case, the fluid should fill about a third of the flask. Fill a small glass tube compactly with bicarbonate of sodium or of potassium. Tie a thread to the tube and hang it in the flask L by pressing the thread between the cork and the neck of the flask. The quantity of bicarbonate taken must be more than sufficient to saturate the acid in the flask; the salt must be free from monocarbonate, but the presence of chloride or sulphate, etc., of sodium, does no harm. In other respects the apparatus is charged in the manner already described in the case where the upright tube in L is straight. After the apparatus has been counterpoised, loosen the cork of the flask L, so that the thread may slacken and let the tube of bicarbonate fall to the bottom of the flask, and at the same instant replace the cork as tightly as possible. Carbonic acid is given off violently at first, then for some time at a uniform rate, afterwards slowly until the evolution ceases. When no more gas is evolved, put the flask L in water of from 50° to 55° C, so hot that the finger cannot long be held in it, and when the renewed evolution of gas thus occasioned has ceased, suck air through the apparatus, as before. When practicable, enough acid should be taken to set free a gramme or two of carbonic acid.

In *Johnson's Modification* of the foregoing method of estimating carbonic acid, the apparatus is weighed full of carbonic acid, both before and after the experiment, so that there is no need of sucking out any gas from the flask. The dessicating material has consequently to dry only as much gas as is yielded by the substance subjected to analysis. The process thus modified is much better than before, and is decidedly to be preferred to the old method in all cases where the substance to be examined dissolves freely and completely in cold acid.

The apparatus may consist of a light, wide-mouthed flask or bottle, from 3 to 3.5 c. m. wide by 6 c. m. high, closed with a caoutchouc stopper, to perforations in which are fitted an upright chloride of calcium tube and the bent end of a tolerably capacious bulb-tube, which serves as a reservoir of acid. A good chloride of calcium tube may be made by blowing two oblong bulbs, each 2.25 c. m. broad, on a glass tube. When finished, the tube may be about 10 c. m. long. The lower bulb is filled with cotton, the upper with small fragments of porous chloride of calcium, and the top of the tube is closed with a rubber connector, plugged with a bit of glass rod. The acid reservoir may be made by blowing an oblong bulb about 3 c. m. wide by 5 c. m. or more long, on a tube of about 7 m. m. internal diameter, and bending the tube in such manner that while the bottom of the bulb, or rather the bent tube immediately adjacent to the bottom of the bulb, shall rest upon the table, close beside the flask, there shall be a short, bent delivery tube to connect the upper, narrower end of the bulb with the rubber stopper, and at the other end of the bulb, a tube pointing upwards and reaching to the level of the top of the flask; this longer tube serves for the introduction of carbonic acid at the beginning of the experiment. The extremity of the upper or delivery tube reaches fairly through the rubber stopper; it is ground off obliquely, so that no drop of liquid can be held in it. At the top of the tube for introducing carbonic acid there is a short rubber connector with a glass-rod plug.

The actual analysis of carbonate of calcium, for example, is made as follows;—Put from half a grm. to a grm. of the substance, best in the form of small fragments, into the flask. Fill the bulbed acid reservoir nearly full of chlorhydric acid of 1.1 sp. gr. Fit the rubber stopper, with its appurtenances, tightly into the mouth of the flask; remove the glass-rod plugs, connect the upright tube with a self-regulating carbonic acid generator, and pass a rather rapid stream of washed carbonic acid through the apparatus during 15 minutes, or until the acid in the bulb is saturated and all the air in the flask has been displaced by the gas. Then plug the top of the chloride of calcium tube, disconnect the apparatus from the source of carbonic acid, plug the upright tube and immediately weigh. Take care to handle the apparatus carefully so that its temperature shall not be changed by the warmth of the hands. After weighing, loosen the plug of the chloride of calcium tube, and, holding the flask by a wooden clamp, incline it so that the acid may flow over upon the carbonate. The decomposition should proceed slowly, so that the escaping gas may be thoroughly dried. As soon as the whole of the carbonate has dissolved, replace the plug in the chloride of calcium tube and weigh the apparatus. — All the joints of the apparatus must be gas-tight; should there be any leak it will be made evident at the final weighing, by a slow but steady loss of weight as the apparatus stands upon the balance. If all the joints are sufficiently tight, the weight will remain the same for at least 15 minutes. Since the temperature of the flask usually rises a little during the solution of the carbonate, it is best, after the decomposition of the latter is completed, to plug the chloride of calcium tube and leave the apparatus at rest for 15 minutes; then connect with the carbonic acid generator as before, and pass dried carbonic acid for a minute before weighing.

When properly executed, the process gives extremely accurate results. It is essential, however, that the operation and the weighings be conducted in an apartment not liable to sudden changes of temperature. A slight change of temperature or of atmospheric pressure between the two weighings greatly impairs the results, or renders them worthless.

In the case of alkali-metals which absorb carbonic acid gas, a small bottle of thick glass and wider mouth should be used in place of the flask above described. The stopper of this bottle should have three holes, the third to carry a narrow tube 3 or 4 inches long enlarged below to a small bulb, to hold the carbonate. This bulb must be so thin that on pushing down the tube within the bottle, it shall be easily crushed to pieces against the bottom of the latter. The carbonate is weighed into the bulb-tube, the latter is wiped clean, corked and fixed in the rubber stopper. The apparatus is filled with carbonic acid and weighed. The bulb is then broken and the process proceeded with precisely as above described. (S. W. Johnson, *Amer. Journ. Sci.*, 1869, **48.** 111, and the American edition of *Fresenius's Quant. Anal.*, New York, 1870, p. 202. The apparatus is figured in both places).

2. *By measuring the carbonic acid gas.* The process is applied to the determination of carbonic acid in soils, manures, mortars, marls, bone-black and the like. Also in the analysis of mineral waters and blood. Like the foregoing method it may be used for estimating the alkalimetric value of the carbonates of potas-

sium and sodium. For the determination of small quantities of carbonic acid, the method by measuring is thought by some chemists to be preferable to either of the other methods.

As described by Pfaff (*Handbuch analyt. Chem.*, 1824, **1.** 505; **2.** pp. 18, 77), the process consisted in putting small fragments of the carbonate into a graduated tube charged with chlorhydric acid, and placed, full of mercury, in a mercury trough. In case the substance to be analyzed was in the state of powder, it was done up to little pellets in paper which had been previously leached with chlorhydric acid. Due allowance was made for temperature and barometric pressure. — The more modern methods are as follows:—

A. *Method of Schulze.* This method may be used for estimating the proportion of carbonate of calcium, or any other carbonate, in rocks, soils and manures, and in general for all cases where the substance to be analyzed is bulky, the volume of gas to be evolved is large, or the evolution of gas attended with frothing. The apparatus required consists of two glass bottles, one for decomposing the carbonate and the other for collecting the gas set free, a tubulated retort to hold the acid by which the decomposition of the carbonate is to be effected and a small metallic water-tank to cool the collecting bottle. — Heat the upper part of the neck of a small tubulated retort in the flame of a blast lamp, and bend the softened glass in such manner that the neck of the retort shall point almost perpendicularly downwards. Obtain a stout, two-necked bottle of about 350 c. c. capacity, for the decomposing jar, and an aspirator-flask of 1 or 1.5 litres capacity, for the collecting bottle. One of the orifices of the decomposing bottle must be tolerably wide, and thick and strong in glass, while the other opening is made in the form of a plain projecting tube. To the upper orifice of the collecting bottle, and to the wider orifice of the decomposing bottle, fit perforated ground glass stoppers. Cement the neck of the bent retort into the perforation of the stopper of the decomposing bottle, and a short, straight glass tube into the perforation of the stopper of the collecting bottle. Connect the two bottles by means of a caoutchouc tube, one end of which is tied to the projecting neck of the decomposing bottle and the other to the straight glass tube in the stopper of the collecting bottle. Place the collecting bottle in a zinc cylinder as tall as the bottle and provided with a lateral opening near the bottom corresponding to the orifice in the side of the bottle. Close the side opening of the bottle with a cork carrying a glass tube bent at a right angle, the longer limb of which projects upwards a little higher than the top of the bottle. The bore of this tube should be about 4 or 5 m. m. in diameter excepting the point, which is narrowed to 2 m. m. By means of a short piece of caoutchouc tubing tied to the lateral tubulures of the collecting bottle and the zinc cylinder, the latter is made tight; it is then filled with water. The collecting bottle is also filled with water, together with enough fatty oil to form a thin layer upon the surface of the water.

In an actual analysis, a weighed quantity of the substance, such as marl or lime stone, to be examined, is placed in the decomposing bottle. The retort is three-fourths filled with dilute chlorhydric acid of 1.05 or 1.07 sp. gr., and the stopper to which the retort is cemented is carefully greased and inserted in the neck of the decomposing bottle. The apparatus is left in a room of tolerably constant temperature until the retort and decomposing jar, as well as the water in the zinc cylinder, have acquired the temperature of the air of the apartment. The stopper of the collecting jar is then smeared with tallow and inserted in its place, and the glass tube in the side orifice of the collecting jar is turned perpendicularly downwards. A beaker is placed beneath this tube to catch the water which flows from the collecting bottle. If the apparatus is tight, only a small quantity of water will flow from the bottle when the tube is turned down. Incline the decomposing bottle to such an extent that acid may slowly flow from the retort upon the carbonate in the decomposing bottle. The carbonic acid evolved will pass over into the collecting jar, while water will flow out of the latter into the beaker to make room for the gas. As soon as the decomposition is finished, prop up the beaker beneath the tube of the collecting jar so that the mouth of the tube shall be immersed in the water, and set the decomposing bottle in a jar of water of the same temperature as that in the zinc cylinder. As soon as all parts of the apparatus have acquired a common temperature, remove the beaker of water, turn the water delivery tube of the collecting jar perpendicularly upwards, take out the stopper of the retort for a moment, again turn the water tube downwards, and collect by itself the small quantity of water which flows from the collecting jar. Measure this water and note the quantity. Measure also the water which was previously collected in the beaker. The difference between the two quantities will give the volume of the carbonic acid under the existing conditions of temperature and barometric pressure,—subject also to a correction for the tension of aqueous vapor at the temperature in question.

The weight of the carbonic acid obtained may readily be calculated from the volume, by referring to the known specific gravities of carbonic acid and air, or it may be determined still more simply by decomposing a known quantity of pure carbonate of calcium in the apparatus. The weight of the volume of carbonic acid obtained from the pure carbonate being known, *a priori*, this volume may be directly compared with any other volume of the gas. By operating in this way, moreover, several slight sources of error incidental to the method of experimenting may be avoided. It is not even necessary to use pure carbonate of calcium for the comparative experiment;—almost any sample of marble whose percentage of carbonic acid has been carefully determined will answer as well. After the value of a marble has once been determined for this purpose a considerable store of the material had better be kept on hand.

The process is simple and is said to yield exceedingly accurate results.

Precautions. The oil employed to cover the water in the collecting bottle must be fresh and indisposed to rancidity. In practice it is found that the oil absorbs carbonic acid so slowly that the loss of gas from this cause during the time required for an experiment is scarcely appreciable. — Care must be taken not to warm the collecting bottle with the hand when the stopper is placed in the bottle. — The chlorhydric acid employed had better be of 1.05 or 1.07 sp. gr., since a stronger acid would affect the tension of the aqueous vapor in the apparatus. — The control experiment with marble or pure carbonate of calcium is of special value in cases,—such as the analysis of artificial manures, soils, peat-ashes, and the like,—where the proportion of carbonic acid in the substance is so small that a comparatively large quantity of liquid has to be mixed with the substance in order that the carbonate contained in it may be wholly decomposed. The liquid of course retains a quantity of carbonic acid in solution, but by repeating the experiment with marble which has been mixed with a quantity of inert sand and as much water as was used in analyzing the substance, this source of error may be so much diminished that accurate results may still be obtained with materials containing no more than 0.5 per cent of carbonate of calcium. In operating with these earthy materials, the decomposing jar must be freely shaken in order that carbonic acid may escape, and the control experiment with the marble must be subjected to the same amount of agitation. The quantity of carbonic acid retained by the water can, for that matter, be determined for any special series of experiments and allowed for, or the carbonic acid may be displaced from the water by means of hydrogen. To this end mix with the carbonate to be analyzed a weighed quantity of zinc powder, each gramme of which has been found capable, by previous experiments, of evolving a certain number of c. c. of hydrogen, when treated with chlorhydric acid. On pouring chlorhydric acid from the retort into the decomposing flask the carbonate will first be decomposed before much of the zinc is acted upon, so that hydrogen will continue to be set free in the liquid long after the carbonate has been completely decomposed. If enough zinc be used, the hydrogen will sweep forward into the collecting jar almost all the carbonic acid which would otherwise have been retained by the liquor. The volume of hydrogen equivalent to the zinc employed is subtracted from the measured volume of carbonic acid and hydrogen. — This method of displacement by hydrogen becomes somewhat more complicated in certain cases, particularly in presence of nitric and oxalic acids, ferric oxide and other substances capable of preventing the evolution of hydrogen. Soils, for example, not only contain substances which combine with and retain nascent hydrogen, but the zinc powder becomes so completely enveloped by the earthy particles that a long time is required in order that the whole of the metal may dissolve. In order to displace carbonic acid by hydrogen therefore, in analyzing soils, manures and the like, it is necessary to treat the zinc with acid in a special vessel and to lead the free hydrogen through the solution of carbonic acid.

A convenient way of arranging the apparatus is to interpose a second two-necked bottle between the decomposing and collecting bottles The orifices of this second bottle are closed with perforated caoutchouc stoppers, through one of which a glass tube passes to the bottom of the bottle. The upper end of this glass tube is connected, by means of a caoutchouc tube, with the outlet of the decomposing bottle. A short, straight glass tube placed in the perforation of the other stopper is connected, by means of another flexible tube, with the top of the collecting bottle. The substance to be analyzed is placed in the new two-necked bottle together with a test tube full of chlorhydric acid of 1.05 sp. gr., while the weighed zinc powder is placed in the old decomposing bottle. After the apparatus has been proved to be tight, the bottle which contains the carbonate is inclined so that the acid in the test tube may flow out into the bottle, and, finally, acid is made to flow from the retort upon the zinc powder. When no more hydrogen is evolved, lift the long tube in the new two-necked bottle, out of the liquid, bring the apparatus to a common temperature and measure the water in the beaker as before: (F. Schulze, *Zeitsch. analyt. Chem.*, 1863, **2.** 289).

B. *Method of L. Meyer*, for determining carbonic acid in mineral waters and in blood. In order to remove carbonic acid (and other gases) from a liquid, Meyer employs a modification of the apparatus of Ludwig[1] based upon the principle of the Torricellian vacuum.

This apparatus may be roughly described as follows:—A stout glass vessel of about 700 c. c. capacity, filled with mercury, is attached to a long vertical tube, also full of mercury, in such manner that on allowing mercury to flow from the bottom of the tube a vacuum can be formed

[1] See Setchenow, *Wien. Akad. Bericht.*, **36.** 293; Schœffer, ibid., **41.** 589, and Heidenhain, *Zeitsch. analyt. Chem.*, 1863, **2.** 120.

in the vessel. The flask which contains the mineral water is connected with the glass vessel by means of a flexible tube closed with a screw compressor, so that on opening the latter the gases contained in the water will be given off to supply the vacuum. As soon as any considerable quantity of gas has escaped from the water the compression cock is again closed and the gas forced over into a collecting tube, standing in a pneumatic trough beside the vacuum jar, by pouring mercury into the latter. The vacuum is then re-established as before, and the compression cock again opened so that a new quantity of carbonic acid may exhale from the water; the glass vessel is thus alternately filled and emptied of mercury as long as any gas continues to escape from the water. The flask which contains the water is heated after a while to facilitate the escape of carbonic acid, and at a certain stage of the process tartaric acid is added to the water, in order to expel any carbonic acid which may have been previously held in combination by a base.

The proportion of carbonic acid in the gas thus set free is then determined by the usual gasometric method described under Hydrate of Potassium.

For details of the method employed for collecting and analyzing the blood of animals the reader is referred to the article of Heidenhain (*Zeitsch. analyt. Chem.*, 1863, **2.** 122). With mineral waters Meyer proceeds as follows:— The water is collected at the spring in ordinary bolt-heads of about 1 litre capacity, whose throats have been drawn out to the width of about 1 c. m. The flasks may be closed securely with thick caoutchouc tubing which has been freed from adhering sulphur by boiling with soda lye or sulphydrate of ammonium, and afterwards soaked in fat upon a water bath. The walls of this tubing should be at least 5 m. m. thick; after the treatment with fat the tubing should be washed first with alcohol, then with water, and dried. — Before forcing the tubing over the neck of the flask the latter should be smeared with a hot mixture of fat and unvulcanized caoutchouc. The rubber tube is then tied tightly to the glass with fine, soft iron wire, and the flask sunk in the water of the spring. When a sufficient quantity of water has run into the flask, the caoutchouc tube is closed under water with a screw compressor, a glass rod smeared with the mixture of fat and rubber is pushed into the outer end of the tube and the tube is tied to the rod with wire, as before. In flasks thus closed, mineral water can be kept for days or weeks without loss of gas. — In order to attach one of these flasks to the vacuum apparatus, take out the glass rod and put in its place a bulb-tube about 10 c. m. long, of 20 or 30 c. c. capacity, attach a second caoutchouc connector to the other end of the bulb-tube, and tie it to the orifice of the glass vessel in which the vacuum is produced. This bulb-tube serves to receive the water which is expelled from the flask through the expansion of the liquid by heat, and also for the reception of the tartaric acid which is added to expel the combined carbonic acid. After the water flask has been connected with the glass vessel, a vacuum established in the latter, and the compression cock opened, the greater part of the carbonic acid escapes from the water immediately with considerable violence. As soon as the first portions of the escaped gas have been forced over into the collecting tube the flask is heated on a water bath, gently at first, but afterwards more strongly, until no more gas is evolved from the water. The collecting tube is then removed and a solution of tartaric acid, which has previously been freed from air by exposure in the vacuum apparatus, is drawn into the large glass vessel through the delivery tube which connects this vessel with the collecting tube and pneumatic trough. By opening the compression cock of the water flask the tartaric acid is then admitted to the water, and the last portions of the carbonic acid are thereby expelled. (Lothar Meyer, *Zeitsch. analyt. Chem.*, 1863, **2.** 237).

C. *Method of E. Dietrich.* Applicable to the determination of carbonic acid in cements, bone-black, precipitates and minerals.

The apparatus required consists of a small, wide-mouthed glass bottle, in which to generate the carbonic acid, and two narrow glass tubes, by means of which to measure the gas. The two tubes are of the same length and caliber, and are held in a vertical position upon a table by means of appropriate rods and clamps. Each of the tubes is capable of holding a little more than 100 c. c. The tube nearest the decomposing bottle (called the measuring tube) is graduated to fifths of cubic centimetres, but the other tube has no graduations,—it is a mere pressure tube. Both tubes are cemented at the bottom into iron caps, provided with short, projecting, vertical pipes, to which caoutchouc tubing may be tied. The top of the graduated tube is in like manner cemented with shellac into an iron cap which has a lateral as well as a vertical projection for the attachment of caoutchouc tubes. But the top of the pressure tube is left open, or only partially closed with a perforated cork. The measuring tube is fastened to its supports in a fixed and immovable position, but the pressure tube is so arranged that it may be readily pushed up or down in a vertical plane.

The two glass tubes are connected at the bottom by means of a caoutchouc tube which must be as long as either of the glass tubes. To free the vulcanized rubber from sulphur, boil it for some time in moderately dilute potash lye. A perforated caoutchouc stopper, carrying a straight glass tube, is fitted to the decomposing bottle, and the bottle is put into connection with the measuring tube by means of a caoutchouc tube, one end of which is tied to the pipe in the stopper of the bottle, and the other end to the lateral projection in the iron cap of the measuring tube. A short piece of caoutchouc tubing, carrying a screw compressor, is attached to the vertical pipe at the top of the measuring tube, so that the tube may be opened or closed at will by turning the screw.

Enough quicksilver is poured into the vertical tubes to fill one of them completely, together with the caoutchouc pipe which connects the two tubes. To prove the tightness of the apparatus, place the stopper in the decomposing bottle, open the screw compressor at the top of the measuring tube, push up the pressure tube until the mercury has risen to the top or 0° mark of the graduated scale, and clamp the pressure tube in that position. Then close the screw compressor on the graduated tube and lower the pressure tube as far as possible. The column of mercury in the measuring tube will instantly fall a little when the pressure is thus removed from it, but if the apparatus be tight the level of the mercury will afterwards remain unchanged as long as the temperature and the pressure of the air remain constant.

In an actual analysis, place a weighed quantity of the carbonate to be examined in the decomposing bottle, together with a small glass tube or bottle two-thirds filled with dilute chlorhydric acid. Cork the bottle tightly, fill the measuring tube with mercury to the zero mark, close the screw compressor, lower the pressure tube and tip the decomposing bottle so that the acid in the tube within it may flow upon the carbonate. When the evolution of gas has ceased, and all parts of the apparatus have again acquired a common temperature, lift the pressure tube until the mercury stands at the same level in it and in the measuring tube, and note the volume of gas in the latter. Correct the observed volume for temperature, barometric pressure, and the tension of aqueous vapor, and calculate the weight of the corrected volume from the known sp. grs. of carbonic acid and air. For Dietrich's table, giving the weights of cubic centimetres of CO_2 under barometric pressures ranging from 720 to 770 m. m., at temperatures from 10° to 25°, see *Zeitsch. analyt. Chem.*, 1865, **4.** 142.

The chief source of error to be taken into account depends upon the fact that a considerable quantity of carbonic acid remains dissolved in the liquid in the decomposing bottle. The quantity of carbonic acid thus dissolved is found to be greater in proportion as the quantity of gas evolved is greater. Thus in a case where 92.8 c. c. of carbonic acid gas were developed, 4.6 c. c. of carbonic acid remained dissolved in the liquid in the decomposing bottle, while in another case where 0.9 c. c. of gas were set free only 0.3 c. c. remained dissolved in the same quantity of liquid. Dietrich (see *Zeitsch. analyt. Chem.*, 1864, **3.** 166 and 1865, **4.** 145) has drawn up tables of corrections to be applied in compensation for this source of error. — As soon as the evolution of carbonic acid ceases, the decomposing bottle should be placed in water of the same temperature as that of the air of the apartment, in order to remove the heat developed by the chemical action. Or, better, the decomposing bottle may be kept in water during the course of the experiment, and, if need be, the gas may be made to flow through a leaden worm sunk in water with the decomposing bottle. — In case more than 100 c. c. of carbonic acid are to be given off from the sample of carbonate under examination, the gas must be measured by portions. To this end, pour the chlorhydric acid, little by little, upon the carbonate, in such manner that the evolution of gas may cease before the measuring tube is completely filled with it. Then slip a screw compressor over the rubber tube which connects the decomposing bottle with the measuring tube, close the tube tightly, and measure the gas which has been evolved; discharge the measured gas, refill the measuring tube with mercury, loosen the compressor attached to the tube of the decomposing flask, and pour another portion of chlorhydric acid upon the carbonate which still remains undecomposed. When a large quantity of carbonic acid is developed in this way, the error depending upon the absorption of gas by the liquid in the decomposing bottle may be readily allowed for. It will be sufficient to measure the chlorhydric acid used, to estimate how much carbonic acid would be absorbed by a similar volume of water at the temperature in question, and add this volume to the measured volume of the evolved gas. Up to 300 or 400 c. c. the errors incidental to measuring do not materially affect the accuracy of the results. (E. Dietrich, *Zeitsch. analyt. Chem.*, 1864, **3.** 162 and figure).

D. *Method of Rumpf.* Seeking to simplify the method of Dietrich, Rumpf (*Zeitsch. analyt. Chem.*, 1867, **6.** 398) has constructed an apparatus which may readily be made from materials to be found in every laboratory. This apparatus consists simply of an ordinary Mohr's burette of from 30 to 50 c. c. capacity, a hydrometer jar, test tube, and small, wide-mouthed bottle, together with a cork and connectors of glass and rubber tubing. The carbonate to be analyzed is decomposed in the bottle and the evolved gas measured in the burette which is inverted for the purpose in mercury contained in the hydrometer jar.

To construct the apparatus, fit a caoutchouc stopper, having three holes, to the decomposing bottle; fix a thermometer in the middle hole of the stopper, and a short straight glass tube in each of the other holes. To one of the projecting glass tubes tie a short rubber connector, and close it with a spring clip or screw compressor. To the other projecting tube tie one end of a thick rubber tube, the other end of which has been previously tied to the point of the burette. — Put a weighed quantity of the carbonate to be analyzed into the decomposing bottle together with a short test tube two-thirds filled with a measured quantity of chlorhydric acid. Replace the stopper, open the spring clip, and sink the burette in the mercury in the hydrometer jar until the last division mark upon the burette is level with the surface of the mercury. Close the spring clip and tip the bottle so that acid may flow out of the test tube upon the carbonate. When the evolution of gas has ceased, take hold of the rubber covered point of the burette, and push the latter into the mercury until the surface of the metal is at the same level inside and without the tube. Read off the number of c. c. of gas in the burette; take the height of the thermometer and barometer and proceed to calculate the weight of the gas, taking care to allow for the tension of the aqueous vapor contained in the gas, and for the volume of gas which remains dissolved in the liquid in the decomposing bottle. (Compare the description of Dietrich's method, above).

E. *Method of Scheibler.* Employed extensively in Germany by sugar manufacturers, for estimating the proportion of carbonate of calcium in bone-black. The process is applicable to the analysis of any carbonate which can be decomposed by cold chlorhydric acid, is easy of execution and yields very accurate results. The rather elaborate apparatus required may be obtained of the dealers in German chemical wares. For a figure and detailed description of it see *Fresenius's Quantitative Analysis*, London, 1865, p. 712, or either of the later editions. The method here alluded to must not be confounded with another process devised by Scheibler for estimating carbonic acid in a mixture of gases by absorption in potash lye. (See Hydrate of Potassium).

F. *Method of Russell.* Applicable to the determination of carbonic acid in carbonates, and to the valuation of oxides, such as binoxide of manganese, which give off carbonic acid when mixed with oxalic and sulphuric acids. — The apparatus required differs but little from that devised by Williamson & Russell for the analysis of gases (see *Proceedings of London Royal Soc.*, **9.** 218; further, *Journ. London Chem. Soc.*, 1868, **21.** 128). The latter consists simply of a tubular pneumatic trough, a eudiometer tube and a pressure tube with clamp-rods to hold the tubes in place.

According to Russell, the pneumatic trough may be made of sheet gutta percha half an inch thick. Seen from above this trough is a pear-shaped box, 6⅛ inches long, in the clear, by 3⅝ inches wide at the broadest part. The sides of the trough are 3.5 inches high, and at its centre there is a well in which the eudiometer and pressure tube may be sunk at will. The broader part of the trough is made circular so that there may be placed in it a tall glass sleeve large enough to envelope both the eudiometer and the pressure tube. During an experiment this outer sleeve is kept full of water, in order that the contents of the eudiometer may be maintained at a constant temperature. The opening of the well is 2.5 inches long by 1⅛ inches broad. One part of the well, devoted to the pressure tube, is 14 inches deep, measured from the bottom of the trough, while the part reserved for the eudiometer has a depth of 19 inches. The narrower part of the trough, not occupied by the water-sleeve has a depression or channel in the middle. This channel is ⅝ of an inch wide and its depth gradually increases from the edge of the trough to 1.75 inch at the side of the well. (See figure in *Journ. London Chem. Soc.*, 1868, **21.** 129). The eudiometer is similar to that of Bunsen (see his *Gasometry*), viz., a straight glass tube closed at the top, some 500 or 600 millim. long by 20 millim. in diameter. It is provided with platinum wires at the top and graduated and calibrated according to Bunsen's plan. The "pressure tube" is simply a straight glass tube of about the same diameter as the eudiometer and closed at the top. Its purpose is to hold a constant quantity of air to be used as a standard of comparison in measuring any gas in the eudiometer. To prepare the pressure tube for use, a very small drop of water is placed at the closed end, and mercury then poured into the tube until most of the air has been displaced. The tube is then inverted in the mercury trough and the height of the mercury in the tube marked off once for all. After this has been done, any effect which the rise or fall of either barometer or thermometer would produce on the bulk of the air in the tube may be exactly counteracted by raising or lowering the tube in the mercury-trough until the mercury again comes to the mark. Before filling the eudiometer with mercury a drop of water is placed in it so that the gas to be measured may always be saturated with aqueous vapor. Any alteration of temperature in the water of the sleeve that surrounds the tubes which may occur during an analysis will not affect the accuracy of the measurements, provided both tubes have undergone the same change of temperature. To ensure an even temperature, the water in the sleeve is agitated from time to time with a wooden stirrer. Before proceeding to measure a gas in the eudiometer the gas is brought to exactly the same tension as that of the standard volume of air in the pressure tube. To do this the two tubes are placed side by side and the eudiometer is raised or lowered until the column of mercury within it is of the same height as that which is required to bring the air in the pressure tube to the original volume. For details relating to the management of the apparatus, see Williamson & Russell's paper in *Journ. London Chem. Soc.*, [2.] **2.** 238.

For estimating carbonic acid or any other gas by measurement, the eudiometer above described is replaced by a somewhat larger graduated tube. This "measuring tube" may be 29 inches long and rather more than 0.75 inch in internal diameter. At the top it is drawn out to a comparatively narrow tube, which is bent at a right angle and left open. The bent part is about half an inch long. A piece of caoutchouc tubing 3.5 inches long is pushed over the projecting bent tube as far as it will go, then bound tightly to the glass and cemented all around with marine glue. The glue is applied as hot as possible so that it may adhere firmly to the glass and make a perfect joint. The rubber tubing should be very thick, the outside diameter being as much as half an inch, while the bore of the tube is only $\frac{1}{8}$ or $\frac{1}{6}$ of an inch. The inside of the tube should be well vulcanized, but the rest of the tube only slightly so.

The carbonate of calcium and acid, or other materials from which the gas is to be generated, may be placed in a small generating flask made by blowing a bulb at the end of a glass tube. The diameter of the glass tube should be such that the rubber tube above described will just fit into it. The rubber tube thus forms a sort of stopper to the flask and connects it in the simplest way with the measuring tube. The manipulations are very simple. A drop of water is placed in the measuring tube and the latter, with the flexible tube attached to it, is put in its position within the water sleeve and clamped, just as a eudiometer would be. To fill the tube with mercury, attach a piece of strong glass tubing, about 2 feet long and slightly tapering at one end, to the flexible tube and sink the measuring tube as far as it will go into the well of the trough. It is not necessary that the well be deep enough to receive the whole of the tube, for by attaching temporarily a thick rubber tube to the end of the glass tube which reaches above the top of the sleeve and applying the mouth to the rubber tube, it is easy to suck up the mercury to a height of several inches. When the mercury has thus been drawn up to within 2 or 3 inches of the top of the mercury tube, pinch the rubber tube for a moment with thumb and finger, and plug it at the end with a bit of glass rod. Elevate the measuring tube so that the flexible tube attached to it shall project above the top of the sleeve, and close this flexible tube with a screw compressor. Withdraw the thick glass tube from the flexible tube, insert the end of the latter into the neck of the generating flask, and loosen the screw compressor so that the flask and the measuring tube may communicate freely one with the other. Bend the flexible tube so that the flask shall be immersed in the water of the sleeve, and fasten the flask in this position by means of a rubber ring slipped over the clamp-rod which holds the measuring tube in place. As soon as the flask and its contents have acquired the temperature of the surrounding water, proceed to measure the volume of air in the flask and tube, as follows:—Lower the measuring tube into the well until the column of mercury in the tube is of the same height as that in the pressure tube, and note the volume of gas in the measuring tube. This reading gives the amount of air in the apparatus before any of the gas to be estimated has been evolved. Again elevate the measuring tube so that the flask may be brought outside the sleeve, and tip the flask so that the ingredients within it may be mixed and made to react upon one another. The flask may be heated at this stage if need be. A second measurement after the reaction in the flask is completed gives the volume of gas produced. To prevent the substances in the flask from coming in contact out of season, one of them is placed in a small tube, which is lowered into the flask in the usual way, and care is taken not to tip the flask prematurely.

Precautions. To avoid leakage the rubber tube which connects the flask with the measuring tube must be slightly wetted. So long as this joint is kept wet it remains perfectly tight, but if it become dry and there is a high column of mercury in the tube it will probably leak. In order to ensure the presence of moisture, a piece of wickyarn is wound round the joining of the flask and rubber tube and the ends of the yarn allowed to dip into the water of the sleeve. It has been proved, by direct experiment, that the varying amount of bending which rubber tubing of the prescribed thickness will undergo when the flask is sunk in the sleeve does not appreciably alter the volume of air in the measuring tube.

It is important that the caoutchouc tube be as short as possible lest, in determining carbonic acid, a considerable error arise from the absorption of the gas by the caoutchouc. In one experiment in which the carbonic acid from 0.124 grm. of marble was collected in a glass tube, to the top of which was attached a piece of the thick rubber tubing some 8 or 9 inches long, it was found that although there was some air in the tube, as well as the carbonic acid which occupied 201.71 divisions, the gas at the end of one hour had diminished 7.7 divisions, at the end of 2 hours, 9.9 divisions, 3 hours, 11.5 divs., 4 hours, 13.3 divs., and by the next day, 30.3 divisions. With a short piece of the rubber tubing, such as that above described, the amount of gas which disappears is about 0.1 of a division of the tube every five minutes. — In an ordinary gas analysis it is sufficient if we know the relative volumes of gas at the different stages of the analysis; but for the present purpose the necessary data for converting these relative volumes into absolute volumes, and for ascertaining their weight, must be determined. The measuring tube is calibrated in the usual way, and a table of volumes drawn up similar to one for an ordinary eudiometer. The volume of mercury which has been used in the calibration is weighed and the temperature noted. The value of one volume in the table, expressed in cubic centimetres, is then determined by the formula:—

$$C = \frac{w \times (1 + 0.0001815\,t)}{13.596\ V}$$

in which w is the weight and t the temperature of the constant quantity of mercury which occupies the volume V, used as the standard in the calibration. 0.0001815 is taken as the coefficient of expansion of mercury, and 13.596 as the specific gravity of that metal at 0° C. It is necessary, moreover, to know the temperature and pressure at which the measurement has been made, in order to find the weight of the gas. When the measurements are made by means of a pressure tube, it is only necessary to ascertain once for all what is the temperature and pressure, at the same moment, of the air in the upper part of the pressure tube; the mercury standing exactly at the mark which indicates the constant volume. When the weight of dry gas occupying one tabular volume of the measuring tube has once been calculated from these data, it is only necessary to multiply any volume of that gas, which may be measured in the tube, by this constant, in order to obtain the weight of the gas.

It has been found in practice that the error which might arise from the retention of carbonic acid, dissolved in the liquid of the flask, may be avoided almost entirely by employing only a slight excess of the dilute acid. The process is liable to another source of error, inasmuch as the tension of the gas in the measuring tube may be slightly altered when a little of the acid is carried over into the tube. But it is found, when operating with a flask of about 120 c. c. capacity, that the alteration of tension is very slight unless the reaction in the flask has been unusually violent. It is well, for that matter, to place a loose plug of cotton-wool in the neck of the flask.

The process yields accurate results, is rapidly executed, and requires only small quantities of material. (Russell, *Journ. London Chem. Soc.*, 1868, **21.** 310).

3. *By noting the weight of the residue, or by analyzing the residue.* Dissolve the carbonate in a slight excess of chlorhydric acid, or other suitable acid, evaporate the solution to dryness, weigh the residue, and calculate the weight of carbonic acid which is equivalent to it. The method is applicable only to the analysis of carbonates free from impurities, and capable of forming fixed and definite chlorides, or other salts. Instead of weighing the dry chloride, the amount of chlorine contained in it may be determined as Chloride of Silver, by titration, and from the weight of chlorine found, the equivalent quantity of metal in the residue, and of carbonic acid in the substance analyzed, may be calculated. According to Mohr (*Titrirmethode*, 1855, **2.** 58) the last named process may be usefully applied in analyzing alkaline carbonates which contain chlorides, and mixtures of carbonates of sodium and chloride of sodium in mineral waters. The chlorine proper to the solution is first determined by titration with nitrate of silver, the mixture is then supersaturated with chlorhydric acid, evaporated to dryness, and again titrated to estimate the chlorine which has taken the place of the carbonic acid.

4. *By absorbing the gas in an alkali.* [Compare Principle I]. This method is convenient of execution, and yields exceedingly accurate results. It is, perhaps, better suited than either of the others for the ordinary requirements of the laboratory. The process may be conducted as follows:—Place a flask of about 300 c. c. capacity, on a piece of wire gauze laid upon a tripod or ring-stand above a lamp, and fit to the flask a caoutchouc stopper provided with two perforations. To one of the holes in the stopper fit a glass tube something more than twice as long as the flask is high; bend the tube twice, near the middle, into the form of an S, in such manner that the bent portion may be wholly outside the cork, while enough of the tube is left straight, at either end, to reach almost to the bottom of the flask; blow a small bulb upon the tube at the middle of the bent portion, *i. e.*, at the centre of the S. To the other hole in the stopper fit a short glass tube, bent at a right angle, to serve as a gas-delivery tube. It is well to blow a bulb on the upright part of the delivery tube, outside the cork, and to grind off the lower extremity obliquely so that drops of water may fall from it. Bring the outer end of the delivery tube into line with a series of absorption tubes arranged as follows:—1st, a bulbed U-tube full of chloride of calcium; 2d, a very small U-tube filled with fragments of glass moistened with 8 or 10 drops of concentrated sulphuric acid and loosely plugged with asbestos at either end; 3d, a U-tube charged with soda-lime and a little chloride of calcium in the ordinary way, and 4th, a U-tube, the inner limb of which is charged with chloride of calcium, and the outer limb with soda-lime or hydrate of potassium. The purpose of this last tube is merely to protect the 3d tube from the moisture and carbonic acid of the air; to the outer end of it attach a piece of rubber tubing 5 or 6 inches long. The 1st tube, which should be comparatively large, is a mere drying tube and is never weighed. The 2d and 3d tubes serve to absorb the carbonic acid, and are weighed together. The purpose of the sulphuric acid in the second tube is merely to show

the rate of flow of the carbonic acid, but since sulphuric acid absorbs a certain amount of carbonic acid, this tube has to be weighed with the others.

After the tubes 2 and 3 have been weighed and tightly connected with the other tubes, place a weighed quantity of the substance to be analyzed in the flask, moisten it with water, cork the flask, and connect its delivery tube with the 1st tube of the series above described. By means of a caoutchouc connector attach a small funnel to the top of the S-tube in the flask, and pour through it a small quantity of mercury to fill the lowermost part of the bend, beneath the bulb, in the S-tube. Then fill the funnel and tube with a mixture of equal volumes of strong nitric acid and water, and suck gently through the caoutchouc tube at the end of the series of absorption tubes, until a little of the acid is drawn past the mercury into the flask. The bulb in the S-tube prevents the mercury from passing over into the flask with the acid.

The carbonic acid set free by the nitric acid is dried by the chloride of calcium in the 1st U-tube, and its rate of flow is made manifest by the bubbles which pass through the sulphuric acid at the bottom of the second U-tube. As soon as the evolution of gas slackens, suck over new portions of the nitric acid into the flask until the carbonate has been completely decomposed. If need be, the flask may be heated gently throughout the process. When no more gas is evolved wash out the S-tube by filling it two or three times with hot water and sucking the water into the flask. Remove the funnel from the top of the S-tube and replace it with a tube filled with soda-lime or hydrate of potassium; then heat the contents of the flask to gentle boiling until the projecting portion of the 1st chloride of calcium tube becomes hot on the side nearest the flask. Extinguish the lamp and, by means of an aspirator fastened to the end of the 4th tube, draw as much air through the apparatus as will amount to at least six times the volume of the flask. After the apparatus has become cold, weigh the 2d and 3d U-tubes. The difference between this second weight of the tubes and the weight before the operation, gives the amount of carbonic acid in the substance analyzed. — The other constituents of the substance may be determined in the nitric acid solution which is left in the flask. By using a definite quantity of standard acid and titrating the excess with a standard alkali, as explained above, in paragraph 1, after the carbonic acid has been expelled, it is easy, in many cases, to determine the quantity of the base with which the carbonic acid was combined. In cases where chlorhydric acid is to be preferred to nitric acid for dissolving the substance, fill one limb of the 1st chloride of calcium tube,—that which is farthest from the flask,—with fragments of pumice stone saturated with anhydrous Sulphate of Copper, and proceed as before.

The 2d tube, charged with sulphuric acid, may be used over and over again for many analyses. So, too, the soda-lime tube may be used repeatedly, subject to the usual rules which apply to this substance. When large quantities of carbonic acid are to be absorbed the soda-lime tube may be replaced by Geissler's potash bulbs, or even by Liebig's bulbs, if the evolution of gas be carefully regulated. (Kolbe, *Annal. Chem. und Pharm.*, **119.** 130, and Fresenius, *Quantitative Analysis*, p. 300).

5. *By absorbing the gas with an alkaline earth.* The carbonic acid generated in the apparatus described above, paragraph 4, might, of course, be collected in the form of Carbonate of Barium or Carbonate of Calcium. As a general rule, however, the absorption by soda-lime is greatly to be preferred.

Principle III. Power of changing the color of blue litmus to violet.

Application. Rough estimation of the free carbonic acid in mineral waters.

Method. Prepare a standard solution of dilute sulphuric acid (see Acidimetry) of such strength that 1 c. c. of it shall contain 10 milligrammes of SO_3; also a solution of litmus, by digesting a quantity of the solid in an equal weight of cold water. There is required also, a concentrated solution of caustic soda containing some carbonic acid, such as may be found upon the shelves of any laboratory.

Measure out about 450 c. c. of distilled water, add to it 1.5 c. c. of the litmus solution and 5 c. c. of the caustic soda. Dilute the mixture with water to the volume of 500 c. c., take out with a pipette 3 several portions of 100 c. c. each, of the liquid, and pour them into three beakers. Place the beakers on a white ground in strong daylight, and pour the standard acid from a burette into each beaker in succession, until its contents appear distinctly violet colored. Towards the close of each operation it is necessary to wait one or two minutes after each addition of acid, in order that the change of color, when it does occur, may become distinctly visible. The experiment in the 1st beaker will give an approximately correct result, so that in the second and third trials the operator will be able to hit the point of coloration with tolerable accuracy. Select the most accurate of the three experiments, to the exclusion of the other two, multiply by 5 the number of c. c. of acid consumed in that case, and record the product of this multiplication as the quantity of acid corresponding to 5 c. c. of the soda solution.

It is to be observed that the color of blue litmus is not changed to violet by bicarbonate of sodium, but only by free carbonic acid; hence, when dilute sulphuric acid is cautiously added to a solution containing carbonate of sodium, no violet coloration will appear until the sodium compound has all been converted into sulphate and bicarbonate, and some carbonic acid has actually been set free from the latter by the addition of a drop of sulphuric acid in excess. It will be noticed, also, that since all the carbonic acid is in the state of a bicarbonate at the close of the neutralization, two molecules of this acid must be reckoned as equivalent to, or as replacing, one molecule of SO_3. Hence, if 1 c. c. of the standard sulphuric acid contains 10 milligrms. of SO_3, the c. c. will correspond to 11 milligrms. of CO_2.

To perform an analysis, add 5 c. c. of the soda solution, or 10 c. c. if need be, to 450 or 500 c. c. of the mineral water under examination, and mix the two liquids. Without heeding any turbidity which may appear in the liquor, measure out as before three separate portions of 100 c. c. each, add to each portion 4 drops of the litmus solution, and pour in the standard sulphuric acid until the liquid exhibits a violet tint. More care and attention should be given to the experiment upon the second portion of liquor than to the first, and still more to the third than to the second. Choose the best of the three experiments, as before; subtract the number of c. c. of acid required to produce the violet coloration in the mixture of mineral water and soda, from the amount of acid which was required to produce the coloration with the soda alone, and multiply the difference by 0.011 (the value of the standard acid in terms of CO_2) in order to obtain the weight of the carbonic acid in grammes.

In most cases the quantity of mineral water taken had better be measured after the admixture of the soda, in order to avoid losing much carbonic acid. If the sample of water to be examined is contained in a bottle, cool it to about 4°, remove the stopper quickly, pour out a little of the water, and add the soda to the rest. — The process is of value for some purposes, but requires a practised eye. (Kersting, *Annal. Chem. und Pharm.*, **94.** 112).

Principle IV. Power of absorbing the heat which radiates from ignited carbonic acid gas.

Tyndall has observed that though carbonic acid gas is one of the weakest absorbents of the heat which radiates from a glowing solid, it has, on the contrary, a remarkable power of absorbing the heat radiated from a flame of burning carbonic oxide. — The fact not only furnishes a delicate qualitative test of the presence of carbonic acid in a mixture of gases, but has been successfully applied to the estimation of carbonic acid in air expired from the lungs. It is not improbable that the process may be found more accurate and convenient than any other for determining the proportion of carbonic acid in air, in cases where a large number of experiments are to be made. (*Zeitsch. analyt. Chem.*, 1868, **7.** 151. For a description of the apparatus employed, see Tyndall's *Heat a Mode of Motion*).

Carbonate of Ammonium.

Principle I. Power of neutralizing acids.

Application. Estimation of carbonate of ammonium in commercial samples.

Method. See Alkalimetry. Mohr (*Titrirmethode*, 1855, **1.** 63) found that the results obtained by supersaturating carbonate of ammonium with oxalic acid and titrating the excess of the latter with caustic soda, were liable to wider variations than those of most other alkalimetric processes. After the expulsion of the carbonic acid the hot acid liquor must be cooled completely before adding any of the standard caustic soda, lest some ammonia be set free by the latter.

Principle II. Volatility.

Application. Estimation of carbonates in natural waters.

Method. Add to 200 c. c. of the filtered water about 0.5 grm. of chloride of ammonium, distill off half the solution and receive the distillate in 10 c. c. of very dilute standard sulphuric acid. Boil the mixture of distillate and acid to expel carbonic acid gas, then allow it to cool, and determine how much of the standard acid still remains free, by titrating with weak standard soda (see Alkalimetry, and above). The process is said to yield good results. (Chevalet, *Bull. Chem. Soc. Paris*, 1868, p. 90).

Carbonate of ammonium is often used as a precipitant, solvent and decomposer of sundry metallic carbonates. It is used also to convert bisulphates of the alkalies to neutral salts. To be fit for use as a reagent it should leave no residue when heated in a platinum dish.

Carbonate of Ammonium and of Magnesium.

See Carbonate of Magnesium and of Ammonium.

Carbonate of Barium.

Principle I. Fixity when not too strongly heated.

Application. Estimation of barium in salts with organic acids.

Method. Heat the salt carefully in a covered platinum crucible until no more fumes are evolved. Then take off the lid, lay it flat upon one side of the triangle which supports the crucible, and lay the crucible flat, or obliquely upon its side with the open end resting on one edge of the cover. Ignite the crucible in this position until the carbon is all consumed and the residue has become perfectly white. Since a portion of the carbonate of barium is apt to be reduced to the condition of oxide by the ignition in contact with carbon, the cold residue must be moistened with a concentrated solution of carbonate of ammonium, in order to reconvert the oxide into carbonate. After adding the carbonate of ammonium evaporate the mixture to dryness on a water bath, heat the dry residue moderately, cool, and weigh. Compare Carbonate of Calcium. — A certain quantity of the substance is almost always lost during the ignition, from fine particles of matter being carried away mechanically by the gaseous products of the distillation. This source of error, however, is the less considerable in proportion as the crucible is heated more slowly. With proper care the process yields satisfactory results.

Properties. Carbonate of barium is unalterable in the air, even when heated to redness. In the intense heat of a forge, however, or blast furnace, it slowly gives up the whole of its carbonic acid (Abich, *Poggendorff's Annalen*, 1831, **23.** pp. 308, 314). The presence of aqueous vapor favors the escape of the carbonic acid. The composition of the salt is as follows:—

Ba	137	77.69
CO_3	60	22.31
	197	100.00

Principle II. Sparing solubility in water.

Applications. The process may be employed for estimating barium in all barium salts which are soluble in water, but is rarely used for this purpose excepting in those cases where the estimation of the metal as Sulphate of Barium would be inadmissible. It is employed also for separating Ba from K, Na, Mg, Mn, etc., and for estimating carbonic acid.

Methods.

Estimation of Barium in solutions of its salts. Mix the moderately dilute solution with ammonia-water in slight excess, add the carbonate of ammonium as long as a precipitate falls, and leave the mixture at rest for 12 hours in a warm place. Collect the precipitate upon a filter, wash it with water, to which a little ammonia has been added, then dry, ignite, and weigh. See Principle I. — Since carbonate of barium is not absolutely insoluble in ordinary water, and is soluble to no inconsiderable extent in aqueous solutions of ammonium salts, the results obtained by this method are usually a little lower than the truth. Care must always be taken to keep the proportion of chloride, nitrate or other salt of ammonium in the solution as small as possible.

It is to be observed that carbonate of barium cannot be precipitated from solutions which contain a citrate or metaphosphate of either of the alkali metals.

Separation of Ba from K, Na and Mg. The process is employed in certain cases for separating barium from potassium, sodium and magnesium, though it is, on the whole, less accurate and convenient than the method described under Sulphate of Barium. It is employed, for example, when barium is to be separated from a mixture of both potassium and sodium, since the mixture of alkalies in the filtrate from carbonate of barium can easily be converted into chlorides, and analyzed as such. But if only one of the alkalies were present, the barium would always be thrown down as Sulphate of Barium, and the alkali weighed as a sulphate. — The process finds application, also, in cases where a mixture of Ba, Sr and Ca is to be separated from K and Na, or from K, Na and Mg. The precipitation of the carbonates of the three metals of the alkaline earths is effected in precisely the same way as if only barium were present. Chloride of Ammonium is added to keep up magnesium in case any of that metal is contained in the substance to be analyzed. But since this chloride of ammonium prevents the precipitation of some barium and calcium, the filtrate from the precipitated carbonates must be subjected to further treatment as follows:—Add to the filtrate a small quantity of dilute sulphuric acid, no more than 3 or 4 drops, together with a few drops of a solution of oxalate of ammonium and let the mixture stand 12 hours in a warm place. A mixed precipitate of sulphate of barium (and strontium) and oxalate of calcium will fall. Collect this precipitate on a small filter and treat it with dilute chlorhydric acid. The oxalate of calcium will dissolve, together with a little oxalate of magnesium which is sometimes thrown down with it, while the sulphate of barium (and strontium) is left as an insoluble residue upon the paper. To recover the magnesium just mentioned, saturate the chlorhydric acid filtrate with ammonia-water, allow the re-precipitated oxalate of calcium to settle, filter, and add the filtrate to the mixture of magnesium and alkalies. Sometimes a little insoluble carbonate of magnesium goes down with the carbonate of barium, etc.; it may be recovered during the process of separating these elements.

To separate Potassium from a mixture of Ba, Ca and Mg, in a solution free from ammonium salts, heat the liquid to strong boiling and add to it a solution of carbonate of sodium, drop by drop, as long as any precipitate falls. Continue to boil until the voluminous precipitate has become compact and granular. It is easy to wash out all the potassium from the precipitate if the boiling has been continued long enough. But the method will be found, in most cases, to be inferior to that of precipitating the potassium directly as Chloroplatinate of Potassium. (Stohmann, *Zeitsch. analyt. Chem.*, 1866, **5.** 307).

Separation of Ba from Mn. See Carbonate of Manganese.

Indirect separation of Ba from Ca and Sr. See Carbonic Acid (Volatility of, — methods by simple ignition, and by ignition with an acid salt).

Estimation of Carbonic Acid.

1. *Absorption in baryta-water.*

Many chemists have estimated carbonic acid by absorbing the gas in baryta-water and weighing the precipitated carbonate, in the manner above described. De Saussure in particular, sought to perfect the process by washing the carbonate of barium with a saturated solution of this substance, in place of water, and did unquestionably obtain accurate and valuable results in that way,—as Boussingault and others have done after him. — The modern methods are as follows:—

A. *Method of Mohr & v. Gilm* (*Wien. Akad. Bericht*, **24.** 279), employed for estimating the proportion of carbonic acid in atmospheric air. By means of an aspirator of at least 30 litres capacity, some 60 litres of the air to be analyzed are slowly drawn through an absorption tube containing a solution of hydrate of barium; the precipitate of carbonate of barium which forms is collected upon a filter out of contact with the air, the tube and precipitate are washed first with distilled water saturated with carbonate of barium, and afterwards with pure water which has been recently boiled. The carbonate of barium upon the filter and in the tube is then dissolved in dilute chlorhydric acid, the solution evaporated to dryness, the residue ignited gently and the amount of chlorine contained in it determined as Chloride of Silver. Or the amount of barium may be determined as Sulphate of Barium. Every atom of Ba, or every two atoms of Cl, will represent a molecule of CO_2.

To prepare the absorbent liquid, dissolve crystallized hydrate of barium in a warm dilute solution of caustic potash, and filter the mixture. Since some carbonate of barium is always precipitated on the addition of the potash lye, the clear solution obtained necessarily contains all the carbonate of barium it is capable of dissolving. — For the absorption tube, choose a tube a metre long, and about 15 m. m. wide. Draw out the upper end of the tube, and at some distance from its lower end bend the tube at an angle of 140°–150°. Fix the tube in such a position that its longer limb shall incline at an angle of 8° or 10° to the horizontal and fill the tube half full of the clear baryta-water. By means of a perforated cork, fit a narrow glass tube into the wider end of the absorption tube, for the admission of air, and connect the other end of the absorption tube with the aspirator. It is well to interpose a couple of little flasks, charged with the baryta-water, between the absorption tube and the aspirator, in order to be sure that the whole of the carbonic acid has been absorbed from the air.

Since, in passing through the absorption tube, the air is compelled to force its way against a column of liquid, it is essential that the aspirator employed should be provided with a small manometer, in order that the volume of air may be accurately measured. The height of the column of mercury in the manometer must be deducted from that observed in the barometer at the time of the experiment.

For filtering the carbonate of barium, v. Gilm uses a double funnel, arranged as follows:—By means of a perforated cork, fit to a sufficiently wide-mouthed bottle a large funnel with wide throat. The rim of the funnel should be ground so that it can be covered tightly with a glass plate. Fit a cork to the upper part of the throat of this funnel, perforate the cork and cut a groove in its side. Through the hole in the cork thrust the tube of a second funnel considerably smaller than the first, and place the filter in this inner funnel. During the operation of filtering, the large funnel is kept closed as much as possible with a glass plate. Air can of course pass freely from the bottle into the large funnel through the slit or groove in the inner cork.

According to observations of A. Mueller (*Zeitsch. analyt. Chem.*, 1862, **1.** pp. 84, 149), this process is open to the objection that filter paper has the power to absorb considerable quantities of baryta out of baryta-water, and to retain it so forcibly that it cannot be washed out.

The precipitated carbonate of barium might of course be decomposed with an acid in an appropriate apparatus, and the Carbonic Acid weighed or measured as such, instead of being treated as above described, or in addition to this treatment.

B. *Methods of Hadfield, of Pettenkofer, and of A. Mueller.* See below, under Principle III (Power of neutralizing acids).

2. *Absorption in an ammoniated solution of chloride of barium.* This method may be employed in the analysis of mineral waters, air, and other gaseous mixtures. It was formerly sometimes used for estimating carbonic acid set free from carbonates by the action of acids, but is now known to be inferior to the methods described in the preceding paragraphs, and to several of the methods described under Carbonic Acid.

Prepare a quantity of an ammoniated solution of chloride of barium, as follows:—Mix an aqueous solution of chloride of barium with an excess of ammonia-water, boil the mixture for a few minutes and filter the hot liquid quickly, in order to remove the carbonate of barium formed by carbonic acid contained in the ammonia-water. After the mixture has once been boiled, take care to protect the liquid as much as possible from contact with carbonic acid of the air. Pour 50 or 80 c. c. of the clear liquid into a light flask of about 300 c. c. capacity, and close the flask tightly with a caoutchouc stopper. Without removing the stopper, weigh the flask together with the barium solution, and pour into the flask enough of the mineral water, or other solution of carbonic acid to be tested, to nearly fill it, then replace the cork immediately, shake the flask, and again weigh it with its contents. The difference between the first and second weighings will give the weight of the mineral water taken. Instead of weighing the mineral water it may be measured as follows:—Measure out the 50 or 80 c. c. of ammonio-barium solution to be placed in the flask, by means of a pipette provided with a rubber ball or soda-lime tube, so that no carbonic acid can enter it from the lungs; then pour the mineral water into the flask, cork the flask, and scratch the glass with a diamond, or paste upon it a bit of paper, with shellac, to mark the height of the liquid. After the carbonate of barium has been precipitated and the experiment finished, fill the flask with water up to the mark, measure the water in a graduated cylinder, and subtract from it the quantity of barium solution in order to obtain the true volume of the mineral water. For another method of measuring the mineral water, see Carbonate of Calcium.

Place the flask in a pan of water, loosen its cork, heat the water to boiling, and keep the flask in the boiling water for an hour or two. Then re-cork the flask, allow its contents to settle out of contact with the air, and proceed to collect the carbonate of barium on a filter, in the manner to be described directly.

It is to be observed that although the liquid in the flask usually becomes turbid as soon as the mineral water is introduced, the mixture must, nevertheless, be heated a long time in order that all the carbonic acid may be thrown down, for at temperatures much below boiling carbonate of barium is soluble to no inconsiderable extent in a solution of chloride of ammonium. The contents of the flask must never be heated to actual boiling, however, for in that case some of the carbonic acid would escape in the form of carbonate of ammonium.

Instead of heating the mixture almost to boiling for an hour or two, it may be left during half a day in a place heated to 80° or 90°.

For filtering, provide a funnel with glass cover, quickly decant the clear liquid from the flask into the filter, and cover the funnel. Nearly fill the flask with warm water, replace the cork, shake the contents of the flask, let the precipitate subside in the flask until the filter has become empty, and again decant the tolerably clear liquor from the flask into the filter. Wash once more by decantation, then transfer the precipitate to the filter, and wash with warm water until the washings no longer give any precipitate when tested with nitrate of silver. In case the last portions of the precipitate cannot be rubbed off the flask, dissolve them in a little dilute chlorhydric acid, mix the solution with pure carbonate of sodium, and collect the precipitate which forms, upon a small separate filter. Dry, ignite and weigh the precipitate as in Principle I.

If the substance analyzed contained no other substance besides carbonic acid capable of precipitating barium or of being precipitated by ammonia, the weight of the carbonic acid may be calculated directly from that of the carbonate of barium. But if, on the other hand, the precipitate is contaminated with phosphate of barium, carbonate of calcium, ferric oxide, or the like, the proportion of Carbonic Acid contained in it must be specially determined.

This method, as well as the analogous method mentioned under Carbonate of Calcium, was formerly often employed, but is now held in comparatively slight esteem. The chief objections to it are found in the solubility of carbonate of barium in water, the difficulty of obtaining an ammoniated barium solution, in the first place, absolutely free from carbonic acid; in the liability of this solution to absorb carbonic acid from the air; in the risk of driving off some carbonic acid in the form of carbonate of ammonium; and in the very decided tendency of carbonate of barium to remain dissolved in a solution of chloride of ammonium. It may readily happen that appreciable quantities of carbonate of barium (or of calcium) may remain dissolved throughout the analysis in case the proportion of chloride of ammonium in the liquid be excessively large or the liquid itself be largely diluted with water. — To guard against the various sources of error, it was customary to operate at one and the same time upon several different portions of the mineral water. — It may here be said that the idea advanced by Kolbe (*Handwœrterbuch der Chem.*, **1.** Supplem., p. 157) and recently defended by Fresenius (*Zeitsch. analyt. Chem.*, 1863, **2.** 49 and 1866, **5.** 321) that the retention of carbonate of barium in solution before boiling, is due to the change of carbonic into carbamic acid, is unsupported by any experimental evidence. The erroneous character of the conception has been shown by Carius (*Annal. Chem. und Pharm.*, 1866, **137.** 108) and by old observations of my own (*Amer. Journ. Sci.*, 1858, **25.** 41; and *Dictionary of Solubilities*).

Instead of drying and weighing the precipitated carbonate of barium, it may be decomposed with an acid and the resulting gas measured (see below), or absorbed in soda-lime and weighed (see below). Or the proportion of carbonic acid may be determined by the Alkalimetric method (see also below, Principle III). But for determining carbonic acid in that way, it had better be precipitated in the form of Carbonate of Calcium.

Properties. Precipitated carbonate of barium is a soft, white powder, soluble in about 48,000 parts of cold water absolutely free from carbonic acid (Bineau), and in 12 or 14,000 of cold water which has been recently boiled to expel most of its carbonic acid. It appears to be but little if any more soluble in hot than in cold water. The aqueous solution has a faint alkaline reaction. The solution in carbonic acid water is also alkaline. — The precipitate is far more readily soluble in neutral solutions of ammonium salts, such as the chloride or nitrate of ammonium, than in water; but is almost completely insoluble in cold water which contains free ammonia and carbonate of ammonium; one part of the precipitate requiring in that case more than 140,000 of the liquid for its solution. When boiled with a solution of chloride of ammonium, carbonate of barium is rapidly dissolved with formation of chloride of barium and evolution of carbonate of ammonium. Carbonate of barium dissolves also to a slight extent in aqueous solutions of most of the salts of potassium and sodium. In cold water saturated with carbonic acid, carbonate of barium dissolves in the proportion of 1 part to about 600 parts of the liquid. (Compare Principle I).

The comparatively large molecular weight as well as the more sparing solubility of carbonate of barium are reasons for preferring it to carbonate of calcium, but the precipitate is apt to be more bulky and less easily filtered than that from the calcium salt.

Principle III. Power of neutralizing acids.

Applications. Estimation of barium and of carbonic acid.

Methods.

A. *Estimation of barium.* Similar to the estimation of calcium described under Carbonate of Calcium.

B. *Estimation of carbonic acid.*

1. *Method of Pettenkofer* (*Annalen Chemie und Pharm.*, 2d Supplement volume, page 23). In this process a definite volume of the air or mineral water to be analyzed is mixed with a measured quantity of baryta-water of known strength; the carbonate of barium which forms is allowed to settle, and a measured portion of the clear supernatant liquid is finally titrated with standard oxalic acid, in order to determine how much of the hydrate of barium in the liquid has remained uncombined with carbonic acid. The difference between the amount of oxalic acid required to neutralize the uncombined hydrate of barium and that required to saturate the baryta-water originally employed, will be equivalent to the amount of carbonic acid in the sample of air or water taken.

To prepare the standard oxalic acid, dissolve 2.8636 grms. of crystallized oxalic acid in water, and dilute to the volume of a litre. 1 c. c. of the liquid will correspond to 1 milligrm. of carbonic acid.

The baryta-water employed must be free from any trace of caustic potash or caustic soda, for it is impossible to titrate baryta-water with oxalic acid in presence of an alkaline oxalate (see Hydrate of Barium).

The baryta-water should be strong or weak accordingly as there is more or less carbonic acid in the air or water to be examined. If the proportion of carbonic acid is comparatively large, it is well to use strong baryta-water prepared by dissolving 21 grms. of crystallized hydrate of barium to the litre, but in case the quantity of carbonic acid to be determined is small, a liquid which contains no more than 7 grms. of the crystallized hydrate to the litre, is to be preferred. Of the stronger baryta-water 1 c. c. will correspond to about 3 milligrms. of carbonic acid and 1 c. c. of the weaker liquid will be equivalent to about 1 milligrm.

To standardize the baryta-water transfer 30 c. c. of it to a small flask and pour in the standard oxalic acid from a burette, little by little, until the liquid is just neutralized. After each addition of the acid, close the flask with the thumb, and shake the liquid. To determine the point of neutralization take up a drop of the liquid upon a glass rod and touch it to a piece of delicate turmeric paper. When a drop of the liquid ceases to produce a brown ring upon the paper the neutralization is known to be complete. In case too many drops of the oxalic acid happen to be added in this first trial, the experiment may be repeated as follows:—Measure off a second 30 c. c. portion of the baryta-water, add to it at once as much of the standard oxalic acid, to within a c. c. or half a c. c., as was used before; then add the acid drop by drop, and test the liquid on turmeric paper after each drop, until the neutralization is complete. A third experiment should agree with the second to 0.1 c. c.

The details of the actual analysis of a mineral water will be found in the description of the lime-water process under Carbonate of Calcium.

In order to determine the proportion of carbonic acid in air, select a bottle of about 6 litres capacity, having a tightly ground glass stopper, and accurately determine its capacity. Dry the bottle thoroughly and by means of a pair of bellows fill it with the air to be analyzed. Pour into the bottle 45 c. c. of the dilute standard baryta-water, and spread the liquid repeatedly over the inner surface of the glass without shaking the bottle any more than is necessary. In the course of about half an hour the whole of the carbonic acid will be absorbed. Then pour the turbid liquid from the bottle into a glass cylinder, close the latter securely from the air, and leave the liquid at rest until it has become clear. By means of a pipette take up 30 c. c. of the clear liquor, transfer it to a flask, and neutralize with the standard oxalic acid. Since only 30 c. c. out of the original 45 c. c. of baryta-water have been employed, the number of c. c. of oxalic acid required to effect neutralization must be multiplied by 1.5. Deduct the product from the number of c. c. required for 45 c. c. of the standard baryta-water. The difference will show how much of the hydrate of barium has been converted into carbonate, and thereby indicate the amount of the carbonic acid.

Instead of measuring the air in a bottle as above described, it may of course be drawn through tubes charged with a measured quantity of standard baryta-water, by means of an aspirator, as described under Principle II, on p. 95. For Pettenkofer's arrangements for effecting the absorption, see his original memoir, as cited above.

2. *Method of A. Mueller* (*Zeitsch. analyt. Chem.*, 1862, **1.** 147, and figure). This method is said to be peculiarly well suited for the determination of carbonic acid in soils, and, in general, for the determination of small quantities of the acid in presence of other volatile substances. It may be employed also for estimating carbonic acid in air, and for experiments on fermentation.

Grind a plate of glass to fit the top of a broad glass cylinder or jar about 100 millim. high and wide. Set a tripod of glass or platinum in the cylinder, and suspend from the tripod a shallow, conical glass vessel of about 40 c. c. capacity. This smaller or "absorption" vessel must be light, and its mouth, though much narrower than the bottom, must be tolerably wide. The top of this vessel also should be ground and fitted with a glass plate. Some forms of ink-stands answer very well for the absorption vessel. There will be needed also a small porcelain crucible of about 15 c. c. capacity.

Prepare standard solutions of nitric acid and of caustic baryta (see Alkalimetry). The baryta solution may be made by mixing a solution of chloride of barium with soda lye. Its strength should be reckoned not in terms of centimetres, but in terms of grammes.

For the analysis weigh out a quantity of the soil, or other substance containing a carbonate, place it in the bottom of the larger glass vessel, and pour upon it 15 or 20 c. c. of water. Place in the porcelain crucible a quantity of some strong non-volatile acid, such as sulphuric, phosphoric, tartaric or lactic acid, more than sufficient to decompose the whole of the carbonate, and set the crucible upon the soil.

Weigh the absorption vessel with its cover, pour into it a quantity of the standard baryta water and again weigh. Remove the glass plate from the absorption vessel, place the latter on the tripod, smear the top of the larger vessel with tallow, close it tightly with its glass plate, and incline the vessel carefully so that some of the acid may flow out of the crucible upon the soil. Repeat the dose of acid from time to time, and then leave the apparatus at rest for a day or two, until the whole of the carbonic acid has been absorbed by the baryta-water. — The rate of evolution of the carbonic acid can be judged of by the appearance of the baryta solution. So long as carbonic acid is generated rapidly, crusts of carbonate of barium will form upon the surface of the liquor as often as the old crust is made to sink by shaking the apparatus. When a crust no longer forms, it is evident that no more carbonic acid is being set free. Wait until the whole of the carbonic acid has been absorbed by the baryta-water, then cover the absorption vessel with its glass plate, and again weigh it with its contents. By means of a syphon and aspirating flask, suck over as much of the clear liquid as possible into a tared flask which contains a weighed quantity of the standard nitric acid, and again weigh the flask after the addition of the baryta solution. Finally titrate the nitric acid still left free in the flask, in order to estimate the amount of baryta which was added to it. The calculation is as follows: —

If w = the weight of carbonic acid absorbed,
a = the weight of the baryta before the absorption,
c = the weight of that portion of the baryta solution taken for titration after the absorption,
b = the weight of baryta in c,
d = the sum of the weights of the baryta solution and the carbonate of barium precipitated in it; and

$$G = \frac{BaO}{CO_2} = 3.477, \text{ then}$$

$$w = \frac{a\,c - b\,d}{G\,(c - b) - b} = \frac{a\,c - b\,d}{3.477\,(c - b) - b}.$$

It is not practicable to operate with a determined volume, instead of a definite weight of the baryta solution, for the weight of the solution varies constantly during the experiment, both from changes of temperature and from the absorption of more or less aqueous vapor by the acid in the jar. Neither can the baryta solution be filtered after the absorption, for filter paper has the power to absorb considerable quantities of baryta, and to abstract it from the solution. — A correction may be applied for the carbonic acid naturally present in the atmospheric air originally contained in the apparatus, or, better, the apparatus may be filled beforehand with air free from carbonic acid.

In case the soil, or other substance analyzed, evolves chlorhydric acid or other volatile acid on being treated with strong acids, the process is modified to the extent that the amount of baryta in the solution is determined before and after the absorption, by precipitating it in the form of Sulphate of Barium.

3. *Old method of absorption in ammoniated baryta-water.* See above (Principle II) and under Carbonate of Calcium.

Principle IV. Decomposition of by solutions of the salts of iron, aluminum, manganese and chromium; by phosphoric and arsenic acids; and by many other saline solutions and acids; the metals or acids in question being at the same time precipitated in the form of hydrates, or of basic salts.

Applications. Separation of Al from Mg, Ca, Zn, Mn, Ni, Co and Fe (see Hydrate of Aluminum). Separation of Fe from Ba, Sr, Ca, Mg, Zn, Mn, Co, Ni and Fe (see Hydrate of Iron). Separation of Cr from Zn, Mn, Ni, Co and Fe (see Hydrate of Chromium). Separation of P_2O_5 from Fe, Al, Ba, Sr, Ca, and all other oxides not precipitable by carbonate of barium (see Phosphate of Iron). Separation of As from Ba, Ca, Sr; Zn, Mn, Ni and Co (see Arseniate of Iron). — For lists of the compounds precipitable by carbonate of barium, see that substance in Dictionary of Solubilities.

For use as a reagent, carbonate of barium

may be prepared as follows: — Dissolve a quantity of crystallized chloride of barium in hot water, filter the solution and heat it to boiling. Prepare a quantity of normal carbonate of ammonium by saturating a solution of the commercial sesquicarbonate with ammonia-water and filtering the mixture after it has been allowed to stand for some time. Add the carbonate of ammonium solution, little by little, to the boiling solution of chloride of barium as long as any precipitate continues to fall. After the mixture has been allowed to settle decant the clear liquor and wash the precipitate 5 or 6 times by decantation with hot water, then throw it upon a filter and wash until the wash water acidulated with nitric acid no longer gives any precipitate when tested with nitrate of silver. Wash the precipitate out of the filter into a beaker and keep it for use, either in the moist state or air-dried.

Principle V. Power of decomposing refractory silicates, aluminates and chromites, when intensely heated.

Applications. Conversion of insoluble silicic acid into the soluble modification, in the analysis of some refractory siliceous minerals. Decomposition of spinel and other native aluminates, and of chrome iron ore, as a preliminary to their solution in acids. Decomposition of silicates as a preliminary to the estimation of alkali-metals.

Method A. For decomposing aluminates and chromites. Mix the finely powdered mineral with from 4 to 6 times its weight of pure, precipitated carbonate of barium, in a platinum crucible. Place the platinum crucible inside a somewhat larger crucible of refractory fire-clay; fill the space between the two crucibles with magnesia, cover the clay crucible and heat it intensely in a Sefström furnace during half or three-quarters of an hour.

Instead of the Sefström furnace, a powerful gas furnace, such as that of Griffin or of Grove, may be used to heat the naked platinum crucible. Or Deville's (*Annales Chim. et Phys.*, 1856, **46.** 182) oil of turpentine furnace may be used. But in any event an intense heat is required to effect the fusion of the mixed aluminate and carbonate of barium. The heat of an ordinary wind furnace is insufficient for the purpose. — After the crucibles have been taken from the fire and allowed to cool, clean the outside of the platinum crucible and press it gently between the fingers to loosen the solid lump within it. Put the lump in a beaker, together with the crucible, if any portion of the fused mass still adheres to it, cover the mass with 10 or 15 times its bulk of water, and add strong chlorhydric or nitric acid, little by little, until solution is complete. Since all the carbonic acid of the carbonate of barium has been expelled by the intense heat, the fused mass will dissolve without effervescence; but care must be taken not to add too large a quantity of chlorhydric acid at any one time, since the chloride of barium formed is only sparingly soluble in acid, and is liable to form an encrustation upon the mass which would impede its solution. (Abich, *Poggendorff's Annalen*, 1831, **23.** pp. 319, 338). The action of fused carbonate of barium, or rather of the oxide of barium, into which the carbonate is converted at a high heat, is exceedingly energetic. Even the most refractory minerals may be readily and completely decomposed by means of it. According to Abich (*loc. cit.*, pp. 339, 341), even chrome iron ore may be completely decomposed by fusing it once or twice during three quarters of an hour with 4 parts of carbonate of barium.

Method B. For decomposing silicates, less carbonate of barium and a lower degree of heat will be required than are necessary for success in the applications of Method A. According to Deville (*Annales Chim. et Phys.*, [3.] **38.** 5), 0.8 part of carbonate of barium is sufficient at a moderate red heat, to reduce 1 part of potash feldspar to the condition of a vitreous transparent mass, decomposable by acids.

The use of a larger proportion of carbonate of barium may even be injurious, since a portion of the potash set free by the caustic baryta formed, might be lost through volatilization. — In other respects the details of the process are similar to those described in A. It has been superseded in great measure, by the process of decomposing with fluorhydric acid. and by L. Smith's process with Carbonate of Calcium and chloride of ammonium. (See further under Silicates, and Oxide of Barium).

L. Smith's (*American Journ. Sci.*, 1853, **16.** 53) old method of fusing silicates with a mixture of carbonate and chloride of barium will be described under the head of Silicates.

Principle VI. Insolubility in an aqueous solution of cyanide of potassium.

Applications. Separation of Ba from Co, Ni and Zn.

Method. See Carbonate of Cobalt.

Basic **Carbonate of Bismuth.**

Principle I. Insolubility in water and in solutions of alkaline carbonates.

Applications. Estimation of bismuth in compounds soluble in nitric acid and free from any admixture of other acids. Separation of Bi from Mn. Separation of Bi from Cu.

Method A. Add a very slight excess of carbonate of ammonium to the bismuth solution, heat the mixture nearly to boiling for a short time, filter, and ignite with the precautions prescribed under Carbonate of Lead. Weigh as Oxide of Bismuth. — The bismuth solution must not be too concentrated. If on diluting it with water, some basic nitrate of bismuth falls, no notice need be taken of it. The process is inapplicable, however, in pressence of sulphuric or chlorhydric acids, since the precipitated carbonate and the ignited oxide would then contain an admixture of basic sulphate or basic chloride of bismuth. The mixture of bismuth solution and carbonate of ammonium must always be heated in order to ensure complete precipitation. The results are in any event a trifle too low, for carbonate of bismuth is not absolutely insoluble in a solution of carbonate of ammonium.

Method B. To separate bismuth from copper, mix the nitric acid solution with an excess of carbonate of ammonium. Most of the copper remains dissolved in the excess of the ammonium salt, but a little of it is retained by the precipitated carbonate of bismuth. The precipitate must therefore be re-dissolved once or twice in nitric acid, and re-precipitated with carbonate of ammonium, in order to remove the last traces of copper. It is well to add some carbonate of ammonium to the water used for washing the precipitate. If these precautions be attended to, it is easy to remove all the copper from the precipitate (R. Schneider, *Journ. prakt. Chem.*, **60.** 311), but a little bismuth always remains dissolved in the carbonate of ammonium and passes into the filtrate (H. Rose, *Pogg. Ann.*, **110.** 430), hence the process is less accurate than that which depends on the insolubility of basic Chloride of Bismuth.

To estimate the copper, heat the ammoniacal filtrate, first by itself and afterwards with caustic lye, and collect the Oxide of Copper which is thrown down.

For the method of separating Bi from Mn, see Carbonate of Manganese.

Properties. When an excess of carbonate of ammonium is added to a nitric acid solution of bismuth, in the cold, a white precipitate of the monocarbonate (Bi_2O_3, CO_2) is immediately thrown down, but the precipitation is incomplete, since a portion of this monocarbonate remains dissolved in the ammonium salt. But on heating the mixture a more difficultly soluble basic salt is formed. Carbonate of potassium also precipitates bismuth completely, but the precipitate in that case retains traces of potash which are hard to wash out. The precipitate obtained by carbonate of ammonium is easily washed. It is as good as insoluble in water, but dissolves readily in acids. When ignited it gives off carbonic acid and is converted into teroxide of bismuth.

Principle II. Insolubility in an aqueous solution of cyanide of potassium.

Applications. Separation of Bi from Cu, Cd, Hg, Ag and Au.

Method. Add to the dilute solution a very slight excess of carbonate of sodium, then add an excess of a solution of cyanide of potassium, heat the mixture for some time, and collect the carbonate of bismuth upon a filter. Since the carbonate thus thrown down always retains some alkali, it must be dissolved in acid and re-precipitated. The cyanide of potassium used must be free from any trace of sulphide. (Fresenius & Haidlen, *Annal. Chem. und Pharm.*, **43.** 129).

Carbonate of Cadmium.

Principle I. Insolubility in water and in carbonate of ammonium.

Applications. Estimation of cadmium in general. Separation of Cd from Mn and Cu.

Method A. Same as that described under Carbonate of Zinc. The precipitate should be collected upon a thin filter to avoid loss through reduction and volatilization, when the precipitate comes to be converted into Oxide of Cadmium.

Method B. In case cadmium is to be separated from copper, add carbonate of ammonium in excess, instead of the carbonate of sodium employed in Method A. Some cadmium will remain dissolved with the copper for a while, but on leaving the mixture exposed to the air all the carbonate of cadmium will gradually be deposited as carbonate of ammonium evaporates. (Stromeyer). The method is said to be more convenient but less accurate than those which depend upon the insolubility of Sulphide of Cadmium in cyanide of potassium, and its solubility in dilute sulphuric acid. It is distinctly inferior also to the method by Sulphocyanide of Copper.

For the method of separating Cd from Mn, see Carbonate of Manganese.

Properties. Carbonate of cadmium is a white precipitate, insoluble in water and the fixed alkaline carbonates; exceedingly sparingly soluble in a solution of carbonate of ammonium, but readily soluble in solutions of the sulphate, nitrate, etc., of ammonium. The water which the precipitate contains is completely expelled by drying, and the carbonic acid by ignition.

Principle II. Solubility in an aqueous solution of cyanide of potassium.

Applications. Separation of Cd from Bi and Pb.

Method. See Carbonate of Bismuth and Sulphide of Cadmium.

Carbonate of Calcium.

Principle I. Fixity when gently heated.

Applications. Estimation of calcium in oxalate of calcium and other compounds of calcium with organic acids.

Method. In case the substance to be operated upon is oxalate of calcium, heat it carefully and gradually in a platinum crucible until the bottom of the crucible has become almost, but not quite, dull red. The crucible should be covered at first, but may be open afterwards. Keep the crucible at this temperature, decidedly below incipient redness, during 8 or 10 minutes, then allow it to become cold, and weigh. — The carbonate of calcium obtained should be white, or only faintly tinged with gray. It should give no alkaline reaction when moistened and subsequently tested with a small slip of turmeric paper. If the turmeric turn brown it will be evident that the precipitate has been overheated and that a portion of it has been changed to caustic lime through escape of carbonic acid. In that event pour as much of a strong aqueous solution of carbonate of ammonium into the crucible as will barely cover the precipitate, evaporate to absolute dryness upon a water bath, heat the dry precipitate gently over a lamp, but not nearly to redness, and again weigh. It is to be observed that a very moderate heat will be sufficient to volatilize the excess of carbonate of ammonium. There is no need of actually igniting the precipitate before weighing it the second time, and consequently no risk of expelling any carbonic acid from the revivified carbonate. The process gives good results when properly conducted, but the operator should on no account fail to apply the test with turmeric paper. — Before proceeding to ignite oxalate of calcium, or carbonate of calcium which has been thrown down as such, take care to remove the precipitate from the filter as completely as possible, and burn the filter thoroughly upon the lid of the crucible out of contact with the precipitate. It is even best not to add the siliceous filter ash to the carbonate in the crucible until after the latter has been heated.

If any other salt than the oxalate is to be converted to the state of carbonate by ignition, proceed as directed under Carbonate of Barium (fixity of), and take special pains in treating the residue with carbonate of ammonium as above described.

Properties. Carbonate of calcium undergoes no change in the air at temperatures below faint redness, but at an intense red heat it gradually loses carbonic acid, especially when exposed to a current of air or steam. When mixed with carbon the decomposition by heat is far more rapid, carbonic acid being reduced and given off in the form of carbonic oxide. Quantities of carbonate of calcium as large as half a gramme may easily be completely converted to quicklime by heating them in an open platinum crucible over an ordinary gas blast lamp; but Fresenius has found that the heat of a Berzelius spirit lamp is insufficient to effect this reduction. The composition of the salt is as follows, both in terms of molecules and per cents.

$$\begin{array}{rcl} CaO & = & 56 \\ CO_2 & = & 44 \\ \hline & & 100 \end{array}$$

Principle II. Sparing solubility in water.

Applications. Estimation of calcium in aqueous solutions of calcium salts. Separation of Ca from Na, K, Mg, Mn. Estimation of carbonic acid in rocks, soils, waters, air, etc. Indirect separation of Ca from Ba and Sr.

Methods.

A. Estimation of Calcium in solutions of its salts. Saturate the moderately dilute solution with ammonia-water, add a solution of carbonate of ammonium in slight excess and let the mixture stand in a warm place for several hours. Collect the precipitate on a filter, wash it with water containing some ammonia, dry and ignite, or rather heat, the precipitate, as directed above (Principle I). The process yields accurate results when the liquid in which the precipitate is formed contains no great quantity of ammonium salts. It is essential that the precipitate be washed with ammoniated water, as will be seen below, under "properties."

B. Separation of Ca from K, Na and Mg. See the similar heading under Carbonate of Barium. Much that is said in that place of the preference to be given to Sulphate of Barium might be said here of Oxalate of Calcium.

C. Separation of Ca from Mn. See Carbonate of Manganese. There is nothing peculiar in the process as applied to the separation of Ca from Mn, excepting that the ignited precipitate must be treated with carbonate of ammonium to revivify the reduced carbonate of calcium, as has been explained above. See Carbonate of Calcium (fixity of). Unless the proportion of calcium in the mixture is large, it will usually be best not to weigh as CaO, Mn_2O_3 + CaO, CO_2,—but to ignite strongly over a blast lamp and weigh as CaO, Mn_2O_3 + CaO. See Oxide of Calcium.

D. Indirect separation of Ca from Ba and Sr. See Carbonic Acid (volatility of,—methods by simple ignition and ignition with an acid salt).

E. Estimation of Carbonic Acid.

1. *Fresenius's method.* Prepare a quantity of dry pulverulent hydrate of calcium by slaking a quantity of recently burnt lime with water. Put a small portion of the hydrate into dilute chlorhydric acid to test whether it is free from carbonic acid. If no effervescence is seen, seal up a number of portions, each of 2 or 3 grms., of the hydrate in small glass tubes for future use. But in case the hydrate is found to contain any carbonic acid place it in a tube of hard glass and ignite it upon a combustion furnace, in a current of air free from carbonic acid; and afterwards seal up several small portions of it as before.

Put 2 or 3 grms. of the pure lime into a light flask of about 300 c. c. capacity, close the flask with a caoutchouc stopper and weigh it together with the stopper and the lime. Pour in enough of the mineral water, or other solution of carbonic acid, to nearly fill the flask, replace the cork, shake the mixture and again weigh. The difference between the two weighings gives the weight of the mineral water. Loosen the stopper and heat the contents of the flask for some time upon a water bath, in order that the amorphous carbonate of calcium at first formed may become crystalline. Without disturbing the sediment at the bottom of the flask, pour the clear liquid upon a small plaited filter and allow the filter to drain; then, without washing either filter or precipitate, throw back the filter with its contents into the flask and determine the carbonic acid by decomposing the carbonate with an acid and collecting the Carbonic Acid in soda-lime, in the manner described on p. 91. — In case the mineral water contains an alkaline bicarbonate, it is well, after filling the flask, to add to the liquid enough chloride of calcium to decompose the bicarbonate. It is unnecessary to make any correction or allowance in this case, for the trifling solubility of carbonate of calcium in water. — The method is accurate, simple and expeditious, and is very much to be preferred to the old method of treating the mineral water with a mixture of chloride of calcium, or chloride of barium and ammonia-water, and afterwards weighing or titrating the precipitate. (Fresenius, *Zeitsch. analyt. Chem.*, 1863, **2.** 56).

In case the mineral water to be examined is contained in a bottle, it may be transferred to the lime flask by means of a syphon, after the bottle and contents have been cooled to about 4°. If the water were poured directly from the bottle into the flask, some free carbonic acid might flow into the latter with the water.

Sometimes the process above given of weighing the mineral water, had better be dispensed with, and only the volume of the water determined.

Thus when water is collected at a spring by opening a large pipette, or a flask provided with two orifices (Mohr, *Annalen der Pharm.*, 1834, **11.** 231 or *Titrirmethode*, 1855, **1.** 115), beneath the surface of the water and then closing and withdrawing the vessel with its contents, it is easy to determine the capacity of the pipette or flask beforehand, and to transfer its contents directly to the lime flask without need of further measurement.

2. *By precipitating with a mixture of chloride of calcium and ammonia.* See the similar heading under Carbonate of Barium, and below under Principle III.

3. *Pettenkofer's method.* See below, and under Carbonate of Barium, also.

Principle III. Power of neutralizing acids.

Applications. Estimation of calcium, of carbonic acid, and of most free acids (see Acidimetry).

Methods.

A. *Estimation of Calcium* (Compare Alkalimetry).

Pour upon the powder or the moist precipitate which is to be examined, a measured quantity of nitric or chlorhydric acid of known strength, taking care to use a little more acid than would be sufficient to dissolve the carbonate. To do this, place the carbonate in a flask and pour the acid slowly upon it from a burette in such manner that no portion of the liquid shall be thrown out of the flask by the escaping carbonic acid. In order to drive out the carbonic acid which remains dissolved in the acid solution, heat the latter carefully until it boils, then add a few drops of litmus, and determine the amount of free acid by titrating with a standard solution of soda. Or, better, omit the boiling and use Cochineal instead of litmus to indicate the point of saturation. By subtracting the amount of acid neutralized by the soda from the quantity of acid taken to dissolve the carbonate, the amount of acid neutralized by the latter will be obtained. The proportion of calcium in the precipitate, that is to say, the amount of calcium equivalent to the acid thus neutralized, may readily be calculated from these data.

B. *The estimation of Carbonic Acid* may evidently be effected in the same way as that of calcium, by the method just described, it being merely necessary to calculate how much carbonic acid is equivalent to the amount of standard acid neutralized by the carbonate.

The manner of applying the process will appear more fully in the following paragraphs.

C. *Pettenkofer's method of estimating carbonic acid by lime water.* Prepare a standard solution of oxalic acid of the strength indicated in the description of the analogous process with baryta-water (see Carbonate of Barium). Standardize a quantity of lime-water with this oxalic acid in the same way the baryta-water is standardized, and proceed with the analysis as follows:—Measure off 100 c. c. of the spring water or other dilute solution of carbonic acid to be analyzed, into a dry flask; add to it 3 c. c. of a highly concentrated solution of chloride of calcium, 2 c. c. of a saturated aqueous solution of chloride of ammonium, and 45 c. c. of the standard lime-water. Close the flask with a caoutchouc stopper, shake its contents, and leave it at rest for 12 hours in order that the amorphous carbonate of calcium, at first thrown down, may become crystalline. The total volume of liquid in the flask amounts to 150 c. c. Take out two portions, each of 50 c. c. of the clear liquid, and by means of the standard oxalic acid and turmeric paper, determine in each portion how much hydrate of calcium still remains free and uncombined with carbonic acid. Precisely as in the operation of standardizing, the experiment upon the first portion of liquid will give an approximation to the truth, and that with the second portion an accurate result. Multiply by 3 the number of c. c. of oxalic acid used in the last experiment and subtract the product from the number of c. c. required to neutralize 45 c. c. of the standard lime-water. The difference will show how much lime has been precipitated by carbonic acid, and, as has been said, each c. c. of the acid corresponds to 1 millig. of CO_2.

The chloride of calcium is added in the case of a mineral or spring water, as supposed above, to decompose any traces of carbonate or other alkaline salt whose acid might be precipitated by lime-water. The purpose of the chloride of ammonium, on the other hand, is to prevent the precipitation of magnesium in case any compound of that metal be present. The addition of chloride of calcium is beneficial, moreover, in case the lime-water happens to contain traces of free caustic alkali, or the carbonic acid water any carbonate of magnesium, for, in the absence of chloride of calcium, an oxalate of either of the alkali metals or of magnesium would react upon the carbonate of calcium, which is almost always contained in carbonic acid waters, to form oxalate of calcium and a carbonate of an alkali or of magnesium, and the latter would immediately combine again with oxalic acid.

If the water under examination contains nothing but carbonic acid, it is merely necessary to add lime-water (or better, dilute baryta-water) to it, and heat the mixture for some time to 70° or 80° to facilitate the change of the amorphous carbonate to the crystalline condition, but in case any chloride of ammonium has been added to the mixture no heat should be applied lest some ammonia be expelled.

This process was formerly recommended by its author for the analysis of waters containing but little carbonic acid, while the analogous method with baryta-water was preferred for analyzing waters highly charged with carbonic acid. In his later papers, however, Pettenkofer urges that the process with baryta-water be always used, to the exclusion of lime-water, for the amorphous carbonate of calcium which forms when lime-water and carbonic acid are first mixed is somewhat soluble in water, and exhibits an alkaline reaction which may easily vitiate the titration. In presence of an excess of lime-water the amorphous carbonate of calcium changes to the insoluble crystalline state comparatively slowly. Baryta, moreover, exhibits a stronger alkaline reaction than an equivalent quantity of lime.

D. *Old method of collecting carbonic acid in a mixture of chloride of calcium and ammonia.* All that has been said under Carbonate of Barium of the preparation of the absorbent solution and the precipitation of the carbonic acid, applies here as well, with the exception that chloride of calcium must be substituted for the chloride of barium. There is, however, in this case no need of bringing the whole of the precipitate upon the filter, nor of rubbing off those portions of it which remain sticking to the flask. After the precipitate has been thoroughly washed, put the funnel which holds the filter into the neck of the flask in which the precipitate was formed, push a glass rod through the point of the filter and wash as much of the precipitate as possible from the filter into the flask, then spread out the filter upon a plate of glass, wash off the last particles of the precipitate into the funnel and flask, and boil the wash water gently for half an hour. The purpose of the boiling is to expel ammonia, some of which is retained by the precipitate even after long continued washing. Pour into the flask a measured quantity of standard nitric or chlorhydric acid, more than sufficient to dissolve the whole of the precipitate, heat the liquid to expel the carbonic acid and titrate the excess of acid with a standard alkali, as described under Alkalimetry. The point of saturation may be indicated either by Litmus solution added to the liquor, or by Turmeric paper as in Pettenkofer's process above described.

According to Mohr, the alkalimetric process with carbonate of calcium is to be preferred to the corresponding process with carbonate of barium, in spite of the fact that carbonate of barium is more nearly insoluble in water than the calcium salt, for precipitated carbonate of calcium is far less bulky than carbonate of barium, and is more readily filtered than the latter, since it is less liable to clog the pores of filter paper. The comparatively high atomic weight of carbonate of barium has no significance in a process where the precipitate is not to be weighed. Like the carbonate of barium process, however, the method of precipitating carbonate of calcium from an ammoniacal solution is exposed to several sources of error, and is no longer held in much esteem.

In case of need, the process may be applied to the estimation of gaseous carbonic acid. In that event, the details of the method remain unchanged, with the exception that the gas is made to flow into a mixture of chloride of calcium and ammonia-water, or better, into ammonia-water to which chloride of calcium is afterwards added.

Properties of Carbonate of Calcium. The precipitate as ordinarily obtained, is a soft, white powder, scarcely at all soluble in absolutely pure, cold water, but appreciably soluble in ordinary distilled or other water which contains traces of carbonic acid gas, and the solution thus obtained exhibits a faint alkaline reaction. It is somewhat more readily soluble in boiling than in cold water. It is far from being insoluble in cold aqueous solutions of ammonium salts, such as the nitrate and chloride, and dissolves readily in hot solutions. The presence of free ammonia and carbonate of ammonium, however, hinders the solvent action of the ordinary ammonium salts. Solutions of the normal salts of potassium and sodium have a tendency to dissolve it, but their action is more feeble than that of ammonium salts. Carbonic acid water dissolves it easily. It cannot be precipitated from solutions which contain citrates or metaphosphates of the alkalies. (See, further, Dictionary of Solubilities).

For use as a reagent, pure carbonate of calcium may be prepared as follows:—Dissolve a quantity of white marble or of ignited stalactite, in chlorhydric acid, neutralize the solution with ammonia-water, heat the mixture to boiling and filter to separate the small quantity of iron and alumina which is thrown down. Heat the filtrate to boiling in a large beaker, and throw into it, one by one, small bits of solid carbonate of ammonium until they cease to dissolve. Wash the heavy crystalline precipitate by decantation with hot water. (Matthiessen).

Principle IV. Power of precipitating iron, aluminum, manganese, chromium, etc. See the Hydrates of those metals. For lists of the substances precipitable by carbonate of calcium see that substance in Dictionary of Solubilities.

Principle V. Insolubility in an aqueous solution of cyanide of potassium.

Applications. Separation of Ca from Co, Ni and Zn.

Method. See Carbonate of Cobalt.

Principle VI? For the use of carbonate of calcium, or rather, of oxide of calcium, in decomposing silicious minerals, see Silicates.

The following account of *Lawrence Smith's method of separating alkalies* from refractory silicates by decomposing the mineral with carbonate of calcium and chloride of ammonium, is taken from Prof. Johnson's edition of *Fresenius's Quant. Analysis*, New York, 1870, p. 303. Though out of place in this connection, the importance of the process is such that early mention should be made of it. — Mix 1 part of the pulverized silicate with 1 part of dry crystallized chloride of ammonium by gentle trituration in a smooth mortar, then add 8 parts of pure, precipitated carbonate of calcium, and mix the whole intimately. Transfer the mixture to a platinum crucible, taking care to rinse the mortar with a little carbonate of calcium. Warm the crucible gradually over a small Bunsen burner until fumes of ammonium salts no longer appear; then heat to full redness, but not too intensely, during 30 or 40 minutes. An ordinary portable furnace or chafing dish provided with a conical sheet-iron cap or chimney 2 or 3 feet high will give heat enough for the purpose, in default of a large Bunsen lamp. — The mass in the crucible should sinter together but not fuse. When cold, it may usually be detached with ease from the platinum. Heat the sintered lump to boiling in a capsule with 100 c. c. of water for several hours, or until it is entirely disintegrated and fallen to powder. In case the lump, from having been overheated, remain partially coherent after long boiling, it may be transferred to a porcelain mortar, ground to fine powder and then boiled as before. Some silicates, notably those containing much protoxide of iron, fuse easily with the proportions of flux above given. It is best when this happens to repeat the ignition on a new quantity of the mineral, using as much as 10 or 12 parts of carbonate of calcium and taking care to bring only the lower three-fourths of the crucible to a red heat. — When completely disintegrated by boiling with water, the sintered mass gives up to the water all the alkalies as chlorides, together with some chloride of calcium and caustic lime. The mixture is filtered and the powder well washed; a quantity (1 or 2 grms.) of carbonate of ammonium in solution is then added to the filtrate and the latter is evaporated to a bulk of about 30 c. c. A little more carbonate of ammonium together with a few drops of ammonia-water is added to ensure a complete separation of the calcium, and the mixture is again filtered. Collect the filtrate and washings in a weighed platinum capsule and evaporate to dryness on a water bath. Place the capsule in a capacious iron cup and heat the latter so that the contents of the platinum capsule within it may be thoroughly dried. Finally, heat the capsule carefully almost to redness to drive out the ammonium salts, cool and weigh. The alkali chlorides thus obtained are nearly pure, though a trifling amount of black residue will usually be seen on dissolving them in a few drops of water. This residue may be removed, if need be, by filtering through a very small filter. — Prof. Smith's process is by far the most convenient and accurate for separating the alkalies from silicates, and is universally applicable except perhaps in presence of boracic acid. (Johnson, *loc. cit.*).

*Bi*Carbonate of Calcium.

Principle. Power of neutralizing acids.

Application. Estimation of bicarbonate of calcium in natural waters.

Method. The method is merely one of Acidimetry, in which a solution of phosphate of copper in chlorhydric acid is made to serve both as the standard acid and as the indicator of the point of saturation. — To prepare the copper solution, mix a solution of cupric chloride with one of ordinary phosphate of sodium, wash the phosphate of copper which is precipitated, mix the washed precipitate with water, and add to the mixture moderately strong chlorhydric acid, drop by drop, until, the chlorhydric acid being slightly in excess, a clear solution is obtained. When such a solution is dropped into a solution of bicarbonate of calcium—or of any carbonate or bicarbonate of either of the alkali–or alkaline–earthy metals—a quantity of the phosphate of copper is precipitated, at first, as the calcium, or other alkaline metal, neutralizes the free chlorhydric acid; but the liquid soon becomes clear again when a further portion of the cupric solution is added, for the precipitate redissolves in the free acid which this solution contains. — The moment at which the turbid solution becomes clear, is taken as the point of saturation, for the quantity of cupric solution required to effect the precipitation and re-solution is proportional to the amount of base in the substance tested, and consequently to the amount of carbonic acid which was combined with the base to form a bicarbonate. The presence of free carbonic acid, that is to say, of an excess of the acid over and above what is necessary to form a bicarbonate, has no influence whatever upon the indications of the process. — The cupric solution is standardized against pure, dry carbonate of sodium. To this end dissolve 0.265 grm. (equal one two-hundredth of an equivalent) of the dry salt in distilled water, dilute the solution to the volume of a litre (see Alkalimetry), and saturate the solution with carbonic acid gas. The copper solution may be made of such strength that 4.4 c. c. of it will saturate, in the manner explained above, 100 c. c. of the standard soda solution. With a liquid of this strength, it will only be necessary to multiply by 22 ÷ 44 (= 0.5) the number of c. c. of the liquid consumed in titrating any 100 c. c. sample of natural water, in order to obtain the amount of carbonic acid in that water expressed in terms of centigrammes. It is well also to employ a burette graduated to fifths of cubic centimetres. — The process is said to be superior to that of Barthélemy (see Carbonate of Mercury), inasmuch as it is applicable to waters contaminated with chlorides and sulphates. It succeeds better with bicarbonates than with the normal carbonates, and may be applied to either of the bicarbonates as well as to bicarbonate of calcium. (Lory, *Chemical News*, **18.** 169).

Carbonate of Cobalt.

Principle I. Solubility in an aqueous solution of cyanide of potassium.

Applications. Separation of Co from Ba, Sr, Ca and Al.

Method. Mix the solution with a slight excess of carbonate of sodium. Add a quantity of cyanide of potassium solution to the mixture of liquid and precipitate, and heat the whole gently until the whole of the carbonate of cobalt has redissolved. Collect the undissolved carbonate of the alkaline earths upon a filter, and precipitate the cobalt in the filtrate as Cobalticyanide of Mercury.

Principle II. Insolubility in water.

Carbonate of cobalt may be precipitated by adding an alkaline carbonate to the solution of a cobalt salt. But the precipitation can with difficulty be made complete even by long continued boiling. The principle is not to be recommended as a means of determining cobalt. (Gibbs & Taylor, *American Journ. Sci.*, 1867, **44.** 214).

Carbonate of Cobalt and of Calcium.

Principle. Insolubility.

Application. Separation of Co from Ni.

Method? Dissolve the mixture of cobalt and nickel in a slight excess of chlorhydric acid and add to the solution 10 parts of chloride of calcium and 10 parts of chloride of ammonium for every 3 parts of the mixed oxides which are contained in it. Mix the solution with 150 parts of cold water and 20 parts of sesquicarbonate of ammonium, previously dissolved in 100 parts of cold water, and gradually heat the mixture to boiling. After the solution has been allowed to cool and settle, collect the precipitate upon a filter and wash with a solution of carbonate of ammonium. To estimate the cobalt, dissolve the mixed precipitate in chlorhydric acid and proceed in the usual way. The nickel may be determined in the filtrate. (L. Thompson, *Zeitsch. analyt. Chem.*, 1864, **3.** 375). Experiments by Winkler, *ibid.*, p. 376), go to show that the process as presented by Thompson, is wholly unfit for quantitative use. Not only is some carbonate of nickel always found in the precipitate, but a quantity of cobalt invariably remains in solution and passes into the filtrate.

Carbonate of Copper.

Principle I. Solubility in an aqueous solution of carbonate of ammonium.

Applications. Separation of Cu from Bi and Cd.

Method. See the Carbonates of Bismuth and of Cadmium.

Principle II. Solubility in an aqueous solution of cyanide of potassium.

Applications. Separation of Cu from Bi and Pb.

Method. See Carbonate of Bismuth and Sulphide of Copper.

Principle III. Insolubility of the basic carbonate.

Applications. Estimation of copper in acid solutions. Separation of Cu from Mn.

Method. Some years since, H. Rose (*Handbuch analyt. Chem.*, 1851, **2.** 188) stated that copper cannot be completely precipitated by means of carbonate of potassium. According to this chemist a certain proportion of the copper remains obstinately in solution, and can only be obtained by evaporating the liquid to dryness and gently igniting the residue. But Gibbs has found that Rose's statement is too strong; the whole of the copper may be precipitated by alkaline carbonates from solutions of sulphate, nitrate or chloride of copper when the latter are sufficiently dilute and are boiled for a long time with the carbonated alkali.

The best method of effecting the precipitation is as follows:—Dilute the copper solution until it contains no more than one gramme of copper to the litre. Add a solution of carbonate of potassium or of sodium, in slight excess, and boil the mixture for half an hour. The boiling proceeds quietly without bumping; the blue-green carbonate soon becomes dark brown and the oxide or basic carbonate finally obtained has a fine granular character which renders it extremely easy to wash. The small portion of the precipitate which usually adheres to the sides of the vessel in which the boiling takes place, must be redissolved in acid and again precipitated; but great care must be taken not to add too large an excess of the alkaline carbonate, lest a solution be formed from which the copper cannot be precipitated by boiling. If the process be well conducted, the original filtrate will be perfectly free from copper. The washed precipitate may be ignited in a current of hydrogen, and the Copper weighed as such; it will be found to be free from alkali. The ignition must be carefully conducted, since the ignited precipitate is so finely divided that particles of it are liable to be carried off in the current of gas. (Gibbs & Taylor, *American Journ. Sci.*, 1868, **44**. 213).

For the method of separating Cu from Mn, see Carbonate of Manganese; as well as the description above given. Test the filtrate from the mixed carbonate with sulphydrate of ammonium, to be sure that all the copper has gone down.

Carbonate of Lead.

Principle I. Insolubility in cold water.

Applications. Estimation of lead in all salts of that metal which are soluble in water, or from which the lead can be dissolved by nitric acid. Separation of Pb from Mn.

Method. Add a slight excess of carbonate of ammonium, together with a small quantity of ammonia-water, to the moderately dilute solution of the lead salt. Heat the mixture for some time, and allow it to settle until the liquid has become clear, then collect the precipitate upon a small, thin filter and wash it with cold water which has been recently boiled to expel carbonic acid. Ignite the dried precipitate in a porcelain crucible and weigh as Oxide of Lead. The filter should be burned by itself upon the cover of the crucible, after the precipitate has been removed from it as completely as possible. It is well also to moisten the cold filter-ash with a drop or two of nitric acid, and to evaporate and ignite before weighing. — The process is a tolerably satisfactory one, though the results obtained are usually somewhat too low, on account of the solubility of the precipitate. It is easy to incur loss also in burning the filter, through reduction of some of the precipitate left upon the paper. The method, however, is better than that which depends on the insolubility of Oxalate of Lead. (Mohr and Fresenius).

Lead may be precipitated as completely by means of the bicarbonate of potassium or sodium as by carbonate of ammonium; normal carbonate of lead being thrown down in both instances. But the precipitation is not complete with the normal alkaline carbonates, and a not insignificant portion of the precipitate may be dissolved in case it is heated with an excess of the normal carbonate of either of the fixed alkalies. (H. Rose).

For the separation of Pb from Mn see Carbonate of Manganese. Add carbonate of ammonium or bicarbonate of sodium to the filtrate from the mixed carbonates of Pb and Mn, to ensure the complete precipitation of the lead.

Properties. Precipitated carbonate of lead is heavy, white and pulverulent. It is almost absolutely insoluble in water which contains no trace of carbonic acid, but dissolves in carbonic acid water. According to Fresenius, 1 part of it dissolves at the ordinary temperature in about 50,000 parts of water which has been boiled. It is more soluble in water charged with ammoniacal salts than in pure water. On ignition it loses its carbonic acid readily. When thrown down by an excess of bicarbonate of sodium the precipitate retains traces of the sodium salt.

Principle II. Insolubility in an aqueous solution of cyanide of potassium.

Applications. Separation of Pb from Cu, Cd, Hg, Ag and Au.

Method. See Carbonate of Bismuth. The carbonate of lead thus precipitated always contains alkali. It may be dissolved in nitric acid and reprecipitated by carbonate of ammonium, as in Principle I, or as Sulphate of Lead. Or it may be reduced to metallic Lead by fusion with cyanide of potassium.

Basic Carbonate of Magnesium.

Principle I. Insolubility in water.

Applications. Estimation of Mg in presence of K, and in a solution of Mg in carbonic acid water. (Method A). — Separation of K from a mixture of Mg, Ba and Ca. (Method B). — Separation of Mg from Mn.

Method A. To estimate magnesium in presence of alkalies, when the quantity of the latter is not to be determined, boil the solution strongly for a long time with an excess of carbonate of potassium, and wash the precipitate with boiling water. The washing must be proceeded with without interruption until a few drops of the filtrate leave only a small residue when evaporated on platinum foil. Though the compound is somewhat soluble in water, it is less so in hot than in cold, and comparatively little of it is taken up by water which is boiling hot and free from carbonic acid. The dried precipitate is ignited strongly and the residue weighed as Oxide of Magnesium. — This method was formerly much employed, but is no longer held in esteem. It is inconvenient and liable to several sources of error. Hence it has been superseded by the method which depends on the insolubility of Phosphate of Magnesium and Ammonium. The normal carbonate of magnesium, which may be supposed to be precipitated at the moment when carbonate of potassium is added, is decomposed, by boiling, into an insoluble basic and a soluble acid carbonate, and in order to decompose the latter completely the liquor has to be boiled for a long time. A similar remark applies to the case where magnesium is to be determined in a carbonic acid water solution. — The mixture must be boiled strongly from first to last, in order to avoid a difficultly soluble compound of carbonate of magnesium and carbonate of potassium, which forms at moderate heats. Carbonate of sodium is somewhat inferior to carbonate of potassium as the precipitant, since an insoluble double carbonate forms more readily with sodium than potassium. Both these double compounds are decomposed by ignition, or rather, after ignition it is easy to wash out the alkaline carbonate from the oxide of magnesium. — Since it is not easy to determine when all the magnesium has been precipitated, and since some of it would inevitably remain dissolved if the boiling were stopped too soon, it is well to evaporate the liquid to dryness in order to be sure that the separation is complete. To this end let the boiled mixture settle, decant the clear liquor into an evaporating dish, best of platinum, throw the precipitate upon a filter and add the filtrate and first portions of wash water to the liquid in the dish. Boil down the liquid rapidly to absolute dryness, heat the residue strongly, and after the dish has become cold treat the residue with hot water. Collect the small portion of the mass that remains undissolved, upon a small filter, and wash it by itself. To avoid loss in the process of drying, place the dish, at last, within a capacious iron cup and stir its contents continually, at a gentle heat, until the mass is completely dry. The liquid must be boiled strongly during the process of evaporation in order to hinder the saline matter from creeping over the edge of the dish. The original method of evaporating the entire mixture of precipitate and liquid to dryness is not to be commended. (v. Bonsdorf, *Pogg. Annal.*, **18.** 128). — In case the magnesium solution contains ammonium salts, carbonate of potassium must be added to destroy them. A large excess of carbonate of potassium is added to the liquid, the mixture is warmed until the odor of ammonia ceases to be perceptible, more carbonate of potassium is added, and the mixture again warmed to make sure that the whole of the ammonia has been expelled, and the liquid is finally boiled as above described. In case the boiled liquid fails to give a strong alkaline reaction with red litmus paper, a new quantity of carbonate of potassium must be added, and the mixture again boiled. (H. Rose, *Handbuch*, 1851, **2.** pp. 33–37, 52).

Method B. Heat the chlorhydric acid solution of magnesium, calcium, barium and the alkalies to actual boiling, add a solution of carbonate of sodium, drop by drop, as long as any precipitate continues to fall, and boil strongly until the voluminous precipitate becomes compact and granular. If the precipitate is boiled long enough it can be washed free from potassium. Acidulate the filtrate with chlorhydric acid, and determine the potassium as Chloroplatinate of Potassium. (Stohmann, *Zeitsch. analyt. Chem.*, 1866, **5.** 307).

For the method of separating Mg from Mn, see Carbonate of Manganese.

According to Bineau, 1 litre of water dissolves 0.06 grm. of three-fourths carbonate of magnesium. According to Chevalet, a litre of water dissolves 0.106 grm. of normal (?) carbonate of magnesium. (*Zeitsch. analyt. Chem.*, 1869, **8.** 91).

Principle II. Solubility in carbonic acid water.

Applications. Separation of Al from Mg and from both Mg and Ca, if the proportion of the latter be very small.

Method. Place the cold, moderately acid, rather dilute solution in a beaker provided with a suitable cover. Add a solution of bicarbonate of potassium or sodium prepared in the cold, as long as any effervescence occurs, or any precipitate falls. Violent effervescence occurs during the precipitation, all the alumina is thrown down in the form of a hydrate, while the magnesium remains dissolved in the liquid charged with carbonic acid. After the mixture has been allowed to stand for 12 hours, decant the clear liquid into a filter and wash the precipitate, first with cold carbonic acid water by decantation, and afterwards with pure water upon the filter. Carbonic acid water for the washing may be prepared by slowly adding to a highly dilute solution of bicarbonate of potassium or sodium a small quantity of chlorhydric acid, insufficient to combine with the whole of the metal. — Magnesium may be determined directly in the filtrate, as Phosphate of Magnesium and Ammonium. The alumina precipitate is often free or almost free from any trace of magnesium (H. Rose) but since it is liable to retain some alkali, it had better be dissolved in chlorhydric acid and the solution treated with ammonia-water to throw down Hydrate of Aluminum. — The process was formerly much employed and would still appear to be valuable.

In case the method is employed for separating both lime and magnesium from aluminum, the solution should be very dilute and a stoppered flask should be employed, instead of a beaker, to effect the precipitation. Satisfactory results can be obtained only when the quantity of lime is very small, since it is liable to be retained by the alumina.

Principle III. Power of neutralizing acids.

Application. Valuation of the commercial carbonate.

Method. Dissolve about a gramme of the substance to be tested in 60 or 70 c. c. of normal nitric acid, taking care to measure the acid and to use an excess of it. Determine the excess of acid by titration with a standard solution of ammonio-sulphate of copper (see Acidimetry), and subtract this excess from the amount of acid taken. The difference will be equivalent to the amount of magnesium in the substance. The results obtained are usually somewhat too low. The process is of technical application only; it has no claim to scientific accuracy. — Instead of the ammonio-sulphate of copper, standard ammonia-water may be employed to determine the excess of acid, but, according to Mohr, the copper solution is to be preferred.

In any event magnesium is less conveniently determined by alkalimetric methods than either of the metals of the alkaline earths. When normal nitric acid and litmus solution are added either to carbonate of magnesium or to calcined magnesia, the red color of the litmus very soon changes to blue and remains so as long as any trace of the magnesium compound is left undissolved. On adding more of the acid, until the color appears bright red, and then titrating backwards with caustic soda until the liquid becomes blue, results are obtained which indicate too little magnesium. (Mohr, *Titrirmethode*, 1855, **1.** pp. 80, 357).

Principle IV. Decomposition of by solutions of ferric salts, while hydrate of iron is precipitated. (Compare the Carbonates of Barium and Calcium).

Application. Separation of Fe_2O_3 from FeO in sulphuric acid solutions.

Method. See Hydrate of Iron. In this process the normal carbonate (*magnesite*) must be employed. The basic carbonate (*magnesia-alba*) will not answer.

*Bi*Carbonate of Magnesium.

(Compare *bi*Carbonate of Calcium).

Carbonate of Magnesium and of Ammonium.

Principle. Sparing solubility in ammoniated water.

Application. Separation of magnesium from the alkalies.

Method. Prepare an exceedingly concentrated solution of the substance to be analyzed, and in case it be acid, neutralize, or slightly supersaturate it with ammonia-water. Add to the liquid a large excess of a solution of normal carbonate of ammonium which contains rather more than one equivalent of oxide of ammonium for each equivalent of carbonic acid. To prepare this solution dissolve 230 grms. of commercial, solid, sesquicarbonate of ammonium in 180 c. c. of ammonia-water of 0.92 sp. gr., and enough water to bring the solution to the volume of a litre. Stir the mixture strongly until the voluminous precipitate which falls at first, on the addition of the carbonate of ammonium has completely re-dissolved. Then leave the mixture at rest for 12 or 24 hours. Collect the crystalline precipitate upon a filter, wash it with the solution of normal carbonate of ammonium, dry, ignite, and weigh the Oxide of Magnesium. — In separating magnesium from sodium, the precipitate can readily be washed free from fixed alkali; but if potassium be present, some carbonate of potassium is retained by the precipitate so strongly that it can only be washed out after the precipitate has been ignited. The ignited magnesia must consequently be washed with hot water in case potassium is present.

It is not absolutely necessary that the precipitate first formed should be completely re-dissolved, since when left to itself in the liquid it gradually contracts and becomes crystalline; but it is always safer to dissolve it by means of an excess of the precipitant and by agitation.

The final precipitate is granular and may be washed readily. It does not adhere firmly to the sides of the beaker. But the chief merit of the process consists in its applicability even in presence of large quantities of salts of ammonium or of the fixed alkalies. It is only necessary in that case to let the mixture stand for 24 hours and to stir it frequently. The process is inapplicable in presence of phosphoric or arsenic acids. (Schaffgotsch, *Pogg. Annal.*, **104.** 482; H. Weber, *Kopp's Jahresbericht*, für 1858, p. 606).

The behavior of the double salt towards solvents has been studied in some detail by Divers, (*Journ. London Chem. Soc.*, **15.** 196; *Zeitsch. analyt. Chem.*, **1.** 474), who appears, however, to have been ignorant of the labors of Schaffgotsch and Weber. According to Divers, the solution of normal carbonate of ammonium may be made to contain 1 part of salt to 6 parts of water. About 4 equivalents of the carbonate of ammonium should be taken for each equivalent of magnesium to be precipitated; and one equivalent or more of chloride of ammonium should be mixed with the magnesium solution before adding the precipitant, in order to prevent the precipitation of any normal carbonate of magnesium. — The composition of the double salt is MgO, CO_2; $(NH_4)_2$ O, $CO_2 + 4H_2O$. According to Schaffgotsch, 1 part of it requires 60,000 parts of a solution of normal or slightly alkaline carbonate of ammonium for its solution. It is somewhat soluble in a solution of chloride of ammonium, when no carbonate of ammonium is present. Divers found 1 part magnesia in 4660 parts of a mother liquor which contained an excess of carbonate of ammonium, some chloride of ammonium and a large quantity of sulphate of ammonium.

Carbonate of Manganese.

Principle. Insolubility in water.

Applications. Estimation of manganese in all salts of that metal which dissolve in water, excepting the salts of some fixed organic acids; and in all compounds from which the manganese can be dissolved out by chlorhydric acid. Separation of Mn from Ca, Ba, Sr, Mg, Al, Fe, Zn, Cd, Pb, Cu, Bi and P_2O_5.

Method. Place the moderately dilute solution in a tolerably large beaker, cover the beaker with a watch-glass-shaped cover, at the centre of which a hole has been bored, and heat the liquid. Pour a solution of carbonate of sodium, drop by drop, through the cover of the beaker until it is slightly in excess, heat the mixture to boiling for a moment and afterwards allow it to settle until the liquid is clear. The original manganese solution should be slightly acid, but not too acid lest some of the liquid be lost through the violent evolution of carbonic acid gas. — In case the manganese solution contains any salt of ammonium, a large excess of the alkaline carbonate must be employed, and the mixture be boiled down nearly to dryness, until no more fumes of ammonia escape, in order to ensure complete precipitation. The formation of dark colored oxide of manganese during the evaporation does no harm. During the evaporation the cover of the beaker must be taken off.

Wash the precipitate by decantation with boiling water, dry the precipitate and ignite and weigh as Manganite of Manganese (MnO, Mn_2O_3). In case the filtrate or wash water from the carbonate of manganese is not absolutely clear, let it stand 18 or 24 hours in a warm place and collect and wash the dirty brown precipitate upon a special filter. When carefully executed the process yields accurate results.

Properties. Recently precipitated manganous carbonate is a white flocculent substance, scarcely at all soluble in pure water, but somewhat soluble in water which contains carbonic acid. It is no more soluble in salts of the fixed alkalies than in water, but dissolves rather easily in ammonium salts. It cannot be completely precipitated in presence of ammonium salts, nor is it precipitable in presence of citric and other fixed organic acids. When left moist in the air, or when washed with water which contains air, the precipitate slowly takes on oxygen and becomes rusty through formation of hydrated sesquioxide of manganese. — When dried, out of contact with the air, the precipitate yields a white powder of the composition $MnO, CO_2 + \frac{1}{2}$ Aq; but when dried in the air the powder is always more or less colored. When heated to redness in the air, the powder becomes black and afterward brown. By long continued ignition it is completely converted into MnO, Mn_2O_3.

To separate either of the metals above enumerated from manganese, precipitate the latter together with the other metal by means of carbonate of sodium, as above described. Wash the mixed precipitate thoroughly, dry, ignite and weigh. The ignited precipitate consists of a mixture of $MO, Mn_2O_3 + x\ MO$, or of $MO, Mn_2O_3 + x\ (MO, CO_2)$, accordingly as the carbonate of M is or is not decomposed by ignition. In case the metal precipitated, together with the manganese, is iron or aluminum, all the manganese will be in the form of Mn_3O_4. — Place a weighed portion of the ignited precipitate in a small flask, together with some pure concentrated chlorhydric acid. Heat the mixture, conduct the chlorine, which is evolved, into a solution of iodide of potassium and estimate the iodine, which is set free, by means of hyposulphite of sodium. Or estimate the Chlorine in any appropriate way. (Compare Manganites, and Manganite of Manganese.) In case the method with iodide of potassium be used, the amount of iodine found will correspond, atom for atom, to that of the chlorine set free from the chlorhydric acid. But for every two atoms of chlorine evolved there must have been present one molecule of sesquioxide of manganese: —

$$Mn_2O_3 + 6HCl = 2MnCl_2 + 3H_2O + 2Cl.$$

To find the weight of the other metal deduct the weight of the sesquioxide of manganese from the total weight of the mixed precipitate, and, if need be, add to the difference the weight of the carbonic acid which was expelled by the manganese during the ignition. That is to say, for the metals Ba, Ca and Sr, which retain carbonic acid on ignition, allow one molecule of CO_2 for every molecule of Mn_2O_3. — As thus far described, excepting the case of iron or aluminum, the method presupposes that more than one molecule of MO is present for every molecule of Mn_2O_3, for if the proportion of MO were less than this the ignited precipitate would contain some MnO, Mn_2O_3, as well as MO, Mn_2O_3. The process may be made available, however, in all cases by the following modification: Weigh out half as much oxide of zinc at there is supposed to be MO and MnO in the mixture, dissolve it in chlorhydric acid, mix the solution with the substance to be analyzed and precipitate the whole with carbonate of sodium. The ignited precipitate will contain the whole of the manganese in the form of Mn_2O_3. (Krieger, *Annal. Chem. und Pharm.*, **87.** 261).

To separate manganese from phosphoric acid. Fuse the weighed substance with carbonate of sodium for some time, boil the fused mass with water, add a little sulphuretted hydrogen water to reduce any manganic acid which may have formed, collect the insoluble carbonate of manganese upon a filter and wash, dry, and weigh, as above.

Carbonate of Mercury. (Mercurous carbonate).

Principle. Insolubility in water and solubility in nitric acid.

Applications. Estimation of carbonic acid in bicarbonates, especially as they occur in natural waters. Separation of bicarbonates of the alkalies from those of the alkaline earths. Use as an indicator in Acidimetry.

Method. Prepare a solution of acid mercurous nitrate by digesting an excess of quicksilver with dilute nitric acid, pouring off the mother liquor from the crystals of the basic salt, and diluting it with 4 or 5 parts of water. By leaving the solution in contact with metallic mercury it may be preserved for a long time. When this solution is added to dilute solutions of bicarbonates of the alkalies or alkaline earths, at temperatures lower than 30°, a precipitate falls which is white at first, but soon changes to yellowish-orange, or sometimes to yellowish-green. This precipitate dissolves in an excess of the acid mercurous nitrate, as well as in sulphuric and nitric acids and in organic liquids, such as urine. Precipitates obtained, under analogous conditions, from the normal carbonates are brown and insoluble in an excess of the mercurous nitrate.

Prepare a standard solution of bicarbonate of potassium by dissolving 0.5 grm. of the bicarbonate in water to the volume of a litre. This quantity of the bicarbonate is equivalent to 0.241 grm. of carbonic acid. Standardize a quantity of the mercurous nitrate by pouring it, drop by drop, from a Gay-Lussac burette into measured portions of the standard solution of bicarbonate of potassium, until the precipitate which forms at first has completely redissolved. Repeat the titration two or three times with different quantities of the bicarbonate solution until the results agree. Then titrate the bicarbonate, or the water, to be tested, in the usual way. (See Acidimetry).

In order to separate alkaline bicarbonates from bicarbonates of the alkaline earths, proceed as above, with one portion of water to be tested. Then boil another portion of 100 c. c. until the carbonates of the alkaline earths have all been thrown down, filter, add water to the filtrate to replace what has evaporated, saturate the filtrate with carbonic acid and titrate with the mercurous solution to determine the proportion of alkaline bicarbonate. Or, add to a measured quantity of the water, a measured volume of standard caustic potash (0.5 grm. to the litre) sufficient to change the bicarbonates of the alkaline earths into neutral salts, let the mixture stand for several days, decant the clear liquid, saturate it with carbonic acid and titrate as before. Subtract, of course, the bicarbonate of potassium which has resulted from the caustic alkali employed.

The presence of sulphates or chlorides in the water interferes materially with the success of the process, but when the proportion of chlorine is not too large the bicarbonates can still be determined approximately, as follows:— Measure out several portions of 100 c. c. each, of the water, acidulate them with nitric acid, and observe about how many drops of the standard mercurous nitrate are required in order to precipitate the chloride and impart a definite gray coloration to the liquid. By trying several experiments it is easy to hit this point with sufficient accuracy. Then add the standard mercurous nitrate to a fresh, non-acidulated 100 c. c. portion of the water, until the yellowish-orange precipitate has disappeared and the liquid has acquired the gray coloration aforesaid. — Since the mercurous nitrate acts upon caoutchouc, it cannot be used with Mohr's burette unless the latter be provided with a glass cock or regulated by a clip at the top. (Barthélemy, *Zeitsch. analyt. Chem.*, 1869, **8.** 91).

Carbonate of Nickel.

Principle I. Solubility in an aqueous solution of cyanide of potassium.

Applications. Separation of Ni from Ba, Sr, Ca and Al.

Method. Same as that described under Carbonate of Cobalt.

Principle II. Insolubility in water.

Method. Same as that described under Carbonate of Copper. The green basic carbonate of nickel, obtained by boiling the dilute solution of a nickel salt with an alkaline carbonate, may be washed much more readily than the precipitate obtained by means of a caustic alkali. According to Genth, the process yields highly satisfactory results. (Gibbs & Taylor, *American Journ. Sci.*, 1867, **44.** 214).

Carbonate of Potassium. [Compare Carbonate of Sodium].

Principle I. Fixity of the salt at moderately high temperatures.

Applications. Estimation of potassium in many potassium salts of organic acids. Valuation of argol and cream of tartar.

Method. Heat the salt gently for a long time in a covered platinum crucible, until absolutely no more visible fumes escape. Wash out the black porous residue from the crucible into a beaker, cover the latter with a watch glass, and add dilute chlorhydric acid to the solution as long as there is any effervescence. Filter the acidulated liquor, to separate the particles of carbon, evaporate the filtrate in a platinum cup and weigh the Chloride of Potassium. Or, instead of chlorhydric acid, use dilute sulphuric acid, and weigh as Sulphate of Potassium. In some cases, as in the examination of tartar, it will be most convenient to estimate the potassium in the carbonate by the method of alkalimetry (see below, Principle II).

The carbonate of potassium in the crucible must never be heated to the point of fusion, and the crucible should be kept covered throughout the operation. If the crucible were left open and the residue roasted until it became white, a considerable quantity of carbonate of potassium would be lost through volatilization. A small proportion of the potassium is usually lost in this way during the process of carbonization, but the process nevertheless yields perfectly satisfactory results when properly conducted. If the process of heating be continued long enough a colorless solution will be obtained when the residue is treated with water. (Braun, *Zeitsch. analyt. Chem.*, 1868, **7.** pp. 149, 150). See also Aug. Vogel (*ibid.* p. 149) for quantitative experiments, showing the large amount of carbonate of potassium which may be lost by volatilization. — It is not well to burn the coaly residue white by throwing in fragments of nitrate of ammonium (as directed by H. Rose), because of the great heat which would be produced by the combustion. Nor is it advisable to use nitric acid instead of chlorhydric, as the solvent of the carbonate.

In some rare cases it may be best to weigh carbonate of potassium as such, though, on account of its tendency to deliquesce, the salt is not well adapted for weighing. In that event, evaporate the solution to dryness in a platinum crucible and ignite the residue at a moderate heat. A few small fragments of solid carbonate of ammonium may be placed in the crucible before the final ignition, in order to convert into carbonate of potassium any traces of caustic potash which may be present. It is to be observed that in igniting carbonate of potassium in contact with carbon, some of the carbonic acid is decomposed, carbonic oxide being set free and caustic potash produced. This decomposition occurs particularly when the carbonate of potassium is heated to fusion in contact with charcoal. The crucible in which carbonate of potassium is weighed should have a tight cover, to hinder the salt from absorbing water from the air during the operation of weighing. All trouble from deliquescence may be avoided, however, by igniting the carbonate with chloride or with sulphate of ammonium and weighing as Chloride or Sulphate of Potassium, as the case may be.

Principle II. Power of neutralizing acids.

Applications. Estimation of potassium in the commercial carbonate; in other words, valuation of pearlash, saleratus, potashes and wood ashes. Valuation of argol and cream of tartar.

Methods. The proportion of pure carbonate of potassium in any given sample of the commercial article may be estimated, either by determining how much of a standard acid can be neutralized by a weighed quantity of the sample (see Alkalimetry), or by determining how much carbonic acid is set free when a weighed quantity of the sample is mixed with a weighed quantity of acid or fused with an acid salt (see Carbonic Acid, volatility of, Methods C and B; also p. 22).

The commercial carbonates of potassium are liable to contain substances insoluble in water,— such as carbonate, silicate, and phosphate of calcium, sand, and dirt, which may be removed by filtration; and neutral salts, such as the alkaline chlorides and sulphates, which have no influence whatsoever in the process of titration. But, besides these harmless ingredients, there may be present certain other compounds,— such as the hydrate, silicate, phosphate, sulphite, sulphide and hyposulphite of potassium,— which must be guarded against in certain cases. [Compare Carbonate of Sodium]. — In many cases it will be sufficient to determine the amount of "available alkali" in the sample, without reference to its precise chemical composition. For example, if the commercial article under examination is to be used for making caustic potash by boiling the aqueous solution with lime, the presence of a small quantity of silicate or of phosphate of potassium will do no harm, any more than that of caustic potash, since both of these compounds will be made caustic by the lime, like the carbonate itself.

Since both silicate and phosphate of potassium have an alkaline reaction and behave towards acids like the carbonate, the method of alkalimetry by neutralization is incompetent to distinguish them from the carbonate; but they do not interfere in any way with those methods of analysis which depend upon the volatilization and quantitative estimation of the Carbonic Acid in the sample. — According to Persoz (*Comptes Rendus*, **53.** 239), the influence of sulphites, sulphides, and hyposulphites may be avoided, so long as the sample is free from carbon or organic matters, by using his method of fusion with bichromate of potassium (see p. 80). Fresenius destroys the sulphur compounds by igniting the weighed sample of carbonate with chlorate of potassium, before proceeding to the treatment with acid. The sulphur compounds are thus changed to inert sulphate of potassium. But if any hyposulphite be present, an equivalent quantity of the carbonate will be destroyed by its oxidation. (See Carbonate of Sodium).

For methods of determining the proportion of caustic and carbonate of potassium in a mixture of the two, like many samples of American potash, see under Alkalimetry.

In case the carbonate of potassium to be examined is contaminated with carbonate or hydrate of sodium, the proportion of potassium in the mixture may be estimated by one of the methods referred to at the close of the article Alkalimetry. According to Fresenius, the following process yields accurate results, and is tolerably expeditious:— Dissolve in water 6.25 grm. of the ignited pearlash, filter the solution into a quarter-litre flask, add acetic acid to slight excess, and warm the liquid to expel carbonic acid. After the carbonic acid has been driven off, add a solution of acetate of lead, drop by drop, to the hot liquor until the formation of a precipitate of sulphate of lead *just ceases*. Allow the mixture to cool, fill the flask with water to the mark, shake the mixture and then let it stand to settle; filter through a dry filter, and transfer 200 c. c. of the filtrate, corresponding to 5 grms. of the pearlash, to a quarter-litre flask. Fill the flask to the mark with strong sulphuretted hydrogen water and shake its contents. If the acetate of lead was carefully added the fluid will now smell of sulphuretted hydrogen and be free from lead. In case the liquid does not smell of sulphuretted hydrogen a stream of that gas must be passed through it. After the sulphide of lead has subsided, filter the liquid through a dry filter; place 50 c. c. of the filtrate, corresponding to 1 grm. of the pearlash, together with 10 c. c. of chlorhydric acid, of 1.1 sp. gr., in a weighed platinum dish, and evaporate the mixture to dryness. Cover the dish, heat the mixed chlorides moderately, and weigh them (see Chloride of Potassium). The weight obtained expresses the amount of chloride of potassium plus chloride of sodium given by 1 grm. of the pearlash. Determine the chlorine by titration (see Chloride of Silver) and calculate the amounts of potassium and of sodium as described under Chloride of Potassium. For a similar calculation see Carbonic Acid, indirect separation of calcium from strontium.

In case it is desired to know how much of any sample of pearlash consists of foreign salts, determine the Water contained in it, by gently heating a weighed quantity (8 or 10 grms.) in a covered platinum dish for a long time, until dew ceases to be deposited upon a piece of cold window glass held over the dish. The loss of weight will give the proportion of water. The difference between the percentage of water plus the percentage of carbonate of potassium, as above determined, and 100, will give the percentage of fixed impurity in the substance analyzed.

In the examination of ashes, or of liquors obtained by leaching ashes, it is well to weigh out or to measure 10 or 12 times as much material as has been directed under Alkalimetry, since the proportion of alkali in ashes is comparatively small.

For use as a reagent, pure carbonate of potassium may be prepared as follows:—Heat a mixture of 10 parts of purified powdered cream of tartar, 10 parts of water, and 1 part of pure strong chlorhydric acid for several hours upon a water bath, stirring the mixture frequently. Place a small filter in the throat of a capacious funnel, pour the mixture into the funnel, and let the liquid portion drain away. Level off the top of the solid matter in the funnel, and press down upon it a disk of compact filter paper turned upwards at its edges. Pour repeated small portions of cold water (iced water is best) upon the filter paper,—in order that all the chloride and phosphate of calcium and any chloride of potassium which has been formed may be removed by percolation,—until a drop of the filtrate ceases to become cloudy when acidulated with nitric acid and tested with nitrate of silver. Then dry the purified tartrate.

Prepare on the other hand, a quantity of pure Nitrate of Potassium, and dry it. Mix 2 parts of the pure tartrate of potassium with 1 part of the pure saltpetre; see to it that the mixture is completely dry; and project the mixture by small portions into a clean, bright, wrought iron pot heated to low redness. As soon as the last portion has deflagrated, heat the contents of the pot strongly until a sample of the carbonate taken from the edge of the mass yields a perfectly colorless solution with water. Then rub up the coaly mass with water, filter, wash the residue slightly and evaporate the filtrate in a porcelain, or better, a silver dish, until it is covered with a permanent crust. Stir the liquor constantly while it cools, and throw the crystalline meal into a funnel as before; allow the meal to drain, wash it somewhat, dry it by heat in a silver or porcelain dish, and preserve the powder in well stoppered bottles. The product should be white, and should give no reaction for silicic, sulphuric, chlorhydric or phosphoric acids when evaporated with chlorhydric acid, or tested with acidulated chloride of barium, nitrate of silver or molybdate of ammonium. It should give no reaction for iron when tested with sulphocyanide of potassium.

A mixture of 13 parts of carbonate of potassium and 10 parts carbonate of sodium is preferable to either of its components for decomposing silicious minerals, by way of fusion, since the mixed carbonates melt at a lower temperature than either of the carbonates taken separately. Most refractory silicates may be decomposed by fusing them with the mixed carbonates over an ordinary Berzelius lamp or simple Bunsen's burner. (See Silicates). — Instead of mixing the pure carbonates directly, the mixture may be prepared either by igniting pure Rochelle salt and lixiviating and evaporating the residue, or by deflagrating a mixture of 20 parts pure bitartrate of potassium, prepared as above, and 9 parts of pure nitrate of sodium, and proceeding as above described.

***Bi*Carbonate of Potassium.** See *bi*Carbonate of Sodium. Compare biCarbonate of Calcium.

Carbonate of Silver.

Principle I. Power of neutralizing acids.

Application. Precise neutralization of nitric acid liquors for the purpose of precipitating phosphate of silver, as a means of separating phosphoric acid from alkalies, alkaline earths, etc.

Method. See Phosphate of Silver. — If a phosphate insoluble in water is dissolved in a slight excess of nitric acid and the solution mixed with nitrate of silver, no precipitate will fall. But, by agitating the solution for a few moments with a slight excess of carbonate of silver, it is easy to neutralize the acid and so cause the complete precipitation of the phosphoric acid (as phosphate of silver) without introducing any new or hurtful reagent. (Chancel, *Comptes Rendus*, 1859, **49.** 997).

For the use of carbonate of silver for decomposing various chlorides, as one step in Grægger's method of estimating combined sulphuric acid, see Chloride of Silver, insolubility of.

Principle II. Solubility in ammonia-water, and insolubility in water.

Applications. Separation of CO_2 from S_2O_2 and H_2SO_4, in the analysis of gunpowder residues.

Method. After all the sulphur which was in the form of an alkaline sulphide has been removed as Sulphide of Cadmium, by digesting the solution with carbonate of cadmium and filtering, heat the filtrate from the sulphide of cadmium, mix it with a neutral solution of nitrate of silver and collect the precipitate upon a filter. The precipitate consists of a mixture of carbonate and sulphide of silver, for

$$K_2O, S_2O_2 + Ag_2O, N_2O_5 = K_2O, SO_3 + Ag_2S + N_2O_5,$$

while the whole of the sulphuric acid will remain in solution, and may be determined in the filtrate as Sulphate of Barium, after the excess of silver has been removed by means of chlorhydric acid. — After washing the precipitate, digest it in ammonia water to dissolve the carbonate of silver; then acidify the solution with nitric acid and estimate the silver as Chloride of Silver. Each equivalent of chloride of silver obtained will correspond to an equivalent of carbonic acid, but from this amount must be subtracted the carbonic acid derived from the carbonate of cadmium previously employed for decomposing the sulphide of potassium, since

$$K_2S + CdO, CO_2 = CdS + K_2O, CO_2.$$

(Werther, *Journ. prakt. Chem.*, **55.** 22).

According to Fedorow (*Zeitsch. analyt. Chem.*, 1870, **9.** 127), the foregoing process is not based upon correct principles, since on precipitating a hyposulphite with a solution of silver, a certain quantity of acid is set free, and a proportional quantity of carbonate of silver dissolved.

For use as a reagent, carbonate of silver may be readily prepared by adding a solution of carbonate of ammonium to one of nitrate of silver. The precipitate may be washed with water by decantation. It is neither necessary to filter nor to dry it, since it acts best when moist.

Carbonate of Sodium.

Principle I. Fixity of the salt when heated.

Applications. Estimation of sodium in the hydrate, bicarbonate, nitrate and chloride of sodium, and in organic salts of that metal. The process is one of far more general applicability than the corresponding method, which depends on the fixity of carbonate of potassium.

Method. In the case of a simple solution of carbonate of sodium, evaporate to dryness in a platinum dish, ignite moderately and weigh. Since the ignited carbonate is not deliquescent it is easy to obtain accurate results in this way, but care must be taken, nevertheless, not to heat the carbonate too strongly, lest some of it be lost by volatilization (compare Carbonate of Potassium); carbonaceous matter and reducing gases, moreover, must be kept out of the crucible. — Bicarbonate of sodium is treated like the carbonate. But the crucible must be kept covered and the heat carefully regulated while the excess of carbonic acid is being driven off. — Caustic soda is treated with an excess of carbonate of ammonium, the mixed solution is evaporated at a gentle heat and the residue ignited.

Compounds of sodium with organic acids are ignited like Carbonate of Potassium, in a covered crucible; the residue is digested in water and the solution filtered and evaporated, with addition of a small quantity of carbonate of ammonium to revivify any caustic soda which may have been formed during the process of carbonization.

Chloride and nitrate of sodium may be converted into the carbonate by mixing their aqueous solution with a moderate excess of oxalic acid, and evaporating the solution to dryness. Water must be added to the dry residue and the process of evaporation several times repeated. All of the chlorhydric acid is driven out, and the nitric acid as well, while a portion of the latter undergoes decomposition. The residue is finally ignited to destroy the excess of oxalic acid, and the carbonate of sodium is weighed.

Properties. The dry carbonate is white and tolerably permanent; when left in the air for any length of time, however, it slowly absorbs a certain quantity of water. It fuses at a strong red heat, and if kept in the fluid state for any length of time, an appreciable quantity of it is lost by volatilization. It scarcely loses weight, however, when ignited moderately, or even when heated to incipient fusion. Carbonate of sodium is much less soluble in dilute ammonia water than in pure water, and is scarcely at all soluble in strong alcohol. The composition of the anhydrous salt is: —

2Na	=	46	=	43.40
C	=	12	=	11.32
O_3	=	48	=	45.28
		106		100.00

Principle II. Power of neutralizing acids.

Applications. Estimation of sodium in the commercial carbonate and bicarbonate. Valuation of soda-ash, black-ball, sal-soda, saleratus, etc. Assay of sulphur in iron- and copper-pyrites (sulphur ore).

Methods. See Carbonate of Potassium.

Soda-ash generally contains, besides carbonate of sodium, a certain proportion of hydrate, sulphate and chloride of sodium; some traces of silicate, aluminate and cyanide of sodium; and not infrequently, appreciable quantities of sulphide, sulphite and hyposulphite of sodium. — The three substances last named are specially objectionable, inasmuch as they neutralize the standard acid and pass for carbonate of sodium, while so far from having any value, considered as alkalies, they are really hurtful. To detect their presence proceed as follows: — Mix a portion of the ash with dilute sulphuric acid, and observe whether any odor of sulphuretted hydrogen can be perceived. Color a quantity of dilute sulphuric acid with a drop or two of a solution of permanganate of potassium, and add to the liquor a quantity of soda-ash, insufficient to neutralize the acid. If the liquor becomes green, or if its purple color is destroyed, either sulphite or hyposulphite of sodium is present; but if the permanganate retains its color, both of these salts must be absent. The presence of hyposulphite of sodium may be made still more manifest by saturating a clear solution of the soda-ash with chlorhydric acid, and noting the odor of sulphurous acid, which is given off, as well as the turbidity due to the precipitation of sulphur. These impurities can be destroyed by igniting the ash with chlorate of potassium, and so converting all the sulphur into the condition of sulphuric acid before proceeding with the titration; though the method is not wholly satisfactory when hyposulphite of sodium is present, inasmuch as some carbonate of sodium will then be decomposed, in accordance with the reaction: —

$$Na_2O, CO_2 + Na_2O, S_2O_2 + 4O = 2(Na_2O, SO_3) + CO_2.$$

The presence of silicate or aluminate of sodium will usually be indicated by the formation of a precipitate when the carbonate is neutralized with acid. As a general rule, however, the quantity of these impurities is so small that they need not be noticed in technical estimations; the hydrate of sodium, also, is often estimated as if it were carbonate, though all of these contaminations have to be estimated and allowed for whenever the proportion of pure carbonate of sodium in the sample is to be determined. — In case it is desired to know the proportion of inert solid matter in the sample, as distinguished from the water which is contained in it, the water must be estimated and allowed for, as directed under Carbonate of Potassium.

To estimate Sulphur in Pyrites, and in the residues left by roasted pyrites, Pelouze (*Comptes Rendus*, **53.** 685) proceeds as follows: — Weigh out about 1 grm. of very finely powdered pyrites, add to it an accurately weighed quantity of pure anhydrous carbonate of sodium (about 5 grms.), and carefully mix the materials with a mixture of 7 grms. of chlorate of potassium and 5 grms. of dry chloride of sodium, which have been weighed roughly upon a coarse balance and rubbed together in a mortar. Heat the mixture in a wrought iron spoon during 8 or 10 minutes, so that it shall gradually be brought to dull redness. Allow the melted mass to cool, treat it 5 or 6 times with hot water, throw the solution upon a filter, boil the residue with water, and wash it upon the filter with boiling water. Bring the filtrate and wash-water to some definite volume (see Alkalimetry), and by means of a standard acid titrate the carbonate of sodium in a measured portion of the liquor. The difference between the weight of carbonate of sodium taken and that found in the matter resulting from the process of oxidation, gives the amount of carbonate of sodium which has been decomposed by the sulphuric acid formed from the sulphur in the pyrites. Each equivalent of carbonate of sodium thus destroyed corresponds with one equivalent of sulphur. The process requires only 30 or 40 minutes for its completion, and yields results which differ only about 1 to 1.5 per cent from the truth. — The whole of the sulphur is converted into sulphate of sodium or potassium, and no sulphurous acid escapes during the process of oxidation. In order to be sure, however, that the washed insoluble residue is free from sulphur, it is well to test it with chlorhydric acid. It is to be observed that any loss of carbonate of sodium through volatilization, projection, or insufficient washing, will tend to make the amount of sulphur appear larger than it really is. The purpose of the chloride of sodium is to moderate the action of the chlorate; the proportion of the chloride may be varied for the rest, according to the quality of the pyrites, and increased so that the oxidation may occur without deflagration, or any appearance of incandescence. In case pyrites-residues are to be analyzed, the chloride of sodium had better be omitted altogether; the mixture may then be made of 5 grms. roasted pyrites, 5 grms. carbonate of sodium, and 5 grms. chlorate of potassium. The presence of quartz, heavy spar or limestone in the pyrites, does no harm. — Instead of using absolutely pure carbonate of sodium it will be sufficient to use that obtained by igniting the commercial bicarbonate, and to determine once for all, by titration, the proportion of pure carbonate contained in it.

Principle III. Power of decomposing or of combining with many silicates, sulphates, chlorides and oxides, when fused therewith.

In decomposing silicates it is usual to mix the finely powdered mineral with 4 times its weight of carbonate of sodium, and to fuse the mixture over a blast lamp or in a furnace. Basic silicate of sodium is formed while the silicates in the mineral are rendered basic and decomposable by acids. (See Silicates).

For the method of decomposing chlorides, see Chloride of Silver; of sulphates, see Sulphate of Barium.

Alumina may be separated from small quantities of the alkaline earths and of manganese, such as often contaminate the Hydrate of Aluminum drawn down by ammonia, by fusing the dry precipitate with 10 or more times its weight of dry carbonate of sodium over a powerful blast lamp, until no more bubbles of carbonic acid are seen to escape from the melted mixture. The alumina precipitate need not be powdered, but the fusion should last from three-quarters of an hour to an hour. Add a little caustic soda to the cold, fused mass, and boil it with water, best in a silver dish, until the soluble matter has all dissolved. In case the solution has a green color, from the presence of manganate of sodium, it should be boiled with a little alcohol to reduce the manganate. Filter off the soluble aluminate of sodium, and wash the precipitate, first with water charged with caustic potash, and then with pure water. (R. Richter, *Journ. prakt. Chem.*, **64.** 378). Compare Aluminate of Sodium.

For use as a reagent, pure carbonate of sodium may be prepared from the commercial bicarbonate, by washing the latter by percolation with ice-water upon a funnel, as described under Carbonate of Potassium, and igniting the dried residue. The process of percolation must be continued until a portion of the filtrate, after having been acidulated with nitric acid, yields no turbidity when tested with the nitrates of silver and of barium. The washed and dried bicarbonate must be ignited gently, best in a dish or crucible of silver or platinum, though in default of these a clear, bright, iron dish, or dish of Berlin porcelain may be employed. When solutions of the salt are needed, or the crystallized salt itself, it is sufficient to recrystallize commercial sal-soda, repeatedly. The solution should yield no insoluble residue of silica when evaporated to dryness after saturation with chlorhydric acid; nor should it give any reaction for iron or for phosphoric, sulphuric or chlorhydric acid, when tested with sulphocyanide of potassium, molybdate of ammonium, chloride of barium or nitrate of silver.

*Bi*Carbonate of Sodium.

[Compare *bi*Carbonate of Calcium].

For the use of bicarbonated alkalies for decomposing and partially dissolving certain sulphates and chlorides insoluble in water, see Sulphate of Lead and Chloride of Lead.

Carbonate of Strontium.

Principle I. Fixity when not too strongly heated.

Applications. Estimation of strontium in organic salts of that metal, and all compounds of strontium which are soluble in water. (Compare below, Principle II).

Method. Same as that described under Carbonate of Barium. — Carbonate of Strontium may be heated to low redness without losing any of its carbonic acid. When heated to intense redness, or moderate whiteness,— as when ignited over a good gas blast lamp, even when not enclosed in a clay cylinder, — it gradually gives off the whole of its carbonic acid. (Compare Carbonic Acid, indirect separation of strontium and calcium). When heated to redness in contact with carbonaceous matter, some carbonic oxide is set free and caustic strontia formed. The latter may be revivified by adding a solution of carbonate of ammonium and evaporating as directed under Carbonate of Calcium. In case the quantity of Carbonate of Strontium in the crucible is small, it may be ignited intensely at the blast lamp and weighed as Oxide of Strontium. In that event the carbonate should be made as compact as possible; it should be precipitated from a hot solution and pressed tightly into the crucible. After the mass has been ignited for some time it is well to turn over the cake in the crucible, in order that all of its surfaces may be thoroughly heated. (Schaffgotsch, *Pogg. Annalen*, **113.** 615.) The composition of carbonate of strontium is:—

Sr =	87.5 =	59.32
CO_3 =	60.0 =	40.68
	147.5	100.00

Principle II. Insolubility in water.

Applications. Estimation of strontium in soluble strontium salts. Separation of Sr from K, Na, Mg and Mn. The precipitation of strontium as carbonate is to be preferred to the estimation of that metal as a sulphate, whenever the addition of alcohol to the solution is inadmissible, and the strontium to be determined is combined with a non-volatile acid.

Method. Same as that described under Carbonate of Barium. The results are accurate, since carbonate of strontium is almost absolutely insoluble in water charged with ammonia and carbonate of ammonium. The presence of ammonium salts, moreover, is less hurtful in the precipitation of carbonate of strontium than in that of carbonate of barium. By direct experiment, Fresenius found 99.82 instead of 100 parts of strontia. As with carbonate of barium, care must be taken that the solution is free from substances likely to prevent the precipitation of the carbonate.

For the separation of manganese and strontium see Carbonate of Manganese.

Properties. Dry, precipitated carbonate of strontium is a white powder soluble in about 18,000 parts of water at the ordinary temperature, and in about 57,000 parts of water, which contains ammonia and carbonate of ammonium. (Fresenius.) It is sensibly less soluble in water than sulphate of strontium. (Dulong; H. Rose.) It is rather easily soluble in solutions of chloride or nitrate of ammonium, but is reprecipitated on adding ammonia and carbonate of ammonium. Carbonic acid water dissolves it, and the solution has a feeble alkaline reaction. It is not precipitated in presence of an alkaline citrate or metaphosphate.

Principle III. Insolubility in a solution of cyanide of potassium.

Applications. Separation of Sr from Co, Ni and Zn.

Method. See Carbonate of Cobalt.

Principle IV. Power of neutralizing acids.

Application. Estimation of strontium in the carbonate.

Methods. Similar to those described under Carbonate of Barium.

Carbonate of Zinc.

Principle I. Insolubility in water.

Applications. Estimation of zinc in all zinc salts which are soluble in water; in those which dissolve in chlorhydric acid, with separation of their acid; and in all those with volatile organic acids. Separation of Zn from

Na, K, Mn and Va. Estimation of CO_2 in mineral waters.

Method A. Similar to that described under Carbonate of Manganese. The precipitation should be made in a tolerably capacious beaker, not more than one-third full of the zinc solution. The liquid should be heated nearly to boiling before adding any of the carbonate of sodium, and be kept at that temperature, or at actual boiling, throughout the experiment. At a certain moment, however, during the addition of the carbonate of sodium, the mixture is liable to suddenly froth violently, from the escape of a torrent of carbonic acid, while a quantity of basic carbonate of zinc is thrown down. The lamp must of course be removed at this instant. It is to prevent the liquid from boiling over when this momentary violent reaction occurs, that the beaker must be large and the source of heat carefully watched and regulated. Boil the mixture freely for a few minutes after all the zinc has been precipitated; then decant the liquid into a filter and wash the precipitate thoroughly with boiling water; at first by decanting two or three times from the beaker, and afterwards upon the filter itself. Dry, ignite in a platinum crucible and weigh as Oxide of Zinc. The carbonate of zinc must be rubbed off from the filter as completely as possible before the latter is burned, and care must be taken to burn the filter in such manner that it cannot reduce any of the oxide of zinc. — In case the zinc solution contains any salt of ammonium the boiling must be continued after the addition of the carbonate of sodium, until all the ammonia is expelled, and the steam no longer turns turmeric paper brown. The precipitation will never be complete until the last trace of ammonia has been driven off. Hence, if the proportion of ammonium salt be large it will be best to boil down the mixture to dryness. (Compare Carbonate of Magnesium). Even when this has to be done the process is much to be preferred to the one depending upon the insolubility of Sulphide of Zinc. Care must be taken to treat the dry residue with boiling water, since cold water would dissolve a part of the zinc in the form of a double carbonate of zinc and sodium. — The process yields excellent results in spite of the fact that a decided trace of zinc always escapes precipitation. On testing the filtrate from the carbonate of zinc with sulphydrate of ammonium a slight precipitate of sulphide of zinc will invariably be obtained, especially when the mixture is left to stand for an hour or two. It is to be inferred, therefore, either that these flocks of sulphide are well-nigh imponderable, or that the precipitated carbonate retains enough sodium, or other foreign substance, to compensate for the lost zinc. It is admitted, however, that the carbonate thrown down, as above described, may be completely freed from alkali by washing with hot water. In case any other substance is to be determined in the filtrate, it will be best to mix the latter with sulphydrate of ammonium, to collect and weigh the Sulphide of Zinc, and add the weight of zinc thus found to that obtained as carbonate.

Precautions. The original solution of zinc should be free from any great excess of acid, in order that no unnecessary effervescence need occur on the addition of carbonate of sodium. The liquor must be heated during the precipitation, since a portion of the zinc would fail to be precipitated at the ordinary temperature, at least from a solution of the sulphate or chloride, and the mixture must finally be boiled to expel the free carbonic acid from the solution, and so prevent it from dissolving some of the carbonate of zinc. But on the other hand, the boiling must not be too long continued lest some of the precipitate be decomposed by the sulphate or chloride of sodium, with formation of carbonate of sodium, and an insoluble basic sulphate or chloride of zinc. These basic salts would not be wholly decomposed on ignition, except at a very high temperature. Even when the carbonate is precipitated from a nitric acid solution, it may retain some nitric acid, but it loses it readily when ignited. (H. Rose).

Instead of operating in the wet way, as above, the dry zinc salt may be heated cautiously to near redness with an excess of carbonate of sodium in a platinum dish, and the residue treated with water, with the precautions above enjoined. This modification of the process is specially to be commended when the compound to be analyzed contains an ammonium salt. (Johnson).

Method B. As a modification of the usual method, E. Jacob (*Zeitsch. analyt. Chem.*, 1865, **4.** 212) has proposed the following:— Heat the somewhat acid zinc solution to 60° or 80° C.; carefully pour into it a solution of carbonate of sodium until, after the lapse of a few seconds, the cloudiness which forms after each addition of the alkali ceases to disappear, and the carbonic acid escapes tranquilly. Then pour in an excess of the carbonate of sodium and boil the mixture.

By proceeding in this way the carbonate of zinc is precipitated completely as a comparatively compact powder, far more easily washed and dried than the light flocculent precipitate obtained by the ordinary process. According to Jacob, no basic salt is formed in his process. It is to be remarked in this connection, that H. Rose has always directed that the zinc solution be mixed with an excess of carbonate of sodium and then heated to boiling. In some of the earlier editions of his Handbuch, he suggests that the zinc solution had better be poured into a warm solution of the alkaline carbonate, since in that case we may be sure that no basic salt would be precipitated.

To separate zinc from vanadic acid, Czudnowicz (*Zeitsch. analyt. Chem.*, 1864, **3.** 379) directs that the acid solution should be heated with oxalic acid, or some other reducing agent, to bring the vanadic acid to the condition of protoxide, and the solution then digested for a long time with carbonate of sodium. In case all the vanadium is in the state of protoxide, the zinc will be precipitated free from any trace of that metal, but in case the reduction were incomplete the carbonate of zinc would be dark-colored. In that event the moist, washed precipitate must again be dissolved in chlorhydric acid, the solution again reduced and the zinc afterwards reprecipitated with carbonate of sodium.

For the method of separating zinc from manganese see Carbonate of Manganese.

With regard to the estimation of carbonic acid in mineral waters, Björklund (*Zeitsch. analyt. Chem.*, 1865, **4.** 229) finds that free carbonic acid can be completely absorbed from an aqueous solution by digesting the latter with oxide of zinc. To apply the method quantitatively, Björklund adds from 4 to 10 grms. of oxide of zinc to a litre of the mineral water, accordingly as there is more or less carbonic acid to be absorbed, and lets the mixture digest during at least 24 hours. The bottle which contains the water should be completely filled with liquid, since the oxide of zinc has little or no power to absorb any carbonic acid which might escape into the air of the space above the liquid in a bottle incompletely filled. Since the oxide of zinc is apt to acquire a crystalline consistency after all the carbonic acid has been absorbed, it is well to wait for this appearance. The mixed oxide and carbonate may then be collected with peculiar ease.

Carbonic acid cannot be absorbed, however, in this way by oxide of zinc, when it exists as a neutral salt in combination with either of the alkali-metals, or with calcium, magnesium or iron; or as a bicarbonate of an alkali. From a solution of carbonate of calcium in carbonic acid water, oxide of zinc absorbs the excess of carbonic acid so that neutral carbonate of calcium is precipitated; a similar remark would apply to the solution of ferrous carbonate in carbonic acid water, were it not that the oxygen which is present in most carbonated waters acts upon a part of the neutral carbonate of iron and converts it into ferric oxide. The acid carbonates of magnesium and manganese behave like those of calcium and iron. In each case the amount of neutral carbonate of calcium, etc., which can be held dissolved by pure water, will remain in solution. In case the water to be examined contains a large proportion of carbonate of ammonium, errors may arise from the solubility of oxide of zinc in that substance.

Properties. As obtained by Method A, recently precipitated basic carbonate of zinc occurs in the form of light, white flocks, as good as insoluble in water. According to Fresenius, one part of the precipitate dissolves in about 45,000 parts of water. It is soluble, however, in carbonic acid water, and readily soluble in solutions of the fixed caustic alkalies, and of carbonate of ammonium, ammonia water and acids. A concentrated solution of zincate of potassium or of sodium is not altered by boiling, but by boiling a dilute solution nearly all the oxide of zinc can be thrown down. In the solutions in ammonia or carbonate of ammonium a precipitate is likewise formed on boiling, especially if the solution be dilute ; such a solution would of course be decomposed with evolution of carbonate of ammonium, when boiled with carbonate of sodium.

As has been already stated, the precipitate obtained by Method A is not normal carbonate of zinc, since more or less carbonic acid is given off during the precipitation, even after all the free acid in the original solution has been neutralized. The proportion of hydrated oxide of zinc in the precipitate varies, however, considerably, according to the degree of concentration of the solutions employed, and the details of the manipulations. — When dried, the precipitate forms a loose white powder, from which all the carbonic acid is readily expelled at a red heat.

Principle II. Solubility in an aqueous solution of cyanide of potassium.

Applications. Separation of Zn from Ba, Sr. Ca, Mg and Al.

Method. Same as that described under Carbonate of Cobalt and Cyanide of Cobalt. Boil the filtrate with chlorhydric acid, to which a little nitric acid has been added, in order to expel the cyanhydric acid, and throw down the zinc as Carbonate, with the precautions enjoined for the case where ammonium salts are present.

Carmine. See Cochineal.

Carthamin and Carthamic Acid. See Safflower.

Casein. [Compare Milk.]

Principle I. Coagulability by acids.

Application. Separation of casein from milk and from alkaline solutions.

Method. On adding a few drops of acetic acid to milk, casein is precipitated, together with fatty matter. To purify the casein, collect the coagulum upon a filter, wash it and redissolve it in a solution of carbonate of sodium; leave the solution at rest for 12 hours, in order that the fatty matter may separate from the alkaline solution of casein. Take off the layer of fat and reprecipitate the casein by adding a few drops of chlorhydric acid to the solution; collect the casein upon a filter and wash, dry and weigh it. (Robin & Verdeil).

According to Lehmann, acetic acid, though formerly almost the only reagent employed for the quantitative determination of casein, by no means effects a thorough precipitation of it, and may even dissolve a very considerable amount of it when added in excess.

Principle II. Coagulability when heated with sulphate of calcium.

Application. Estimation of casein in milk.

Method. Stir into the milk about one-fifth its weight of finely powdered gypsum, and heat the mixture to 100°. Perfect coagulation ensues, and there is obtained, on evaporating, a brittle residue, which may be readily pulverized, and from which ether and alcohol easily remove fat, milk, sugar, and most of the salts. (Haidlen, *Annalen Chem. und Pharm.*, **45.** 273.)

Catechu. See Tannin.

Cellulose.

Principle I. Insolubility in dilute acids, alkalies, alcohol and ether.

Method. Weigh out 3 grms. of the thoroughly dried and finely powdered substance (hay or straw) to be analyzed, and place it in a porcelain dish, together with 200 c. c. of dilute sulphuric acid, containing 1.25 per cent of oil of vitriol. Boil the liquid for half an hour, and continually add small quantities of water to replace that which evaporates Then allow the mixture to settle, and draw off the clear acid liquor with a syphon and a pipette into a beaker. Pour 200 c. c. of water upon the residue and boil for half an hour, taking care to replace, as before, the water which evaporates. Allow the mixture to settle, draw off the clear liquor into the same beaker as before, add another quantity of 200 c. c. of water to the residue and again boil, settle and decant. — The residue is next boiled for half an hour with a mixture of 50 c. c. of five per cent potash lye (80 grms. of fused hydrate of potassium to the litre), and 150 c. c. of water; and afterwards with two successive portions of water, each of 200 c. c., for half an hour in each case. The decanted liquors are collected in a second beaker.

The solid matter which remains in the dish after the several boilings, is thrown upon a weighed filter; to it is added any sediment which may have been deposited in the beaker which has received the alkaline liquids, and the contents of the filter are then washed with water, until the washings no longer exhibit an alkaline reaction. The sediment in the beaker which received the acid liquors decanted from the cellulose, is then thrown upon the same filter, and the contents of the latter washed successively with water, alcohol and ether, as long as these solvents continue to dissolve anything. Before throwing the sediments from the alkaline and acid liquors upon the filter, the clear supernatant liquids above these sediments must be carefully decanted and thrown away. (Henneberg, *Die landwirthschaftliche Versuchs-Stationen*, 1864, **6.** 497). The final residue left after washing with ether, is cellulose, contaminated with ash and with nitrogen, for which corrections must be made. The ash is readily determined by incinerating a weighed portion of the crude cellulose, while the nitrogen is assumed to belong to some albuminoid, the amount of which is calculable. Since all the albuminoids contain, on the average, about 16 per cent of nitrogen, it will be sufficient to determine the percentage of nitrogen in the crude cellulose, and to multiply this quantity by $100 \div 16 = 6.25$ in order to obtain the amount of albuminoid to be subtracted. — Even with these corrections the estimation of cellulose in this way is not absolutely accurate, as may usually be seen from the appearance of the product. According to v. Hofmeister, crude cellulose prepared in this way from pease was perfectly white, while that from wheat-bran was brown, and that from rape-cake almost black. In order to obtain comparable results, care must be taken to operate under the precise conditions prescribed. If the solvents are too concentrated, or the temperature at which they act is too high, some of the cellulose will be dissolved, while, if the reagents are too much diluted, a quantity of other matter will escape solution.

Principle II. Power of resisting oxidizing agents.

Method. Place one part (from 2 to 4 grms.) of the dry pulverized substance, which has been previously extracted with water, alcohol and ether, in a glass-stoppered bottle, together with 0.8 part of chlorate of potassium and 12 parts of nitric acid of 1.10 sp. gr., and digest the mixture at a temperature not exceeding 18° for 14 days. At the end of this time add a quantity of water to the contents of the bottle, filter the mixture and wash the residue first with cold, and afterwards with hot, water. — As soon as all the acid and soluble matters have been washed out, throw the contents of the filter into a beaker and heat them to 74° for three quarters of an hour, with weak ammonia water, made by mixing 1 part of commercial ammonia with 50 parts of water. Finally collect the insoluble matter upon a weighed filter and wash it, first with dilute ammonia, as long as the filtrate passes off colored, then with cold and hot water, then with alcohol, and at last with ether. The cellulose obtained in this way, though free from chlorine, still contains a small quantity of ash and nitrogen, for which corrections must be made, as in Method 1; it is purer, however, than that obtained by Method 1, and the quantity obtained is larger—usually from 0.5 to 1.50 per cent larger — than that obtained by Method 1. Method 2 yields more correct results than Method 1; they appear to vary only about one per cent from the truth. (F. Schulze, 1857; Kühn & others, 1866, see *Henneberg's Journal für Landwirthschaft*, 1866, pp. 289–297, and *Johnson's "How Crops Grow,"* pp. 60, 61).

Chloric Acid. (Compare Chlorates).

Principle. Power of neutralizing alkalies.

Application. Separation of chloric acid from any other free acid.

Method. Measure out two portions of the liquid. In one portion determine the total amount of acid by titration with a standard alkali (see Acidimetry), and in the other portion estimate the other acid in any appropriate way. The difference will give the amount of chloric acid.

Chlorates.

Principle. Oxidizing power of.

Applications. Estimation of chloric acid, free or combined. Separation of chloric acid from other acids and from metals. Estimation of chlorine and chlorides. Oxidation of organic matters, S, H, As, Sb, Cr, etc., as a preliminary to the estimation of these substances and various others, as explained under Carbon, and the other elements enumerated.

Method 1. *Depending upon the action of a chlorate upon chlorhydric acid.*

A. A weighed quantity of the chlorate to be analyzed is placed in a flask, provided with a suitable delivery tube (compare Chromic Acid, reduction of by HCl), together with an excess of strong chlorhydric acid. The mixture is heated gently at first, but afterwards to boiling, and the Chlorine set free is absorbed and estimated in some appropriate way. One way, for example, is to conduct the gas into a solution of iodide of potassium, and to estimate the iodine which the chlorine liberates, by means of a standard solution of hyposulphite of sodium, or by sulphurous acid. (See Iodine). Another way is to estimate the chlorine by means of arsenite of sodium as described under Arsenious Acid, p. 45. — The reaction between chlorhydric acid and a chlorate may occur in accordance with either or all of the following equations:—

$$Cl_2O_5 + 2HCl = Cl_2O + Cl_2O_3 + H_2O.$$
$$Cl_2O_5 + 4HCl = 3Cl_2O + 2H_2O.$$
$$Cl_2O_5 + 6HCl = 2Cl_2O + 4Cl + 3H_2O.$$
$$Cl_2O_5 + 8HCl = Cl_2O + 8Cl + 4H_2O.$$
$$Cl_2O_5 + 10HCl = 12Cl + 5H_2O.$$

But no matter what products of decomposition are formed, they all agree in this, that when brought in contact with iodide of potassium, they will set free 6 equivs. of iodine for each equivalent of chloric acid in the original chlorate. The process yields accurate results. (Bunsen, *Annalen Chem. und Pharm.*, **86.** 282).

In case an aqueous solution of chloric acid is to be examined, mix a weighed or measured portion of it with a slight excess of soda lye, evaporate to dryness upon a water bath, and proceed with the residue as above described.

When metals have to be separated from chloric acid they may be estimated in the residue of the decomposition above described, or a special portion of the substance may be boiled with strong chlohydric acid on purpose.

B. The same principle is often employed for the oxidation of organic matter, sulphur, arsenious acid, etc., as a preliminary step in various processes of analysis. To this end small crystals of a chlorate, usually chlorate of potassium, are from time to time thrown into the hot, strong chlorhydric acid solution of the substance, until the oxidation is complete. A mixture of chlorhydric acid and a chlorate is however a far less powerful oxidizing agent than a mixture of nitric acid and the chlorate. (See below).

Method 2. *Action of a chlorate upon a ferrous salt.*

A. *By estimating the iron of the residual ferrous salt.* Mix a weighed quantity of the chlorate with an excess of a standard solution of protoSulphate of Iron, or of Sulphate of Iron and Ammonia in a glass-stoppered bottle. Acidulate the mixture with sulphuric acid and let it stand for some time in a warm place. Finally heat the mixture almost to boiling and proceed to determine by means of permanganate of potassium, how much of the Ferrous Salt remains unoxidized. One equivalent of a chlorate will oxidize 12 equivalents of the ferrous salt. The results obtained by this method are apt to be somewhat too high, since a portion of the ferrous salt may be oxidized by the air in the bottle when the mixture comes to be heated. (Mohr, *Titrirmethode*, 1855, **1.** 238 and **2.** 128).

B. *By estimating the ferric salt formed.*

Place a roughly weighed quantity of fine iron wire in a quarter-litre flask, together with a weighed quantity (0.1 to 0.15 grm.) of the chlorate to be examined. Pour upon the mixture from 30 to 50 c.c. of chlorhydric acid of 1.12 sp. gr., connect the flask with a carbonic acid generator, and warm it gently. The quantity of iron taken should be at least ten times greater than the weight of the chlorate. After the lapse of about twenty minutes, gradually raise the temperature of the flask until the liquid begins to boil. Then allow the flask to cool, and estimate the Ferric Salt formed by titration with hyposulphite of sodium. The process is said to be more accurate than that described in § A. (Braun, *Zeitsch. analyt. Chem.*, 1867, **6.** pp. 61, 43, 47). — To estimate chloric acid in presence of nitric or nitrous acid, determine the amount of efficient oxygen, as above, (compare Nitric Acid), and calculate therefrom the amounts of chloric and of nitric acid, by the method of indirect analysis. (See Carbonic Acid, p. 80). (Braun, *loc. cit.*, p. 62).

Instead of hyposulphite of sodium, Braun, (*Journ. prakt. Chem.*, **81.** 421) formerly used protoChloride of Tin to estimate the ferric salt.

In case a mixture of a chlorate and some other salt is to be examined, estimate the chlorate as above in one portion of the mixture, and the other acid in another portion of the material. — In case free chloric acid is to be estimated in an aqueous solution, saturate a measured quantity of the solution with soda lye, evaporate on a water bath and operate as above described upon the residue.

C. *By estimating the chlorine in the residual chloride.* When an alkaline solution of chloric acid is heated with freshly precipitated ferrous oxide, decomposition occurs in accordance with the following equation :—

$$K_2O, Cl_2O_5 + 12FeO = 2KCl + 6Fe_2O_3.$$

Hence chloric acid may be estimated by combining it with potassium or sodium, mixing the solution with a quantity of pure ferrous sulphate, supersaturating with potash lye free from chlorine, and heating the mixture to boiling. The reaction occurs immediately, so that it is only necessary to filter and acidulate the liquid before proceeding to determine the chlorine as Chloride of Silver. In presence of nitric acid, this method yields better results than that depending upon the reduction of chloric acid by nascent hydrogen (see below No. 3). (Stelling, *Zeitsch. analyt. Chem.*, 1867, **6.** 32). Chloric and nitric acids may evidently be estimated in presence of one another in this way by the method of indirect analysis (Compare Carbonic Acid, p. 80).

Method 3. *Reduction of the chlorate by nascent hydrogen.* Dissolve a weighed quantity of the chlorate in a small quantity of water, place a piece of zinc in the solution, and afterwards add some pure dilute sulphuric acid, and allow the mixture to stand for some time. If no more than 0.1 grm. of chlorate of potassium be taken, half an hour will be sufficient to complete the reduction of the chloric to chlohydric acid. Take out the zinc, wash it carefully, and proceed to determine the chlorhydric acid in the solution by precipitation as Chloride of Silver. From the amount of the latter, calculate the corresponding weight of chloric acid.

An aqueous solution of free chloric acid may be treated as if it were the solution of a chlorate. — In case the chlorate to be examined is contaminated with an alkaline chloride, weigh out two portions of it. In one portion determine the chlorine directly as Chloride of Silver, but treat the other with zinc and sulphuric acid, as above described, before adding the silver solution. The chloric acid may finally be calculated from the difference in the weights of the two precipitates of chloride of silver. (Sestini, *Zeitsch, analyt. Chem,,* **1**, 500.)

Method 4. *Reduction of the chlorate by sulphuretted hydrogen,* Pass sulphuretted hydrogen gas into the warm solution until its odor persists. Add drop by drop a solution of ferrous sulphate to destroy the excess of sulphuretted hydrogen, acidulate the mixture with nitric acid, filter if need be, and estimate the chlohydric acid as Chloride of Silver, and from the amount of this substance calculate that of chloric acid. The process is applicable to the estimation of free chloric acid in an aqueous solution.

Method 5. *Action of a chlorate upon nitrous acid, or a nitrite.* All the oxides of chlorine, excepting perchloric acid, are immediately reduced with formation of chlorhydric and nitric acids, when nitrous acid is added to their dilute aqueous solutions. A solution of nitrite of lead, prepared by leading carbonic acid into water in which basic nitrite of lead (4PbO, H_2O, N_2O_3) is suspended, may be employed to effect the reduction. Such a solution may be kept for a long time in glass stoppered bottles, which are completely full of it.

The analysis may be made either in the gravimetric or the volumetric way. In the first case, the substance to be analyzed is treated with a slight excess of the cold dilute solution of the nitrite of lead, then acidulated with nitric acid, and warmed. The chlorhydric acid is then precipitated as Chloride of Silver, and the amount of chloric acid (or other oxide of chlorine) calculated from the weight of the precipitate.

In the second case, there will be needed a normal solution (see Alkalimetry) of chlorate of potassium, containing 0.01227 grm., (= 0.0001 equiv.) of the chlorate in each cubic centimetre, and a solution of nitrate of silver of approximately known value, say 17 grms. of the nitrate in 100 c.c. The solution of nitrite of lead, moreover, should be prepared by weighing out roughly a quantity of basic nitrite of lead, rubbing it up with water, and passing a continuous stream of carbonic acid through the mixture, until the basic salt has almost entirely disappeared and a yellow solution of the normal salt has been obtained which is no longer rendered cloudy by carbonic acid. During the passage of the carbonic acid the liquid should be gently warmed. — The value of the solution of nitrite of lead is determined by means of the normal solution of chlorate of potassium, and is then used for estimating the amount of chloric or chlorous acid in sample of unknown composition.

The details of the process are as follows: Place the highly dilute solution of chlorate of potassium in a glass stoppered bottle, together with an excess of the nitrate of silver solution to serve as an indicator, and acidulate strongly with nitric acid. Heat the closed bottle in a water bath, and add the solution of nitrite of lead from a burette, until the last drop fails to precipitate any chlorate of silver. The bottle must be shaken frequently to facilitate the deposition of the chloride of silver. There is no special difficulty in obtaining good results, provided the solution of nitrite of lead is sufficiently dilute, and the solution under examination is so dilute that no appreciable quantity of chlorous acid gas can escape from it. It is well to dilute the 8 or 16 c.c. of normal solution of chlorate taken, to the volume of 250 c.c, Since the solution of nitrite of lead is slowly decomposed by the action of atmospheric oxygen, it must be standardized afresh for each new series of determinations. It may be kept, however, for a week without much change. The presence of perchlorates has no influence upon the results. (Toussaint, *Annalen Chem. und Pharm.*, **137**. 114).

Method 6. *Action of a chlorate upon protochloride of tin.*

Mix a weighed quantity of the chlorate with strong chlorhydric acid, and a measured quantity of a standard solution of stannous chloride, the latter, of course, to be in excess, and estimate by means of a standard solution of bichromate of potassium how much of the stannous salt has been left undecomposed. (Streng). According to Mohr the method is less commendable than those depending upon the action of chlorhydric acid and of a ferrous salt upon chlorates.

Method 7. *Action of a chlorate upon nitric acid.* When a chlorate, chlorate of potassium for example, is heated with moderately strong nitric acid, a mixture of great oxidizing power is obtained, well fitted for the oxidation of compounds of sulphur, arsenic, chromium, carbon, and the like, as a preliminary to the estimation of these and other substances by the ordinary methods of analysis. The reaction between nitric acid and chlorate of potassium may be roughly formulated as follows:—

$$8KClO_3 + 6HNO_3 = 6KNO_3 + 2HClO_4 + 6Cl + 13O + 3H_2O.$$

The oxidizing power of such a mixture is much greater than that of mixtures of sulphuric or chlorhydric acid and a chlorate (compare the equations above, under Method 1). — In most cases it will be sufficient to cover the substance to be oxidized with nitric acid of 39° B, to heat the mixture and to throw in one by one crystals of chlorate of potassium during 15 or 20 minutes, or until the oxidation is complete. When a metallic sulphide is to be treated, however, it is best to mix it first of all in the condition of fine powder, with as much or rather more chlorate of potassium, before adding any nitric acid. On heating such a mixture much of the sulphur will be oxidized before it has time to cohere into lumps. The last portions of sulphur will, however, have to be destroyed in almost every instance, by throwing fresh crystals of the chlorate into the nitric acid after the original mixture has ceased to act. The operation may be conveniently effected in a porcelain dish covered with an inverted glass funnel with bent stem. It is safer to heat the mixture upon a water bath, though in that event the oxidation proceeds more slowly than when the acid mixture is actually boiled. (Storer, Pearson, and others, *American Journ. Sci.*, 1869, **48**. 190 *et seq*).

Principle II. Decomposition of by heat. [Compare Carbon].

Applications. Separation of chloric acid from metals, and from nitric and nitrous acids. Estimation of chloric acid in chlorates, in presence of water or the chloride of an alkali metal.

Method A. Ignite a weighed quantity of the chlorate until the whole of its oxygen is expelled and weigh the residual chloride. If it be desirable the oxygen may be collected and measured. — The process is applicable only to those chlorates which are completely decomposed by ignition into oxygen and a metalic chloride. There is an objection to it, even in the case of chlorate of potassium, that a portion of the chlorate is carried off as dust in the escaping oxygen. Traces of chlorine also escape with the oxygen.

Method B. In case the chlorate to be analyzed is contaminated with a chloride, weigh out 2 portions of it. In one, estimate chlorhydric acid directly by precipitation as Chloride of Silver. Ignite the other portion cautiously, to decompose the chlorate, dissolve the residue in water and estimate the total chlorhydric acid as Chloride of Silver. From the difference between the weights of the two precipitates calculate the amount of chloric acid. — When a metal is to be separated from chloric acid the latter may be destroyed in a special portion of the substance, by careful ignition before proceeding with the analysis.

Method C. To separate chloric from nitric or nitrous acids ignite a weighed quantity of the mixture with repeated doses of chloride of ammonium, estimate the amount of chlorine in the residue as Chloride of Silver, and calculate the amount of chloric acid by the method of indirect analysis. (Braun, *Zeitsch. analyt. Chem.*, 1867, **6**. 62).

Chlorate of Potassium. See Chlorates, above.

For use as a reagent, the commercial salt is usually sufficiently pure. It may readily be purified, if need be, by recrystallization from a hot solution. For the analysis of Carbon compounds the fused chlorate is often required. It is prepared by heating the commercial crystals in a porcelain dish until the mass is fairly fused, but no longer, and pouring the liquid into a platinum dish or upon porcelain. Break the hot sheet into small fragments and keep them in a tight bottle for use.

Chlorhydric Acid. [See finding list for Chlorine, in the Appendix].

Principle I. Volatility.

Applications. Estimation of HCl in certain cases, separation of Cl from K, Na, Ba, Sr, Ca, Mg, Pt and other metals which form non-volatile chlorides.

Method A. Place a weighed quantity of the powdered chloride to be examined, in a platinum or light porcelain capsule, cover it with pure sulphuric acid somewhat diluted with water, and heat the mixture until all the chlorhydric acid and the excess of the sulphuric acid has been expelled. Ignite and weigh the residual sulphate. Calculate the chlorine from the loss of weight. Care must be taken to heat the mixture gently at first, in order to avoid frothing, but a strong heat will subsequently be needed to expel the free sulphuric acid. — This method will not serve for the analysis of the chlorides of silver, lead, mercury and tin, since these substances are either undecomposable by sulphuric acid or only imperfectly decomposed by that agent.

Method B. To separate chlorine from platinum, heat the compound in stream of hydrogen gas, lead the chlorhydric acid, which is formed, into a solution of nitrate of silver and weigh the Chloride of Silver. (v. Bonsdorff).

Principle II. Power of neutralizing alkalies.

Application. Valuation of the aqueous acid.

Method. See Acidimetry, method by neutralization.

Principle III. Comparative weight of a given volume of the aqueous solution in proportion as it contains more acid.

Application. Valuation of the commercial acid.

Method. See Acidimetry, method by taking the specific gravity.

Principle IV. Power of expelling boracic acid from its compounds with metals.

Application. Indirect analysis of alkaline borates. Separation of boracic acid from several of the metals.

Method. Dissolve a weighed quantity of the borate in water, add an excess of chlorhydric acid, and evaporate the mixture to dryness upon a water bath. Take care to add a few drops of chlorhydric acid to the residue before it is quite dry, and continue to heat the residue until no more chlorhydric acid vapors escape. Determine the chlorine in the residue as Chloride of Silver, calculate therefrom the amount of metal, and the boracic acid by the difference. For the analysis of borax this process yields very satisfactory results. (Schweizer).

Principle V. Power of reducing high oxides. See the various high oxides and oxygenated compounds, notably Chromic Acid, biChromate of Potassium, binOxide of Manganese, Chlorates, etc., etc. Compare also Arsenious Acid.

For use as a reagent, pure chlorhydric acid may be prepared, after Fresenius, as follows: Pour a cooled mixture of 7 parts of concentrated sulphuric acid, and 2 parts of water, over 4 parts of chloride of sodium in a retort. Set the retort upon a sand bath with its neck slightly raised, and by means of a tube bent at an obtuse angle connect the neck of the retort with a bottle containing 3 or 4 parts of water. Keep the receiving bottle cool and heat the retort as long as much gas continues to be evolved from it. The abduction tube should touch, but scarcely more than touch, the surface of the water in the bottle. As soon as the abduction tube begins to get hot change the receiving bottle, since the aqueous vapor charged with gas which is now going forward would weaken the strong acid already obtained. The action of fuming chlorhydric acid, is, in general, far more energetic than that of the diluted acid. In order to obtain an acid absolutely free from arsenic and chlorine, care must be taken to use purified Sulphuric Acid.

The pure acid should neither color a mixture of starch paste and iodide of potassium (chlorine), nor discolor a liquid made slightly blue with starch and iodine (sulphurous acid), nor should it when highly diluted give any cloudiness or coloration with chloride of barium, sulphuretted hydrogen, or sulphocyanide of potassium.

Chlorides.

Most of the highly volatile chlorides, such as the compounds of Cl, with P, S, Se, Te, As, Sb, Cr, Ti, Sn, etc., have to be dissolved in water, as a step preliminary to their analysis. It is to be remembered that their solution, or rather decomposition by water, is usually of such nature that the hydrogen of the water goes to the chlorine, while the oxygen unites with the other element to form an oxygen acid. But if we can determine the amount of oxygen in the acid formed, the composition of the original chloride can be easily calculated; for example

$$PCl_3 + 3H_2O = 3HCl + H_3PO_3$$
$$SiCl_4 + 2H_2O = 4HCl + SiO_2.$$

Or instead of estimating the oxygen acid, the chlorhydric acid may be estimated as Chloride of Silver. Only a few of the volatile chlorides enumerated deposit anything on being mixed with water, though some sulphur and selenium is set free from the chlorides of those elements, and a little phosphorus, previously held in solution is dropped by terchloride of phosphorus. Chloride of tellurium leaves metallic tellurium as well as tellurous acid. (H. Rose). When liquid, these volatile chlorides may be weighed out for analysis in little glass bulbs such as have been described under Carbon, analysis of volatile liquids, p. 67. The weighing bulbs should have long stems and the points should be drawn out very fine so that a liquid may be weighed even in an unsealed bulb without losing anything through evaporation. The filled and weighed bulb is finally dropped into a glass-stoppered bottle, containing some cold water; the stopper is inserted and held in tightly, and the bottle shaken, first to break the bulb, and again sometime afterwards to absorb the last traces of chlorhydric gas. These precautions are necessary because of the heat developed by the action of water upon the chloride. — Solid volatile chlorides like that of phosphorus may be weighed in a small glass-stoppered flask or tube, and the latter subsequently opened and carefully thrown into a bottle of water.

Principle II. Decomposition of by the protoxides of various metals. See the Chlorides of these metals. The value of the principle depends upon the fixity of the chlorides. This principle comes into play also when Ammonia is volatilized by means of lime or either of the fixed alkalies, and when Hydrate of Aluminum or Hydrate of Chromium is precipitated by adding chloride of ammonium to an alkaline solution of these metals.

Principle III. Power of reducing certain high oxides. See Arsenic Acid, Chromic Acid and Chloride of Antimony.

Principle IV. Power of converting salts of other acids into Chlorides. Compare Sulphate of Potassium.

Principle V. Power of retaining various metals in solution, or rather of forming soluble double salts with the compounds of these metals.

Applications. The most important perhaps of the applications of this principle may be seen under Arseniate of Magnesium and Ammonium, and under Phosphate of Magnesium and Ammonium. It is applied almost always in the analysis of compounds which contain magnesium; and frequently serves also to prevent the precipitation of small quantities of manganese and nickel (See Hydrate of Iron), of tartrate of calcium, etc.

For use as a reagent, a solution of the pure crystallized commercial salt in 8 parts of water is ordinarily employed. It should be neutral to test papers, should give no coloration of iron when tested with sulphide of ammonium and should leave no residue when heated upon platinum foil. For methods employed by Stas for preparing chloride of ammonium of an exceptional degree of purity, see *Zeitsch. analyt. Chem.*, 1867, **6.** 423.

Chloride of Ammonium.

Principle I. Fixity of the salt at 100°.

Applications. Estimation of ammonium in ammonia gas, ammonia water, and the compounds of ammonium with weak volatile acids, such as carbonic and sulphydric acids. Estimation of free chlorine.

Method A. To estimate ammonia collect the gas in an excess of chlorhydric acid of 1.13 sp. gr.; or in case ammonia water or an ammonium salt be under examination, supersaturate cautiously with the chlorhydric acid. Evaporate the liquid to dryness upon a water bath, and dry the residue at 100° in an air bath until the weight remains constant.

In case carbonate or sulphide of ammonium is to be examined, it may be placed in an inclined flask before adding the acid. In the case of sulphide of ammonium, the liquid should be filtered to remove sulphur after the treatment with acid. The process yields accurate results; for though solutions of chloride of ammonium suffer a certain amount of decomposition when boiled, so that alkaline vapor is exhaled and the residual liquid has an acid reaction, as Emmet (*American Journ. Sci.*, (*1.*) **18.** 255) and Fittig (*Annalen Chem. und Pharm.*, **128.** 189) have shown the proportion of material thus lost by volatilization is very trifling in a properly conducted analysis.

Method B. To estimate free chlorine, see under Chlorine, power of decomposing ammonia.

*ter*Chloride of Antimony.

Principle. Volatility of the chloride.

Applications. Separation of Sb from Na, K, Ba, Sr, Ca. Estimation of Sb, by loss.

The *Method* is described under Arsenic Acid (reduction of by chloride of ammonium). It is well after the first ignition with the chloride of ammonium to stir a fresh portion of this salt into the crucible and again ignite, and to repeat the process, if need be, until the weight of the residual chloride remains constant. In separating antimony from the alkali metals a double ignition will usually be found to be sufficient, and as a general rule the antimoniates of the true alkaline earths are so completely decomposed by a single ignition with the chloride that no trace of antimony is left in the residue. From magnesium on the other hand it is difficult to expel all the antimony, even by repeated ignitions with the chloride. It is impossible, moreover, to separate antimony from aluminum in this way since a large portion of the latter volatilizes with the chloride of ammonium. As regards the alkali metals and those of the alkaline earths, the process yields very accurate results (H. Rose).

*quinqui*Chloride of Antimony.

Principle. Volatility of the chloride.

Applications. Separation of Sb from Co, Ni, Pb, Cu, Ag, Au and Pt.

Method. Heat the mixed sulphides in an atmosphere of chlorine gas, in a bulb tube of hard glass, in a manner analagous to that described under Antimony Compounds. If it be desired to estimate the antimony directly it may be caught for further examination in a couple of flasks or U-tubes attached to the bulb tube and charged with a mixture of tartaric and chlorhydric acids. — Instead of the mixed sulphides, alloys of antimony and the metals above enumerated may be heated in the atmosphere of chlorine. The alloy should be in the form of powder if that be possible, or as finely divided as may be. It should be but gently heated when the chlorine first acts upon it. In case any portion of the alloy begins to glow from the action of the chlorine the lamp should be instantly removed. In case lead be present the bulb must not be too strongly heated lest some chloride of lead be volatilized. As applied to the analysis of alloys the process is not specially accurate, since the alloys are decomposed with difficulty, and small portions of the alloy are liable to be protected from decomposition by the residual metallic chloride, especially when this is fusible, or when the alloy cannot be finely powdered. When applied to the mixed sulphides, however, the process yields excellent results. Compare Chloride of Sulphur.

The process somewhat modified, may be applied to the analysis of antimoniates or antimonites.— such as those of copper, for example,— which are insoluble in acids and which cannot be conveniently decomposed by fusion with carbonate of sodium and sulphur (see Sulphide of Antimony). In that event the substance is placed in a bulb-tube and heated in the first place in an atmosphere of hydrogen, so as to reduce it to the condition of a metallic alloy or mixture, and subsequently in an atmosphere of chlorine in the same tube. — It is possible to analyze in this way the product known as *Kupferglimmer* (Copper mica), consisting of teroxide of antimony combined with copper and nickel, which occurs as an impurity in some varieties of copper, notably in that of Goslar. This substance is not acted upon by acids or the most powerful reagents ; and it cannot well be treated with carbonate of sodium and sulphur on account of the presence of nickel and copper ; but it is easily decomposed by hydrogen at a not very high heat, and the porous alloy formed is readily acted upon by chlorine. For the analysis of this alloy the method now in question is to be preferred to the process of dissolving in aqua regia and precipitation by sulphide of ammonium. (H. Rose).

Chloride of Arsenic.

Principle. Volatility of. (Compare Arsenic Acid, reduction of by chloride of ammonium).

Applications. Estimation of arsenic by loss. Separation of As from Na, K, (Method A), Co, Ni, Pb, Cu, Ag, Au and Pt, (Method B), less perfectly from the metals of the alkaline earths. (Method A).

Method A. By heating with chloride of ammonium. See Arsenic Acid. To separate arsenic from barium, strontium and calcium in this way, the mixture in the crucible has generally to be treated with five successive portions of chloride of ammonium. Arseniate of magnesium cannot be thoroughly decomposed even by repeated mixture and ignition with the ammonium salt. (H. Rose).

Method B. Heat the mixed sulphides in a current of chlorine gas in a bulb tube of hard glass and weigh the residual metallic chloride or the metallic gold or platinum. (Compare Antimony Compounds and Chloride of Sulphur). Arsenical alloys, even when very finely divided can only be decomposed with extreme slowness when heated in an atmosphere of chlorine.

Chloride of Barium.

Principle. Power of precipitating carbonates, sulphates and chromates. Compare Carbonate of Barium, Chromate of Barium and Sulphate of Barium.

Pure chloride of barium, for use as a reagent, may be prepared by passing a stream of chlorhydric acid gas through an aqueous solution of commercial chloride of barium as long as a precipitate continues to form. Almost the whole of the chloride of barium in the solution may be separated in this way as a crystalline powder free from calcium and strontium. Throw the precipitate upon a filter, and wash it with repeated small portions of pure chlorhydric acid until a sample of the washings, diluted with water and precipitated with sulphuric acid, gives a filtrate that leaves no residue when evaporated on platinum foil. (Fresenius). The aqueous solution (1 part $Ba\ Cl_2$ to 10 parts of water) should be neutral to test papers, and should neither be colored nor rendered cloudy on the addition of sulphuretted hydrogen or sulphide of ammonium.

Chloride of Bismuth. (normal = $Bi\ Cl_3$).

Principle. Volatility of.

Applications. Estimation of Bi, by loss. Separation of Bi from Pb, Cu and Ag.

Method. Heat the finely divided alloy in a current of chlorine gas in a bulb-tube of hard glass (Compare Antimony Compounds) and weigh the residual metallic chloride. Care must be taken to heat the mixture until the whole of the bismuth has been expelled, but if lead be present the heat must be regulated so that no chloride of lead shall be volatilized. If it be desirable to estimate the bismuth directly, it may be caught for further treatment, in a couple of little flasks or U tubes attached to the bulb tube, and charged with chlorhydric acid. From this solution the bismuth may be precipitated as basic Chloride of Bismuth.

Chloride of Bismuth. (basic = $BiCl_3$; $2Bi_2O_3 + H_2O$).

Principle. Insolubility in water and dilute acids.

Applications. Separation of bismuth from other metals, such as Na, K, Ba, Ca, Sr, Mn, Cd, Zn, Co, Ni, Cu, Hg, but not from Fe or Al.

Method. Mix the nitric acid solution of bismuth with chlorhydric acid, or better with chloride of ammonium and dilute the mixture with a very large quantity of water. Since a larger quantity of water will be needed in proportion as the solution contains more free acid it is impossible to determine beforehand just how much water should be added. The precipitate at first formed must consequently be allowed to settle and portions of the clear liquor above it must be poured off and tested with new quantities of water. So long as any cloudiness is thus produced, fresh portions of water must be added to the main mixture of liquid and precipitate. — In order to avoid the necessity of adding a large volume of water care must be taken in the first place not to employ a large excess of acid in dissolving the bismuth compound. If the solution be already strongly acid it may be nearly neutralized with ammonia before adding the water. In case the solution to be analyzed is very dilute it should be concentrated by evaporation so as to expel much of its free acid. There is no danger in thus evaporating if only nitric acid be present, but if the solution contains chlorhydric acid or aqua regia some chloride of bismuth may be lost by volatilization unless care be taken to prevent this result. It is to be remarked in that connection that the chloride of bismuth does not begin to volatilize until the larger part of the free chlorhydric acid in the solution has been expelled. Instead of evaporating the liquor, the free chlorhydric acid may be neutralized by ammonia; — to this end add ammonia water until the solution has only a faint acid reaction, before adding the excess of water. — Collect the precipitated basic chloride of bismuth upon a weighed filter, and wash it at the ordinary temperature with water very slightly acidulated with chlorhydric acid until the filtrate yields no residue when evaporated on platinum. Acidulated water is used instead of pure water for washing because the latter would dissolve out some chlorine from the precipitate. From the weight of the precipitate, dried at 100°, calculate the weight of the bismuth.

In case the original bismuth solution contains sulphuric acid, the precipitated chloride will retain a slight trace of it in the form of a basic sulphate, but the quantity is so small that it can hardly have any influence upon the accuracy of the result, especially as the equivalent weight of the basic chloride of bismuth does not differ very much from that of the basic sulphate. Some phosphoric acid is dragged down in like manner by the basic chloride of bismuth, in case there is any of it in the original solution. But it is easy in either event to determine the amount of bismuth in the precipitate by melting it with cyanide of potassium. (See Bismuth, fixity of). — It is not possible to estimate bismuth in the basic chloride by the method of reduction with hydrogen, since a large quantity of chloride of bismuth would be lost by volatilization. (H. Rose).

Properties. The precipitate is as good as absolutely insoluble in water and dilute acids. Bismuth may be so completely precipitated in this way that no trace of the metal can be detected in the filtrate from the basic chloride.

Chloride of Calcium.

Principle I. Fixity when heated.

Applications. Estimation of chlorine in organic compounds. (Method A). Separation of Ca, from Mg, As, Sb, Bi, Hg, and Sn. (Method B).

Method A. Choose a combustion tube about 40 c. m. long, and close it at one end as if it were a test tube. Place a layer 6 c. m. long of Oxide of Calcium, free from chlorine, at the end of the tube, then the weighed substance to be analyzed (in small glass bulbs if it be volatile), and another 6 c. m. layer of the lime. Mix these materials with a wire, as described under Carbon, p. 61, and afterwards fill the tube almost completely with lime. Hold the tube in a horizontal position, and rap it gently against the table until its contents have settled so as to leave a free passage for the escape of the gases which are to be evolved, then place it in a combustion furnace and heat the mixture as in an ordinary combustion (see Carbon). When all the organic matter has been decomposed, dissolve the contents of the tube in dilute nitric acid, and estimate the chlorine as Chloride of Silver. According to Kolbe, the products of the combustion may be removed from the tube as follows:— As soon as the decomposition is complete, remove the fire from the tube, cork the latter, and in case charcoal has been used as the fuel, remove every particle of ash from the tube. Immerse the hot tube, sealed end downwards, into a beaker two-thirds full of distilled water, so that it may break into numberless pieces and its contents fall out to be readily acted upon by the acid. — When compounds rich in nitrogen are treated in this way, some cyanide of calcium is liable to be formed. For the method of freeing chloride of silver from cyanide, see Cyanide of Silver.

Method B. See the Chlorides of the several metals above enumerated (volatility of).

Principle II. Power of absorbing water. [Compare Sulphuric Acid, Sulphate of Copper, Hydrate of Calcium, and Phosphoric Acid].

Applications. Estimation of water and of hydrogen. Separation of hydrogen from carbon, etc.

Method. Tubes of light, but strong glass, either U-shaped, or horizontal with a bulb near one end, are filled with small lumps of the dry porous chloride, and their large ends closed with perforated corks, provided with short abduction tubes. The corks should be covered with a thin, smooth coating of sealing wax, to prevent them from absorbing moisture from the air, or giving up moisture to the air according to the state of hydration of the latter.

In filling the tubes, small, loose tufts of cotton should be placed so as to prevent particles of the chloride from falling into the narrow abduction tubes, by means of which the tube is to be connected with other pieces of apparatus. The abduction tubes should be wiped clean and dry by means of little rolls of filter paper, and when not in use their ends should be kept covered with little caps, formed by plugging short pieces of rubber tubing with bits of glass rod.

Sometimes the tube is filled with fused chloride of calcium, instead of the porous chloride. But in either event the filled tube is weighed, after it has been wiped clean, and left at rest for some time in a place of tolerably constant temperature, — then attached to the apparatus from which moisture is to be derived and finally weighed again, together with the water which has been absorbed by the chloride. (See Carbon, p. 62, and Carbonic Acid, p. 79).

Chloride of calcium has the great merit of not absorbing any carbonic acid when a mixture of that substance and aqueous vapor is passed through it; but it is to be remarked that as a mere dessicating agent, it is distinctly less efficient than Sulphuric Acid. Pettenkofer observed, some years since, that air which had been dried by chloride of calcium, still gave up moisture to strong sulphuric acid; and Fresenius (*Zeitsch. analyt. Chem.*, 1865, **4.** 177 *et seq.*) has determined by careful experiments that neither fused chloride of calcium, nor the ordinary dry porous variety, — which in his experiments contained as much as 21 per cent. of water, — is capable of thoroughly drying moist air. Air which has been dried as completely as possible by chloride of calcium, always gives up moisture to strong sulphuric acid. The fused chloride does not dry air any more completely than the porous chloride, but acts rather more quickly than the latter; a shorter column of it will suffice to remove all the moisture that can be removed by chloride of calcium. The experiments of Fresenius show, moreover, that dry air, — that which has been dried by means of sulphuric acid, — on being passed over ordinary porous chloride of calcium, or fused chloride which has been made to absorb a little water, is capable of removing a very small quantity of moisture from the chloride.

The inability of chloride of calcium to remove all the moisture from air, is no objection to the use of this agent in those cases where the substance to be analyzed for water is heated in a current of air. For if the incoming as well as the outgoing air be dried by chloride of calcium, it will be brought to a normal condition of dryness in both instances, and the whole of the water of the substance to be analyzed will be retained in the chloride of calcium tube. Errors would evidently be introduced, however, in case the incoming air were dried by sulphuric acid, and the outgoing by chloride of calcium, or if this arrangement were precisely reversed.

It is in the estimation of Carbonic Acid, by loss, that the inefficiency of chloride of calcium is specially conspicuous.

For use as a reagent pure chloride of calcium may be prepared as follows: — Mix 1 part of ordinary commercial chlorhydric acid with 6 parts of water, and place the liquid in a tall stone jar; add chalk or powdered marble to the acid by small quantities, until the last portion refuses to dissolve. Then add a little slaked lime, and sulphuretted hydrogen water, until a filtered sample of the liquid gives no reaction with sulphide of ammonium. Put the solution in a stoppered bottle and let it stand for 12 hours in a warm place; then filter, exactly neutralize the filtrate, concentrate by evaporation, and crystallize. Let the crystals drain and dissolve 1 part of them in about 5 parts of water for use. — A solution of chloride of calcium should be perfectly neutral, and should neither be colored nor precipitated by sulphide of ammonium. It should not evolve ammonia when heated with slaked lime.

To obtain porous chloride of calcium, fit for absorbing water in the analysis of Carbon compounds, evaporate a neutral solution to dryness in a clean iron dish or pan, and expose the residue for several hours to a heat of about 200° upon a sand bath. The white porous mass thus obtained consists of $CaCl_2 + 2H_2O$.

The fused chloride, such as is used for drying air and other gases needed for quantitative operations, may be obtained by evaporating any solution of the chloride to dryness, and fusing the residue in a Hessian crucible. Pour out the fused mass upon stone or metal, break the sheet into pieces and preserve the fragments in well-stoppered bottles.

Chloride of Chromium.

Principle. Colorific power.

Applications. Estimation of Cr and of Co.

Methods. As regards the estimation of chromium, see the general subject of Colorimetry. Since the color of chloride of chromium is complementary to that of sulphate of cobalt, the quantity of the latter may be estimated by titrating with a standard solution of chloride of chromium, until the colors have precisely neutralized one another. (Himly & Dehms). According to Dehms (*Zeitsch. analyt. Chem.*, 1864, **3.** 222) the process yields useful results, though it is inferior to the corresponding one in which sulphate of nickel is employed. With the chrome salt a difficulty is encountered in the fact that the point where the colors neutralize one another is concealed by a yellowish tint, which then appears in the liquid.

Chloride of Cobalt.

Principle I. Fixity when heated.

Applications. Separation of Co from As, Sb, Bi, Hg and Sn.

Method. See the Chlorides of the several metals.

Principle II. Colorific powers.

Application. Estimation of cobalt.

Method. See Colorimetry, A. Mueller's process, and *Zeitsch. analyt. Chem.*, 1866, **5.** 426.

Instead of a glass plate of a color complementary to that of the cobalt solution, Winkler uses a solution of nickel.

Principle III. Change of color from blue to red by means of water.

Application. Volumetric estimation of water in solutions of organic compounds.

Method. Ignite a quantity of the blue anhydrous chloride gently, grind it in a warm mortar, and treat it with absolute alcohol in a perfectly dry flask. Allow the mixture to stand 24 hours, then separate the blue solution from the undissolved chloride by filtering it under a tubulated bell glass, sealed with concentrated sulphuric acid below and carrying a chloride of calcium tube at the tubulure. This saturated solution, containing 23.67 per cent of chloride of cobalt, is next diluted with absolute alcohol until the solution contains no more than 5 or 10 per cent of the chloride. To standardize the dilute solution, weigh out about 5 grms. of water in a beaker, place the latter upon a white ground, and pour the cobalt solution into it from a burette until the liquid is violet colored, or even shows a suspicion of blue. Some experience is required in order to be sure of the precise tint at which to stop. It is well, therefore, to retain the liquid which has been used in the operation of standardizing as a point of comparison for future experiments. — In an actual experiment, the estimation of water in glycerin, for example, it is simply necessary to weigh out 5 or 10 grms. of the material, and pour upon it the standard cobalt solution until the predetermined violet tint is reached. Though the results obtained by this process are not absolutely accurate, they still furnish tolerably close approximations to the truth. The process will be found useful in many cases, though of course, it can be applied only to substances soluble in alcohol, and to those which have no chemical action upon chloride of cobalt. Operating with a solution of chloride of cobalt, each c. c. of which represented 0.220 grm. of water, Winkler found in glycerin 13.6 and 52.8 per cent of water instead of 12 and 52 per cent; in aldehyde 73.9 per cent water instead 75 per cent, and in a solution of cyanhydric acid 91.9 per cent instead of 90.8. — The process was not found to be fully applicable in the case of alcohol. Good results were obtained with spirit, which contained more than 25 per cent of water, but when less than 25 per cent of water was present the blue color appeared too soon, and the results were worse and worse in proportion as there was less water in the alcohol examined. (Winkler, *Journ. prakt. Chem.*, **91.** 209).

Principle IV. Reducibility by hydrogen.

Applications. Separation of Co from Ba, Sr, Ca, and Mn.

Method. See Cobalt, fixity when heated.

*Di*Chloride of Copper.

Principle. Power of reducing ferric chloride:—

$$CuCl + FeCl_3 = FeCl_2 + CuCl_2.$$

Application. Volumetric estimation of iron.

Method. There are needed three solutions: —one of dichloride of copper, one of sesquichloride of iron and one of sulphocyanide of potassium (1 part of the salt in about 9 parts of water). To prepare the dichloride of copper, dissolve sheet copper in aqua regia, evaporate to destroy the excess of nitric acid, dissolve the residue in water acidulated with chlorhydric acid, transfer the solution to a flask and mix with it a quantity of chloride of sodium about equal to the weight of the dichloride of copper to be obtained. Place also in the flask some strips of sheet copper and heat the mixture to boiling until the liquid has become almost colorless; then cork the flask, allow its contents to cool and dilute with water acidulated with chlorhydric acid to such an extent that one c. c. of the liquid can reduce about 6 milligrams of iron from the state of sesqui- to protochloride. The purpose of the chloride of sodium is to prevent the deposition of any solid dichloride of copper when the liquid is boiled with the copper strips. — The solution of dichloride may be kept for use with but little change, by placing it in a tightly stoppered bottle in which has been placed a spiral of stout copper wire reaching from the bottom of the bottle almost to its top. Or the liquid may be kept in the same way as protoChloride of Tin. But it is best in either event to standardize it anew before each new series of experiments.

The solution of ferric chloride needed for standardizing the copper solution is prepared by dissolving 10.03 grms. of fine iron piano wire (equivalent to 10 grms. of pure iron) in chlorhydric acid, with addition of a little chlorate of potassium. After heating the liquid to expel chlorine it is diluted to the volume of a litre.

In an actual analysis, the iron solution to be tested should be strongly acid and highly dilute. Only 4 or 5 drops of the solution of sulphocyanide of potassium, which is to serve as an indicator, should be added to the liquid in order that the difficultly soluble sulphocyanide of copper may not be formed too soon. When once formed this precipitate only dissolves with difficulty. If the indicator be added in correct proportion, as above, the decoloration of the red sulphocyanide of iron will be sharply defined, and only after all the ferric chloride has been reduced to a ferrous salt will the next drop of dichloride of copper produce a soft permanent cloudiness due to precipitation of disulphocyanide of copper.

is well to dilute a solution of ferric chloride which contains from 0.1 to 0.2 grms. of iron to the volume of 500 c. c. or more, not that the dilution is absolutely essential to the success of the experiment, but that the indicator acts more plainly and satisfactorily in solutions thus diluted. The presence of metals such as cobalt, nickel and copper, the salts of which give colored solutions, do not interfere with the process when the liquor is sufficiently dilute. Arsenic acid moreover does not interfere with the process. Attempts to use other substances as indicators, such as molybdate of ammonium, iodide of potassium and starch, diniodide of copper, etc., gave less satisfactory results than those obtained with the sulphocyanide ; and without an indicator the method is not exact.

The merit of dichloride of copper as compared with that of protochloride of tin is found in the fact that it acts readily in the cold. (Winkler, *Journ. prakt. Chem.*, **95.** 417). Hoch & Clemm (*Zeitsch. analyt. Chem.*, 1866, **5.** 328), object to this process that the decoloration of the sulphocyanide of iron and the precipitation of sulphocyanide of copper do not occur simultaneously. According to circumstances, either one or the other of these phenomena will first appear. If the appearance of the cloudiness due to sulphocyanide of copper be regarded as indicative of the point of complete reduction it is essential to success that the conditions of each and every experiment should be similar. For a given quantity of iron the consumption of dichloride of copper will increase accordingly as the liquid is more highly diluted. It will decrease on the other hand if the proportion of sulphocyanide of potassium be increased. The quantity of free acid alone does not appear to exert an influence upon the indicator, provided only that the requisite quantity of acid be present.

On the other hand, if the destruction of the coloration due to sulphocyanide of iron be made to serve as the indicator, it is no easier to obtain good results, since under some circumstances the decoloration is complete some little time before the appearance of any cloudiness in the liquor, while at other times a decided cloudiness will appear before the liquid is completely decolorized. The premature decoloration of the liquid appears to depend upon the facts that the blood red color obtained by adding sulphocyanide of potassium to a solution of ferric chloride is in the first place weakened or diluted by the addition of chloride of copper, and finally changed to green by a great excess of the latter ; and that this tendency to obscuration is increased by the presence of ferrous chloride.

*proto***Chloride of Copper** [basic].

Principle. Fixity when heated.

Applications. Estimation of chlorine in organic compounds. Separation of Cl from C and H. (Method A). Separation of Cu from As, Sb, Bi, Hg and Sn. (Method B.)

Method A. Bend a combustion tube and pack a portion of it with asbestos, as described under Carbon, p. 72. About 10 inches of the posterior horizontal part of the tube should be thus packed with purified asbestos ; a space of about 2 inches being left vacant in front of the column. Place a short plug of asbestos in front of the two-inch vacant space, then a column 4 or 5 inches long of an intimate mixture of asbestos and brown oxide of copper,—prepared by precipitating a copper salt with potash lye and igniting the washed precipitate over the flame of a Bunsen burner,—and last of all, another plug of asbestos to keep the oxide in place. — Place the tube upon a gas combustion furnace and push an air bath over that portion of the anterior part of the tube which contains the mixture of asbestos and oxide of copper. This air bath is a simple box or chamber of sheet iron provided with three holes. One hole, at the top of the box, serves for the insertion of a thermometer. The other two holes which are oppoite one another in the sides of the box admit of the tube being thrust completely through the box so that a given portion of its length (that which contains the oxide of copper) may be enclosed in a confined volume of air and so subjected to a definite temperature. The bulb of the thermometer is placed in a central position, in the interior of the bath, close by the side of the combustion tube The bath is heated by a Bunsen burner placed under its front end. It should be kept at a temperature of from 250° to 350° throughout the analysis. — To the front of the combustion tube attach weighed chloride of calcium tubes and potash bulbs as prescribed under Carbon.

The substance to be analyzed is introduced at the posterior end of the tube and is burned by means of a current of oxygen gas, in the 10 inch column of asbestos, precisely as has been described under carbon. (See Warren's method, p. 72). All the chlorine is absorbed and retained by the hot oxide of copper in the anterior end of the tube while the water and carbonic acid go forward to be absorbed by chloride of calcium and caustic potash in the usual way. After the combustion is finished and the tube has become cold the chloride of copper, together with the excess of oxide, is dissolved in dilute nitric acid and the chlorine is estimated as Chloride of Silver. The solution of the chloride and oxide of copper is readily and completely effected by immersing the front part of the combustion tube, in a vertical position, in a cylinder charged with the dilute acid. Since the quantity of oxide of copper required to absorb the chlorine is very small, there is no difficulty in dissolving it, or in the subsequent stages of the chlorine solution of the chloride and oxide of copper is readily and completely effected by immersing the front part of the combustion tube, in a vertical position, in a cylinder charged with the dilute acid. Since the quantity of oxide of copper required to absorb the chlorine is very small, there is no difficulty in dissolving it, or in the subsequent stages of the chlorine determination. Four or five grammes of oxide of copper are amply sufficient for a single analysis, but it is to be remarked that the oxide must be mixed with enough asbestos to fill 4 or 5 inches of the combustion tube. A shorter column than this sometimes fails to retain all the chlorine. The temperature of the air bath had better be maintained between 250° and 350°. Temperatures of 150° or 160° have been found to be too low, though the soft brown oxide of copper above prescribed will absorb chlorine at lower temperatures than a harder oxide. Oxide of copper is found on the whole to serve better than oxide of lead or oxide of zinc as an absorber of chlorine, though in the case of some difficultly combustible chlorine compounds, which cannot be burned completely by oxygen gas, oxide of zinc is to be preferred (See Chloride of Zinc). It is probable that bromine and iodine compounds can be analyzed in this way as well as chlorine compounds. (Warren, *Proc. American Acad.*, 1866, **7.**84; *Zeitsch. analyt. Chem.*, **5.** 174).

Method B. See Chloride of Antimony, Chloride of Sulphur, and the Chlorides of the other metals enumerated above. The chloride of copper left in the bulb tube cannot be wei hed directly since it is a mixture of mono- and basic-chloride. The residue must be dissol ed in chlorhydric acid and the copper estimated in some appropriate way.

*proto***Chloride of Iron.** [Compare Ferrous Salts].

Principle. Volatility of.

Applications. Separation of Fe from Al, Cr, Gl, and most of the rare earths.

Method. A quantity of the mixed precipitate of sesquiOxide of Iron, alumina and chrome is weighed out in a small porcelain boat; the boat is placed in a porcelain tube, and there heated to redness in a current of dry hydrogen gas, as long as any water is formed. Dry chlorhydric acid gas is then passed through the hot tube, and afterwards hydrogen again. The whole of the iron volatilizes as protochloride, while the alumina (or the chrome) is left in a state of purity to be weighed. The iron is either estimated from the loss, or collected and determined. In the latter case a tubulated receiver is attached to the porcelain tube, and at the close of the experiment moist chlorhydric acid gas is conducted into the tube and the receiver. To this end a flask of dilute chlorhydric acid is attached to the end of the porcelain tube, and its contents are heated to boiling. During this operation the tubulure of the receiver is pointed downwards into a beaker, and the tube is slightly inclined. (Deville, *Annales Chim. et Phys.*, **38**).

The foregoing process has been modified by J. P. Cooke (*American Journ. Sci.*, 1866, **42.** 78) as follows: — Instead of the porcelain tube, take a platinum tube about 6 inches by 0.4 inch in diameter, the neck of a small platinum still, for example, will answer perfectly; and instead of the porcelain boat. take one of platinum as large as the tube will admit, and about 1.5 inch long. Mount the tube horizontally upon any convenient stand, and close one end with a two-holed rubber stopper. Through one hole of the stopper thrust the delivery tube of a self-regulating hydrogen generator, and through the other the delivery tube of a small flask fitted to generate chlorhydric acid gas. To the other end of the tube attach by means of a caoutchouc connector, a small glass adapter, so curved that the end may dip into a beaker of water. Weigh out in the boat a quantity of the finely powdered oxides to be analyzed; place the boat in the tube, pass a slow current of hydrogen, and heat the tube at the point where the boat is, by means of a single Bunsen burner. In the course of half an hour the whole of the oxide of iron will be reduced to the metallic state. Then replace the stream of hydrogen by a much more rapid current of chlorhydric acid gas, and instead of the Bunsen burner, heat the tube with a blast lamp. The reduced iron is rapidly changed to chloride, which is carried forward in the current of gas and dissolves in the water of the beaker. After a few minutes the action ceases, and the chlorhydric acid is to be replaced by hydrogen. The lamp is then withdrawn, and the apparatus allowed to cool. In case the alumina or other earth in the boat is not perfectly white, it should be replaced in the tube and the process repeated. — For generating the chlorhydric acid, the flask should be charged with coarse salt and sulphuric acid, which has been diluted with about one-third its volume of water and allowed to cool. This mixture when gently heated, gives a constant flow of gas which stops almost immediately when the lamp is withdrawn. Both the chlorhydric acid and the hydrogen should be made to pass through a bottle charged with oil of vitriol, before being admitted to the platinum tube. It is not well to try to use a small porcelain tube in place of the platinum tube, since the lamps employed could not heat it sufficiently, and there would be risk of breaking the porcelain tube if it were heated by a lamp. According to Cooke, Deville's process affords exceedingly accurate results.

Johnson (see his N. Y. edition of Fresenius) proceeds as follows: — Weigh out a quantity of the powdered mixed oxides in a platinum crucible. Ignite over a Bunsen burner in a stream of hydrogen gas for about an hour, or as long as water forms. Then ignite over a blast lamp in a current of mixed hydrogen and chlorhydric acid gases. The whole of the iron volatilizes as protochloride, and is determined by the loss, while the alumina and chrome are left in a state of purity.

*Sesqui***Chloride of Iron.** [Compare Ferric Salts].

Principle I. Volatility.

Application. Estimation of carbon in cast iron.

Method. Weigh out a quantity of the iron in a porcelain boat, place the boat in a hard glass tube and heat it to dull redness in a stream of chlorine, until no more chloride of iron is formed. Care must be taken to dry the chlorine completely by passing it through pumice stone wet with sulphuric acid; otherwise some carbon will be lost through the formation of a volatile hydrocarbon. When dry chlorine is used, the whole of the carbon will remain in the boat. It is to be burned in oxygen, and the carbonic acid collected and weighed, see Carbon. An estimation of carbon in iron may be made in this way in 2 hours. The process is excellent, and yields remarkably concordant results. (Berzelius; Wœhler; Tosh, *Chemical News*, 1867, **16.** 94). According to M. Buchner, this process affords results as accurate as those obtained by Berzelius's method which depends on the insolubility of Carbon in chloride of copper. It is commended also by B. Kerl. — H. Rose (edition of 1851, p. 753) has objected that the process is inconvenient since it is difficult to obtain chlorine absolutely free from oxygen or air, and if oxygen be allowed to go forward in the chlorine, some of the carbon will be oxidized by it and so lost. There is a tendency, moreover, on the part of the chloride of iron to stop the tube.

Properties. Fears are sometimes expressed lest small quantities of ferric chloride may be lost on evaporating its solution to dryness, as in the analysis of silicates, or on boiling a chlorhydric acid solution of it. But Fresenius (*Zeitsch. analyt. Chem.*, 1867, **6.** 92) has shown by experiments that no appreciable loss occurs, either on long continued boiling of the concentrated chlorhydric acid solution, or on evaporating solutions to dryness, either upon the water bath or at a somewhat higher temperature, such as is employed in the analysis of silicates.

Principle II. Colorific power.

Application. Estimation of iron.

Method. See Colorimetry, Dehm's later process.

For use as a reagent ferric chloride may be prepared as follows: — Mix 1 part of pure chlorhydric acid with 10 parts of water; place the mixture in a flask together with a quantity of small iron tacks, and heat it until no more hydrogen is given off, the iron being all the while in excess. Filter the liquid into another flask and pass chlorine gas through it with frequent shaking, until a drop of the liquid no longer gives a blue precipitate when tested with ferricyanide of potassium. Heat the liquid to expel the excess of chlorine, and dilute with as much water as may be necessary. A solution of ferric chloride should contain no excess of acid. A small sample of it stirred with a rod which has just been dipped in ammonia water will give a precipitate at once if the acid be not in excess. It should not give a blue precipitate with ferricyanide of potassium.

To prepare a standard solution of ferric chloride, Fresenius (*Zeitsch. analyt. Chem.*, 1862, **1.** 27) proceeds as follows: — Weigh out 10.03 grms. of fine iron wire (piano wire), that is to say, as much as will contain 10 grms. of absolutely pure iron, dissolve it in chlorhydric acid in a long necked flask set slanting, add enough chlorate of potassium to oxidize the iron, and expel completely the excess of chlorine by long continued gentle boiling. Finally dilute the solution with water to the volume of 1 litre.

Chloride of Lead.

Principle I. Fixity when heated.

Applications. Separation of Pb from As, Sb, Bi, Hg and Sn.

Method. See the Chlorides of the several metals enumerated, and Chloride of Sulphur. Care must be taken not to ignite the mixed sulphides and chlorides too strongly, lest some of the chloride be lost through volatilization. In case it is desired to remove the chloride of lead from the bulb tube in order to weigh the latter, heat the chloride in a stream of hydrogen and dissolve the metallic lead in nitric acid. — For Warren's experiments on the use of this principle for estimating chlorine in organic compounds, see *Proc. Amer. Acad.*, 1866, **7.** pp. 88, 89; compare protoChloride of Copper (basic).

Principle II. Insolubility in a mixture of alcohol and ether.

Applications. Separation of Pb from Cr, Hg, Cu and Bi. Estimation of Pb in the nitrate, chromate, iodide, and bromide of that metal.

Method. Mix the solution with an excess of chlorhydric acid, evaporate to dryness upon a water bath, and add to the residue a large quantity of absolute alcohol, mixed with some ether. Let the precipitate settle, then collect it upon a filter, and wash with the mixture of alcohol and ether. Dry the precipitate at a moderate heat in a porcelain crucible, after having burned the filter upon the lid. The precipitate must not be heated to redness, lest some of the chloride of lead be lost through volatilization. In case the precipitate be small it had better be collected on a weighed filter. The alcoholic solution should be tested with sulphuric acid, and in case a precipitate of sulphate of lead falls it should be collected on a filter and weighed. No free chlorhydric acid should be permitted to remain in the liquid before adding the alcohol and ether, for chloride of lead is not insoluble in alcohol acidulated with chlorhydric acid. In the case where chromate of lead has been reduced by chlorhydric acid and alcohol, there will necessarily be some free chlorhydric acid in the liquid; the filtrate must, therefore, be treated with sulphuric acid, as a matter of course. But on the other hand, according to H. Rose, it is not advisable to add sulphuric acid immediately after the reduction of the chromate, with the view of determining the whole of the lead as sulphate, all at once. The method is not often employed, though there are some cases in which it may be resorted to. (Compare *Zeitsch. analyt. Chem.*, 1866, **5**. 227).

Principle III. Insolubility in ammonia water.

Application. Separation of Pb from Ag.

Method. See Chloride of Silver. Melt the residual insoluble basic chloride of lead with cyanide of potassium, in order to obtain the amount of metallic Lead.

Principle IV. Solubility in an aqueous solution of nitrate of ammonium.

Application. Separation of small quantities of lead from a large excess of silver, as in the analysis of coin.

Method. Convert the mixture of lead and silver to the condition of nitrates, acidulate the solution with nitric acid, heat it upon a water bath and pour into it a solution of chloride of ammonium until no more chloride of silver falls, taking care that no unnecessary excess of the chloride of ammonium is added. The chloride of silver is thus precipitated in presence of a large excess of nitrate of ammonium, in which the lead remains dissolved, since chloride of lead is readily soluble in such a solution. The chloride of silver is collected upon a filter, and washed with hot water, the lumps of chloride of silver being broken down as much as possible with a glass rod. The filtrate and wash water is evaporated in a porcelain dish to the consistence of a thin syrup, and then transferred to a flask of hard German glass, in which the evaporation is continued until the nitrate of ammonium is entirely destroyed. This operation requires care in the application of the heat, lest violent decomposition of the nitrate of ammonium ensue, and portions of the substance be thrown out of the flask. Wash the residue out of the flask into a porcelain dish, using at last a little nitric acid to remove the particles which adhere to the glass. Add a slight excess of pure sulphuric acid, and estimate the lead as Sulphate of Lead. The process gives satisfactory results. The presence of a small amount of copper, and of traces of silver dissolved by the nitrate of ammonium does no harm. Gold is got rid of as explained under Sulphate of Lead. (Eliot & Storer, *Proc. American Acad.*, 1860, **5**. 55).

Principle V. Decomposition of by solutions of bicarbonates of the alkali metals.

Application. Estimation of chlorine in chloride of lead.

Method. [Compare Sulphate of Lead]. Digest the precipitate at the ordinary temperature in an aqueous solution of pure bicarbonate of sodium for a couple of days. Collect the carbonate of lead upon a filter, wash this carbonate thoroughly, and estimate chlorine in the filtrate as Chloride of Silver. Since the Carbonate of Lead contains some alkali, it cannot be weighed directly. It must be dissolved in acid, and the lead reprecipitated in some appropriate way. — According to Warren (*Proc. American Acad.*, 1866, **7**. 88) it is a very difficult matter to decompose chloride of lead, which has been over heated or converted to the state of oxychloride, with a solution of bicarbonate of sodium. Even after treatment for more than two weeks, using fresh portions of the bicarbonate, and occasionally agitating the mixture, a part of the chloride of lead still remained undecomposed.

Properties. The precipitate occurs either in the form of small glistening needles, or as a white powder unalterable in the air. One part of it is soluble in 135 parts of water at the ordinary temperature, and in much less hot water. It dissolves somewhat less readily in water acidulated with chlorhydric or nitric acid, but is easily soluble in strong chlorhydric acid from which it is partially reprecipitated on addition of water. It is exceedingly sparingly soluble in alcohol of 70 or 80 per cent. and is insoluble in absolute alcohol. It fuses at a temperature below redness without decomposition, but at higher temperatures volatilizes slowly. In contact with air it suffers partial decomposition at high temperature, some chlorine escaping and a mixture of oxide and chloride of lead being left. Its composition is

Pb	207 . . .	74.46
Cl_2	71 . . .	25.54
	278	100.00

Chloride of Lithium.

Principle. Solubility in a mixture of alcohol and ether.

Applications. Separation of Li from Na and K.

Method. Weigh out a quantity of the powdered mixed chlorides, dried at 120°, into a flask provided with a cork or stopper; pour into the flask a quantity of a mixture of equal volumes of absolute alcohol and ether, close the flask and let the mixture stand for a number of hours (24 at least) shaking it occasionally. The chloride of lithium dissolves while the chlorides of potassium and sodium remain insoluble. Decant the mixture into a filter, wash the residue with the mixture of alcohol and ether, and weigh it.

Evaporate the filtrate to dryness in a tared dish, convert the residue into Sulphate of Lithium and weigh, or estimate the lithium from the loas. Pure alcohol cannot be used instead of the mixture of alcohol and ether since chloride of potassium is not insoluble in alcohol as it is in the mixture. As regards the separation of lithium from potassium, the process is less accurate than that depending upon the insolubility of Chloroplatinate of Potassium. It is apt to give too much lithium, since neither of the alkaline chlorides are absolutely insoluble in the mixed alcohol and ether. A higher degree of accuracy may be obtained if the chloride of lithium, left on evaporating the filtrate, be treated a second time with a mixture of alcohol and ether to which a drop of chlorhydric acid has been added. The residue thus obtained is of course added to the first residue. — The original mixture of chlorides to be analyzed should not be ignited, since in that event a small quantity of caustic lithia is formed, which, by absorption of carbonic acid, would form an insoluble carbonate. In case this happens a drop of chlorhydric acid must be added to destroy the carbonate. (Rammelsberg, *Poggendorffs Annal.*, **66.** 79). In case the process is to be applied to the analysis of sulphates, these may be converted into chlorides by mixing the solution with a slight excess of chloride of calcium, adding a double volume of alcohol to the mixture, and washing the precipitated sulphate of calcium with diluted alcohol The excess of calcium employed is then removed from the filtrate by means of oxalic acid and the second filtrate is evaporated to dryness and carefully heated. It is well to redissolve the precipitated sulphate of calcium in hot dilute chlorhydric acid, to evaporate the solution almost to dryness, and exhaust the residue with diluted alcohol. The traces of alkaline chlorides thus obtained are added to the rest. It is to be remarked that sulphate of lithium cannot be completely changed to the chloride by ignition with chloride of ammonium like the other alkaline sulphates. (H. Rose).

Properties. Chloride of lithium deliquesces so easily in the air that it is very difficult to weigh it accurately. It cannot be ignited owing to its volatility and to its tendency to decompose when moisture is present.

Chloride of Magnesium.

Principle I. Decomposition of by hot oxygen or air.

Application. Separation of Mg from Ca, especially when the former is largely in excess of the latter.

Method. A combustion tube is bent upwards at both ends, and at one end (the posterior) is again bent outwards and drawn out so that it can be attached to a chloride of calcium tube and a set of Liebig's potash bulbs. The combustion tube is then placed in position, in an oil bath, as if it were a U-tube, and an aspirator and tell-tale bottle are attached to its anterior end. Before the attachment of the aspirator, a chlorhydric acid solution of the substance to be analyzed is poured into the combustion tube, care being taken that a free channel is left above the liquid for the passage of air. — A slow stream of air is drawn through the tube by means of the aspirator, and the oil bath is heated to 100° at first and afterwards to 200°. At first the air takes up aqueous vapor and chlorhydric acid but afterwards free chlorine. After the lapse of several hours the oil bath should be removed and a gas flame substituted for it. After long continued ignition, all but a faint trace of the chloride of magnesium will be decomposed while the chloride of calcium suffers no change. As soon as the decomposition of the magnesium salt is thus complete, the contents of the tube are repeatedly leached with hot water, and the filtrate after evaporation is mixed with chloride of ammonium and carbonate of ammonium, for the precipitation of Carbonate of Calcium. In order to remove the last traces of magnesium the washed carbonate of calcium should be redissolved in chlorhydric acid and again precipitated as before. The oxide of magnesium which remains sticking to the tube is to be dissolved in chlorhydric acid and thrown down with the rest as Phosphate of Magnesium and Ammonium. In the opinion of its author the process is preferable for separating small quantities of calcium from a large quantity of magnesium to those which depend upon the insolubility of Sulphate of Calcium and Oxalate of Calcium. (Oeffinger, *Zeitsch*, *analyt. Chem.*, 1869, **8.** 457).

Principle II. Decomposition of by oxalic acid.

Application. Separation of Mg from K and Na.

Method. Add to the solution of the mixed chlorides enough oxalic acid to convert all the metals present, viewed as potassium, into quadroxalate. Add more water if need be to ensure complete solution, evaporate to dryness in a platinum capsule and ignite. During the evaporation the alkaline chlorides are partially, and the chloride of magnesium completely converted into oxalates, and by the final ignition the oxalate of magnesium is changed to magnesia, and the small quantity of oxalates of the alkali metals to carbonates. Treat the residue with repeated small portions of boiling water, collect the Oxide of Magnesium on a filter, and in case the filtrate is at all turbid, evaporate it to dryness and again filter to separate any magnesia which may have been deposited. Finally add chlorhydric acid to the filtrate and estimate the alkalies as chlorides. See Chloride of Potassium.

Chloride of Manganese.

Principle. Fixity when heated in an atmosphere of hydrogen.

Applications. Separation of Mn from Co and Ni.

Method. See Cobalt, and Chloride of Cobalt. The turbid solution of chloride of manganese, decanted from the reduced cobalt (or nickel) is acidulated with chlorhydric acid, and concentrated somewhat by evaporation, after which the manganese is precipitated as Carbonate of Manganese.

*di*Chloride of Mercury. (Calomel).

[Compare Mercuric Compounds].

Principle I. Insolubility in water, and fixity at 100.°

Applications. Estimation of mercury in mercurous salts. Separation of Hg from Pb, Cd, Cu. Estimation of free chlorine. (Method D).

Methods: — Gravimetric.

Method A. Mix the cold, highly dilute solution with chlorhydric acid or with a solution of chloride of sodium as long as any precipitate forms. Let the mixture settle; collect the precipitated dichloride of mercury upon a weighed filter, dry at 100° and weigh. The results are accurate. In case the substance to be analyzed is a solid, insoluble in water, it may either be treated directly with dilute chlorhydric acid in the cold, or it may be dissolved in highly dilute nitric acid, and the solution mixed with a large quantity of water, after most of the acid has been neutralized with carbonate of sodium, before proceeding with the precipitation. Care must be taken not to change any of the mercurous salt to the condition of a mercuric salt. If lead be present, the precipitate must be washed with special care with water of 60° or 70° until the washings are no longer colored on addition of sulphuretted hydrogen.

Volumetric.

Method B. [Compare Principle III, below].

Pour from a burette a standard solution (one-tenth normal) of chloride of sodium into the cold solution of the mercurous salt acidulated with nitric acid, until a precipitate is no longer produced, and a slight excess of the chloride of sodium has been added. It is easy to determine when an excess of the chloride of sodium is present, though from the fact that, unlike chloride of silver, the precipitate does not ball or become granular, it would be well-nigh impossible to stop at the precise moment when precipitation ceases. Collect the precipitated dichloride upon a filter; wash the precipitate thoroughly, taking care not to use any more water than is absolutely necessary, and determine how much chloride of sodium has gone into the filtrate by precipitating the chlorine as Chloride of Silver, with a standard silver solution. By subtracting the excess of chloride of sodium thus found, from the total amount used, it will be seen how much was consumed in precipitating the mercurous salt, and the amount of the latter may then be readily calculated. — The process yields accurate results, though from the need of filtering and washing it has no great advantage over the gravimetric method. It is well, sometimes, to weigh the precipitated dichloride, and to titrate the excess of chloride of sodium also, for the sake of control.

Method C. Precipitate dichloride of mercury in a glass stoppered bottle. (Compare Method A., gravimetric, above). Let the precipitate settle; collect it on a filter without heeding the particles which adhere to the sides of the bottle, and wash it as well as the bottle. Then push a glass rod through the apex of the filter, and rinse the precipitate back into the bottle. Pour into the bottle a sufficient quantity of a solution of iodide of potassium, together with a measured quantity of a standard Iodine solution, put in the stopper and shake the bottle until the precipitate has all dissolved.

For every gramme of dichloride of mercury about 2.5 grms. of iodide of potassium and 100 c. c. of one-tenth normal iodine solution may be taken. Since an excess of iodine is used, the solution will be colored brown. Estimate the excess of Iodine by means of hyposulphite of sodium. (Hempel, *Annalen Chem. und Pharm.*, **110.** 176).

Method D. To estimate free chlorine in chlorine water, put a weighed quantity of metallic mercury in the liquid and leave the mixture for a long time, until the odor of the chlorine has wholly disappeared. The mercury must be in excess from first to last. Collect upon a weighed filter the dichloride of mercury which is formed, together with the excess of metallic mercury, dry at a temperature below 100° and weigh. The increase in weight of the mercury gives the weight of chlorine in the liquid. A small proportion of the mercury is liable to be lost by volatilization, even when the temperature is kept below 100°, so that the process is rather less accurate than the corresponding method, depending on the insolubility of Chloride of Silver. (H. Rose).

For Schulze's method of estimating chlorine in organic compounds, see protOxide of Mercury.

Principle II. Volatility when strongly heated.

Applications. Separation of mercury from several of the lighter metals, and from the metals of the alkalies, alkaline earths and earths.

Method. The substance to be analyzed is either ignited in a crucible and the mercury estimated from the loss, or it is heated in a combustion tube drawn out and curved in front so as to dip into a beaker of water, by which the chloride is retained.

Principle III. Conversion of to mercuric chloride by chlorine: — $HgCl + Cl = HgCl_2$.

Application. Estimation of chlorine in chlorine water. Valuation of bleaching powder, and of binoxide of manganese.

Method A. Put a weighed quantity of dichloride of mercury in the liquid which contains free chlorine. Shake the mixture frequently, until the odor of the chlorine has completely disappeared. Then collect upon a weighed filter the dichloride, which still remains undissolved, and dry and weigh it. The difference between this last weight and the weight of dichloride originally taken, gives the amount which has been dissolved, and each equivalent of the dichloride thus lost corresponds to an atom of chlorine in the liquid. Considerable time is required for the complete removal by the dichloride of chlorine from its solution.

Method B. One of the earliest methods of Chlorimetry depends upon the principle now in question. As applied by Balland (*Journ. de Pharm.*, 1838, **24.** 105), a solution of mercuric nitrate is poured into the solution of bleaching powder, until a precipitate persists even on agitation. But the process is not accurate. Marezeau (*Pogg. Ann.*, **22.** 273), on the other hand, poured a mixture of water and a weighed quantity of the bleaching powder to be tested, into a liquid holding in suspension a known quantity of dichloride of mercury, until the latter was all dissolved and the liquid had become perfectly clear. The necessary solutions were prepared as follows:— Weigh out 5 grms. of pure, dry chloride of sodium, dissolve in water and dilute to the volume of 400 c. c. Measure off in a pipette 50 c. c. of a solution of mercurous nitrate, or more or less, according to the degree of concentration of the liquor, place the solution in an eight ounce glass-stoppered bottle, and dilute to the volume of about 5 ounces. Slowly pour the chloride of sodium solution from a burette into the bottle, until the last drop ceases to produce a precipitate. During this operation keep the bottle in hot water, and shake it after each addition of the chloride of sodium, to promote the settling of the precipitate. In case too much chloride of sodium happens to be added at first, add a few more c. c. of the mercurous solution, and again add chloride of sodium. Read off how much chloride of sodium has been used, and calculate therefrom how much mercury is contained in each c. c. of the mercurous solution. Then dilute a quantity of the latter to such an extent that each c. c. of the diluted liquid shall correspond either to 0.01 grm. of chlorine, or to 0.005 grm.

For the actual analysis, grind up 5 grms. of bleaching powder with water, to a definite volume of milky solution, in the usual way (see under Arsenious Acid). Measure out, by means of a pipette, 100 volumes of the solution of the mercurous nitrate, corresponding to 0.5 grm. of chlorine. Place the liquid in a beaker, dilute with water, and add an excess of a solution of chloride of sodium, together with a small quantity of chlorhydric acid free from chlorine or sulphurous acid. Cautiously pour the bleaching powder liquor from a burette into the mixture in the beaker, until the liquid just becomes clear. The mixture must be stirred incessantly, and must be kept acid from first to last. Read off the number of c. c. of bleaching powder liquor used, and proceed to calculate the amount of chlorine contained in them, from the known value of the mercurous solution used.

Method C. To determine the value of binoxide of manganese, and of other high oxygen compounds, place a weighed quantity of the oxide, together with some strong chlorhydric acid in a flask, fitted with a perforated cork carrying a delivery tube bent at two right angles. Let the outer long limb of the delivery tube reach into another flask containing water, in which a weighed quantity of very finely powdered dichloride of mercury has been suspended. Warm the mixture of oxide and acid, as occasion may require, so that an exceedingly slow development of chlorine shall occur, taking care, meanwhile, to shake the absorption flask frequently, to prevent the dichloride of mercury from becoming impacted at the bottom of the flask. For each atom of chlorine absorbed, one equivalent of the dichloride goes into solution. At the end, boil the liquid in the flask which contained the manganese, in order to sweep forward the last traces of chlorine. Then collect upon a weighed filter the dichloride of mercury which remains undissolved, and wash, dry and weigh it. Subtract the weight thus found from the weight of dichloride taken, and from the difference reckon the corresponding quantity of binoxide. It is to be observed that the absorption of chlorine by the suspended dichloride is slow, much slower than in the corresponding process, where the Chlorine is made to act on sulphurous acid, hence care must be exercised to prevent the escape of any of the chlorine.

Properties. The precipitated dichloride is a heavy white powder almost absolutely insoluble in cold water. It is slowly decomposed by boiling water. It does not dissolve at the ordinary temperature in water acidulated with chlorhydric acid; at a higher temperature some of it dissolves, and on boiling with access of air the whole of the dichloride is gradually converted into soluble protochloride. Boiling concentrated chlorhydric acid decomposes it rather rapidly into metallic mercury and protochloride. Boiling nitric acid decomposes and dissolves it. Chlorine water and aqua regia convert it into soluble protochloride even in the cold. Solutions of the alkaline chlorides decompose it slightly in the cold, and to a considerable extent when heated.

The dichloride is permanent in the air and may be dried at 100° without loss. It volatilizes without previous fusion at a temperature below redness. Its composition is:—

Hg	200	84.93
Cl	35.5	15.07
	235.5	100.00

*proto***Chloride of Mercury.** (Corrosive sublimate). [Compare Mercuric Compounds.]

Principle I. Volatility.

Applications. Separation of Hg from Pb, Cu, Ag, and, in general, from all metals which form non-volatile chlorides (Method A). Analysis of chloro-mercurates (Method B). Separation of Mg from alkali metals and from chlorine (Method C).

Method A. Precipitate the mixed solution with sulphuretted hydrogen (see Sulphide of Mercury, etc.), collect the precipitate upon a weighed filter, dry at 100° and weigh. Introduce a weighed portion of the precipitate into a weighed bulb tube of hard glass, or into a porcelain boat, which is afterwards placed in a piece of hard combustion tubing; pass a slow current of dry chlorine through the tube, and heat the mixed sulphides in the boat or bulb, at first gently, but afterwards to faint redness. A couple of small flasks charged with water should be attached to the bulb tube to receive the volatile matters which are expelled from it, and the tube should be bent so as to project into the first flask. The chloride of sulphur which is formed distils over completely into the flasks, but a part of the chloride of mercury is apt to condense in the colder part of the tube. In that event, cut off the soiled portion of the bulb tube, and wash out the chloride from it into the flask. Or, if a porcelain boat has been used, withdraw the boat from the combustion tube and wash out the latter. The piece of bulb tube which has been cut off is dried and weighed together with the rest of the bulb tube and its contents. If the residue contains only a single metal, the weight of the latter can be calculated at once from the weight of its chloride, but if there are several metals present the chlorides must be reduced in a stream of hydrogen, and the metallic residue thus obtained dissolved in nitric acid as a preliminary to their separation by appropriate processes. The solution of chloride of mercury is mixed with some chloride of ammonium, heated to expel chlorine, then filtered to separate the sulphur and subjected to analysis. See di-Chloride of Mercury or Sulphide of Mercury. Compare Chloride of Sulphur. The process yields accurate results. — If it be deemed sufficient to estimate the mercury by the loss, as may be done in case only one metal is mixed with the mercury, the apparatus may be simplified as follows:—Place a weighed quantity of the mixed sulphides carefully dried at 100°, in a porcelain boat, place the boat in a wide glass tube and heat it in a current of chlorine until all the sulphur and mercury have been expelled. From the weight of the residual chloride, calculate how much of the corresponding sulphide was contained in the matter analyzed, and in the entire precipitate. The difference between the latter quantity and the weight of the precipitate is sulphide of mercury.

Method B. To analyze the double compounds formed by the union of mercuric chloride with other metallic chlorides—the so called chloromercurates—v. Bonsdorff (*Poggendorff's Annalen*, **17.** 115) proceeds as follows:—Select a glass tube about 0.5 inch in diameter, close it at one end and blow out that end into the form of a little flask; at a point several inches above this flask blow a small bulb, or rather an enlargement of the tube; above this protuberance draw out the tube so that its diameter may be about 0.25 inch; there cut it off, and smooth the edges. The length of the finished apparatus should be about 6 inches. Weigh the dried apparatus, put a quantity of the substance to be analyzed in the bottom of the flask, again weigh, and close the end of the tube partially by means of a loosely fitting cork. Clamp the apparatus in a nearly horizontal position, and warm the flask gently with a lamp. The water of crystallization in the compound will be first driven off, and will collect in the protuberance on the neck of the flask. It may be removed with slips of filter paper or expelled by heating it very gently, in which event no residue, or a scarcely perceptible residue of mercuric chloride will be left upon the under side of the bulb. After the first instalment of water has been thus removed replace the cork and again heat the contents of the flask gently, to be sure that all the water is expelled, then allow the apparatus to cool, and weigh it. The loss of weight represents the water of crystallization in the substance. Again heat the flask, loosely corked as before, until the whole of the mercuric chloride is driven into the neck of the flask near the bulb, then cut the apparatus in two at a point below the sublimate, and weigh each portion, first with its contents, and afterwards without them. Since the mercuric chloride might be contaminated with a few milligrammes of water, it had better be left to dry over sulphuric acid before weighing it.

In case the chloride left in the flask, is of such character that it can be heated in the air without change, the flask had better be ignited after the first weighing, and again weighed to make sure that all the mercuric chloride has been expelled. When carefully conducted the process yields good results. There is scarcely any risk of losing chloride of mercury during the evaporation of the water, or its expulsion from the flask.

Method C. To separate magnesium from alkalies and from chlorine, place the solution of the mixed chlorides in a crucible, evaporate the solution to dryness, and in case any chloride of ammonium be present ignite the residue. The substance to be analyzed must contain no acid other than chlorhydric. Mix the residue with a little warm water, and stir into the mixture a quantity of pure elutriated oxide of mercury — about three times as much as there is of the mixed chlorides. Again evaporate to thorough dryness upon a water bath, with frequent stirring; then cover the crucible and heat it to redness until all the chloride of mercury, which has been formed by the decomposition of the chloride of magnesium, is expelled. In the case of separating magnesium from chlorine the ignition must be continued until the excess of oxide of mercury employed has all been expelled; but in separating magnesium from the alkalies it is enough to drive off the chloride of mercury by heat, and to separate the oxide of mercury, together with the oxide of magnesium, by filtration from the undecomposed alkaline chlorides. In that case it will finally be expelled when the Oxide of Magnesium comes to be ignited.

The method is convenient and yields satisfactory results. Care must be taken to add only a little more than enough oxide of mercury to decompose the chloride of magnesium and to avoid inhaling the fumes of corrosive sublimate which are given off during the ignition. The alkaline chlorides obtained should tbe tested for magnesium; a trace of it will usually be found in them. (Berzelius, in his *Jahresbericht der Chemie*, 1842, **21.** 142.)

Principle II. Oxidizing power. See Mercuric Compounds.

Chloride of Nickel.

Principle I. Fixity when heated.

Applications. Separation of Ni from As, Sb, Bi, Hg and Sn.

Method. See the Chlorides of the metals above enumerated.

Principle II. Colorific power.

Application. Estimation of nickel.

Method. See Colorimetry, A. Mueller's process, and *Zeitsch. analyt, Chem,*, 1866, **5.** 426. Instead of a glass plate of a color complementary to that of the nickel solution, Winkler uses a solution of cobalt.

Principle III. Reducibility by hydrogen.

Applications. Separation of Ni from Ba, Sr, Ca aud Mn.

Method. See Nickel, fixity of.

Chloride of Phosphorus.

Principle. Volatility.

Application. Estimation of phosphorus in phosphide of iron.

Method. Heat a glass tube at about 30 c. m. from its anterior end in such manner that about 15 c. m. of the tube shall be made narrower. Bend the narrowed portion downward and then again horizontal, in a plane parallel to that occupied by the rest of the tube. Finally draw out to a point and turn upwards the anterior end of the tube, leaving 10 c. m. or so. of the tube horizontal and of the original diameter. This anterior horizontal space serves as a reservoir for a few c. c. of water, as will be directly explained. — Into the long, posterior wide portion of the tube push an asbestos plug, then a layer of 12 or 15 c. m. of coarsely powdered chloride of potassium, then a second small loose plug of asbestos, and place the tube upon a wire gauze support. Behind the second asbestos plug place a porcelain boat charged with lumps of the phosphide of iron to be analyzed, then a third plug of asbestos, and close the tube with a perforated cork carrying a short delivery tube. Charge the anterior horizontal part of the tube with water, as above indicated, and by means of a caoutchouc connector attach to that end of the tube another tube charged with moistened bits of porcelain. The purpose of the water and moist porcelain is to retain chloride of phosphorus; that of the chloride of potassium is to retain chloride of iron, and to destroy a compound of the chlorides of iron and phosphorus.

Heat the chloride of potassium with a lamp and pass a current of dry air through the tube to remove all traces of moisture. Then pass in a stream of chlorine gas, and as soon as all the air has been expelled from the tube, heat the porcelain boat with a second lamp. Immediately a red liquid will begin to condense in the tube, and to diffuse itself among the lumps of chloride of potassium. The latter must only be heated in the neighborhood of the boat, and no more strongly than will suffice to melt the double salt of chloride of potassium and chloride of iron which is formed, and thus prevent the tube from becoming stopped. Towards the close of the operation the heat may be increased, but never to dull redness, since at that temperature chloride of phosphorus would be decomposed by the oxygen in the silicic acid of the glass with formation of phosphoric acid. All the chloride of phosphorus is driven forward by means of a gentle heat into the anterior reservoir of water, and the operation is considered to be finished when no more sublimate can be seen. The reservoir-tube is finally detached by cutting the narrowed part of the tube, then emptied into a porcelain dish and washed out with water. The tube charged with bits of porcelain is also washed out into the same porcelain dish. Nitric acid is added to the liquid, and the mixture is evaporated. The chlorhydric acid decomposes without effervescence, as the liquor becomes concentrated, and the phosphoric acid is finally determined as Phosphate of Silver.

It is best to maintain a constant stream of chlorine, so that this gas may always be present in slight excess. To this end a self-regulating apparatus provided with a stop-cock should be employed. A telltale bottle charged with water, and placed before the tube which contains the bits of porcelain will indicate the rate of flow of the chlorine. (Schloesing, *Zeitsch. analyt. Chem.*, 1868, **7.** 474).

*Bi*Chloride of Platinum.

Principle I. Colorific power. See Colorimetry, A. Müller's process, and *Zeitsch. analyt. Chem.*, 1864, **3.** 407.

Principle II. Power of precipitating potassium and ammonium. See the Chloroplatinates of these metals.

To prepare bichloride of platinum for use as a reagent, cut up a quantity of worn out foil or wire into very small pieces, boil them for some time in a porcelain dish with nitric acid alone, to remove impurities, then decant the nitric acid and boil the metal with repeated small portions of aqua regia until it has all, or nearly all, dissolved. Decant each portion of the aqua regia, after it has acted upon the platinum for some little time, into another porcelain dish and evaporate the mixed solutions almost to dryness upon a water bath, then add some chlorhydric acid and again evaporate until no odor of chlorine or of chlorhydric acid can be perceived. On allowing the dish to cool the chloride will solidify to a crystalline mass, which should be dissolved in a not too large quantity of water, and the solution kept for use. It is well to know approximately how strong the solution is, and to label in that sense the bottle which contains it. A good strength is 0.5 grm. of platinum to each c. c. of solution. It is important that the evaporation be made upon a water bath lest some of the salt suffer decomposition and protochloride of platinum be formed. The solution when evaporated to dryness upon a water bath should leave a residue completely soluble in ordinary alcohol. According to Rose, it is best to operate in porcelain vessels and to avoid glass, in order that as little alkali as possible may be dissolved out to contaminate the platinum solution.

A still better way is to bring the platinum in the first place into the condition of powder, so fine that it can be readily dissolved by aqua regia. To this end fuse 5 parts of metallic zinc in a clay crucible, and little by little throw into the melted metal 1 part of platinum scraps. Stir the alloy with a pipe stem plugged with dry clay, and pour it into water. Treat the drops of alloy with chlorhydric acid somewhat diluted, until there is no longer any effervescence, then boil for a time with fresh chlorhydric acid to remove the last traces of zinc, wash with water, boil with nitric acid, again wash, and finally dissolve in aqua regia, added all at once, and evaporate the solution as above.

Chloride of Potassium.

[Compare Chloride of Sodium].

Principle. Fixity at moderately high temperatures.

Applications. Estimation of potassium in the hydrate, sulphate, chromate, chlorate (which see), silicate and sulphide of that metal, as well as in the potassium salts of weak volatile acids, — such as carbonic acid and various organic acids. Separation of K from Na, both direct and indirect. Estimation of chlorine in organic compounds.

Method. Evaporate the clear solution in a platinum dish almost to dryness, then transfer the liquor to a platinum crucible, taking care to use as little water as possible in washing the dish, and evaporate to dryness upon a water bath. Transfer the crucible from the water bath to the middle of a small iron pot, fit to serve as an air bath, and heat the pot to such an extent that the chloride of potassium may be exposed to a temperature somewhat higher than that of boiling water. After some time transfer the crucible to a ring-stand and heat it almost, but not quite, to dull redness, by means of a lamp, and weigh. The crucible should be kept covered during the ignition, and the temperature carefully regulated, so that none of the chloride shall be lost through volatilization. By heating the salt in an air bath, as above described, before the final ignition, the last traces of water are expelled, and all chance of loss through decrepitation avoided. When carefully executed the process yields very accurate results. — Instead of weighing the chloride, the chlorine contained in it may be determined by titration as Chloride of Silver. This method saves time where many estimations have to be made, but is of no advantage in the case of a single determination.

The presence of free acid in the original solution does no harm. It is, in fact, often necessary to evaporate the compound to be analyzed with an excess of chlorhydric acid, in order to expel some other acid. In the case of nitrate of potassium the evaporation must be repeated with 5 or 6 new portions of chlorhydric acid till the weight of the ignited chloride remains constant. In order to avoid frothing when carbonate of potassium is to be converted into chloride—as in the analysis of residues obtained by igniting the potassium salts of nonvolatile organic acids—it is best to treat the carbonate with a solution of chloride of ammonium instead of chlorhydric acid, and to evaporate the mixture to dryness and ignite the residue. A slight excess of chloride of ammonium should be used, and the mixture had better be evaporated and ignited in a tolerably large dish to expel the chloride of ammonium before transferring the chloride of potassium to the crucible. By heating the dish moderately at first, and afterwards almost to redness, the chloride of ammonium can all be expelled without loss of any of the potassium salt, for the ammonium salt changes at once from the solid to the gaseous state without becoming liquid, and so long as there is any chloride of ammonium left to evaporate, none of the chloride of potassium can volatilize. Whenever chloride of ammonium has to be expelled in this way, the residue should be taken up in the least possible quantity of water and filtered, in order to separate particles of dirt and carbon, which almost always present themselves. — In no event should any chloride of ammonium be allowed to get into the platinum crucible with the solution of chloride of potassium, for it would inevitably cause loss of the latter by creeping over the rim. Even when the quantity of chloride of ammonium in the solution is very small it must be expelled by ignition in a dish before the potassium salt can be safely transferred to the crucible. It is to be observed that chloride of ammonium can be more easily driven off from chloride of potassium than sulphate of ammonium from sulphate of potassium. For the conversion of sulphate of potassium to the state of chloride by means of chloride of ammonium, see Sulphate of Potassium; for its conversion by means of baryta water see Sulphate of Barium, and for its conversion by means of chloride of strontium see Sulphate of Strontium.

In the preliminary steps to the separation of potassium and sodium, both by the direct method (see Chloroplatinate of Potassium) and the indirect method (see Chloride of Silver), the process is conducted precisely as above described, *i. e.*, as if nothing but chloride of potassium were present. It is important, however, to remember that the mixed chlorides ought never to be weighed until their purity has been proved by dissolving them in water, and testing the solution with carbonate of ammonium and ammonia. No precipitate should be produced by these reagents, and the solution should moreover be clear before they are added. Filter, if need be, and again evaporate. As obtained by L. Smith's process (see Carbonate of Calcium) from a silicate, the chlorides are peculiarly liable to leave a black residue on being treated with water.

Estimation of chlorine in organic compounds. In acid organic compounds, such as chlorospiroylic acid, for example, chlorine may be determined by dissolving the compound in an excess of dilute potash lye, evaporating to dryness, and igniting the residue, by which means the whole of the chlorine is obtained as chloride of potassium. Dissolve the residue in water, and estimate the chlorine as Chloride of Silver. (Lœwig).

To separate potassium (or sodium) from chromic acid, mix the compound with about twice its weight of dry powdered chloride of ammonium, ignite the mixture carefully until the whole of the ammonium salt is expelled, dissolve the chloride of potassium in water, separate it from the oxide of chromium by filtration, and proceed as above. A. Mitscherlich (*Journ. prakt. Chem.*, **83.** 459) has shown that no loss of the alkaline chloride is occasioned by the escape of chloride of ammonium, not even when 20 parts of the latter are ignited with one of the former.

When a mixture of chloride of potassium and oxalate of ammonium is ignited to expel the latter, an appreciable quantity of the chloride of potassium is decomposed with evolution of chlorhydric acid, and more or less carbonate of potassium is left mixed with the residual chloride. If free oxalic acid be ignited with chloride of potassium, a still more considerable quantity of the carbonate will be formed. It is even possible to change the whole of the chloride of potassium into carbonate by igniting with oxalic acid. But the carbonate of potassium thus formed can readily be destroyed, after it has been formed, by means of chloride of ammonium, as above described; and if chloride of ammonium be added to the original solution of oxalic acid and chloride of potassium, before evaporating it, the decomposition of the latter may be prevented.

Principle II. Insolubility in absolute alcohol.

Application. Separation of K from Li.

Method. See Chloride of Lithium.

Principle III. Power of absorbing heat while dissolving in water.

Application. Estimation of chloride of sodium in impure chloride of potassium. Rough valuation of potashes and pearlash.

Method. It is a matter of experience that chloride of potassium in the act of dissolving in water cools the liquid far more than a similar quantity of chloride of sodium can. Thus while 50 grms. of chloride of potassium in dissolving in 200 grms. of water contained in a glass, weighing 185 grms., and capable of holding 320 grms. of water, will lower the temperature 11.4°, the same amount of chloride of sodium, under similar circumstances, will reduce the temperature only 1.9°. From these data, according to Gay-Lussac (*Annales Chim. et Phys.*, [2.] **12.** 14 and **39.** 356), the reduction of temperature produced by any mixture of chloride of potassium and chloride of sodium, dissolved in a similar quantity of water, may be reckoned; and conversely from the observation of the temperature in any given case, the proportion of chloride of potassium in the mixture can be estimated with tolerable accuracy. Let the quantity of chloride of potassium in 50 grms. of the mixture be called x, and that of the chloride of sodium $50 - x$. Then the reduction of temperature effected by x parts of chloride of potassium will be $\frac{11.4 \times x}{50}$, and that produced by $50 - x$ parts of chloride of sodium will be $\frac{1.9\,(50 - x)}{50}$. The observed reduction of temperature which may be called t, is the sum of these two quotients; consequently

$$t = \frac{11.4 \times x}{50} + \frac{1.9\,(50-x)}{50},$$

and the quantity x of chloride of potassium in 50 parts of the mixture may be obtained by reduction as follows:

$$x = 50\left(\frac{t-1.9}{11.4-1.9}\right) = \frac{50t-95}{9.5}.$$

To avoid the trouble of calculating each single result, Gay-Lussac has constructed a table by means of the above formula, which gives the per cent of chloride of potassium in all possible mixtures for each tenth of a degree of temperature between 1.9° and 11.4°. (See the original memoir of Gay-Lussac, or *Handwœrterbuch der Chemie*, **1.** 893).

It is essential that the mixture of the chlorides should be finely powdered, in order that the salts may dissolve as rapidly as possible, and that the observations of temperature be made with a highly sensitive thermometer, marking tenths of degrees. — The details of the process are as follows:—Weigh out 200 grms. of water in a French glass flask of 185 grms. weight and 320 c c. capacity, place the thermometer in the water and bring the apparatus to the temperature of 20.4°. Take hold of the flask by its neck, so that the warmth of the hand shall not influence the temperature of the water, pour in the 50 grms. of the dry, powdered chlorides, and twist the flask about rapidly while observing the thermometer. The observation is finished when the chlorides have dissolved, and the thermometer has sunk to the lowest point. The difference between 20.4° (the original temperature) and the degree to which the mercury has fallen, will be equal to t, as above stated. Since a single experiment occupies only about 10 minutes, the process has a certain value for technical purposes, such as the valuation of the chloride of potassium used in the manufacture of potash- from soda-saltpetre, and for making alum.

It is to be observed that Gay-Lussac's table is true only for the case where the 50 grms. of mixture and 200 grms. of water are placed in a glass flask of the given weight and capacity. With a different sized flask a different rate of thermometric reduction would obtain. Instead of trying to fulfill all these conditions, it would probably be easier for each experimenter to choose a flask for himself at random, and to determine once for all the reductions of temperature produced in it by 50 grms. of each of the pure chlorides when dissolved separately in 200 grms. of water. A formula could then be made out at once (with the new values in place of 11.4 and 1.9), and a new table calculated to be used in place of the table of Gay-Lussac. The temperature of 20.4° chosen by Gay-Lussac as the normal, may as well be retained.

The process may be applied, of course, to the determination of sodium in carbonate of potassium, contaminated with that substance. It will be necessary only to convert the mixed carbonates to chlorides, and then proceed as above.

Similar methods are evidently applicable to the analysis of other mixtures of salts, which absorb or evolve heat on being dissolved in water. The chief objection to the process in any case is the disturbing influence which the presence of other salts may exert upon the fall or rise of the thermometer.

Properties. Chloride of potassium is readily soluble in water, but much less soluble in dilute chlorhydric acid. It is almost completely insoluble in absolute alcohol, and but slightly soluble in spirit. It is permanent in the air at the ordinary temperature, fuses at a moderate red heat without change, and volatilizes in white fumes at a higher temperature. The volatilization proceeds more slowly the more effectually access of air is prevented (Fresenius). When the aqueous solution is evaporated, a small quantity of water is obstinately held mechanically enclosed in the seemingly dry residue. Great care must be exercised to expel this moisture before the salt is ignited, as has been already explained. When evaporated with an excess of nitric acid it is readily and completely converted into the nitrate. When evaporated repeatedly with an excess of oxalic acid it is converted into oxalate of potassium. The composition of chloride of potassium is,

K	39.1	52.41
Cl	35.5	47.59
	74.6	100.00

Chloride of Roseo Cobalt.

Principle. Sparing solubility in cold dilute chlorhydric acid, chloride of ammonium and alcohol.

The proposition of Terreil (*Comptes Rendus*, **62.** 139) to separate Co from Ni and Mn, by precipitating the former as chloride of roseo cobalt, by the action of permanganate of potassium upon a boiling ammoniacal solution of the three metals, has been shown to be valueless by Braun and Fresenius (*Zeitsch. analyt. Chem.*, 1866, **5.** pp. 114–116). Some of the cobalt escapes the oxidizing action of the permanganate, and the chloride formed is not wholly insoluble in dilute chlorhydric acid.

Chloride of Silver.

Principle I. Fixity when heated in the air.

Application. Weighing of chloride of silver as obtained by the gravimetric method under Principle II. Retention of chlorine in the analysis of some organic compounds. See Carbon, p. 68.

Method. See under Principle II, and Properties, below.

Principle II. Insolubility in water and dilute nitric acid.

Applications. Estimation of silver, of combined chlorine, and of chlorhydric acid. Separation of Ag from K, Na, NH_4, Ca, Sr, Ba, Mg, Al, Cr. Fe. P, S. Se, Te, As, Sb, Sn, Co, Ni. Mn, Zn, Cd, Cn, Pb, Bi, Hg, Ur, Au, and Pt. Separation of chlorhydric acid from sulphuric, phosphoric, boracic, silicic, fluorhydric, chloric, nitric, oxalic, arsenious, arsenic, and chromic acids. Direct and indirect separation of K from Na. Assay of coins and other alloys of silver.

Gravimetric Method.

Method. The moderately dilute solution of a silver salt, of chlorhydric acid, or of a chloride, is slightly acidulated with nitric acid, either in the cold or at temperatures no higher than 60° C., and then mixed with dilute chlorhydric acid, or with a solution of nitrate of silver, as the case may require, as long as any precipitate continues to be formed. Care must be taken not to add any considerable excess of chlorhydric acid, or of chloride of sodium, when these substances are employed as precipitants, for chloride of silver is somewhat soluble in them. The presence of all other substances capable of dissolving chloride of silver must of course be carefully avoided. (See properties, below, and Dictionary of Solubilities). — Precipitated chloride of silver may be washed either by decantation, or upon a filter. The method by decantation is best when the amount of precipitate is large, and when nothing but silver (or chlorine) is to be determined in the solution; but when the precipitate is small, and when other substances besides silver or chlorine are to be determined, a filter should be employed, lest too large a proportion of the precipitate be lost, or too great a bulk of liquid be obtained.

In case it is proposed to wash by decantation, place the acidulated solution to be analyzed in a tall conical flask, with long neck and narrow mouth, immerse the flask in water, and maintain the latter at a temperature of about 60° C. The flask should be provided with a smooth and sufficiently large cork, or with a finely-ground glass stopper. Add the precipitant little by little, while occasionally twisting the flask until the last drops produce no further cloudiness, then cork the flask and shake it vigorously until the chloride of silver has united into large coherent lumps, and the supernatant liquid has become tolerably clear. Remove the cork and wash from it and from the neck of the flask any particles of chloride which may have adhered there, then leave the flask at rest in the warm water during several hours, until the precipitate has separated completely, and the liquid above it is absolutely clear and transparent. Where a high degree of accuracy is required the mixture may be left to settle for 12 hours in a moderately warm, dark place.

Decant the clear liquid from the flask into a beaker, as completely as may be practicable, taking care not to pour out any of the chloride; then fill the flask with hot water and again decant as soon as the mixture has settled, and repeat the process until a small portion of the water tested with chlorhydric acid, or with nitrate of silver, as the case may be, does not become cloudy. It is well to acidulate with nitric acid the water used for washing the chloride of silver, since the settling of the latter is thereby slightly promoted.

Transfer the washed precipitate from the flask to a weighed porcelain crucible as follows: Set a tolerably large evaporating dish, full of water upon the table; fill the flask which contains the chloride of silver completely full of water and invert the porcelain crucible over the mouth of the upright flask, so that the latter shall be in a measure closed by the crucible. Hold the crucible as tightly as possible against the mouth of the flask and suddenly invert the latter, so that the crucible and the mouth of the flask shall be completely sunk in the water of the evaporating dish. Lift the flask a little, so that its rim shall be just within the crucible, and rap the flask so that the last particles of the chloride of silver shall fall down into the crucible. Finally slip the mouth of the flask over the rim of the crucible, still keeping the mouth under water, and lift the crucible out of the dish. Decant the clear water from the crucible, dry the chloride thoroughly upon a water bath, and heat it over a lamp until it begins to fuse at the edges. When the crucible has become cold, weigh it with its contents. — To facilitate the settling of the precipitate in the crucible it is well to heat it upon the water bath before attempting to decant the last portions of liquid. The last drops of water may be removed by holding the crucible in an inclined position and touching the water with little rolls of filter paper. In case any particles of chloride of silver are seen in the beaker into which the liquid from the flask is decanted, leave the mixture at rest in a warm, dark place, for a number of hours, decant the clear liquid, collect the residual chloride of silver upon a small filter, as below, and add its weight to that of the main precipitate.

The precipitate obtained by adding nitrate of silver to a solution of chlorhydric acid, settles much more quickly than that obtained by adding chlorhydric acid to a solution of nitrate of silver.

Since chloride of silver is decomposed by light, it should be kept in the dark as much as possible during the processes of precipitation and decantation. A sleeve of cloth to cover the flask will be sufficient for the purpose, and the operator need not be afraid to uncover the flask as often as he may please, to observe the progress of the precipitation. Direct sunlight must of course be carefully avoided. — In order to remove the fused chloride of silver from the crucible, put a bit of zinc upon the chloride and fill the crucible with highly dilute chlorhydric acid. Some of the chloride will uickly be reduced to the condition of metallic ilver, and its adhesion to the flask destroyed.

In case the precipitate is collected upon a filer, the latter should be as small as possible, nd the precipitate should be left to settle for ome hours in a warm, dark place, before proceeding with the filtration. The precipitate s washed with hot water, thoroughly dried, nd ignited in a porcelain crucible, as above. Before igniting, the chloride of silver must be emoved as completely as possible from the aper, and the latter burned upon the lid of he crucible. Since some of the chloride is lways reduced in this operation, the filter ash hould be moistened with two or three drops f dilute nitric acid and gently warmed, then reated with a drop or two of chlorhydric cid and thoroughly dried, in order to reconert the metal into chloride. Or the reduced ilver may be weighed by itself, as such. Or he crucible may be discarded altogether, and he chloride collected and weighed, on a tared ilter dried at 100°. — It will often be ound convenient to combine the method of washing by decantation with that of filtration. The great bulk of the precipitate may be washed by decantation and dried in a porcelain crucible, as above, while the decanted iquor is passed through a filter, and the chloride thus collected incinerated by itself.

Exceptions and special precautions. In separating silver from *mercury*, care must be taken to keep the chloride of silver from being contaminated with difficultly soluble salts of mercury. In case the precipitation be effected in presence of a nitric acid solution of mercuric oxide, some basic nitrate of mercury is liable to go down with the chloride of silver, and the proportion is larger, accordingly as the amount of free nitric acid in the solution is less. On the other hand, if there be present a nitric acid solution of mercurous oxide, some mercurous chloride will go down with the silver precipitate. See also, below.

To avoid these contaminations, the original solution may be mixed with acetate of sodium, or acetate of ammonium, and the silver precipitated with a solution of chloride of sodium. Or chlorhydric acid, in not too great excess, may be added to the diluted original solution, and the precipitate heated with a mixture of nitric and chlorhydric acid, after the clear supernatant liquor has been decanted. Or, instead of treating the precipitate with the mixed acids, the basic mercuric nitrate may be decomposed and removed from it by means of a solution of acetate of sodium or acetate of ammonium, to which a very small quantity of nitric acid has been added.

In case of having to deal with a solution which contains mercurous oxide, the impure precipitated chloride of silver may be heated with aqua regia to convert the insoluble mercurous chloride into soluble mercuric chloride. It would be difficult to transform the mercurous to a mercuric salt by heating the original solution with nitric acid, in case the solution were very dilute. — According to Levol, no thorough separation of the precipitate from the liquid can be obtained unless an alkaline acetate be added to the original solution. In the absence of an acetate some chloride of silver would, moreover, remain dissolved in the mercuric nitrate (Wackenroder). — In separating silver from *lead*, also, acetate of sodium should be added to the original solution. The mixture is then heated and precipitated by the addition of rather dilute chlorhydric acid, taking care to add no more of the acid than is absolutely necessary. In this case, the purpose of the acetate of sodium is to retain chloride of lead in solution (Anthon).

In separating silver from *bismuth*, care should be taken to have enough free nitric acid present to prevent the precipitation of any basic chloride of bismuth. For the sake of certainty, it is best to heat the precipitated chloride of silver with dilute nitric acid, after the clear supernatant liquor has been decanted, in order to dissolve any chloride of bismuth which may have gone down. In both cases the precipitate must be washed thoroughly with water acidulated with nitric acid, before any pure water is added to it. Instead of proceeding in this way, some tartaric acid may be added to the original solution, then ammonia, or some other alkali in excess, and finally chlorhydric acid in slight excess. Nothing but chloride of silver will go down from the mixture, even if water be also added.

To separate silver from *gold* or *platinum*, in alloys, treat the alloy with cold dilute aqua regia, and after this agent has ceased to act dilute the liquid largely, and collect the undissolved chloride of silver upon a filter. This method is applicable only to alloys which contain less than 15 per cent of silver. If a larger proportion of silver be present, the chloride of silver formed by the action of the first portions of aqua regia will envelop the rest of the alloy so completely as to protect some of it from the action of the acids. — To estimate *free chlorine*, put a weighed quantity of finely divided silver into the liquid which contains the chlorine, and leave it there, taking care to shake the mixture occasionally, until the odor of the chlorine has completely disappeared. Collect, wash, dry, ignite and weigh the mixed precipitate of silver and chloride of silver. The increase of weight of the silver gives the amount of chlorine in the liquid. To prepare the finely divided silver precipitate some nitrate of silver with ferrous sulphate, or reduce some chloride of silver with zinc.

For the separation of *chlorine from the metals*, in chlorides soluble in water, or nitric acid, no special precautions are necessary, excepting in the cases of the chlorides of antimony and platinum, mercuric chloride, stannic chloride, and green protochloride of chromium.

When a solution of either of the *chlorides of antimony* or of *mercuric chloride* is mixed with nitrate of silver, the chloride of silver thrown down will carry with it a certain quantity of the other metal. To avoid this contamination the mercury or the antimony may be first thrown down by means of sulphuretted hydrogen, the excess of the latter destroyed by adding a few drops of a solution of ferric sulphate, and the chlorine subsequently determined in the filtrate. It will not do to drive off the excess of sulphuretted hydrogen by heat, for a part of the chlorhydric acid would escape with it. In the case of antimony it will be necessary to add some tartaric acid to the solution to prevent the separation of a basic salt (see Sulphide of Antimony). — When nitrate of silver is added to a solution of *chloride of platinum*, insoluble chloroplatinate of silver ($2AgCl, PtCl_4$) is thrown down, together with the chloride of silver. Various methods have been proposed to avoid this difficulty. v. Bonsdorff, for example, heated chloride of platinum and its compounds in a current of hydrogen, and led the chlorhydric acid gas into a solution of nitrate of silver. Other analysts have melted the platinum chlorides with carbonate of sodium, and estimated chlorine in the solution obtained by treating the fused mass with water.

Instead of these methods Topsöe (*Zeitsch. analyt. Chem.*, 1870, **9.** 32), directs that an excess of zinc filings be added to the tolerably dilute solution of the Platinum compound, in order to precipitate the platinum in the metallic state. The chlorine (or bromine or iodine in case these elements be present) may then be determined in the filtrate in the usual way. After the development of hydrogen has ceased in the aqueous solution, the liquid is saturated with ammonia and heated upon a water bath before filtering off the platinum. This method is said to yield excellent results.

Or the platinum compound may be converted into Sulphite of protoxide of Platinum, a compound which does not interfere with the precipitation of chloride of silver (see Iodide of Silver). — From a solution of *stannic chloride* nitrate of silver would precipitate, besides chloride of silver, a quantity of binoxide of tin and oxide of silver. To avoid this result, precipitate Hydrate of Tin in the first place by adding sulphate of sodium or nitrate of ammonium to the liquid, and estimate the chlorine in the filtrate. (Lœwenthal, *Journ. prakt. Chem.*, **56.** 371). — From a solution of the green *protochloride of chromium* nitrate of silver cannot throw down all the chlorine. The chromium must consequently first be precipitated as Hydrate of Chromium, by means of ammonia, and the precipitate removed by filtration before the estimation of the chlorine can be proceeded with. (Peligot). In estimating chlorine in presence of a *fluoride* it is best to throw down the fluorine, as Fluoride of Calcium, by adding nitrate of calcium before proceeding to precipitate the chloride of silver; though when the substance is soluble in water it is perfectly possible to estimate the chlorine first by adding nitric acid and nitrate of silver in the usual way. An insoluble chloro-fluoride would have to be fused with carbonate of sodium and silicic acid. — To estimate chlorine in a *silicate* dissolve in dilute nitric acid, if that can be done, and if no gelatinization occurs add nitrate of silver to the highly dilute solution, without warming it at any time. Before proceeding to estimate Silicic Acid in the filtrate, add dilute chlorhydric acid to it in the cold, and so remove the excess of silver.

In case the silicate gelatinizes with the nitric acid, dilute the jelly and allow the mixture to settle; filter, wash the precipitated silicic acid and add nitrate of silver to the filtrate. If the silicate to be examined cannot be decomposed by nitric acid, fuse it with a mixture of carbonate of sodium and carbonate of potassium in a platinum crucible, boil the fused mass with water, remove the dissolved Silicic Acid by means of carbonate of ammonium, and precipitate the filtrate with nitrate of silver after acidulating with nitric acid. (H. Rose).

To separate chlorine from nitric and chloric acids, in case the metal with which the chlorine, etc., is combined, is capable of uniting with phosphoric acid to form an insoluble compound, boil the solution with recently precipitated, thoroughly washed tribasic phosphate of silver. Chloride of silver and phosphate of the metal with which the chlorine was previously combined, will separate as insoluble powders, together with the excess of the phosphate of silver, while the nitrates and chlorates remain in solution (Chenevix; Lassaigne, *Journ. de Pharm.*, **16.** 289).

Estimation of Chlorine in organic compounds. To estimate chlorine (or bromine or iodine) in organic compounds, or to separate chlorine (or bromine or iodine) from cyanogen, seal up the substance in a glass tube, together with some nitric acid of 1.2 sp. gr., and a slight excess of nitrate of silver, as below. In most cases the oxidation of the organic matter is easy and rapid, while the chloride (iodide or bromide) of silver separates out. As regards cyanogen, the separation depends on the rapid decomposition of cyanide of silver under the conditions above described. Organic substances, such as the members of the aromatic series, which do not oxidize easily, may be destroyed by putting some bichromate of potassium in the tube with the acid and the silver salt. But in that event the highly diluted acid liquid must be heated to destroy the chromate of silver which is formed, before proceeding to collect the chloride of silver. The free nitric acid is then neutralized, almost, but not quite, completely, by adding pure carbonate of sodium, and the insoluble chloride of sodium is collected and weighed, together with the fragments of the glass bulb, the weight of the latter being subtracted from the total weight of the precipitate in order to obtain the true weight of the chloride of silver. Since there is always an excess of nitrate of silver present in the liquid above mentioned, the solvent action of the nitrate of sodium upon the chloride of silver has but little significance. — The details of the operation are as follows:—Prepare some little weighing bulbs, somewhat similar to those described under Carbon, on p. 67. Each of these bulbs should have two capillary stems, one upon either side of the bulb. These stems should be about 1 m. m. wide at their ends, very thin in glass, and bent or curled to one side at the ends. In one of these bulbs seal up from 0.15 to 0.4 grm. of the liquid to be analyzed, taking care to leave as little air as possible in the bulb. Put the loaded bulb in a tube of difficultly fusible glass from 10 to 12 m. m. wide and closed at one end to a round end. Fill the tube about half full of nitric acid of 1.2 specific gravity, with addition of nitrate of silver, as above, then draw out the upper end of the tube at the blast lamp to a thick, capillary tube, heat the nitric acid to boiling, and when all the air has been expelled seal the capillary tube at the lamp. As soon as the nitric acid has become cold, shake the tube until the points of the weighing bulb are broken off so as to leave openings of about 1 m. m. through which the nitric acid can gain easy access to the contents of the bulb. Then place the glass tube in an iron tube closed at one end, and set the latter in an oblique position in a sheet iron box or air bath, and heat the latter with a gas flame to 120° or 140°. In order to open the tube at the close of the operation, let the glass become perfectly cold, warm the capillary point gently, in order to expel any liquid which may have collected there, and finally heat the extreme point of the tube to redness, so that the gases within the tube may escape through the softened glass without violence. Then cut the tube in two below the capillary portion, and wash out its liquid and solid contents. The object in placing nitrate of silver in the tube is to prevent the formation of any chlorine gas. In case there was no nitrate of silver in the tube, some gaseous chlorine would be formed as well as chlorhydric acid, but in that event, the point of the tube may be broken off beneath the surface of a dilute solution of sulphite of sodium contained in a glass cylinder. For each part of the organic substance, from 8 to 10 parts of the sulphite would be needed, and as much water as would amount to 40 times the volume of the nitric acid. The point of the glass tube must be broken off carefully, by pressing it against the side of the cylinder, after a light scratch has been made upon it with a glass-knife of hardened steel. The gases within the tube will then escape in so fine a stream that all the Chlorine will be readily changed to chlorhydric acid. Before opening the tube completely, let some of the sulphite solution enter it to destroy the chlorine there. Heat the mixture until the whole of the sulphurous acid has been expelled, add enough pure carbonate of sodium to nearly neutralize the acid, and precipitate the chlorhydric acid with nitrate of silver. — The amount of nitric acid to be sealed up with the organic substance in the tube ranges from 20 to 60 times the weight of the substance; it varies according to the quantity of oxygen which the substance is capable of consuming. Nitric acid of 1.2 specific gravity is best suited for the requirements of the case; when a stronger acid is used the tubes are more liable to explode, and with a weaker acid a higher temperature and longer continued heating will be required. With acid of 1.4 specific gravity, the conversion of the organic matter into carbonic acid and water, is rapid at a temperature not much superior to 100°; with acid of from 1.12 to 1.2 specific gravity, the mixture must be heated from 1 to 3 hours to form 120° to 150°. Some substances, such as naphthalin and phenylalcohol, are completely decomposed only after 6 or 8 hour's heating, to 150° or 180°, in case no bichromate of potassium has been added. The process is applicable also for the estimation of Sulphur and Phosphorus, as will be explained hereafter. (Carius, *Liebig and Kopp's, Jahresbericht der Chemie*, 1860, **13.** 668, and 1861, **14.** 833; Kraut, *Zeitsch. analyt. Chem.*, 1863, **2.** 243). The process would be an excellent one were the sealed tubes less liable then they are to be broken by explosions.

B. For the steps preliminary to the addition of the silver solution in other methods of estimating chlorine in organic compounds, see Chloride of Copper. Chloride of Calcium, Chloride of Potassium, Chloride of Sodium and Chloride of Zinc.

For separating magnesium from the alkali-metals, Sonnenschein (*Poggendorf's Annalen*, **74.** 313) evaporates the solution of the mixed chlorides to dryness, ignites moderately to expel chloride of ammonium, treats the cooled residue with water, and boils the solution with an excess of carbonate of silver, until the liquid exhibits a strong alkaline reaction. It is well to continue to boil for 10 minutes, with constant stirring. The mixture is filtered as hot as possible, and the mixed precipitate of chloride of silver, carbonate of magnesium and the excess of the carbonate of silver, is washed with hot water. A little chlorhydric acid is added to the filtrate to throw down a trace of silver which goes into solution, the mixture is again filtered, and the alkaline chlorides determined as explained under Chloride of Potassium. The matter upon the filter is digested in chlorhydric acid, and from the solution thus obtained magnesium is precipitated as Phosphate of Magnesium and Ammonium. It is to be observed that a small portion of the chloride of magnesium is changed to oxide by the first ignition. Oxide of Silver might be used instead of the carbonate.

According to Fresenius, this process is not to be commended, since the filtrate invariably contains something more than mere traces of magnesium.

B. Volumetric Method.

Instead of weighing the precipitated chloride of Silver, as in A, its amount may be determined with great accuracy by employing a standard solution of chloride of sodium or of nitrate of silver, as the case may be, and noting how much of this solution is required to precisely precipitate a weighed quantity of the silver compound, or of the chloride under examination.

For the estimation of silver, three standard solutions may be prepared as follows:— 1st. A strong solution of chloride of sodium, by dissolving 5.4145 grms. of pure chloride of sodium in distilled water, to the volume of 1 litre at 16°.

If the salt be absolutely pure, 100 c. c. of this solution will contain a quantity of chloride of sodium equivalent to 1 grm. of metallic silver, and 1 c. c. to 0.01 grm. Clean, well crystallized, native rock salt is pure enough for this purpose. It should be coarsely powdered and dried by a moderate ignition, but should not be fused.

2d. A decimal solution of the chloride, by diluting 50 c. c. of the aforesaid strong solution to the volume of half a litre, at 16°. (Compare Alkalimetry, p. 18.) Each c. c. of this weak solution will correspond to 0.001 grm. of silver.

3d. A decimal silver solution, by dissolving 0.5 grm. of chemically pure silver, in 2 or 3 c. c. of pure nitric acid of 1.2 specific gravity, and diluting the acid liquor to the volume of half a litre, measured at 16°. Each c. c. of this solution will contain 0.001 grm. of silver.

Each of the three solutions should be kept in a glass-stoppered bottle, and each solution should always be shaken before any portion of it is poured out for use. The bottle which contains the silver solution should be kept in the dark, by enclosing it in a sleeve of cloth, for example.

To fix precisely the value of the chloride of sodium solutions, weigh out exactly a little more than one gramme of chemically pure silver (from 1.001 to 1.003 grm.) and put it in a white glass bottle of a little more than 200 c. c. capacity. This *test bottle* should be provided with a well ground glass stopper, running to a point below. Pour into the bottle 5 c. c. of perfectly pure nitric acid of 1.2 specific gravity, and set the bottle in an inclined position upon a water or sand bath, and heat it gently until the whole of the silver has dissolved. By means of a bent glass tube, blow the red fumes out of the upper part of the bottle, and after the bottle has been allowed to cool somewhat, set it in a stream of water, the temperature of which is not far from 16°. As soon as the bottle has been reduced to the temperature of the water, wipe it dry and place it in a box or case made of pasteboard blackened internally, and reaching to the neck of the bottle. Next measure out 100 c. c. of the strong solution of chloride of sodium in a pipette and let it flow in upon the silver solution in the test bottle. In order to measure the salt solution with due accuracy, the pipette should be fixed firmly in a vertical position, in a support after it has been filled above the mark, before the excess is allowed to flow out. Moisten the stopper of the test bottle with water, press it firmly into the bottle and shake the latter violently until the chloride of silver coheres in lumps and the liquid is left perfectly clear. During this process of shaking, the bottle should be kept in its pasteboard case and its neck covered with a cap of black cloth, in order that the chloride of silver may be protected from the action of light. When the chloride of silver has completely settled, turn and incline the bottle so as to wash down any of the chloride which has remained adhering to its upper part, then remove the stopper and add the decimal chloride of sodium solution, little by little, as long as any precipitate continues to be formed. The decimal solution should be poured from a burette graduated to tenths of cubic centimetres, and the bottle had better be held in an inclined position so that the drops may fall against the lower part of its neck. The portions of the decimal solution first added may be as large as 0.5 c. c., but afterwards smaller quantities should be added, in proportion as the precipitate produced is smaller, and towards the close, no more than two drops of the solution should be added at once. The operation is finished when the last two drops fail to produce a precipitate. After each addition of the decimal solution, the bottle should be lifted out of its case so that the amount of precipitate produced by that addition may be observed. The bottle should then be shaken until the liquid becomes clear, before any more of the decimal solution is added. Towards the close of the operation the height of liquid in the burette should be read off and noted before each addition. The reading before the last addition (which fails to produce any precipitate) is taken as the correct reading. Suppose 3 c. c. of the decimal chloride of silver solution have been used besides the 100 c. c. of the strong solution, and that the amount of pure silver weighed out was equal to 1.002 grms., then 100.1 c. c. of our chloride of sodium will be equal to 1 grm. of silver, for:—

$$1.002 : 1 = 100.3 : x (= 100,0998).$$

If by any chance, too much of the decimal chloride of sodium solution has been added so that the exact point of the cessation of precipitation cannot be made out, add 2 or 3 c. c. of the decimal silver solution, above described, and proceed again with the decimal salt solution until the point of saturation is exactly hit. The amount of silver thus added in the decimal solution must of course be added to the weight of silver originally taken.

The value of the standard solution should be marked upon the label of the bottle which contains it. It is the fundamental number upon which all the actual assays or analyses of silver made by means of that solution will depend. If at any time there is reason to apprehend that the strength of the solution may have altered, its value should be determined anew, against a new quantity of pure silver.

For the actual *assay of a silver alloy*, weigh out as much of the alloy as will contain about one gramme of silver, or a few milligrammes more than one gramme, dissolve it in 5 or 6 c. c. of nitric acid in a test bottle as above described, and proceed precisely as in the foregoing paragraphs. In order to obtain absolutely accurate results, it is necessary to know approximately beforehand, what proportion of silver is contained in the alloys. In the case of coins such as those of the United States and France, which contain 10 per cent of copper to 90 per cent of silver, it will be sufficient to weigh out a quantity equal to about 1.115 or 1.12 grm., and an analogous remark would apply to English coins, in which the proportion of copper is only 7.5 per cent. But with alloys of unknown composition, an approximate estimation of the silver value should be made before proceeding to the final assay. To this end weigh out 0.5 grm. of the alloy,—or 1 grm., in the case of alloys poor in silver,—dissolve it in from 3 to 6 c. c. of nitric acid, and add the strong solution of chloride of sodium, until the last drops produce no precipitate. The operation is conducted in the usual way, as above described, taking care to add smaller and smaller portions of the salt solution towards the close. The reading of the burette previous to the last addition is of course taken as the true reading.

Suppose 0.5 grm. of alloy was taken, and that 25 c. c. of the chloride of sodium solution of the above mentioned strength have been consumed, the amount of silver in the alloy will appear from the following proportion,

$$100.1 : 25 :: 1 : x = (0.2497),$$

and the quantity of alloy to be weighed out will be found by the proportion

$$0.2497 : 1.003 :: 5 : y = (2.008).$$

To dissolve this 2 grms. and more of alloy, a comparatively large quantity of nitric acid will be needed, say 10 c. c.

It need hardly be added that the results obtained by this preliminary trial are already tolerably close approximations to the truth. If the experiment be carefully conducted, the quantity of silver present in the alloy can be determined in this way, to within $\frac{1}{1000}$ or $\frac{1}{500}$. It is only where the highest possible degree of accuracy is required that the second titration with the decimal solutions need be resorted to.

It is to be remarked that alloys of silver and copper are never of absolutely homogeneous composition throughout their mass, excepting the alloy of $\frac{718.93}{1000}$ silver. The sheets of rolled metal from which coins are stamped, have often been found to show 1.5 to 1.7 in a thousand more silver in the middle than at the edges.

The following example will serve to illustrate the manner in which the results of an assay are calculated. Suppose that 1.116 grms. of American coin have been taken, and that 5 c. c. of the decimal solution (= 0.5 c. c. of the strong solution) of chloride of sodium have been consumed in addition to the 100 c. c. of strong solution first added. Suppose moreover, that the strength of the salt solution is as above stated, *i. e.*, 100.1 c. c. = 1 grm. silver, then

$$100.1 : 100.5 :: 1 : x$$

where x will equal 1.004, the amount of silver in the alloy. The same result may be arrived at in the following way:—

There was required for precipitating the silver in the alloy	100.5 c.c.
For 1 grm. of pure silver . . . . . . the chloride of sodium solution.	100.1 c.c. of
Difference =	0.4 c.c

There was consequently 0.004 grm. of silver more than one grm. present in the alloy, on the presumption that 0.1 c. c. of the strong, or 1 c. c. of the weak, chloride of sodium solution corresponds to 0.001 grm. of silver.

To dissolve alloys which contain sulphur, and those which consist of gold and silver, with a small proportion of tin, Levol (*Annales Chim. et Phys.*, (3.) **44.** 347) employs about 25 grms. of strong sulphuric acid; the mixture is boiled until the alloy has dissolved, and the liquid then treated in the usual way. But since concentrated sulphuric acid does not dissolve the whole of the silver when there is much copper present, Mascazzini (*Chem. Centralblatt*, 1857, p. 300) prefers to treat the weighed alloy, which may contain small quantities of lead, tin and antimony, besides gold, with the least possible quantity of nitric acid, as long as red vapors are formed, before adding the sulphuric acid. After the addition of the latter, the liquid is boiled until the gold has settled well together; the mixture is then cooled, diluted with water, and titrated with chloride of sodium, as above.

If the substance to be analyzed contains

any mercury, some acetate of sodium should be added to the nitric acid solution before proceeding with the titration (see the description of the gravimetric method above).

The reason why the value of the standard solution of chloride of sodium cannot be obtained directly from the weight of salt taken, is that chloride of silver is not absolutely insoluble in a solution of nitrate of sodium. And such a solution is of course formed by the reaction of chloride of sodium upon nitrate of silver. The presence of nitric acid does not hinder, or prevent, this solution. Mulder has shown, for example, that if a solution of $\frac{1}{20}$ milligrm. of silver be treated with 50 c. c. of nitric acid of 1.2 sp. gr., and the solution thus formed be mixed with enough chloride of sodium to precisely neutralize the nitrate of silver, the precipitate which forms at first will disappear. But the further addition of a trace either of chloride of sodium or of nitrate of silver, to the clear liquid thus obtained, will occasion the formation of a persistent precipitate. The existence of the clear solution seems to depend upon that of a certain equilibrium between the affinities of nitric acid and chlorine in presence of water, for sodium and silver. But this equilibrium is at once destroyed on the addition of an excess of either chloride of sodium or nitrate of silver. It follows from the foregoing that one equivalent of chloride of silver dissolved in water is not quite sufficient to precipitate one equivalent of silver dissolved in nitric acid. Unless a little more than one equivalent of chloride of sodium be used, we encounter the so-called "neutral point," at which both nitrate of silver and chloride of sodium can produce precipitates. Hence, if we wish to stop at the point of permanent precipitation, the value of the standard solution must be determined by experiment, as above explained.[1]

The bearings of the neutral point upon the accuracy of the silver assay, have been carefully studied by Mulder (in his *Silber-Probirmethode*, Leipzig, 1859). From what has been said already, it follows that if to a silver solution there be added first a strong solution of chloride of sodium, and then a decimal solution, drop by drop, until no more precipitate appears, a small precipitate will be again produced on adding a decimal silver solution to the mixture; but if the decimal silver solution be now added, drop by drop, until the last drop produces no turbidity, the addition of some of the decimal solution of chloride of sodium will again produce a small precipitate. By taking note of the number of drops of the two decimal solutions which are consumed in passing from one limit to the other, it will be seen that the same number of each is used. Let us suppose that, like Mulder, we have a dropping apparatus, of such size that 20 of its drops are equal to 1 c. c.; that to a silver solution there has been added a decimal chloride of sodium solution, until precipitation has ceased, and that 20 drops of decimal silver solution have been used in reaching the point at which that liquid ceased to produce turbidity; it will then be found that 20 drops of decimal solution of chloride of sodium must again be added in order to arrive at the point at which this liquid ceases to react. If only 10 drops were added, instead of the 20 which are required, we would be involved in the neutral point.

Instead of stopping at the point enjoined in these pages, at which chloride of sodium has just ceased to precipitate the silver, the operator might stop either at the neutral point, or at the point at which silver solution has just ceased to precipitate chloride of sodium. But whichever point be chosen, that point must always be adhered to. It would be wrong, for example, to stop at one point when fixing the value of the chloride of sodium solution, and at another in performing an analysis. According to Mulder, the difference obtained by using first the first point, and then the second is about 0.0005 grm. of silver for 1 grm. of silver at 16°. By employing first the first point, and then the third, as was permitted in the original process of Gay-Lussac, the difference is increased to 0.001 grm. — For the ordinary purposes of the laboratory it is usual to consider only the first point. If by any chance that point is overstepped by the addition of too much of the decimal solution of chloride of sodium, 2 or 3 c. c. of the decimal silver solution should be added all at once, and the end point again sought for by adding the decimal solution of chloride of sodium, until precipitation ceases. The silver thus added must of course be subtracted from the amount finally obtained, as has been already stated. The foregoing process was devised by Gay-Lussac, *Instruction sur l'essai des matières d'argent par la voie humide*, Paris, 1832. It has been studied with great care by Mulder, *Die Silber-Probirmethode*, Leipzig, 1859, whose book should be consulted by all persons specially interested in the subject. The account of the method here given has been copied from Fresenius's work on Quantitative Analysis. For a modification of the method, proposed by Mohr, see below.

For the estimation of Chlorhydric Acid, or of Chlorine, in combination with a metal, a standard solution of nitrate of silver may be added to the liquid under examination, until a precipitate is no longer seen to form (compare the description of the silver assay above), or an indicator (see Acidimetry) may be employed to show the point at which the precipitation ceases. Thus Levol (*Journ. prakt. Chem.*, **60.** 384) has proposed to add to the

[1] It is to be remarked that Stas, in a preliminary note (*Comptes Rendus*, **67.** 1107; *Zeitsch. analyt. Chem.*, 1869, **8.** 466), has called attention to the fact, that by using bromide of sodium as the precipitant, in place of chloride of sodium, silver may be completely precipitated without encountering any neutral point.

perfectly neutral solution of the chloride to be tested 0.1 volume of a saturated solution of phosphate of sodium, and to stop adding nitrate of silver at the moment when a persistent yellow precipitate of phosphate of silver appears. Since no permanent precipitate of the phosphate can form until the whole of the chlorine has been removed from the solution, the formation of a yellow precipitate will indicate the fact that all the chlorine has been thrown down. But according to Mohr (*Titrirmethode*, 1856, **2.** 13) the color of the phosphate of silver is too feeble to serve as a really useful indicator. Long practice with solutions of known value would be required to enable the operator to stop at precisely the right moment. There would always be a tendency to overstep the mark. In consequence of this defect, Mohr (*loc. cit.*) has proposed to substitute normal chromate of potassium for the phosphate of sodium of Levol, and the idea has been found to be one of much merit.

For the ordinary work of a laboratory, a perfectly neutral decimal solution of nitrate of silver (1 litre = 0.1 equiv. HCl) will be well suited. To prepare this solution, dissolve from 18.75 to 18.80 grms. of pure fused nitrate of silver in 1100 c. c. of water, and filter the liquid, if it is not already clear. Weigh out carefully into four separate beakers four portions of pure dry chloride of sodium, each of from 0.1 to 0.18 grm., dissolve them in 20 or 30 c. c. of water, and add 3 drops of a cold, saturated solution of pure, yellow chromate of potassium to the contents of each beaker. The silver solution, which is a little too concentrated, is now allowed to drop slowly from a burette into one of the beakers, while the liquid in the latter is constantly stirred. Each drop of the silver solution produces a red spot of chromate of silver where it touches the yellow liquid in the beaker; but the red color disappears instantly on stirring, owing to the decomposition of the chromate of silver by the chloride of sodium. At last, however, a slight red coloration will persist, at the moment after all the chlorine in the solution has combined with silver. Note down the readings of the burette, and without throwing away the contents of this first beaker repeat the experiment with the salt solutions in beakers Nos. 2 and 3, taking care to stop as nearly as possible at the moment when the same shade of red as in No. 1, is manifested. Note the readings of the burette as before, and proceed to calculate in each case how much of the silver solution would have been required for 0.1 equivalent of NaCl (*i. e.*, for 5.846 grm.). Thus, if 0.11 grm. of chloride of sodium was weighed out, and 18.7 c. c. of the silver solution consumed, then

$$0.11 : 5.846 :: 18.7 : x = (993.8).$$

Suppose the second and third trials give the value of x as 995 and 993 respectively; then the mean of the three numbers (= 993.9) would be taken as the true result, and it is only necessary to measure out that number of c. c. of the silver solution and to dilute it to the volume of a litre in order to obtain the standard solution which is desired. Or a dry litre flask may be filled to its mark with the solution, and 6.14 c. c. of water then added; for if 993.9 c. c. of the solution require 6.1 c. c. of water to make a litre, then 1000 c. c. will need to be diluted with 6.14 c. c. No matter how the mixture is made, take care to close the litre flask with a caoutchouc stopper, and shake it thoroughly before transferring the solution to the bottle in which it is to be kept. It is well to control these operations by testing the value of the finished solution. To this end empty the burette and rinse it with the new solution. Then fill it, and proceed to determine how much chlorine is contained in the portion of salt which was weighed out into beaker No. 4. If the standard solution be correct, the number of c. c. of it used multiplied by 0.005846 will exactly equal the weight of salt taken.

The analysis of any sample of chlorhydric acid or of any metallic chloride is conducted precisely like the experiment in beaker No. 4, care being taken in all cases to hit that identical shade of red which the operator had determined upon for himself. On this account, it is best to employ for an analysis a solution of the same bulk and about as strong as that used for standardizing the silver solution. It is best never to operate with excessively dilute solutions or with warm solutions, since chromate of silver is soluble to an appreciable extent, especially in hot water. — The quantity of silver solution consumed in producing the coloration is extremely small, varying in amount between 0.05 and 0.1 c. c.; no very considerable inaccuracy will be introduced by means of it even in cases where it stands in a widely different proportion to the quantity of chlorine from that which obtained in the solutions used for standardizing. If the amount of silver used for producing a visible coloration were always the same, it would simply be necessary to subtract this amount from each and every observation, in order to obtain an absolutely correct result; but since in point of fact more chromate of silver is required to produce coloration in presence of a large mass of chloride of silver, than when less of the chloride is present, no such deduction can be safely made. In spite of this hindrance, the process gives very satisfactory results. — It is to be observed that the solution containing the chloride to be analyzed must be neutral, for chromate of silver is soluble in free acids. If need be it must be neutralized by adding nitric acid or carbonate of sodium, before proceeding with the titration. It had better be a trifle alkaline than at all acid. In case the red coloration at the close is so strongly marked as to give reason to fear that too much of the silver solution has been added, it will be well to add 1 c. c. of a decimal solution of chloride of sodium and afterwards the silver solution again, drop by drop. The amount of chlorine in the c. c. of salt solution thus added must of course be allowed for and deducted in the calculation.

Attempts to use arseniate of sodium as the indicator in chlorine determinations gave Mohr (*Titrirmethode*, 1856, **2.** pp. 13, 14) far less satisfactory results than those obtained with the chromate. Since arseniate of silver has a dark brownish red color it is much more readily seen than the yellow phosphate, but its color is always far less conspicuous than that of the blood red chromate of silver.

Indirect Separation of K from Na. See to it that the mixed Chloride of Potassium and Chloride of Sodium to be treated is as pure as possible. To this end, redissolve the mixed chlorides after they have been once ignited, add a few drops of carbonate of ammonium and ammonia, and filter before again evaporating and igniting. Weigh the mixed chlorides, dissolve them in water and proceed to determine the amount of chlorine, by means of a decimal solution of nitrate of silver, using chromate of potassium as the indicator, as above described. Calculate the amounts of potassium and sodium, as explained below.

The process is easily and rapidly executed and yields perfectly satisfactory results when the mixed chlorides are free from impurities. It answers best for the analysis of mixtures which contain tolerably large quantites of both the metals. As in all similar indirect methods, the errors of observation are multiplied in calculating the results. (F. Mohr, *Zeitsch. analyt. Chem.*, 1868, **7.** 173).

According to experiments of Collier (*American Jour. Sci.*, 1864, **37.** 344), the process is equal in accuracy to the method of direct analysis by means of Chloroplatinate of Potassium.

The calculation is as follows (Compare Carbonic Acid, indirect separation of Ca from Sr):—Suppose the mixed chlorides weighed 3 grms. and there was found 1.6888 grm. of chlorine. Calculate how much KCl this chlorine would indicate if there were no sodium present: — thus

Atomic wt. of Cl 35.5	:	Molecular weight of KCl 74.6	: :	Wt. of Cl found 1.6888	:	x (= 3.5497).

The difference between $x = 3.5497$ and the weight (3 grms.) of the mixed chlorides actually found is proportional to the weight of chloride of sodium in the mixture. Thus,

Difference between the molec. weights KCl and NaCl 16.1	:	Molec. wt. of NaCl 58.5	: :	Difference as above 0.5497	:	Wt. of NaCl in mixture y (=1.9974).

Whence the short rule, multiply the quantity of chlorine in the mixture by 74.6 ÷ 35.5 (= 2.1014), deduct from the product the sum of the chlorides, and multiply the remainder by 58.5 ÷ 16.1(= 3.6336). The product will express the quantity of chloride of sodium in the mixed chloride. Or better, calculate from the formulæ of Collier, in which the atomic weight of chlorine is taken as 35.46, and that of potassium as 39.11 (instead of 35.5 and 39.1): —

W = weight of the mixed chlorides; C = weight of the total chlorine.

$$NaCl = C \times 7.6311 - W \times 3.6288$$
$$KCl = W \times 4.6288 - C \times 7.6311.$$

The following formulæ by Bosse (*Otto's Lehrbuch der Chemie*, 3 Aufl, **2.** 928) may be employed to find directly the amounts of potassium and sodium in the mixed chlorides : —

W = the weight of the mixed chlorides; C = weight of the chlorine; x = weight of the potassium; and y = weight of the sodium.

$$x = \frac{[(W - C) \cdot 1.54] - C}{0.63}.$$

$$y = \frac{C - [(W - C) \, 0.91]}{0.63}.$$

$$1.54 = Cl \div Na; \quad 0.91 = Cl \div K;$$
$$0.63 = (Cl \div Na) - (Cl \div K).$$

This method has been recommended by Anthon (*Dingler's polytech. Journ.*, **71.** 286) for estimating the amount of chloride of sodium in commercial chloride of potassium. Anthon (*loc. cit.*) has, moreover, constructed a table from which the percentage amount of chloride of potassium in any mixture may be obtained at a glance from the number of c. c. of silver solution used. But Fresenius (*Zeitsch. analyt. Chem.*, 1862, **1**, 110) urges that a process so dependent upon the purity of the mixed chlorides can hardly be employed with safety, for the analysis of commercial products. It admits of calculation that a commercial chloride of potassium containing 88 per cent of KCl, 10 per cent NaCl and 2 per cent of foreign salts, such as sulphates or nitrates, would require as many c. c. of decimal silver solution as correspond to only 3 per cent NaCl instead of the 10 per cent actually present; the presence of only 2 per cent of foreign impurity being sufficient to introduce an error of 7 per cent in the process. For the application of the method to the estimation of Carbonic Acid, see p. 91.

Mohr's method of estimating Silver, is a mere application of the foregoing process of determining chlorine. A measured quantity of standard chloride of sodium solution, more than sufficient to precipitate the whole of the silver, is added to the substance to be analyzed, and the excess of chlorine thus employed is then estimated by means of standard nitrate of silver, using chromate of potassium as the indicator. Since chromate of silver is somewhat soluble in water, care must be taken not to operate with dilute or with warm solutions. Though probably somewhat less accurate than the ordinary method of estimating silver, this process may nevertheless be sometimes found useful in the laboratory, especially in cases where the proportion of silver in the substance to be examined is wholly unknown. — The details of the process, as applied to the assay of silver coin are as follows:—Weigh out a sample of the alloy (say 1.08 grm.) and dissolve it in a flask in the least possible quantity of warm nitric acid. Pour from a burette into the solution of the coin, a decided excess of a decimal solution of chloride of sodium. This salt solution may be added by portions of 10 c. c. each until there is manifestly an excess of it in the mixture. The free acid and the copper in the solution must be got rid of before proceeding to determine the excess of chlorine which has been used. To this end heat the contents of the flask to boiling, throw in crystals of carbonate of sodium, free from chlorine as long as any effervescence occurs, and continue to boil until the oxide of copper has turned black. Then pour the mixture of liquid and precipitate into a narrow, graduated cylinder of 150 c. c. capacity, rinse the flask with water and fill the cylinder to the 0 mark with water, taking care to measure at about 16°. Close the mouth of the cylinder with a sheet of greased vulcanized rubber, held in place by the palm of the hand, and shake the mixture thoroughly. Let the solid matter settle; by means of a pipette transfer 50 c. c. of the clear supernatant liquid to a beaker; add 2 or 3 drops of a solution of yellow chromate of potassium and estimate the amount of chlorine by means of standard nitrate of silver as explained above. Multiply the number thus obtained by three, and subtract the chloride of sodium thus found from the amount of that substance first added to the solution of the alloy; the difference will be equivalent to the silver in the alloy. — The rule just given, to multiply by three, is not absolutely correct, since it is not a clear solution of chloride of sodium, but a mixture of liquid and solid matter, which is diluted to 150 c. c. The 50 c. c. taken out do not represent absolutely a third of the solution; but the error is trifling in any event, and is wholly insignificant when only a slight excess of chloride of sodium is added to the solution of the alloy.

Instead of the mixing cylinder, a marked flask of 150 c. c. capacity may be used. It is necessary to boil thoroughly after adding the carbonate of sodium, for carbonate of copper is somewhat soluble in water and if any of it be left undecomposed, a coloration will be produced by it on the addition of the chromate of potassium. When the operation is properly conducted, however, the liquid is left so free from copper that not the slightest precipitate or coloration will be perceived on adding to it sulphide of ammonium. The precipitate produced by the carbonate of sodium consists of chloride of silver as well as of oxide of copper, but the chloride of silver is not in the least acted upon by the boiling carbonate of sodium, nor does the oxide of copper carry down any chlorine admixed or in combination. Before transferring the mixture to the measuring cylinder, it should be cooled by immersing the flask in cold water. Instead of taking the red color of chromate of silver as the final point, the appearance of that color may be regarded only as a first approximation to the truth. After enough of the decimal solution of nitrate of silver has been added to produce the persistent red coloration, a decimal solution of chloride of sodium may be made to fall drop by drop upon the mixed precipitate of chloride and chromate of silver as long as any lighter shade of color can be distinguished at the place where the drop first touches the cloudy liquid. In this way the excess of silver required to produce the red coloration may be annulled and eliminated. The chloride of sodium thus added must of course be considered, and allowed for in calculating the results of the experiment. If the operator prefers to work backwards with chloride of sodium, as just indicated, the preliminary experiments made to familiarize him with the proper color of chromate of silver may be omitted. It will only be necessary to prepare the decimal solutions, by dissolving 5.846 grm. of pure, dried chloride of sodium in water, and diluting to the volume of a litre, and by dissolving 1.08 grm. of pure silver (or an equivalent amount of pure, fused nitrate of silver) in nitric acid, evaporating to absolute dryness to ensure neutrality, and dissolving in water to the volume of a litre. (Mohr, *Titrirmethode*, 1856, **2.** 54).

Volumetric estimation of combined sulphuric acid. A process devised by Græger (*Zeitsch. analyt. Chem.*, 1867, **6.** 443), for determining sulphuric acid in sulphates, depends essentially upon the principle now under consideration. It is as follows.—Dissolve the sulphate (of an alkali metal) in water, add a solution of chloride of barium in slight excess, and then, without filtering, digest the mixture for a short time with an excess of carbonate of silver. By the action of this agent, the metallic chloride which was formed by the action of the chloride of barium upon the sulphate is decomposed, together with the excess of chloride of barium employed, and converted into chloride of silver and a carbonate of the alkali metal. The chloride of silver, admixed with the excess of carbonate of silver and the carbonate of barium resulting from the decomposition of the excess of chloride of barium, are separated by filtration, and the amount of carbonated alkali in the filtrate is determined by means of standard nitric acid (see Akalimetry or Carbonate of Potassium). From the amount of alkali found, calculate an equivalent quantity of sulphuric acid. A trace of carbonate of barium always remains dissolved in the solution, and goes to increase the amount of alkali. — In case the sulphate to be analyzed is mixed with a carbonate and a chloride of an alkali, three separate measured portions of the solution must be treated: The first portion is titrated at once with standard nitric acid; in the second portion, acidulated with nitric acid, the chloride is determined as Chloride of Silver, by means of standard nitrate of silver and chromate of potassium, while the third portion is neutralized with nitric acid, and treated with chloride of barium and carbonate of silver, as above described. The amounts of alkali found in the first portion, and that equivalent to the chlorine found in the second portion, must of course be substracted from that found in the third portion, before proceeding to calculate the sulphuric acid.

According to Græger the process is applicable to all sulphates whose bases can be completely precipitated by means of alkaline carbonates, and particularly for the analysis of mineral waters. It is only necessary to decompose the sulphate by means of a measured quantity of standard carbonate of sodium, and to proceed as above. — In order to gain an idea of how much carbonate of silver should be used in any given instance, it is well to estimate the chlorine by titration, in a small measured portion of the liquid, after the action of the chloride of barium, and to calculate from this result how much of the silver salt will be needed. A slight excess of it should always be employed. In using this process it should be remembered that carbonate of silver is slightly soluble in aqueous solutions of the alkaline carbonates.

For Carius's method of estimating chlorine in organic compounds, see Nitric Acid, oxidizing power of, and Iodate of Silver.

Principle III. Reduction of to metallic silver by hot hydrogen.

Application. Estimation of silver in chloride of silver and of chlorine in presence of bromine.

Method. See Bromide of Silver.

Principle IV. Reduction of by metallic zinc.

Method. See Silver and Silver Compounds.

Principle V. Reduction of, by hot alkaline carbonates.

Applications. Separation of chlorine from silver. Estimation of chlorhydric acid in certain cases.

Method. Mix the chloride of silver with 3 times its weight of a mixture of carbonate of sodium and carbonate of potassium, and ignite the mixture in a porcelain crucible until the materials begin to agglutinate. On treating the cold mass with water, the metallic silver will be left undissolved, while the alkaline chloride goes into solution. The chlorine may then be determined in the manner described above under Principle II.

Principle VI. Fixity when heated in chlorine.

Applications. Estimation of bromine and iodine in presence of chlorine. Separation of Ag from Sb, As, Bi and Hg.

Methods. See Bromide of Silver, Chloride of Sulphur, and the Chlorides of the metals above enumerated (volatility of). In case it is desired to remove the chloride of silver from the bulb tube, in order to weigh the latter, reduce the chloride by heating it in a stream of hydrogen, and dissolve the metal in nitric acid. — According to Stas, chloride of silver fused in chlorine gas absorbs traces of chlorine, and does not give up the whole of the gas on cooling. In very delicate experiments, therefore, the chloride must at the last be heated in a stream of carbonic acid to expel this adherent chlorine. Stas found that about 100 grms. of chloride of silver lost 7.13 milligrammes on expelling the absorbed chlorine.

Principle VII. Solubility in ammonia-water.

Applications. Separation of silver from lead. Separation of chloride of silver from sulphide of silver in estimating chlorhydric acid.

Method. To separate silver from lead, add chlorhydric acid to the mixed solution, and immediately afterwards a large excess of ammonia-water. Basic Chloride of Lead will remain undissolved. Filter and acidulate the filtrate with nitric acid to throw down the chloride of silver, which collect and weigh, as above. In order that the chloride of silver may dissolve freely, it is important that the ammonia-water be added immediately after the precipitation. The mixture should be kept from the light. The process does not yield very accurate results, since more or less of the chloride of silver remains dissolved in the nitrate of ammonium, and is lost in the filtrate. — In separating chlorine from metals, it often happens that the latter must be thrown down by sulphuretted hydrogen before the estimation of the chlorine can be proceeded with. It is possible in that event, though not as a rule advisable, to add nitrate of silver directly to the filtrate from the metallic sulphide, and to dissolve out, by means of ammonia-water, the chloride of silver from the mixed precipitate of chloride and sulphide of silver. The process is involved, however, and the determination of chlorine less trustworthy than it would be if the sulphuretted hydrogen were removed in the usual way, by means of ferric sulphate, before adding the nitrate of silver.

Principle VIII. Conversion into bromide of silver by Bromide of Potassium (Wittstein's Method). See Iodide of Silver.

Properties of Chloride of Silver. When recently precipitated, chloride of silver is white; but on exposure to light it changes to violet, and finally to black. A small amount of chlorine is lost during this transformation, and some dichloride of silver formed. Though the change is superficial, it is attended with an appreciable loss of weight. When the chloride which has been darkened by light is treated with ammonia-water, there is left undissolved a small quantity of metallic silver, resulting from the decomposition of the dichloride.

On drying, the chloride becomes pulverulent, and yellow-colored if strongly heated. It fuses at 260° to a transparent yellow liquid, which solidifies to a white or yellowish horn-like mass on cooling. At high heats it volatilizes unchanged. It can be easily reduced to the metallic state by means of zinc and iron, and by hot hydrogen, coal gas and carbonic oxide, but not by ignition with charcoal.

The fresh precipitate has a peculiar curdy consistency, and its particles tend to cohere to large lumps when the liquid in which they are suspended is agitated. In order to obtain perfect coherence of the particles, and, as a consequence, a clear supernatant liquor, it is best to have a solution of silver slightly in excess in the liquid.

The precipitate is exceedingly sparingly soluble in water, and in dilute nitric and sulphuric acids, but dissolves rather easily in chlorhydric acid, especially when the latter is hot and concentrated. From the solution in strong chlorhydric acid, it is precipitated almost completely on the addition of water. On being kept in contact with water for some hours, even in the dark, chloride of silver decomposes very slightly, and becomes gray, more chlorine than silver being dissolved; in hot water this change is more rapid, but it is slight in any event. — Chloride of silver is soluble in solutions of all the metallic chlorides which are soluble in water; also, to a certain extent, in solutions of various nitrates, especially when the solutions are hot and concentrated. It is readily soluble in various cyanides and hyposulphites. Concentrated solutions of caustic potash and soda decompose it, especially when they are hot. When recently precipitated, ammonia-water dissolves it easily, but after having been kept for a long time and boiled with water, it dissolves with difficulty (see also above). Solutions of the carbonates of potassium and sodium only decompose traces of it even on boiling. Chloride of silver is perceptibly soluble in warm tartaric acid, but is less soluble in the cold. Both bromide and iodide of potassium transform it completely to Bromide or Iodide of Silver, as the case may be. Its composition is

Ag =	108	75.26
Cl =	35.5	24.74
	143.5	100.00

Chloride of Sodium.

Principle I. Fixity at moderately high temperatures.

Applications. Estimation of sodium in the hydrate, chromate, chlorate, silicate, sulphide and carbonate of that metal, and in its salts with various organic acids. Separation of Cl from Pt.

Method. See Chloride of Potassium. In separating chlorine from platinum, mix the solution of the platinum compound with a solution of carbonate of sodium, evaporate the mixture to dryness upon a water bath and fuse the residue in a platinum capsule. Dissolve out the chloride of sodium with water and estimate the chlorine as Chloride of Silver (*Zeitsch. analyt. Chem.*, 1870, **9**. pp. 30, 31).

Estimation of chlorine in organic compounds. [Compare Chloride of Potassium]. In many easily decomposable organic compounds, such for example as the substitution products of acids, chlorine (bromine or iodine) may be determined by leaving the substance for several hours in contact with water and sodium amalgam, until decomposition is complete, then acidulating with nitric acid and estimating the chlorine as Chloride of Silver (Kekulé).

Principle II. Sparing solubility in alcohol.

Applications. Separation of Cl from Br and from I.

Method. See Bromide of Sodium, solubility of.

Principle III. Solubility in water.

Applications. Estimation of water in organic solutions, particularly in beer and in milk. Valuation of beer and of milk.

Methods.

A. Fuchs's hallymetric beer test. Since 100 parts of water at temperatures between 15° and 40° dissolve almost precisely 36 parts of pure chloride of sodium, it is possible to estimate the amount of water in solutions of many organic bodies by adding a weighed quantity of salt in excess to the aqueous liquid and afterwards determining the amount of salt which remains undissolved, time enough being allowed of course for the water to become saturated with the salt. Suppose, for example, it be found that 315 grains of salt have been dissolved by a given sample of beer then

$$36 : 100 :: 316 : x \; (= 875 =$$

the amount of water in the sample). — The chief difficulty in the process is to estimate quickly and accurately the salt which remains undissolved. For this purpose Fuchs (*Dingler's polytech. Jour.*, 1836, **62**, 309) employs a measuring tube about 20 c. m. long. At the top this tube is 4 c. m. wide, but just below the middle it is contracted to such an extent that the lower half of the tube is no wider than 1 c. m. To make such a tube, two tubes of the different sizes must be fused together, the larger tube being drawn down sufficiently to correspond with the narrower one. The lower, narrow part of the apparatus is graduated so that each fine mark shall represent one grain and each heavy mark five grains of moist chloride of sodium of the degree of fineness and compactness to be described directly. The first tube of this kind must be graduated by means of powdered salt wet with a saturated solution of salt, but after one tube has been made copies of it may be multiplied to any extent, by means of quicksilver. — For calibrating this *hallymeter* as well as for the subsequent beer tests, Fuchs employs chemically pure salt of such a degree of fineness that it will pass through a sieve the holes in which are 0.0673 Paris lines broad and 0.0757 Paris lines long, the thickness of the brass wire being 0,0458 Paris lines. To graduate the tube weigh out 600 grains of water, which would be enough to dissolve 216 grains of salt, add to it 221 grains of salt or 5 grains more than it can dissolve, and as soon as solution is complete bring the mixture of salt and liquor into the tube, taking special care that no particle of the solid salt is lost. The solution may be effected in a small flask, and the salt transferred to the measuring tube by closing the flask with the thumb, inverting it and shaking the liquid until all the salt has fallen upon the thumb. On now holding the flask over the measuring tube and loosening the thumb the whole of the salt will be washed from the thumb into the tube, It will be observed that the upper part of the measuring tube which serves as a mere reservoir for liquids is made large enough to hold a considerable quantity. As soon as the salt is in the tube rap the latter repeatedly against the table in the manner described below, in order to bring the particles of salt to the proper degree of compactness; make a fine mark upon the tube and number it "5." Then empty the tube, wash it with water, dry it with blotting paper and repeat the operation so as to make other marks corresponding to 10, 15, 20, etc. grains, of salt. In each case the necessary excess of salt, over and above what 600 grains of water can dissolve must be weighed out. It will not answer to add fresh portions of powdered salt to the liquid in the measuring tube, for the powder is not of uniform fineness, and since in actual practice the finer particles of the powder dissolve first while the coarser particles are left to be measured, these conditions must be insisted on in the operation of calibration. For the beer test, it will not be necessary to make more than 7 or 8 of these coarse divisions. The subsidiary divisions, each representing 1 grain of salt may be filled in by means of a dividing machine ; care having of course been taken in the first place to choose a tube of uniform bore for that part of the apparatus which is to be graduated,

In testing beer two experiments are made; the 1st, to determine the amount of water and carbonic acid, and the 2d to determine the amount of "extractive matters" in the beer. For the first experiment, weigh out (but do not measure) 1000 grains of the beer in a small flask, add to it 330 grains of the powdered and sifted salt, cover the flask with a glass plate and counterpoise it upon a tolerably delicate balance. In order to facilitate solution and to expel carbonic acid, heat the flask carefully, but not above 38° lest some of the alcohol in the beer be expelled. It is well to set the flask in a dish of water, which is kept at 37° or 38°, and to shake it or rather twirl it about from time to time. Only 5 or 6 minutes are really needed to complete the solution. At the expiration of that time, cool the flask by immersing it in cold water, blow a little air into the upper part of the flask to expel carbonic acid, wipe it dry and weigh it, the loss of weight represents approximately the carbonic acid; in good Bavarian beer it will amount to about 1.5 grains. Then pour the contents of the flask into the measuring tube, taking care as before to lose no particle of the salt. If any of the salt clings to the side of the flask it may be washed out by pouring back some of the liquor from the upper part of the measuring tube or by means of a saturated solution of chloride of sodium. — Special care must be taken both in graduating the tube and in measuring the undissolved salt, in an actual experiment, to make the salt settle down into the smallest possible space in the measuring tube. To effect this, clasp the tube at the middle with the thumb and curved forefinger of the left hand so that it may be supported in a vertical position and be free to move up and down, then with the thumb and finger of the right hand take hold of the very bottom of the graduated tube, lift the apparatus to a distance of about 0.25 inch above the table and let it drop back so that the bottom of the tube may strike the table with a shock. Repeat this operation many times, working with such rapidity that the tube shall receive about 100 blows per minute. Take care to hold the tube upright and to have the blows follow one another with precision and regularity. After about two minutes thrust a wire into the tube and without actually stirring up the salt move it about gently to expel any little bubbles of air which it may have entangled. Then remove the wire and continue to tap the tube against the table as before as long as the salt continues to sink upon itself. This operation of settling usually consumes about 15 minutes. When the salt has ceased to settle note the mark and fraction of a mark upon the tube to which its surface corresponds and proceed to calculate how much water the dissolved salt represents, in the manner indicated above.

For the second experiment, which is to determine the extractive matter in the beer, weigh out 1000 grains of the liquid in a flask and evaporate it in the flask to about one-half in order to be sure that the whole of the alcohol has been expelled. It is well to heat the flask upon a sheet iron plate by means of a lamp, but the heat must be very gentle at first, lest the liquid boil over through the too rapid escape of carbonic acid. If this catastrophe threaten the flask should be immediately removed from the fire and gently agitated or twirled around. A rather larger flask should be taken for this experiment than for the first.

After the first tempestuous escape of carbonic acid is over the liquid will boil gently and may be left to itself. When half the original liquid has evaporated remove the flask from the fire, immerse it in cold water in order to cool its contents, wipe it dry and by means of filter paper remove any drops of water which may have condensed in its neck. Then weigh it with its contents. Next pour into it 180 grains of the powdered salt and proceed to determine how much of this salt is dissolved by the liquid, precisely as in the first experiment By subtracting the weight of water thus found from the weight of the concentrated liquid weighed, the weight of the so called extractive matters is obtained. By subtracting water, extractive matters and carbonic acid from the original 1000 grains of beer the weight of the alcohol is obtained. — According to Fuchs, the results obtained by this process are satisfactory and useful for the comparison of different kinds of beer. The substances here classed as extractive, exert little or no influence upon the solubility of the salt in water. The chief merit of the method of course consists in its rapidity; the complete examination of a sample of beer requiring only about two hours time. Tables giving the amount of alcohol and of extractive matter in 1000 grains of beer for each grain of undissolved salt will be found in Fuchs's memoir.

B. Reichelt's milk test. Weigh out 1000 grains (= 62.5 grms.) of the milk to be tested, add to it 324 grains (= 20.25 grms.) of chloride of sodium and 240 grains of a solution of litmus saturated with common salt. The chief purpose of the litmus solution is to color the milk so that the undissolved salt can be seen and measured, but it is found also that the addition of this salt solution improves the consistency of the milk and facilitates the deposition of the undissolved salt. After time enough has been allowed for the salt to dissolve in the milk, the mixture is transferred to the measuring tube precisely as in Method A, see above, and the experiment finished as there described. In one experiment, 4 grains of salt were found to have remained undissolved. 1000 grains of the milk had consequently dissolved 324 — 4 = 320 grains which are equivalent to 888.89 grains of water, for

$$36 : 100 = 320 : 888.89.$$

The percentage of solid matter in the milk was consequently 100 — 88.90 = 11.10. An experiment made for the sake of control by the method of evaporation gave 11.03 solid matter and 88.97 per cent water. The method seems to be as accurate as that in which the milk is mixed with a known weight of quartz sand and evaporated at 105°—110°, and is of course far more expeditious (Reichelt, *Wagner's Jahresbericht Chem. Tech.*, 1859, **5.** 443).

Principle IV. Decomposition of by nitric acid.

Application. Indirect separation of K from Na.

Method. The chlorides of the alkali metals are easily changed to nitrates by means of nitric acid. If enough nitric acid be taken a single evaporation to dryness will expel the last trace of chlorine. The converse process of changing nitrates to chlorides by means of chlorhydric acid, is far more difficult.

The process consequently consists in simply treating a weighed quantity of the mixed chlorides with nitric acid in an evaporating dish loosely covered with a watch glass, evaporating to dryness and weighing the residue. The calculation is as follows:—

Call the weight of the mixed chlorides S, and that of the mixed nitrates, obtained therefrom, R; the weight of the KCl in the mixture a and that of the NaCl b, then $S = a + b$; $a = S - b$; and $b = S - a$, but a grms. of KCl will yield $1.355 \times a$ grms of KNO_3 and b grms. of NaCl will yield $1.453 \times b$ grms. of $NaNO_3$, for

$$\begin{matrix}\text{Molec. wt.}\\ \text{of KCl}\\ 74.6\end{matrix} : \begin{matrix}\text{Molec wt.}\\ \text{of } KNO_3\\ 101.1\end{matrix} :: \begin{matrix}\text{Grm KCl}\\ 1\end{matrix} : \begin{matrix}\text{Grm. } KNO_3\\ x\ (= 1.355),\end{matrix}$$

and

$$\begin{matrix}\text{Molec. wt.}\\ \text{NaCl } 58.5\end{matrix} : \begin{matrix}\text{Molec. wt.}\\ NaNO_3\ 85\end{matrix} :: \begin{matrix}\text{Grm. NaCl}\\ 1\end{matrix} : \begin{matrix}\text{Grm. } NaNO_3\\ 1.453\end{matrix}$$

hence

$$1.355 \times a + 1.453 \times b = R.$$

By substituting for b, in this equation, its value $(S - a)$, we have

$$1.355 \times a + 1.453 \times (S - a) = R,$$

whence

$$1.453\ S - R = 0.098\ a,$$

and

$$a = \frac{1.453\ S - R}{1.098}$$

or

Chloride of Potassium $= a = 14.83\ S - 10.2\ R.$
Chloride of Sodium $= S - (14.83\ S - 10.2\ R)$; or
Chloride of Sodium $= b = 10.2\ R - 14.83\ S.$

Reckoning backwards,

The Nitrate of Potassium × 0.7375 = the Chloride of Potassium; and the Nitrate of Sodium × 0.6877 = the Chloride of Sodium. (Mohr, *Zeitsch. analyt. Chem.*, 1868, **7.** 175.)

Properties. Chloride of sodium is much less readily soluble in chlorhydric acid than in water. It is nearly insoluble in absolute alcohol and only sparingly soluble in spirit. It fuses at a red heat without decomposition but volatilizes in white fumes at a white heat; or at a bright red heat, when heated in an open vessel. It is less volatile, however, than chloride of potassium. In fusing salt, care must be taken to prevent the flame from touching it, for in that case some chlorhydric acid would escape and a little carbonate of sodium be formed through the action of carbonic acid and water resulting from the fire. When exposed to moist air even pure salt will slowly absorb a little water. Crystals of salt decrepitate on being heated owing to the escape of a little moisture which they retain. To avoid loss from this cause, careful drying is essential (See Chloride of Potassium). On being evaporated with oxalic or nitric acids, or ignited with oxalate of ammonium, chloride of sodium behaves like chloride of potassium. It composition is

Na	= 23	39.31
Cl	= 35.5	60.69
	58.5	100

For use as a reagent, perfectly pure native rock salt is well suited. Or, pure salt may be prepared, after Margueritte, as follows:—Pass chlorhydric acid gas into a concentrated solution of common salt as long as any of it continues to be absorbed. Throw the small crystals of salt, which are deposited, into a funnel, let them drain thoroughly, wash with chlorhydric acid and finally dry them in a porcelain dish to drive off the last traces of free acid.

In drying salt before weighing it for the preparation of standard liquors, it should not be fused, but ignited moderately, powdered roughly while still warm and placed in a weighing tube which admits of being closed. Pure salt will dissolve in water to a clear liquid and the solution will not become cloudy when tested with oxalate of ammonium, phosphate of sodium and chloride of barium.

Chloride of Sulphur.

Principle. Volatility of.

Application. Separation and estimation of sulphur in metallic sulphides, especially those of complex composition.

Method. Weigh the powdered sulphide in a tolerably long, narrow, glass tube, closed at one end, so that the powder may be poured therefrom into a bulb tube without soiling the stems of the latter. Attach to one end of the bulb tube an apparatus for generating dry chlorine and to the other a couple of large U tubes or small flasks to serve as receivers of the chloride of sulphur. Fill each of the flasks about one quarter full with water, or if antimony be present, with a solution of tartaric acid in dilute chlorhydric acid. [Compare the description of Lindt's arrangement for absorption in soda lye, below]. One end of the bulb tube should be bent so as to reach nearly to the surface of the water in the first flask or U tube. Beyond the last flask there should be an abduction tube reaching to the chimney or into a large bottle loosely filled with pellets of paper wet with alcohol, so that the excess of chlorine may not incommode the operator.

Before attaching the bulb tube to the chlorine generator wait until the latter and the washing bottles connected with it are completely full of chlorine. Pass a slow current of chlorine through the bulb tube in the cold, until no further alteration of the sulphides is observed. Then heat the bulb very gently, taking care also to keep the outlet of the bulb tube warm so that no volatile metallic chloride shall condense in it and stop it. By the action of the chlorine the sulphides are completely changed to chlorides, most of which remain in the bulb, though volatile chlorides, such as those of arsenic, antimony, mercury, etc., go forward with the chloride of sulphur. The chloride of sulphur on coming in contact with the water in the flasks is decomposed to chlorhydric and hyposulphurous acids, with separation of some free sulphur. The hyposulphurous acid decomposes again to sulphur and sulphurous acid and the latter is converted into sulphuric acid, by means of the chlorine with which the water and the flasks is filled, so that the final products are sulphuric acid and more or less free sulphur. The current of chlorine should be slow from first to last, and the flasks as well as the rest of the apparatus should be full of chlorine before the contents of the bulb tube are heated.

At the end of the operation the sulphur is collected upon a tared filter, washed, dried and weighed, and the sulphuric acid is estimated in the filtrate as Sulphate of Barium. The metals of any volatile chlorides which have gone forward, may be estimated in the filtrate from the Sulphate of Barium. The chloride which remains in the bulb tube is weighed as such if it be chloride of silver, chloride of lead or any other fixed chloride, or, if need be, it is dissolved in chlorhydric acid, aqua regia or other suitable solvent and subjected to analysis. — In heating the bulb tube it is not necessary to wait until the whole of the ferric chloride is expelled, in case any be present; but care should be taken to drive all the chloride of sulphur and other volatile chlorides as far forward towards the water in the flask as may be practicable. When the bulb has become cold cut off its bent stem, and in case the latter contains a portion of the volatile metallic chlorides, close it by inverting upon it a moistened glass tube closed at one end and leave it at rest for 24 hours. When left in this way the chlorides will slowly absorb moisture and dissolve without evolution of heat. Finally rinse out the bent part of the bulb tube with dilute chlorhydric acid, and gently heat the contents of the flasks or U tubes to expel the chlorine, before proceeding to collect the sulphur. The sulphur is apt to separate in liquid drops. If need be, time should be allowed for these drops to become solid before filtering. — Compounds of the sulphides of arsenic, antimony or tin, with basic sulphides, may be decomposed in this way in the course of a few hours, but simple sulphides such as those of lead, silver, copper, cobalt, manganese, etc., require a much longer time and need to be heated far more strongly. Several days would be required to completely decompose a few grains of sulphide of lead. (Berzelius and Rose).

Instead of absorbing the chloride of sulphur in water as above, Lindt (*Zeitsch. analyt. Chem.*, 1865, **4.** 370) employs soda lye. He heats the sulphide, moreover, in a porcelain boat instead of a bulb tube. The apparatus is arranged as follows:— Blow two bulbs upon a tolerably long glass tube, the diameter of which is sufficient for the admission of a small porcelain boat. One of the bulbs may be close to one end of the tube, but the other bulb should be six or eight inches from the end. Bend the tube close to the last mentioned bulb at a right angle, taking care to leave the outer six or seven inches of the tube horizontal, for in that part of the tube the boat is subsequently placed. At a point 2 or 3 inches below the bulb in question, bend the tube again to a slightly acute angle. At this stage of operations there will be the outer six or seven inches of horizontal tube, say at the left of the bulb, then a short vertical tube, near the top of which the bulb is situated, and to the right of the bulb a length of 10 or 12 inches of tube sloping upwards somewhat towards the right hand, and bearing a bulb at its outer end. Bend the tube once more close to this outer bulb so that the bulb shall point upwards and be in the same plane with the other bulb or rather a little higher than the latter. Fix the tube in a wooden clamp so that the outer six or seven inches shall be horizontal, pour enough soda lye into the outer bulb to fill the lower sloping part of the tube, push the loaded porcelain boat into the horizontal part of the tube and connect that extremity of the tube with a chlorine generator. Attach also a caoutchouc tube to the outer bulb to lead away the excess of chlorine.

The chloride of sulphur on coming in contact with the soda lye is immediately decomposed with formation of sulphide, hyposulphite, chloride and hypochlorite of sodium. The decomposition of the metallic sulphide is soon completed, but it is well to continue to pass chlorine through the apparatus for a couple of hours, in order that some chlorate of sodium may be formed. After the porcelain boat has been removed, wash out the liquid from the tube into a porcelain dish, evaporate to dryness and ignite strongly to decompose the chlorate of sodium. No violent action occurs during this ignition and at its close all the sulphur will be found to be in the condition of Sulphate of Sodium. Determine it as Sulphate of Barium.

*proto***Chloride of Tin.**

Principle. Reducing power of.

Applications. Volumetric estimation of Sn, Fe, Hg, Cu, Pb, Mn, Co, Ni, Cl, Br, I, and chromic, chloric, sulphurous and ferricyanhydric acids. Indirect gravimetric estimation of protoxide of tin in presence of the binoxide.

Methods. See *bi*Chromate of Potassium, Chromic Acid, and the other acids above enumerated; also Permanganate of Potassium, Ferric Salts and Iodine.

The essential points of the process are either that a solution of stannous chloride is converted into stannic chloride, by means of oxidizing solutions of known strength, or that an excess of a standard solution of stannous chloride is added to the solution to be reduced and analyzed, and the portion of the tin salt which has escaped oxidation determined. The oxidizing solution may be *bi*Chromate of Potassium (after Streng and Penny), Permanganate of Potassium (after Schlagdenhauffen), Iodine in alkaline solution (after Lenssen), or Iodine, in iodide of potassium solution (after Fresenius).

1. *Chloride of tin against biChromate of Potassium.*

As will be seen under biChromate of Potassium, the oxygen dissolved in the water in which the analyses are made has much influence upon the correctness of the result. The action of this dissolved oxygen is peculiar. It differs totally according to circumstances. Thus while ferrous salts can be titrated with permanganate of potassium without a particle of the oxygen in the water taking part in the oxidation, it is wholly different when stannous salts (or sulphurous acid) are treated with the permanganate. In the latter case almost the whole of the dissolved oxygen suddenly unites with the stannous oxide the moment the stannous salt and permanganate are mixed. The dissolved oxygen seems, in fact, to be thrown into an active state at the moment when the permanganate acts upon a stannous salt, or upon sulphurous acid, while it remains wholly passive when other reducing agents are treated in its presence with the permanganate. What is true of the permanganate is true of chromic acid, and of several other reagents.

As regards the stannous salts, experiments have shown that the oxygen dissolved in water is rendered active, and that it unites with the tin in the case of the following oxidizing agents:—Chromic acid and chromates, permanganate of potassium, chlorous and hypochloric acids, peroxide of hydrogen and ozone. But that it remains inactive, and has no influence upon the reaction when the tin salt is oxidized with bromine, chlorine, hypochlorous acid, iodic acid, ferric or cupric chloride, or a solution of peroxide of lead (Lenssen and Lœwenthal, *Journ. prakt. Chem.*, 1862, **86**. pp. 193, 205, 214). — It has been repeatedly shown that the influence of the dissolved oxygen in the titration of chloride of tin may be annulled or avoided by adding a certain quantity of ferric chloride to the solution. Lenssen and Lœwenthal have corroborated this observation, and explain the fact by supposing that $SnCl_2$ and $2FeCl_3$ decompose one another to form $SnCl_4$ and $2FeCl_2$, and that the last named compound is really the substance that is titrated. By resorting to this artifice it becomes possible, therefore, to titrate stannous oxide with any reagent with which ferrous oxide can be titrated.

The affinity of chloride of tin for active dissolved oxygen is so strong that, in the case of a mixture of iodine, chromic acid and dissolved oxygen, the tin salt will unite with the latter before proceeding to reduce the chromic acid and iodine. The dissolved oxygen becomes active only in presence of the stannous salt and chromic acid. When the tin salt is replaced by hyposulphite of sodium, the oxygen remains inactive. In case the stannous salt to be oxidized by chromic acid is mixed with so much iodhydric acid that the chromic acid is completely decomposed, there will be a direct titration of the stannous salt with iodine, and the dissolved oxygen will have no active action. But if the proportion of iodhydric acid is so small that a certain quantity of chromic acid can exist in its presence, some of the dissolved oxygen will become active, since this activity is induced by the contact of chromic acid and the stannous salt. It appears, however, that the presence even of small quantities of iodhydric acid tends to diminish the quantity of oxygen which becomes active, while the amount of the latter increases as the proportion of chromic acid or of the tin salt is increased, or as the water contains more oxygen in solution. Similar remarks apply to the oxidation of stannous chloride by permanganate of potassium in presence of chlorhydric acid. When chlorous or hypochloric acid is treated with the tin salt the dissolved oxygen becomes active, no matter whether iodhydric acid be present or not. But as regards peroxide of hydrogen it appears that the presence of iodhydric acid prevents the activity of the dissolved oxygen. (Lenssen and Lœwenthal, *loc. cit.*, pp. 193–215).

A. *For the estimation of Chromic Acid* see under that head.

B. *For the estimation of Tin* see biChromate of Potassium.

C. *To estimate Mercury*, dissolve the substance to be analyzed in chlorhydric acid, pour into it from a burette a slight excess of a standard solution of protoChloride of Tin, heat the mixture until the mercury is reduced to the metallic state, and the metal has collected into a ball. Decant the clear liquid into a flask, wash the metallic mercury by decantation with water which has been boiled to expel air, and estimate the excess of protochloride of tin by means of one-tenth normal solution of the bichromate, as explained under Chromic Acid. The presence of nitric acid is inadmissible. To analyze mercurous nitrate in this way it must be first precipitated as Chloride by means of chlorhydric acid, and washed before adding the solution of tin. Mercuric nitrate can be decomposed by boiling with chlorhydric acid (Mohr, *Titrirmethode*, 1855, 1. 274).

D. *To estimate Iron* see Ferric Salts.

E. *To estimate Copper* see under Copper Salts, reduction of by iodide of potassium.

F. *To estimate Lead, Manganese, Cobalt or Nickel*, precipitate the metal as a peroxide by means of a solution of bleaching powder, reduce the washed oxide with a standard solution of chloride of tin and titrate the excess of the latter with a standard solution of bichromate. (Streng.)

2. *Chloride of Tin against Mercury Salts.* Compare C, above. To estimate stannous oxide when mixed with stannic oxide, determine the total amount of tin in one portion of the mixture; then dissolve another portion in chlorhydric acid, taking care to exclude the air, and drop the solution into a large excess of mercuric chloride kept constantly stirred and somewhat warm. Allow the precipitated *di*Chloride of Mercury to stand for a long time, then collect and weigh it in the usual way. One molecule of $HgCl_2$ corresponds to one of $SnCl_2$ (H. Rose).

3. For the use of stannous chloride as an absorbent of free chlorine in Mitscherlich's method of organic analysis, see Chloroplatinate of Potassium, power of decomposing organic substances.

For use as a reagent, stannous chloride may be prepared as follows:– Fuse a quantity of metallic tin in a small porcelain dish, remove the dish from the fire and rub the melted metal with a pestle until it becomes solid. Transfer the metal thus powdered to a flask, throw upon it a strip of platinum foil or the cover of a platinum crucible, pour in a quantity of strong chlorhydric acid and boil the mixture under a hood until the evolution of hydrogen from the surface of the platinum is observed to have become feeble. Take care always to have the tin in excess. The presence of the platinum greatly facilitates the solution of the tin. Since chloride of tin must always contain some free acid to be fit for use, it is best not to push the boiling until the evolution of hydrogen has entirely ceased.

In order to prevent the solution from absorbing oxygen from the air, it should be kept in a well stoppered bottle containing a number of small pieces of metallic tin. This remark applies however, only to solutions which have not been standardized. To keep a standard solution Fresenius recommends the following arrangement:— Place the bottle which is to contain the tin solution at the very corner of a table and fit to it a caoutchouc stopper with two holes. To one hole of the stopper fit a syphon, made by bending a glass tube at two right angles. The shorter arm of this syphon is straight and reaches almost to the bottom of the bottle while the outer arm reaches below the table to a point considerably lower than the bottom of the bottle, it is curved upwards at the end and closed with a piece of rubber tubing compressed by a spring clip. By means of another tube bent at two right angles and fitted to the second hole in the stopper the bottle is connected with another smaller though still tolerably large bottle, and this in turn with a couple of U tubes. The U tubes as well as the bottom of the bottle are filled with fragments of pumice stone saturated with a strongly alkaline solution of pyrogallic acid. When everything is ready, put a piece of glass tubing in the rubber connector at the outer end of the syphon, open the clip and suck at the tube until the syphon is full of the stannous chloride. Then close the clip and remove the sucking tube. At the outer extremity of the last U tube there is a cork carrying a tube of rather fine bore so that the outer air can communicate freely with the interior of the apparatus; but as the pyrogallic solution absorbs oxygen greedily, the vessels soon come to contain nothing but nitrogen. — To fill a pipette or Mohr's burette with the solution, open the clip so that a few drops of the solution may run to waste, then insert the point of the pipette in the rubber tube and open the clip. Since the outer leg of the syphon reaches far below the bottom of the bottle the liquid will flow into the pipette without any difficulty. In case any of the liquid is to be transferred to a beaker, a ∩-shaped tube may be inserted in the rubber connector. In a bottle thus protected the tin solution may be preserved unchanged, for any reasonable length of time. It is well to prepare the pyrogallate of potassium in the vessels themselves some time before the apparatus is put together, by mixing concentrated potash lye with a solution of pyrogallic acid.

As a substitute for the foregoing arrangement, Mohr (*Zeitsch. analyt. Chem.*, 1869, **8.** 113) simply pours a quantity of petroleum into the bottle which contains the chloride of tin, so that a layer of petroleum about 1 c. m. thick shall float upon the surface of the liquid. The bottle stands at the corner of the table as before and is in like manner provided with an upturned syphon tube for filling the burette, but at the top of the bottle there is now needed nothing more than a perforated cork carrying a small tube for the admission of air when the outlet tube is put in action.

For the methods of standardizing a solution of stannous chloride see biChromate of Potassium, Iodine, and the other substances against which it is used.

*bi*Chloride of Tin.

Principle. Volatility.

Applications. Separation of Sn from Co, Ni, Pb, Cu, Ag, Au, Pt.

Methods. See *ter-* and *quinqui*Chloride of Antimony.

Chloride of Zinc.

Principle. Fixity when not too strongly heated.

Applications. Separation of Zn from S, As, Sb, Sn and Hg (Method A). Estimation of Chlorine in organic compounds (Method B).

Method A. When a mixture containing sulphide of zinc is decomposed by chlorine, in the manner described under Chloride of Sulphur or Chloride of Antimony, the whole of the chloride of zinc will remain behind with the non-volatile chlorides in the bulb tube, provided the heat to which the sulphide is subjected has not been too strong. But if the heat has been too high, a very small quantity of chloride of zinc will go forward with the volatile chlorides. (H. Rose).

Method B. Compare *proto*Chloride of Copper (fixity of). In the analysis of difficultly combustible organic substances, such as chloroform, which cannot be completely burnt in oxygen gas, by Warren's method, oxide of zinc is a good material to complete the combustion and absorb the chlorine. Oxide of copper cannot be used in this case, since at the high temperatures employed some dichloride of copper would be formed, which being insoluble in dilute acids would interfere with the determination of the chlorine. — The apparatus required is described under Chloride of Copper, and the details of the method are similar to those there indicated except that there is no oxide of copper used, and that the whole of the asbestos in the tube—both the anterior and posterior columns, is mixed with oxide of zinc. The asbestos and oxide of zinc are simply rubbed together in a mortar, and then packed into the tube. About 3 grms. of the oxide will be enough for the asbestos of the posterior column and 1 grm. for that of the anterior column. At the close of the combustion all but a trace of the chlorine will be found combined with the zinc of the posterior column. The air bath may be kept at 160°, or it may be dispensed with altogether. The process yields excellent results (Warren, *Proc. American Acad.*, 1866, **7.** pp. 87, 89).

Chlorimetry.

A term applied to the valuation of bleaching powder and analogous compounds — the so-called "chlorides of lime, potash and soda, etc.," which are mixtures of the hypochlorites, chlorides and hydrates of the respective metals. The bleaching or disinfecting value of these substances depends upon the amount of chlorine set free when they are treated with an acid, thus: —

$$CaO, Cl_2O; CaCl_2 + 2H_2SO_4 = 2CaSO_4 + 2H_2O + 4Cl,$$

and this chlorine may be readily estimated by a variety of methods. See, for example, Arsenious Acid (oxidation by chlorine); Chlorine (oxidizing power of) and power of expelling iodine from iodide of potassium, and Ferrocyanide of Potassium; older methods of chlorimetry are described under *di*Chloride of Mercury and Indigo.

Chlorine is commonly determined as Chloride of Silver, or as Chlorine, by some one of the processes depending upon its oxidizing power to be described immediately. [See the finding list in Appendix.]

Principle 1. Oxidizing power of.

Applications. Valuation of bleaching powder. Estimation of chlorine, hypochlorous acid, sulphurous acid and hyposulphurous acid. Separation of Mn from K, Na, Ba, Sr, Ca,

Mg, Al, Zn, Fe and Ni (see No. 7); of Pb from K, Na, Ba, Sr, Ca, Mg, Ni, Zn and Cu (see No. 7); of Co from Ni, etc. Analysis of many high oxides, acids and oxygenated compounds, which evolve chlorine when heated with strong chlorhydric acid (see No. 4). Valuation of binoxide of manganese and other high oxygen compounds.

Methods.

1. For methods depending upon the action of chlorine upon Antimony, Arsenic, diChloride of Mercury, sesquiOxide of Chromium, and Indigo, see these several substances.

2. *Action of chlorine upon ferrous salts.* This principle, first suggested by Dalton, was reduced to practice by Graham, and introduced into Germany by Otto. It is often called Otto's process by German writers, in spite of repeated explanations by this chemist, that the process was not invented by himself.

A. Provide a small, tolerably long-necked flask, to which has been fitted a two holed caoutchouc stopper, carrying a couple of short glass tubes, each bent at a right angle; weigh out about 0.15 grm. of clean, fine iron wire (piano wire), and place it in the flask, together with a quantity of pure chlorhydric acid. Insert the stopper, clamp the flask in an inclined position, and pass a slow current of washed carbonic acid through the flask. Heat the acid until all the iron has dissolved, keep up the current of carbonic acid until the flask has become cold, then dilute the iron solution to the volume of 200 c.c., and pour upon it, from a 50 c. c. burette, the freshly shaken solution of bleaching powder until the whole of the ferrous salt has been converted into a ferric salt. To determine the point when the oxidation is completed, place a number of drops of a solution of ferricyanide of potassium upon a white porcelain plate, and, as the operation draws to its close, take up upon a thin glass rod a little of the liquid under examination, add it to one of the drops of the ferricyanide, and observe whether a blue precipitate is produced. Repeat this test after each fresh addition of two drops of the bleaching powder solution until the last addition of liquid fails to yield any blue precipitate. Then note how many c. c. of the bleaching powder solution have been expended, and calculate the chlorine by the proportion

$$\frac{\text{Wt. of 2 atoms}}{\text{of Fe 56}} : \frac{\text{At. wt. of}}{\text{Cl 35.5}} :: \frac{\text{Wt. of Fe}}{\text{taken}} : x.$$

The turbid solution of bleaching powder required should be prepared in the manner described under Arsenious Acid. — Instead of ferrous chloride the inventor of the process employed pure crystallized ferrous sulphate, obtained by dropping a hot, fresh solution of clean iron in dilute sulphuric acid into twice its volume of alcohol. The crystalline precipitate thus obtained was washed with alcohol, dried between sheets of filter paper, and kept for use in tightly corked bottles. 0.7831 grm. of the sulphate corresponds to 0.1 grm. of chlorine, for

$$2Fe, SO_4 + H_2SO_4 + 2Cl = Fe_2O_3, 3SO_3 + 2HCl,$$

and

$$\frac{\text{Wt. of 2 Atoms}}{\text{of Cl 71}} : \frac{\text{Wt. of 2 molecs.}}{(FeSO_4+7Aq)\ 556} :: 0.1 : x\ (= 0.7831).$$

To avoid calculations, the operator may weigh out 3.1324 grms. (= 0.7831 × 4) of the sulphate and dissolve it, with addition of a few drops of dilute sulphuric acid, to the volume of 200 c. c. 50 c. c. of this solution (corresponding to 0.1 grm. of chlorine) may then be taken out with a pipette and diluted with 150 to 200 c. c. of water, and a sufficient quantity of chlorhydric acid for the titration Were it not for the inconvenience of weighing out a definite quantity of iron wire, it would be well to start with 0.6308 grm. of pure metallic iron,—or, on the assumption that fine wire contains 99.7 per cent of pure iron, take 0.6327 grm. of piano wire,—and to dissolve it with the precautions above enjoined. The solution diluted to 200 c. c. would correspond to 0.4 grm. of chlorine, like the solution of the sulphate. — This method gives very satisfactory results. Since it requires no standard solutions, it is particularly well adapted for occasional examinations of bleaching powder, and to serve as a control for other processes of chlorimetry. As a process for every day use, however, it has far less significance than was formerly the case.

B. Instead of the foregoing process, the following modification may be used. Weigh out about 0.3 grm. of piano wire, dissolve in chlorhydric acid, with the precautions above given, dilute the highly acid solution to 200 or 300 c. c., and slowly add to the liquid, from a burette, 50 c.c. of the turbid solution of bleaching powder, while constantly stirring the iron solution. Lastly, determine how much iron remains unoxidized by means of a standard solution of biChromate of Potassium, or of Permanganate of Potassium. By subtracting the weight of iron thus found from the weight of iron taken, the weight of iron equivalent to the chlorine in the bleaching powder used will be obtained. The process yields accurate results, particularly when applied to the valuation of chlorine water. Hypochlorites must be added very slowly to the iron solution, since they give off chlorine very easily on coming in contact with the acid liquor, so that there is danger of losing some of the gas. According to Mohr (*Titrirmethode*, 1856, **2.** 132), an acid solution of a ferrous salt can never absorb instantly all the chlorine which is brought in contact with it. The odor of chlorine can be perceived, moreover, in liquors which still contain ferrous oxide. A certain amount of time seems to be required for the ferrous salt to absorb the oxygen or chlorine necessary to convert it to a ferric salt. — In examining chlorine water, the latter should be left in contact with the ferrous salt in a stoppered bottle for a short time, before proceeding with the titration. In measuring out chlorine water for analysis attach a tube, fitted with fragments of moist, caustic potash to the top of the pipette.

C. The bleaching powder solution is mixed with an excess of a solution of ferrous chloride of known strength and the amount of ferric chloride formed is determined by means of protochloride of tin (see Ferric Salts), or by hyposulphite of sodium (see Ferric Salts). Each equivalent of $FeCl_3$ corresponds to one equivalent of Cl, for

$$FeCl_2 + Cl = FeCl_3.$$

D. The bleaching powder solution is heated with a mixture of alkali and freshly precipitated ferrous oxide and the chlorine estimated as Chloride of Silver, precisely as in the analysis of Chlorates (by estimating the chlorine in the residual chloride). The process of course gives only the total chlorine in the substance analyzed (Stelling, *Zeitsch. analyt. Chem.*, 1867, **6.** 33).

E. The bleaching powder solution is made to act upon a solution of ferrous chloride and the amount of ferric salt formed is estimated by determining how much metallic copper the solution will dissolve off from a weighed sheet or bar of that metal. See Ferric Salts (Runge).

F. To estimate the value of binoxide of manganese, Otto heated a weighed quantity of the powdered mineral with strong chlorhydric acid in a flask provided with a proper delivery tube, led the chlorine evolved into the solution of a weighed quantity of ferrous sulphate and determined how much of the latter had been converted to ferric sulphate by the chlorine.

Another old way of valuing the binoxide was to heat it with chlorhydric acid as above, and absorb the chlorine in milk of lime. The amount of chlorine in the product was then determined by some one of the ordinary methods of chlorimetry.

3. *Action of chlorine upon ferrocyanide of potassium.* Mix the solution of bleaching powder or chlorine water with an excess of a standard solution of Ferrocyanide of Potassium, acidulate the mixture strongly with chlorhydric acid and determine the excess of the ferrocyanide by means of a standard solution of biChromate of Potassium. The operation is finished when a drop of liquid touched to a drop of highly dilute ferric chloride, upon a white plate, no longer occasions any blue or green coloration (E. Davy. *Phil. Mag.*, (4), **21.** 214). According to Davy, bichromate of potassium is to be preferred to permanganate of potassium for estimating the excess of ferrocyanide, but Fresenius (*Zeitsch. analyt. Chem.*, 1863, **2.** 93) urges that the titration may be made more conveniently with the permanganate. — For Gay-Lussac's method of using ferrocyanide of potassium in chlorimetry see *Annales Chim. et Phys.*, **60.** 225.

4. *Action of chlorine upon arsenious acid and arsenites.* Compare Arsenious Acid.

As applied to the valuation of bleaching powder this process has already been described under arsenious acid. It may be employed also for the valuation of binOxide of Manganese, Chromate of Potassium, Chlorate of Potassium and many other oxygen compounds capable of evolving chlorine when heated with chlorhydric acid. But in these cases special precautions must be taken in generating and collecting the chlorine, [Compare Principle II, below]. — A simple form of apparatus for absorbing chlorine in arsenite of sodium has been described by Mohr (*Titrirmethode*, 1855, **1.** 313). It may be prepared as follows:— Fit a two holed cork or caoutchouc stopper to an ordinary flask of about one litre capacity. In one hole of the stopper place a small narrow funnel full of rather small fragments of glass or porcelain and in the other hole insert a tolerably wide glass tube, which shall reach almost to the bottom of the flask and also project three or four inches above the stopper. This tube though straight in the main should be bent slightly at a point above the stopper, so as to incline somewhat to one side and not interfere with the funnel. Slip the ring of a spring clip over the projecting part of this tube so that the clip may hang there ready for use. This large flask is to contain the solution of arsenite of sodium and serve as the receiver of the gaseous chlorine. To connect it with the decomposing flask, bend another glass tube, of similar bore to the first, twice—at one acute and one obtuse angle. The little flask which is to serve to generate the chlorine is closed with a perforated cork of the best quality. Through the hole of this cork, push that portion of the bent tube which is below the obtuse angle; and by means of a rubber connector attach the other end of the bent tube,—that under the acute angle—to the straight tube which projects from the absorption flask. Take care to arrange the rubber connector so that it can be closed by means of the spring clip when necessary. It is well to grind off obliquely the lower end of the bent tube, which falls within the decomposing flask, to facilitate the dropping of liquid from it. When everything is ready a measured quantity of a standard solution of arsenite of sodium (see below) is placed in the large absorption flask together with a considerable excess of pure carbonate of sodium. The latter is poured through the chips of glass in the funnel so that they may be left moistened with it. The oxide to be examined is placed in the small flask together with a quantity of concentrated chlorhydric acid, and the mixture is boiled until the whole of the chlorine has been set free and driven over into the absorption flask. It is safe to conclude that all the chlorine has gone over when the tube which projects from the absorption flask feels distinctly hot to the hand, and the peculiar noise of steam condensing in the liquid of the absorption flask is heard. During the earlier part of the decomposition most of the acid vapor generated in the decomposing flask will condense in the sloping tube which connects it with the absorption flask and the liquid thus formed will flow back into the decomposing flask. But as soon as the sloping tube has become hot some chlorhydric acid will pass over into the large flask, and by acting upon the carbonate of sodium will occasion a strong effervescence. To stop the process, remove the lamp from the decomposing flask and at the same instant close the rubber connector with the spring clip. Without opening the clip, disconnect the bent tube from the absorption flask, shake the latter and allow it to cool. Wash the glass chips with water as well as the lower part of the tube through which the chlorine was admitted to the flask. Pour a little starch paste into the flask and with a standard solution of Iodine determine how much arsenious acid remains undecomposed. No odor of chlorine should be perceived when the flask is opened. It is to be observed that the straight tube in the absorption flask should dip slightly beneath the surface of the solution of arsenite of sodium. With regard to the quantities of materials to be used, it may be said that for 0.2 to 0.5 grm. of bichromate of potassium Mohr places from 40 to 100 c. c. of arsenite of sodium solution in the absorption flask. For the preparation of the arsenical solution see Arsenious Acid. — See Arsenious Acid also for the application of the process to the estimation of low oxides such as As_2O_3 and FeO which are capable of being oxidized by nascent chlorine.

At first sight it would seem to be easy to apply the process to the estimation of chlorine in ordinary chlorides by heating the latter in the decomposing flask with a high oxide (like MnO_2 or K_2O, $2CrO_3$) and sulphuric acid, but the experiments of Mohr (*Titrirmethode*, 1855, **1.** 317) go to show that the whole of the chlorine cannot be expelled from the chlorides in this way by a single distillation. Possibly pure permanganate of potassium might do better than the oxydizing agents above mentioned. A still less satisfactory result was obtained on distilling in the dry way a mixture of chloride of sodium, binoxide of manganese and bisulphate of potassium.

The method now in question has suffered somewhat from a prejudice in favor of the somewhat analogous method of Bunsen (see below, Principle II) in which iodide of potassium is used to absorb the chlorine. It has, however, the distinct merit of requiring no large quantity of expensive materials and is to be commended in cases where the amount of chlorine to be estimated is comparatively large. Fresenius has urged also that the indirect character of the process must not be lost sight of. In Bunsen's process, with the iodide of potassium, we determine directly the quantity of iodine which has been set free by the chlorine, while in the process now under consideration it is not the amount of arsenious acid which has been oxidized that is determined, but the portion which has been left unacted upon. This point has special significance according to Fresenius, in cases where the amount of chlorine turns out to be less than was expected and where there is consequently an undue excess of undecomposed arsenite to be estimated. The errors of observation and those due to slight changes in the composition of the standard arsenite solution go to impair the accuracy of the determination of the chlorine.

5. *Action of chlorine upon stannous chloride.* Mix the chlorine water, or less conveniently the solution of bleaching powder, with an excess of a standard solution of protochloride of tin, and determine the excess of the latter by means of a standard solution of biChromate of Potassium. For the objections to the use of protochloride of tin, see biChromate of Potassium. According to Mohr it is hardly possible to avoid the loss of some chlorine when a solution of bleaching powder is added to the acid solution of chloride of tin.

6. *Action of chlorine upon sulphites and hyposulphites.*

A. *To estimate sulphurous or hyposulphurous acids.* Saturate the solution of the substance to be examined with chlorine gas and warm the mixture. Precipitate and weigh the sulphuric acid, as Sulphate of Barium. The process is specially applicable for the analysis of sulphites which are wholly free from any contamination of sulphuric acid.

B. *To estimate chlorine in aqueous solutions.* Mix the liquid under examination, which must be free from sulphuric acid, with a slight excess of hyposulphite of sodium in a stoppered bottle, and leave the bottle for a short time in a warm place until the odor of chlorine can no longer be detected. Then heat the liquid to boiling with an excess of chlorhydric acid in order to destroy the last portions of hyposulphite of sodium, filter off the sulphur and determine sulphuric acid in the filtrate as Sulphate of Barium. Each equivalent of sulphuric acid found will correspond to 2 equivalents of chlorine. About 0.5 grm. of hyposulphite of sodium will be enough to take for 30 grms. of chlorine water (Wicke, *Annalen Chem. und Pharm.*, **99.** 99).

C. *To estimate the total chlorine in a mixture of free chlorine and chlorhydric acid or a chloride*, mix a weighed quantity of the liquid with an excess of sulphurous acid, acidify the mixture with nitric acid and throw down the whole of the chlorine as Chloride of Silver. In another portion of the original liquid estimate the free chlorine by means of iodide of potassium or in some other appropriate way.

D. *For estimating the value of bleaching powder*, Fordos & Gelis proposed to substitute a standard solution of hyposulphite of sodium for the Arsenious Acid employed by Gay-Lussac. The standard solution was prepared by dissolving 2.77 grms. of the hyposulphite in water and diluting to the volume of 1 litre. Such a solution neutralizes its own volume of chlorine and is equivalent to the arsenical solution of Gay-Lussac. For an analysis, 10 c. c. of the standard solution of hyposulphite were placed in a beaker, diluted with 100 c. c, of water, acidulated, and colored blue with a few drops of a solution of indigo. The solution of bleaching powder was then poured into the beaker from a burette, in the manner described under Arsenious Acid. The blue color is but slowly destroyed so that the operator has plenty of time to observe the change. The process has the great disadvantage that a fresh supply of the standard solution must be prepared for each set of experiments. The reason for diluting the standard solution of hyposulphite before adding the bleaching powder is to avoid the precipitation of sulphur. It was urged as an advantage of the process that the hyposulphite would be acted upon by any chlorite in the powder while arsenious acid would not be oxidized thereby.

D. A gravimetric method of estimating the value of bleaching powder based upon the oxidation of hyposulphite of sodium by the chlorine of the powder has been proposed by Duflos, and afterwards by Nœllner (*Annalen Chem. und Pharm.*, **95.** 113). The oxidation was supposed to be effected in accordance with the following equation:—

$$Na_2O, S_2O_2 + 5H_2O + 8Cl = 2H_2SO_4 + 2NaCl + 6HCl.$$

According to Nœllner, 1 grm. of the bleaching powder is ground up with 2 grms. of hyposulphite of sodium in a mortar, the mixture is washed into a flask and the solution warmed upon a steam bath. The excess of the hyposulphite is decomposed with chlorhydric acid and the liquid warmed until it has become clear, and the sulphur has collected in drops. The liquid is then filtered, and the sulphuric acid determined as Sulphate of Barium in the usual way. Manifest objections to the process will be found in the liability of hyposulphite of sodium to contain some sulphuric acid, and of chlorhydric acid to contain a little chlorine. The operations of washing and weighing the sulphate of barium are, moreover, tedious, and not wholly free from difficulty. The accuracy of the method has, moreover, been called in question by Knop (*Pharm. Centralblatt*, 1855, p. 656), and disproved by Fresenius. The latter found that the results of the process differed considerably, accordingly as more or less hyposulphite of sodium was used.

E. *To estimate the value of binoxide of manganese and the like*, treat a weighed quantity of the finely powdered dry mineral with chlorhydric acid, lead the chlorine gas which is evolved into a mixture of sulphurous acid and chloride of barium, collect and weigh the Sulphate of Barium formed, and from its amount calculate that of the binoxide. The details of the process are as follows:—Prepare a not too concentrated aqueous solution of chloride of barium, acidulate it with chlorhydric acid which has been so much diluted that it will not throw down any of the chloride of barium, and lead sulphurous acid into the liquid until the latter smells strongly of the gas. Put the liquid aside in a tightly stoppered bottle until it has deposited any precipitate due to the presence of a little sulphuric acid. On the other hand fit to a flask, by means of a perforated cork, a delivery tube bent at two right angles. Place in the flask a weighed quantity of the binoxide of manganese, cover it with strong chlorhydric acid, cork the flask and support it in such position that the long, outer limb of the delivery tube shall reach into another flask charged with the mixture of chloride of barium and sulphurous acid. This second flask should be partially closed by means of a loosely fitting cork. Heat the mixture of manganese and acid moderately, until no more chlorine is given off, or until the oxide of manganese has all dissolved, and the brown color of the liquid in the flask has disappeared, then boil the liquid for a short time to drive the last traces of chlorine out of the flask, taking care that it does not boil over. Finally tip the flask so that the outer limb of the delivery tube shall be lifted out of the chloride of barium solution, remove the lamp, wash the end of the delivery tube and cork the absorption flask. The liquid in the bottle must still smell strongly of sulphurous acid at the close of the experiment. — To collect the sulphate of barium the mixture may either be left to settle and then be filtered, out of contact with the air, or the contents of the absorption flask may be boiled to expel the excess of sulphurous acid, as soon as the chlorine has all been absorbed. It is a defect of this method that access of air can easily convert some of the sulphurous acid into sulphuric, but on the other hand it is applicable in cases where the manganese is contaminated with oxide of iron, carbonates of the alkaline earths, etc. An error in weighing the sulphate of barium, moreover, is diminished as regards the binoxide. When carefully conducted, the process yields tolerable results. (Duflos; Ebelmen.)

7. *Action of chlorine upon salts of manganese, cobalt and lead.* [Compare Bromine.]

A. To separate *manganese* from the alkali metals, saturate the chlorhydric acid solution of the substance with chlorine gas, and precipitate Hydrate of Manganese by means of carbonate of barium, or less completely by ammonia water. The process may be employed also, in certain cases, for separating manganese from the alkaline earths, though it is inferior to the method in which manganese is thrown down as binoxide, see below, because portions of the alkaline earths are liable to fall in combination with the hydrate of manganese. — For separating manganese and *cobalt* from nickel, place the chlorhydric acid solution in a capacious flask and dilute with water, in such proportion that there shall be about 1 litre of solution for 2 grms. of the mixed oxides. Saturate the liquid with chlorine gas, and see to it that the upper part of the flask above the liquid is full of chlorine. Add an excess of elutriated carbonate of calcium, leave the mixture in the cold for 18 or 24 hours, shaking it as often as may be convenient, and finally filter the mixture. The whole of the nickel will pass into the filter, together with some traces of cobalt, while the hydrates of manganese and cobalt, together with the excess of carbonate of calcium, will be left upon the filter (H. Rose, *Pogg. Annalen*, **71.** 545, and **110.** 412). Instead of chlorine, D. Smith uses a dilute solution of bleaching powder, which has been so completely decomposed with sulphuric acid that no particle of hypochlorite remains in it. If there were any hypochlorite present, some nickel would go down with the oxides of manganese and cobalt. Bromine is a still better oxidizing agent, as Henry has shown.

To separate manganese from iron and nickel, and from the metals of the alkaline earths, pass a current of chlorine through a dilute acetic acid solution of the substance, or better, through a mixture of the chlorides of the metals and acetate of sodium. If the substance is in the form of a chlorhydric acid solution mix this solution with enough acetate of sodium to decompose the chlorides and bind the free chlorhydric acid. The presence of acetic acid does not prevent the precipitation of the *bin*Oxide of Manganese. Since cobalt would be partially precipitated in this way as well as manganese, it must be excluded from the solution (Schiel, *American Journ. Sci.*, 1853, **15.** 275; Compare Rivot, Beudant & Daguin, *Comptes Rendus*, 1853, p. 835).

According to Fresenius the solution should be heated to 50° or 60° during the passage of the chlorine. Instead of chlorine gas, a solution of hypochlorous acid or of hypochlorite of sodium may be used, but in case the latter is employed take special care to keep the liquid acid, with acetic acid. The process is held in good esteem, but Fresenius finds that the binoxide of manganese precipitated in this way retains a certain amount of alkali. He recommends that it be redissolved in chlorhydric acid and thrown down, as Carbonate of Manganese.

H. Rose directs that the solution to be oxidized should be concentrated, or at all events not too dilute. In case it is strongly acid neutralize it almost completely with carbonate of sodium before adding the acetate of sodium. Heat the solution almost to boiling in a beaker, and stir into it strong chlorine water [or better, Bromine] until the mixture smells strongly of chlorine after thorough stirring. In case the chlorine water contains much chlorhydric acid neutralize the latter with a little carbonate of sodium. The operation is known to be complete when the liquid, after the deposition of the black or dark brown precipitate, remains colored red, from the presence of permanganic acid. But this coloration cannot be obtained unless the liquid has been made almost neutral with carbonate of sodium before adding the acetate. To destroy the permanganate add a very small quantity of alcohol to the still warm solution after the precipitate has been allowed to settle for the most part. Wash the precipitated oxide thoroughly with hot water. It is of varying composition as regards the proportion of oxygen, and is liable to retain a certain quantity of soda, so that after having once been ignited and reduced to manganite of manganese it may, on further ignition, increase in weight through absorption of oxygen to form manganate of sodium. — In case the solution to be analyzed contains any large proportion of salts of ammonium, these compounds must be destroyed by boiling with carbonate of sodium before proceeding with the oxidation, for the manganese cannot all be thrown down by chlorine in presence of ammonium salts. For this reason, ammonia water cannot be used instead of carbonate of sodium for neutralizing the original acid solution.

B. To separate *lead* from alkalies, alkaline earths, magnesium, zinc and nickel, dissolve the mixture in acetic acid, heat the liquid to 50° or 60°, and lead chlorine gas into the liquid as long as any binOxide of Lead continues to fall. The precipitation is soon finished. In case the original solution contains free chlorhydric acid, nearly neutralize it with carbonate of sodium and add acetate of sodium to decompose the chlorides. The process is not applicable to the separation of lead from iron, cobalt and manganese. (Rivot, Beudant & Daguin, *Journ. prakt. Chem.*, **61.** 136.) According to H. Rose, though chlorine, when made to act upon chloride of lead and sulphate of lead suspended in water, can convert them into the binoxide, there will always be traces of chloride of lead, and in presence of sulphuric acid, a comparatively large proportion of sulphate of lead in the binoxide obtained as above. The binoxide, therefore, cannot be weighed as such, but must be converted into sulphide of lead. — The binoxide of lead, moreover, always carries down larger or smaller quantities of the other oxides from which it is to be separated, notably oxide of zinc, and must consequently be redissolved and again treated, to effect a complete separation.

C. *Cobalt* can be precipitated in the same

way as manganese from hot solutions slightly acidulated with acetic acid. It is well to add from time to time a few drops of carbonate of sodium and chlorine water, until the solution smells of chlorine. For the treatment of the precipitate see Hydrate of Cobalt.

8. *Action of chlorine upon copper.* Chlorine can be estimated indirectly in aqueous solutions by placing in the latter a weighed quantity of some metal like copper, which unites with chlorine to form a soluble chloride, noting the loss of weight of the metal, and calculating therefrom the amount of chlorine. This principle has been applied to the valuation of bleaching powder (Runge) and of binoxide of manganese (Fickentscher, *Journ. prakt. Chem.*, **17.** 173). The process, as regards binoxide of manganese, is as follows: — Weigh out 3 or 4 grms. of finely powdered, dry binoxide of manganese, and place it in a flask to which has been fitted a perforated cork carrying a narrow glass tube. Pour some water upon the powder, and place in the flask a strip of brightly polished, not too thin, sheet copper, which has been accurately weighed. It should weigh about 30 grms. Then pour into the flask as much strong chlorhydric acid as may be necessary to dissolve the manganese, and put the cork into the neck of the flask. Let the mixture stand for some time in the cold, then heat it until the manganese has dissolved, and finally boil the liquid for some time. At the moment of removing the lamp close the tube in the cork with a pellet of wax. During this process of dissolving the manganese no trace of chlorine should be evolved. This accident may occur in case the mixture is heated too quickly, or when very strong chlorhydric acid is employed. After the flask has become cold take out the strip of copper, wash it thoroughly with water, dry it and weigh it. From the loss of weight which the copper has undergone, calculate the amount of real binoxide in the manganese examined.

The dichloride of copper formed dissolves in part in the excess of chlorhydric acid, and is, in part, deposited as a white powder. In case any of it should adhere firmly to the metallic copper, it may be dissolved off with dilute chlorhydric acid, to be followed immediately by water. Or, better, the separation of the dichloride may be entirely prevented when, instead of water, a concentrated solution of common salt is placed in the flask with the manganese, for dichloride of copper is soluble in chloride of sodium. — The method gives tolerably accurate results, and is useful sometimes for technical purposes, though it suffers from the fact that a little metallic copper will dissolve in chlorhydric acid, even when air is carefully excluded. It is not applicable for the analysis of binoxide of manganese which contains any considerable amount of ferric oxide, since the ferric chloride at first formed would be reduced to ferrous chloride at the expense of the copper; unless, indeed, the amount of ferric oxide be determined in a separate portion of the mineral, and subtracted from the first result. To this end weigh out as much of the sample as was taken for the first trial, boil it in an open flask with an excess of chlorhydric acid until no more chlorine is evolved, then place a weighed strip of copper in the flask, close the latter with a perforated cork, and proceed as in the first experiment. This process has the merit that it may be applied without trouble to the valuation of manganese which is contaminated with carbonates of the alkaline earths.

9. *Action of chlorine upon organic coloring matters.* See the various coloring matters. The old method of chlorimetry depending upon the decoloration of Indigo will be described under that head. A method of testing the value of indigo, devised by Schlumberger, is as follows: — Powder the several samples of indigo which are to be compared, as described under Indigo, weigh out precisely 1 grm. of each sample and 1 grm. of pure indigotine, obtained by collecting the sublimate of indigo vats, and washing it first with acidulated water and then with pure water; dissolve the samples in fuming sulphuric acid, as described under biChromate of Potassium (action upon coloring matters), and dilute the solution to the volume of 1 litre. By means of a marked pipette transfer 50 c. c. of the solution to a glass cylinder, and pour into the blue liquid, from a graduated pipette, successive 2.5 c. c. portions of a solution of bleaching powder marking 1° B, until the blue color is totally destroyed. To determine the excess of chlorine thus added, pour a quantity of the indigo solution in question into a burette, and pour it thence into the liquid in the cylinder until a greenish tint pervades the liquid. It is well to verify the results of the first trial by means of a second experiment. The calculation is as follows: — Suppose that 5 c. c. of the chlorine solution have decolorized 59 c. c. of the solution of indigo, viz.. the 50 c. c. first taken and 9 c. c. subsequently added from the burette. On the other hand, let it be admitted that of the solution of pure indigotine, 46 c. c. corresponded to 5 c. c. of the chlorine solution. Then :

Since 46 c. c. of indigotine solution = 59 c. c. of the sample, these 59 c. c. will contain 0.046 grm. of indigotine, whence the proportion
59 : 0.046 = 1000 : x (= the pure indigotine in 1 grm. of the sample).

The process was formerly esteemed, but is inconvenient because of the instability of the solution of bleaching powder. It is of course necessary to determine the value of this solution against pure indigotine in each series of experiments. It is impossible, moreover, to avoid losing some chlorine at the moment when the bleaching powder solution comes in contact with the acid liquid. A better method is described under biChromate of Potassium.

(Schützenberger's *Matières Colorantes*, 1867, **2**. 558.)

Principle II. Power of decomposing iodide of potassium and other metallic iodides.

Applications. Estimation of chlorine and of hypochlorous acid; of iodine and iodhydric acid. Valuation of bleaching powder. Separation of chlorine, bromine and iodine from one another. Analysis of many oxides, acids and oxygenated compounds which evolve chlorine when heated with strong chlorhydric acid [Compare Arsenious Acid, Iodine, biChromate of Potassium, Permanganate of Potassium, binOxide of Manganese, and the various high Oxides].

Method A. Bring the chlorine, either in the gaseous form or in solution, into contact with an excess of a solution of one part of iodide of potassium in ten parts of water, and determine how much iodine has been set free by the chlorine, by means of a standard solution of hyposulphite of sodium or sulphurous acid (see Iodine). Each atom of iodine found corresponds to one atom of chlorine. Chlorine water may be simply allowed to flow from the pipette into an excess of the iodide of potassium solution. If the latter be not present in sufficient excess, a persistent black precipitate of iodine will be formed. Chlorine gas, on the other hand, as obtained by the action of concentrated chlorhydric acid upon any high oxide (compare Arsenious Acid, oxidation of by nascent chlorine), must be set free in a flask or retort, and carefully led into the solution of iodide of potassium in such manner that none of it shall be lost. To this end provide a small flask, free from any projecting rim or lip, and by means of a sufficiently wide caoutchouc tube connect it with an empty, bulbed chloride of calcium tube of the same diameter as the neck of the flask. The caoutchouc connecter should be thoroughly boiled in dilute potash lye, and afterwards well washed with water to free it from adhering sulphur. When everything is ready, clamp the flask in an inclined position, so that any liquid condensing in the bulb of the empty chloride of calcium tube could run back into the flask, and by means of narrow glass tubes and short rubber connecters freed from sulphur, connect the outer end of the chloride of calcium tube with a small flask containing iodide of potassium, to which is attached a U-tube likewise charged with the iodide. Both the vessels which contain iodide of potassium should be placed in a dish of cold water, and the tube which delivers the chlorine gas should not be allowed to reach quite to the surface of the solution of iodide in the receiving flask. A weighed quantity of the substance to be examined (such as binOxide of Manganese, Chromate of Potassium, or other oxygen compound capable of generating chlorine with chlorhydric acid) is placed in the decomposing flask, together with a quantity of concentrated chlorhydric acid, the flask is then connected with the remainder of the apparatus and its contents heated — at last to boiling — until the whole of the chlorine has been driven over into the iodide of potassium. Take care to shake the flask which contains the latter occasionally. The empty chloride of calcium tube serves to condense a good part of the acid vapors, and to return them to the decomposing flask. Finally pour the iodide of potassium solution into a beaker, together with the rinsings of the vessels which contained it, and estimate the free Iodine by means of a standard solution of hyposulphite of sodium. — The process is exceedingly accurate, and is to be recommended in all cases where the quantity of chlorine to be estimated is but small. It is less convenient, however, when any considerable quantity of chlorine has to be set free.

To estimate the value of bleaching powder by this process, pour 10 c. c. of the turbid solution of bleaching powder, prepared as described under Arsenious Acid, p. 44, into a beaker, add about 6 c. c. of a solution of Iodide of Potassium (1 part of the salt to 10 parts of water), dilute the mixture with about 100 c. c. of water, acidulate with chlorhydric acid and estimate the Iodine, which is set free, by means of hyposulphite of sodium or sulphurous acid. Each equivalent of iodine found represents an equivalent of chlorine. This method yields excellent results, but is less employed than the processes for which only cheap materials are required (see Arsenious Acid and Chlorine, above). (Bunsen.)

Method B. When chlorine water, or a solution of hypochlorite of sodium, is added to the solution of a metallic iodide, the iodine expelled from combination with the metal, unites with chlorine to form pentachloride of iodine. Hence, by employing a standard solution of chlorine and a proper indicator, to show when the last particle of iodine has been combined with chlorine, the amount of iodine in any compound can readily be determined. Golfier-Besseyre employed starch paste as the indicator, but A. and F. Dupré (*Annalen Chem. und Pharm.*, **94**. 365) have improved upon that suggestion, and employ as the indicator chloroform, or bisulphide of carbon, both of which substances are colored intensely violet by free iodine, as well as by all compounds of iodine and chlorine which contain less than 5 atoms of the latter to one of the former element. — There are two ways of proceeding: —*The first way* is as follows: Prepare some highly dilute chlorine water and determine its strength by means of a solution of iodide of potassium, as explained above, under Method A. Weigh or measure out a quantity of the liquid to be examined; pour it into a glass stoppered bottle, and add to the mixture a few grms. of pure chloroform, or of recently distilled bisulphide of carbon, free from sulphur and sulphuretted hydrogen. The quantity of material taken for analysis ought not to contain much more than 10 milligrms. of iodine. Pour the standard chlorine solution from a burette, drop by drop, into the bottle, shaking the latter sharply after each addition, until the violet color of the chloroform or the bisulphide just disappears; 6 atoms of chlorine consumed correspond to 1 of iodine. Or, still more simply, determine the value of the dilute chlorine water, in the first place by making it act upon a known quantity of iodide of potassium,— say 10 c. c. of a standard solution containing 0.001 grm. iodine in each cubic centimetre. It may then be employed at once upon the liquid to be examined; the amount of chlorine consumed upon the known weight of iodide of potassium being to the iodine therein contained, as the quantity used in the analysis of the substance under examination is to x.

The point of disappearance of the violet color can be hit with great precision. In cases where so much iodine is present that it can itself color the aqueous liquid perceptibly, it is best not to add any chloroform or bisulphide until the brown iodine coloration first produced in the liquid has been nearly destroyed by the further addition of chlorine.

Since this way of proceeding is inadmissable in presence of substances liable to be acted upon by free chlorine or iodine,—such, for example, as the organic matters with which the mother liquors which contain iodine are usually contaminated; — the *second way* has often to be employed. It is as follows: Add chloroform or bisulphide of carbon to the liquid under examination, and then dilute chlorine water of unknown strength until the liquid is just decolorized, and all the iodine has been converted into ICl_5. Then add a moderate excess of a solution of iodide of potassium. By the reaction of this iodide of potassium upon the chloride of iodine, 6 atoms of iodine will be liberated and go into solution,

$$5KI + ICl_5 = 5KCl + 6I.$$

Determine this Iodine by means of hyposulphite of sodium or sulphurous acid, and divide the amount obtained by six. The quotient will express the amount of iodine in the liquid analyzed. — The process in both its modifications yields excellent results. It is particularly well adapted for the determination of very small quantities of iodine.

Still a *third modification* of the process must be resorted to for estimating iodine in presence of bromides. The Dupré's find that when the solution of an iodide contains 1 part or more of bromide of potassium in 1500 parts of water, protobromide of iodine (IBr) will be formed on the addition of chlorine water. If, on the other hand, the solution contains less than 1 part of bromide of potassium in 1500 parts of water there will be formed in addition to the protobromide of iodine, varying proportions of the higher bromides of that element. If the solution contains only 1 part of bromide of potassium to 13000 parts of water, pentabromide of iodine will alone be formed. In presence of bisulphide of carbon the combination of iodine with bromine to form IBr is marked by a change of the violet color of the liquid to yellowish brown, while the formation of IBr_5 is marked by the change from violet to white. — In presence of a bromide, consequently, proceed as follows:— Try, in the first place, whether the color will change from violet to white by adding some bisulphide of carbon, and then gradually chlorine water, to a portion of the liquid to be analyzed. In case the color does not thus change dilute the liquid to the necessary degree, and to make quite sure add half as much again water as would be strictly required. Then proceed, as above described, as if no bromine were present. The process yields satisfactory results, and is particularly useful for determining small quantities of iodine in liquids highly charged with chlorides; and containing not too small quantities of bromides. [Compare the analogous principle below, in which chlorine is made to decompose bromide of potassium.]

The foregoing process may be applied to the separation of chlorine, bromine and iodine, from one another as follows: — In one portion of the liquid throw down Bromide, Chloride and Iodide of Silver all together, collect and weigh the mixed precipitate, and by means of hot hydrogen (see Bromide of Silver) determine how much silver is contained in it. In another portion of the solution determine the amount of iodine by the method now in question, and calculate the amounts of iodide of silver and of silver corresponding to the quantity of iodine found. Deduct this calculated iodide of silver from the weight of the mixed precipitate of bromide, chloride and iodide of silver, and the calculated amount of silver from the amount of that metal in the mixed precipitate. The remainders are respectively the joint amount of chloride and bromide of silver, and the quantity of silver contained therein. For the method of calculating the results see Bromide of Silver. The quantity of bromine in the mixture must not be too small, lest inexact results be obtained from the indirect method on which its estimation depends.

To estimate a small quantity of an iodide in the presence of a large quantity of a chloride, add dilute chlorine water of unknown strength to the liquid, drop by drop, with constant shaking, till the violet coloration of the bisulphide has just vanished, and all the iodine has been converted into ICl_5. Separate the aqueous solution from the bisulphide, add iodide of potassium in sufficient excess, and determine the Iodine, which is set free, by means of hyposulphite of sodium, or in some other appropriate way. Every six parts of iodine found correspond to one part of iodine in the original substance. The process gives good results. To avoid the trouble of pouring off the aqueous liquid from the bisulphide and of washing the latter, the mixture may be poured into a rather narrow measuring cylinder after the chlorine has been made to act to decoloration, and the volume of the aqueous solution noted. A definite portion of it may then be measured out in a pipette for the experiment.

Principle III. Power of decomposing metallic bromides.

Applications. Estimation of bromine and bromhydric acid. Separation of bromine, chlorine and iodine from one another.

Method A. This method depends upon the volatility of bromine as well as upon the principle now in question. It was proposed for estimating bromine for technical purposes in mother liquors. When chlorine is made to act upon the aqueous solution of a metallic bromide, each atom of it liberates an atom of bromine, and the latter imparts a yellow color to the solution. But on boiling the liquid the bromine escapes and the yellow tint disappears. Hence by using a standard solution of chlorine, and noting the point of disappearance of the yellow coloration, the amount of bromide in a liquid can be estimated volumetrically. The chlorine water employed must be highly dilute. It should be standardized just before use, either by making it act upon a solution of bromide of sodium of known strength, acidulated with a few drops of chlorhydric acid in the manner described below; or by means of iodide of potassium and hyposulphite of sodium as explained above, under Principle II. — The liquid to be examined is placed in a flask and heated nearly to boiling; some of the standardized chlorine water is then added to it from a burette covered with black paper, and the mixture heated for about three minutes, or until the yellow tint which appeared on the addition of the chlorine has disappeared, Then let the mixture cool for 2 minutes, drop into it another portion of the chlorine water and again heat it. Proceed in this way until the last addition of chlorine fails to impart a yellow color to the liquid. In case the experiment should not be finished under several hours, the strength of the chlorine water ought to be redetermined at the end of the operation, and the calculation of the result based upon the mean of the two trials,

The process is inapplicable in presence of iodine, the protoxides of iron and manganese and organic matters. Mother liquors colored yellow by organic matter may be evaporated to dryness, after adding some carbonate of sodium and the residue gently ignited. The residue may then be treated with water, and the solution filtered to fit it for analysis. In this case as in others, where the liquid is alkaline, the solution must be slightly acidulated with chlorhydric acid before adding the chlorine water. The purpose of the carbonate of sodium added before the evaporation, is to prevent the loss of bromine from decomposition of bromide or chloride of magnesium by heat. (Figuier, *Ann. Chim. et Phys.*, **33.** 303).

Method B. The bromine liberated by means of chlorine is dissolved in ether, and the color of this ethereal solution is compared with that of other ethereal solutions containing known quantities of bromine. The process is applicable for the valuation of mother liquors, the amount of bromine in which is known approximately beforehand. (Heine, *Journ. prakt. Chem.*, **36.** 184.) Satisfactory results obtained by this method have been reported by Fehling (*Journ. prakt. Chem.*, **45.** 269). The brine examined by this chemist could at the most contain no more than 0.02 grm. bromine in 60 grms. of liquid. Hence he prepared ten different standard liquids by adding to as many 60 grm. portions of a saturated solution of common salt, definite quantities of bromide of potassium, increasing regularly from 0.002 to 0.02 grm. of bromine. Equal volumes of ether were added to each solution, and then chlorine water until there was no further change observed in the color of the ether. It is of the highest importance to hit this point exactly, since too little as well as too much chlorine makes the liquid appear lighter than it should. To avoid all chance of error Fehling prepared three samples of each of his standard liquors, and chose the darkest colored in each case for the comparison. Several samples each of 60 grms. of the brine to be examined are now weighed or measured out, the same volume of ether as before is added, then the chlorine water, and the color of the ethereal solution obtained is compared with that of the several ethereal solutions which were obtained from the standard liquors. The mean of several good experiments is taken as the correct result. Direct sunlight must be avoided, and the operations conducted with expedition. According to Fresenius it is well to use chloroform or bisulphide of carbon instead of the ether.

Method C. The chlorine is added in such quantity that it shall combine with the liberated bromine. Since chloride of bromine merely communicates a yellowish tinge to chloroform, while free bromine colors it yellow or orange, the point at which the latter color disappears, on the continued addition of a standard solution of chlorine can be made to mark the amount of bromine in a solution. The process is applicable for the analysis of bromides of the alkali-metals in neutral solutions — and especially for the determination of small quantities of bromine in the mother liquors from brine and kelp. It is as follows: — Place the solution to be examined in a glass-stoppered bottle, together with as much pure chloroform as would fill a hazel-nut and pour dilute standard chlorine water upon the mixture, from a burette covered with black paper. On shaking the bottle the chloroform becomes yellow, then, on further addition of chlorine, orange, then yellow again, and lastly yellowish white, at the moment when 2 atoms of chlorine have been expended for each atom of bromine in the liquid.

$$KBr + 2Cl = KCl + BrCl.$$

The chlorine water is standardized with iodide of potassium and hyposulphite of sodium as explained above, under Principle II; its strength should be made to conform somewhat with the amount of bromine to be determined, and adjusted so that about 100 c.c. of it may be used in an experiment. Considerable practice is required before the operator can be sure of the end reaction. It is well on that account to place the bottle on a sheet of white paper, and to have at hand a dilute solution of yellow chromate of potassium of the required tint, with which to compare the color of the chloroform. The process yields good results; for example, 0.018 instead of 0.0185; 0.055 instead of 0.059, and 0.0112 instead of 0.01. If the liquid to be examined contains organic matter, add caustic soda to alkaline reaction, evaporate to dryness, ignite the residue, best in a silver dish, dissolve the residue in water, filter, neutralize exactly with chlorhydric acid and test as above. (Reimann, *Annalen Chem. und Pharm.*, **115.** 140).

Bromine may be determined by this process in presence of iodine by adding in the first place, standard chlorine water until the violet color of the chloroform due to iodine is just destroyed, and ICl_5 (see Principle II., above,) or IBr_5 has been formed; and afterwards adding more of the chlorine water until the whole of the bromine is converted to BrCl. In the first step 6 atoms of chlorine are expended for each atom of iodine, while of the second quantity of chlorine, every 2 atoms represent one atom of bromine. Suppose there were 5 molecules of KBr and 1 of KI in the mixture to be examined; then,

$$KI + 5KBr + 6Cl = 6KCl + IBr_5.$$

and

$$IBr_5 + 10Cl = ICl_5 + 5BrCl.$$

Method. To separate bromine from chlorine, precipitate from one portion of the solution both these elements together as Bromide of Silver and Chloride of Silver, either by the volumetric or gravimetric process. In another portion estimate the bromine by one of the processes above described, and calculate the chlorine from the difference. According to Fresenius, this method affords an expeditious means of examining the mother liquors of brine.

Principle IV. Power of decomposing ammonia.

Applications. Estimation of free chlorine. Separation of free chlorine from combined chlorine. Valuation of bleaching powder.

Method A. To estimate chlorine gas lead it carefully into dilute ammonia water. A part of the ammonia is decomposed, and nitrogen evolved, while chloride of ammonia is formed.

$$4NH_3 + 3Cl = 3NH_4Cl + N.$$

Care must be taken to bring in the chlorine as slowly as possible in order that time may be allowed for the decomposition, and that no chlorine shall go to waste with the nitrogen. It is well to place the ammonia water in two or three Woulfe bottles or connected flasks, and still better to have nothing but water in the last flask to catch the vapors of chloride of ammonium which are formed. The ammonia water must be in excess everywhere, not only that all the chlorine may be retained, but that no chloride of nitrogen shall be formed. When the development of chlorine has ceased, drive out those portions of it which remain in the decomposing vessel by means of a current of carbonic acid. The solution of Chloride of Ammonium may then be evaporated and weighed as described under that head; or the residue of the evaporation may be dissolved in water, the solution acidulated with nitric acid, and the amount of chlorine in the chloride determined as Chloride of Silver, by titration. From this result the weight of the chloride of ammonium is then calculated.

In case chlorine water is to be estimated it may be carefully mixed with an excess of dilute ammonia water, and the mixture then evaporated. — This process was formerly esteemed, but since Schoenbein (*Journ. prakt. Chem.*, **84.** 386) has shown that a little chlorate of ammonium is formed by the action of chlorine on ammonia water, it can no longer be regarded as irreproachable. Fresenius regards it as inferior to the process depending upon the action of Chlorine upon sulphurous acid. Experiments by Haarhaus (*Zeitsch. analyt. Chem.*, 1863, **2.** 59) indicated that of 100 parts of free chlorine decomposed by ammonia, 98.6 parts were converted into chloride of ammonium and 1.4 parts to chlorate of ammonium. The proportion of chlorate formed would be likely to vary however with the degree of concentration of the ammonia water. But it could doubtless be got rid of by the use of an appropriate reducing agent.

Method B. In liquids which contain free chlorine plus chlorhydric acid, or a metallic chloride, the amount of combined chlorine may be determined as follows: — Mix a weighed quantity of the liquid with an excess of dilute ammonia water, and estimate the total chlorine as Chloride of Silver. In another weighed portion of the original liquid, estimate the free Chlorine by means of iodide of potassium, or in some other appropriate way and subtract this amount from the weight of the total chlorine; the difference will represent the combined chlorine. — It would be wholly inadmissable to mix the original solution with nitrate of silver, since only five-sixths of the free chlorine would go down as chloride of silver,

$$6Ag_2O + 12Cl = 10AgCl + Ag_2O, Cl_2O_5.$$

(Weltzien, *Annalen Chem. und Pharm.*, **91.** 45.)

Method C. To estimate the total chlorine in bleaching powder mix the latter with ammonia water, acidulate the liquor with nitric acid, and by means of a standard solution of silver estimate the chlorine of the chloride of calcium which is formed. See Chloride of Silver. The reaction which occurs between the ammonia and the hypochlorite may be formulated as follows: —

$$4NH_3 + 3 (CaO, Cl_2O + CaCl_2) = 6CaCl_2 + 6H_2O + 4N.$$

By estimating the total chlorine in this way in one portion of the powder, and subjecting another portion to one of the processes of Chlorimetry, the amount of inactive chlorine in the sample will be found from the difference between the two results. (Kolb, *Comptes Rendus*, **65.** pp. 530 — 534). To destroy any chlorate of calcium which the bleaching powder may contain, treat the dilute mixture of bleaching powder and ammonia water, after the reaction between the two has ceased, with zinc and sulphuric acid. The nascent hydrogen evolved will reduce the chlorate to the condition of chloride. (Fordos & Gelis). Compare Chlorates, reduction of by hydrogen.

Method D. Instead of weighing the chloride of ammonium, as in Method A, the gaseous nitrogen may be collected and measured, as has been proposed by Henry & Plisson. The apparatus required consists of a flask of the capacity of 300 or 400 c. c., to the cork of which is fitted a funnel with glass stop-cock and a delivery tube. The bent tube leads to a pneumatic trough, and there delivers gas from the flask into a graduated tube of the capacity of 80 c. c. 10 grms. of the bleaching powder and 250 c. c. of water are placed in the flask, and 100 c. c. of ammonia water diluted with an equal bulk of water, are gradually added through the funnel. The contents of the flask are gradually heated and the nitrogen is collected in the graduated tube, which is clamped in a vertical position. Both the graduated tube and the pneumatic trough are filled with an alkaline liquid. When no more nitrogen is given off, the flask is filled with water through the funnel in order to drive forward all the nitrogen. The Oxygen from the air originally contained in the flask is then determined by absorption with pyrogallic acid, or in any other appropriate way, and the volume of air to which it corresponds is subtracted from the contents of the graduated tube. The process yields accurate results when carefully conducted, but is less convenient than the methods in ordinary use.

Principle V. Volatility.

Application. Separation of free chlorine from free chlorhydric acid.

Method. Add to the solution as much sulphate of potassium as may be required to react upon the whole of the chlorhydric acid, for the formation of chloride of potassium, and bisulphate of potassium. Evaporate the mixture carefully in the dark to expel the free chlorine, and estimate the chlorine of the chloride of potassium by precipitation as Chloride of Silver. (Kœne.)

For use as a reagent, chlorine gas may be prepared as follows: — Fit up a flask with a delivery tube, and connect it with a washing bottle charged with concentrated sulphuric acid, and a cylinder full of chloride of calcium. Place in the flask a mixture of 18 parts by weight, of coarsely powdered chloride of sodium, and 15 parts of finely powdered binoxide of manganese, of good quality, and pour in a perfectly cold mixture of 45 parts of oil of vitriol, and 21 parts of water. On shaking the flask a steady evolution of chlorine will set it at once. When the flow of chlorine shows signs of slackening, heat the flask very gently (Wiggers). — Hager (*Zeitsch. analyt. Chem* 1867, **6.** 421) puts chlorate of potassium and chlorhydric acid of 25 per cent in a small flask, in the proportion of 2. 5 grms., of the chlorate to 25 grms. of the acid. The mixture must not be heated, nor shaken, but simply left to itself. — For the arrangements for developing chlorine devised by Mitscherlich and Brugnatelli, see *Zeitsch. analyt. Chem.*, 1867, **6.** pp. 137, 393.

To prepare chlorine water, simply conduct the gas into a bottle of cold water until the latter is saturated, taking care to lead off the excess of chlorine into a chimney, or into a large bottle filled with pellets of paper which have been moistened with alcohol. Since chlorine water is rapidly decomposed by light, it should be kept in a dark place, or in bottles covered with black paper.

Chloroplatinate of Ammonium.

Principle. Insolubility in alcohol.

Applications. Separation of NH_4 from Na, Li, Ba, Sr, Ca, Mg. Estimation of NH_4 in salts of ammonium. Estimation of platinum. Separation of Pt from Mn, Fe, Ni, Co, Zn, Cd, Ur, Hg, Cu, Bi.

Methods. Same as those described under Chloroplatinate of Potassium. In estimating ammonium it is well to control the results of the first weighing as follows : — Ignite the precipitate weighed at 100°, in a platinum crucible, weigh the metallic platinum which is left and calculate the ammonium which corresponds to it. If the precipitate is pure, the results obtained from the two weighings will agree. To ignite the precipitate roll it up in the filter, place the ball in the crucible, cover the latter and heat it moderately for a comparatively long time, then remove the cover, place the crucible upon its side, lay the lid in front of the crucible and burn the carbon of the filter at a gradually increased heat.

Unless care be taken to heat the precipitate very gradually, particles of the precipitate will be carried off by the escaping chloride of ammonia and some of the platinum be lost. In case the amount of precipitate is very small it is best to collect it at once upon an unweighed filter, and to ignite the dried mass and weigh as platinum. So too in estimating platinum, or separating it from other metals (see below). A little platinum will almost always adhere to the platinum crucible in which the ignition is made, so that the crucible is heavier after the experiment than before, but a platinum crucible is nevertheless to be preferred to one of porcelain for effecting the decomposition of the chloroplatinate. The process yields satisfactory results, though owing to the sparing solubility of the chloroplatinate and the liability of losing a little platinum while igniting, they are usually a little too low.

In case both potassium and ammonium are to be separated from the other metals, ignite the weighed precipitate of mixed chloroplatinates of potassium and ammonium with the prescribed precautions, until the chloroplatinates are fully decomposed, treat the residue with water and in the filtrate determine the Chloride of Potassium. Collect also and weigh the platinum for the sake of control. Calculate the chloride of potassium found into chloroplatinate of potassium, and subtract the latter from the weight of the mixed precipitate in order to obtain the weight of the chloroplatinate of ammonium. The method is seldom employed. In case the precipitate is large, it will be well to finish the reduction in a stream of hydrogen, or to ignite the precipitate with oxalic acid. In case the latter is employed take care to acidulate the filtrate with chlorhydric acid.

In estimating platinum by adding chloride of ammonium to its solution, it is not safe to weigh the precipitated chloroplatinate of ammonium, since it is impossible to wash out with alcohol a certain excess of chloride of ammonium (1 or 2 per cent of the weight of the precipitate) which goes down with the chloroplatinate. It is essential in that case to ignite the precipitate, and weigh the metallic platinum (Fresenius). — In separating platinum from other metals, wash the precipitate first with a saturated solution of chloride of ammonium in order to remove the other metals, and finish the washing with alcohol.

Properties. The precipitated compound is either in the form of a heavy lemon-yellow powder, or in small, bright yellow octahedral crystals. It is difficultly soluble in cold water, but more readily soluble in hot water; very sparingly soluble in absolute alcohol, but more readily in spirit. At 15° or 20°, 1 part of it dissolves in 26500 parts of alcohol of 97.5 per cent; in 1400 parts of alcohol of 76 per cent, and in 670 parts of alcohol of 55 per cent. In alcohol acidulated with chlorohydric acid it is far more readily soluble (Fresenius). It undergoes no change in the air, or when heated to 100°. When ignited it gives off chlorine and chloride of ammonium, and metallic platinum is left in the form of a porous sponge.

Its composition is . —

$2NH_4Cl$	107	23.97
$PtCl_4$	339.4	76.03
	446.4	100.00
N_2	28	6.27
H_8	8	1.79
Cl_6	213	47.72
Pt	197.4	44.22
	446.4	100.00

Chloroplatinate of Potassium.

Principle. Insolubility in alcohol, and ether. Also insolubility in a solution of chloride of ammonium.

Applications. Estimation of K and Pt. Separation of K from Na, Li, Ba, Mg, (compare Method D,) Sr, Ca, As, P, B, and S, and from other elements enumerated under Method B. Separation of Pt from Mg, Zn, Cd, Mn, Fe, Co, Ni, Ur, Hg, Cu, Bi, and Au.

Method A. To separate potassium from sodium and lithium, weigh the mixed chlorides after they have been ignited to expel acid and ammoniacal salts as explained under Chloride of Potassium, dissolve them in a small quantity of water, add a decided excess of a pure, concentrated, neutral solution of biChloride of Platinum, prepared as described under that head. Evaporate the mixture almost, but not completely, to dryness on a water bath, add a quantity of alcohol of 0.86 or 0.87 specific gravity, cover the dish with a glass plate and let the mixture stand for several hours with occasional stirring. After the precipitate has settled, the supernatant liquid should exhibit a deep yellow color, due to the presence of dissolved chloroplatinate of sodium. If it is not decidedly yellow, there is reason to apprehend that too little of the bichloride of platinum was used. Pour the clear supernatant liquid upon a weighed filter, but before transferring the precipitate to the filter examine it carefully as to its purity. If the precipitate is a heavy yellow powder, exhibiting nothing but small octahedral crystals when sufficiently magnified, it is pure chloroplatinate of potassium, but if tesseral crystals of a dark orange color, transparent by transmitted light, and of relatively large size, are visible, then the precipitate is contaminated with chloroplatinate of lithium (Jensch). If on the other hand white particles of chloride of sodium are to be seen mixed with the yellow powder, then the quantity of bichloride of platinum used was insufficient, since some of the chloride of sodium has escaped being converted into chloroplatinate of sodium. In this event add some water to the precipitate in the dish to dissolve the chloride of sodium, then a new quantity of bichloride of platinum, evaporate nearly to dryness and proceed as before. The risk of adding too little of the platinum solution may be avoided by employing a solution of approximately known strength, and calculating how much of it will be necessary to transform the whole of the weighed quantity of chlorides taken into chloroplatinates.

When the yellow precipitate is of proper appearance, transfer it from the dish to the filter, by means of the filtrate rather than with alcohol, wash it with alcohol of 0.86 specific gravity, and dry at 100° until it ceases to lose weight; or, better, wash with a mixture of 4 parts of such alcohol and 1 part ether. Such a mixture of alcohol and one-fifth or one-sixth ether — in which the chloroplatinate of potassium is as good as insoluble — is used by many chemists. It was first proposed by H. Rose. (*Berzelius's Lehrbuch der Chemie*, 4 *Aufl.*, **10**. 73). In order to be certain that no potassium is left in the solution, add some water to the filtrate, and some bichloride of platinum, together with some chloride of sodium, in case the proportion already present is supposed to be small. Evaporate the mixture nearly to dryness upon a water bath no hotter than 75°, and treat the concentrated liquor in the manner already described. In case any precipitate of chloroplatinate of potassium is formed, collect it upon a special filter and add its weight to that of the principal precipitate. The object in adding chloride of sodium is to prevent the decomposition to which bichloride of platinum is subject when its alcoholic solution is evaporated. The temperature is kept at 75° during the evaporation of the alcoholic liquor for a similar reason. (Compare A. Mitscherlich's precautions, below). — In evaporating the original mixture of bichloride of platinum and potassium solution upon the water bath, care must always be taken not to push the process to dryness, or so far as to expel any water of crystallization from the chloroplatinate of sodium which is to remain dissolved.

When properly executed the process yields good results, though there is usually a small loss of potassium due to the sparing solubility of the precipitate in alcohol. On the other hand, it must be remembered that the air of a laboratory often contains ammonium salts which may go to form chloroplatinate of ammonium, and so increase the weight of the potassium compound. According to Fresenius, repeated experiments have shown that in this method of separating potassium from sodium, the amount of potassium found is always a little less than it should be. When the process is properly conducted, however, the loss of potash ought not to exceed 1 per cent. — In order to avoid the possibility of decomposing any of the bichloride of platinum by evaporating the alcoholic solution, A. Mitscherlich (*Journ. prakt. Chem.*, 1861, **83**. 460) prefers to add bichloride of platinum to a concentrated aqueous solution of the mixed chlorides of potassium and sodium and without adding any alcohol, to collect and wash the precipitate upon an unweighed filter. He then evaporates the filtrate and wash water upon a water bath, dissolves the residue in a very small quantity of water, adds alcohol, collects the new precipitate of chloroplatinate of potassium upon a second unweighed filter and washes it with alcohol. After having been dried, the precipitates are carefully removed from the filters, dried at 100° and weighed, while the two filters are burned, the ashes leached with water and the chloride of potassium thus obtained determined by itself.

The quantity of sodium in the substance analyzed is usually estimated "by the difference," *i.e.*, by subtracting from the weight of the mixed chlorides taken that of the chloride of potassium, which corresponds to the chloroplatinate obtained. Some chemists, however, prefer to add oxalic acid to the filtrate from the chloroplatinate of potassium, to evaporate to dryness, to ignite, to take up with water the Chloride of Sodium, and to weigh it as such. If this is to be done, it is still essential to make the special secondary trial for potassium as above described, else any potassium left in the filtrate will be weighed as chloride of sodium. Instead of estimating the sodium as chloride, sulphuric acid may be added to the filtrate from chloplatinate of potassium, the mixture evaporated to dryness and ignited, the residue extracted with water and evaporated, and weighed as Sulphate of Sodium. (A. Mitscherlich, *loc. cit.*) The best way is to spare no pains in determining the potassium, and to take the sodium from the difference.

By good rights, the compound of potassium and sodium to be examined should be a chloride, though it is possible to separate sodium from potassium in this way, even in presence of boracic, phosphoric, sulphuric and other acids soluble in alcohol, as will be explained below, under Method C. For the methods of converting other salts to the state of chlorides, see Chloride of Potassium, and the methods of separating acids from metals, as given in the finding lists of the appendix.

When the amount of chloplatinate of potassium to be collected and weighed is very small, it is better not to attempt to collect it upon a weighed filter. Instead of that it should be collected upon a very small unweighed filter, and ignited in a crucible after it has been washed and dried. The ignition, which may be made in a platinum crucible, must be conducted with special care, in the manner described under Chloroplatinate of Ammonium. The residue left after the ignition is carefully treated with hot water, the solution of chloride of potassium thus obtained is decanted off from the heavy powder of metallic platinum, and the latter is washed, dried, ignited and weighed in the crucible. Each atom of platinum found corresponds to two atoms of potassium. A small quantity of platinum will remain adhering to the crucible, so that the weight of the latter will be greater after the experiment than before, but the accuracy of the results is not impaired, and no serious harm is done to the crucible. The decanted liquid will be colorless if the decomposition of the chloplatinate is complete; no tinge of yellow is permissible. The metallic platinum should be washed until the washings cease to give a reaction for chlorine when tested with nitrate of silver. — Quantities of the chloroplatinate which weigh less than 0.03 grm. can be safely decomposed by ignition in this way, but with larger quantities it is not easy to obtain good results. The melted chloride of potassium so envelops the carbon of the filter that the latter cannot be thoroughly burned, not even after it has been washed with water. It is, moreover, difficult to complete the decomposition of the chloroplatinate; to facilitate the decomposition a small quantity of oxalic acid may be made to act upon the substance in the crucible. If this be done, let the crucible become cold after the process has been pushed so far that the filter has been burned to ashes, throw in a minute fragment of pure oxalic acid, cover the crucible and ignite gently at first, but afterwards at a strong red heat. By the aid of oxalic acid it is possible to decompose the chloplatinate in a porcelain crucible. A still better way is to ignite the precipitate in a current of hydrogen at a temperature so low that the chloride of potassium cannot fuse. This reduction may be effected in a crucible of platinum, into which hydrogen is admitted through a hole in the cover. But in order to avoid losing any particles of the compound in the current of gas, the chloroplatinate should be partially decomposed by carefully heating it by itself before hydrogen is admitted to the crucible. After the hydrogen has acted, wash out the chloride of potassium and ignite the residue in the air to destroy the last particles of carbon from the filter.

Method B. A modification of the foregoing method has been proposed by Finkener (*Poggendorff's Annalen*, 1866, **129.** 637). It depends in part upon the insolubility of the chloroplatinate in a strong solution of chloride of ammonium, and is applicable to the separation of potassium from sodium when this metal is present in the form of a sulphate. — The details are as follows: — Dissolve the mixed sulphates in a small quantity of water, add some chlorhydric acid, and enough bichloride of platinum to color the liquid intensely yellow; dilute the mixture with so much water, that the whole of the chloroplatinate of potassium will dissolve on boiling, and evaporate the solution upon a water bath to a small volume, so that a pasty, but not solid residue will be left when the dish is allowed to cool. Care must be taken not to leave the dish upon the water bath after the liquid has once been sufficiently evaporated. Stir into the cold residue, at first by small portions, 15 or 20 times its bulk of a mixture of 2 volumes alcohol of 0.8 specific gravity, and 1 volume of ether; collect the mixed precipitate of chloroplatinate of potassium and sulphate of sodium upon a filter and wash it with a mixture of alcohol and ether, such as was poured upon the residue in the dish, until the washings are colorless. Then wash the matter in the filter with a cold, saturated, aqueous solution of chloride of ammonium, until the washings no longer give any reaction when tested for sulphuric acid. If the filter be filled completely full once or twice with the solution of chloride of ammonium, so as to thoroughly saturate the paper and the precipitate, it will be easy to wash out the whole of the sulphate of sodium. The chloroplatinate of ammonium formed by the action of the chloride of ammonium upon any portions of bichloride of platinum which the precipitate may have retained does no harm. — To determine the potassium in the precipitate, dry the filter, with its contents, in a large porcelain crucible, at a temperature somewhat above 100°, and heat the crucible until the filter has charred, but has not yet begun to glimmer, then lead a current of hydrogen into the crucible, and continue to heat it as long as the escaping gas yields a white cloud when tested with ammonia. The reduction by hydrogen is tolerably rapid at temperatures as low as 240°. Treat the residue with hot water, separate the chloride of potassium from the metallic platinum and charred paper by filtration, wash thoroughly, evaporate to dryness and ignite, to expel any trace of chloride of ammonium; then weigh the Chloride of Potassium directly, or estimate the chlorine in it by titration with a standard solution of nitrate of silver (see Chloride of Silver).

In case the precipitate of chloroplatinate of potassium is small it may be dissolved off the filter with boiling water, the solution evaporated to dryness in a porcelain crucible, the residue ignited to expel chloride of ammonium, and then reduced by means of hydrogen. The chlorine in the residue may then be estimated volumetrically, as Chloride of Silver, without further trouble. — The sodium, which will be found for the most part in the chloride of ammonium washings, may be recovered by evaporating these, together with the alcoholic filtrate, but the operation is very troublesome. The process gives tolerably satisfactory results in spite of the fact that the chloride of ammonium decomposes and dissolves a little of the potassium precipitate, for a trace of sodium is always retained by the potassium precipitate. For a quantitative discussion of the several sources of error see Finkener's memoir. — The process can be applied for separating potassium from lithium, as well as from sodium. In case lithium and sodium are both present, the first filtrate, *i.e.*, the one obtained before washing with chloride of ammonium, may be treated to remove the excess of platinum, and the lithium may then be precipitated as Phosphate of Lithium.

Potassium may still be estimated by this method in presence of arseniate, borate, and phosphate of sodium, or the sulphates of Mg, Zn, Mn, Fe, Ni, Co, Cu and Al. It is only necessary to mix a little chlorhydric acid with the first portions of chloride of ammonium used for washing the mixed precipitate of chloroplatinate and sulphate. When the mixture to be analyzed contains no sulphate, but only nitrates, borates, phosphates, arseniates, etc., enough sulphuric acid must be added to convert the whole of the metals present into sulphates. The presence of a moderate excess of sulphuric acid does no harm. — It is not safe to attempt to reckon the potassium from the quantity of platinum left after the ignition and reduction of the precipitate. The result might readily come out too high, especially if much sulphate of sodium were present, or a large excess of bichloride of platinum were used, or the evaporation so conducted that comparatively large crystals of sulphate of sodium should form and enclose some of the platinic mother liquor. The best way in case the potassium is to be calculated from the weight of the platinum is to omit altogether the washing with chloride of ammonium, and to proceed as in Method C.

According to Stohmann (*Zeitsch. analyt. Chem.*, 1866, **5**. 307) there is no difficulty in applying Method A for the separation of potassium from barium, calcium and magnesium, since the chloroplatinates of all these metals are soluble in alcohol. In case a mixture of sulphates and chlorides of potassium, calcium and magnesium is to be analyzed, add chloride of barium, drop by drop, to the boiling solution, until all the sulphuric acid is precipitated, filter, and treat the filtrate, or a measured portion of it, with bichloride of platinum in the usual way. A good deal of bichloride of platinum will be required in order to convert the whole of the foreign metals into chloroplatinates, but it is readily recovered from the filtrate.

Method C. To estimate potassium in any of its compounds proceed as follows: — If the potassium salt contain only a volatile acid, like acetic, nitric or cyanhydric acid, the solution may simply be mixed with an excess of chlorhydric acid, evaporated to dryness to convert the salt to Chloride of Potassium, and treated with bichloride of platinum, as in Method A. — If the substance to be analyzed is a bromide or iodide of potassium, treat it with chlorine water, evaporate to dryness, redissolve the residue in chlorine water, and again evaporate to expel the last traces of bromine or iodine. — Instead of proceeding with the analysis precisely as in Method A, some chemists prefer to omit the evaporation after the addition of the bichloride, and to wash with a mixture of alcohol and ether, as has been already remarked. H. Rose directs that about 1 grm. of chloride of potassium be taken, that it be dissolved in 15 c. c. of water, and that after the addition of the requisite quantity of bichloride of platinum a mixture of 75 c. c. absolute alcohol and 15 c. c. of ether be added to the liquid. In order to prevent the evaporation of the ether place the mixture under a bell glass set upon a plate of ground glass which has been smeared with glycerine, and leave it at rest for 12 hours, then filter, wash with a mixture of alcohol and ether and weigh.

A modification of this process has been devised by Teschemacher and D. Smith (*Zeitsch. analyt. Chem.*, 1869, **8**. 90) for the rapid commercial analysis of saltpetre and other potassium salts: — Weigh out 500 grains of the salt, dissolve in water and dilute the solution to the volume of 5000 fluid grains; measure out 500 fluid grains of the liquid, and dilute again to 5000 grains. Take 1000 fluid grains of the diluted liquid, add to it 50 fluid grains of chlorhydric acid, if the salt be not already a chloride, and wash the mixture into a dish. Heat the liquid, which should amount to about 1500 grains in all, almost to boiling, add to it a large excess of a solution of bichloride of platinum, so that as much as 20 grains of platinum may be present, and evaporate so far upon a water bath that the liquid would stiffen to a pasty mass if the dish were removed from the bath and allowed to cool. Without allowing it thus to stiffen, quickly pour upon the concentrated liquid 500 or 600 fluid grains of rectified methyl alcohol of 85 per cent, mix the materials by giving a rotary motion to the dish, cover the dish, and let the mixture stand for 5 minutes. The chloroplatinate of potassium will separate in the form of large crystalline scales, from which the supernatant liquid can be readily and completely decanted. Pour the liquid into a comparatively large filter — large enough to hold 400 or 500 fluid grains, and wash the crystals in the dish twice by decantation with methyl alcohol; then by means of a wash bottle, and without touching them with a rod lest they be broken, wash the crystals out of the dish into the filter, and wash them there thoroughly with methyl alcohol. After the filter and contents have been dried, the precipitate can be removed from the filter so completely that the latter can be burned to ashes by itself, and the ashes and dried precipitate weighed together in a crucible. — The process yields excellent results. In describing it, Teschemacher and Smith take occasion to deny the allegations of Chalmers and Tatlock (*loc. cit.*, p. 88) that on account of the frequent impurity of the bichloride of platinum employed in laboratories the processes depending upon the principle now in question are unreliable, and apt to yield results by which the potassium is estimated 1 or 2 per cent higher than the truth.

In case the potassium be combined with a non-volatile acid, such as phosphoric, arsenic, boracic or sulphuric acid, which is soluble in alcohol, or in mixed alcohol and ether, make a strong aqueous solution of the salt, add some chlorhydric acid, then an excess of bichloride of platinum, and a quantity of the strongest alcohol, or of mixed alcohol and ether, let the mixture stand for a number of hours and filter, etc., as in Method A. According to Finkener, for 1 grm. of the potassium salt 30 c. c. of chlorhydric acid of 1.05 specific gravity, 150 c. c. of anhydrous alcohol, and 25 c. c. of ether may be taken. After the mixture has stood under a bell glass for several hours, it may be collected on a weighed filter and washed with a mixture of chlorhydric acid, alcohol and ether, in the above given proportions, until the filtrate is colorless, and afterwards with alcohol and ether. The method can be applied for separating potassium from sodium and lithium, as well as for estimating potassium. Compare Method B.

Method D. To separate potassium from magnesium and sodium, Scheerer (*Annalen Chem. und Pharm.*, **112**. 117) weighs the mixture in the form of anhydrous sulphates, divides the aqueous solution into two measured portions, and precipitates chloroplatinate of potassium in one portion of the liquid, and Phosphate of Magnesium and Ammonium in the other. By calculating both the potassium and the magnesium found as sulphates, and subtracting the sum from the weight of the original mixed sulphates, the weight of the sulphate of sodium will be obtained. — Instead of dividing the liquid, Rube (*Journ. prakt. Chem.*, **94**. 117) precipitates chloroplatinate of potassium from the whole of the dissolved sulphates, and evaporates the filtrate a second time to make sure of the whole of the potassium. He then adds to the filtrate a solution of chloride of ammonium to remove the excess of platinum, filters off the chloroplatinate of ammonium, and throws down Phosphate of Magnesium and Ammonium in the filtrate.

Method E. To estimate platinum in platinum compounds, and to separate platinum from the metals above enumerated: — Add caustic potash to the solution until the free acid, if any there be, is nearly neutralized, then add a slight excess of a strong solution of chloride of potassium, and a comparatively large quantity of absolute alcohol. In case the solution be very dilute, it should be concentrated by evaporation before adding the alcohol. After the mixture has been allowed to stand for 24 hours, collect the precipitate upon a weighed filter, as directed in Method A. — If no metal other than platinum be present, wash at once with alcohol of 70 per cent, but if other metals are contained in the mixture wash the precipitate first with a saturated solution of chloride of potassium to remove these metals, and afterwards with the alcohol. After the precipitate has been dried and weighed at 100°, reduce it to metallic platinum by means of hot hydrogen gas. See Platinum Compounds. — The weight of platinum in the precipitate cannot be calculated from the weight of the latter, since it is impossible to wash out all the chloride of potassium by means of spirit without dissolving a portion of the chloroplatinate as well. The process is preferable to that which depends upon the insolubility of Chloroplatinate of Ammonium, since the latter is rather more soluble in alcohol than chloroplatinate of potassium, and because with the potassium salt there is less risk of losing material in the subsequent process of reducing to metallic platinum.

Properties. The precipitate occurs either as a lemon colored powder, or as small reddish-yellow octahedrons. It is difficultly soluble in cold water, but more readily in hot water. It is almost completely insoluble in absolute alcohol, and still less soluble in a mixture of alcohol and ether; in ordinary alcohol it is somewhat, though still sparingly, soluble. According to Fresenius, 1 part of it dissolves in 12,000 parts of absolute alcohol, 3,800 parts of alcohol of 76 per cent, and 1,000 parts of alcohol of 55 per cent. In alcohol acidulated with chlorhydric acid it is considerably more soluble. It dissolves readily in caustic potash. According to Andrews it retains a little water (0.0055 of its weight) even when heated to temperatures considerably above 100°. When left in the air, or heated to 100° it undergoes no change, but at an intense red heat it suffers decomposition, some chlorine being expelled, and a mixture of metallic platinum and chloride of potassium being left. This decomposition is never complete, however, if any considerable quantity of the compound is ignited. Even after long continued fusion of the mass, a little of the chloroplatinate will always remain undecomposed. To reduce this last trace, the substance must be ignited in an atmosphere of hydrogen, or with oxalic acid, as has been described above. (See also Platinum Compounds.)

The composition of the chloroplatinate is:—

K_2 —	78.2 —	16.01
Pt —	197.4 —	40.40
Cl_6 —	213.0 —	43.59
	488.6	100.00

2KCl —	149.2 —	30.54
$PtCl_4$ —	339.4 —	64.66
	488.6	100.00

Principle II. Power of decomposing organic substances.

Application. Estimation of carbon, hydrogen and nitrogen directly, in one and the same quantity of material.

Method. A weighed quantity of the substance to be analyzed is shaken into a porcelain tube which is filled with a mixture (containing about 8 grms. of platinum) of pumice stone and chloroplatinate of potassium, and the air is expelled from the apparatus by a stream of tolerably pure nitrogen. To the front of the tube there is attached first a tube charged with anhydrous phosphoric acid, then three sets of bulb tubes charged with solutions of nitrate of lead, stannous chloride, and caustic potash respectively. The lead and tin solutions are as concentrated as possible, while the potash lye is such as is ordinarily employed in estimating Carbon. In case there is no hydrogen in the substance to be analyzed, the phosphoric acid tube is omitted. — After the apparatus has been proved to be tight the porcelain tube is heated to redness in the places where none of the organic substance has been placed; afterwards the mixture of substance and chloroplatinate is gradually heated, so that the substance may be slowly decomposed. The chloroplatinate of potassium is employed as a substitute for free chlorine, on the ground of its being an easily decomposable chlorine compound. It has the merit of being free from water, and inalterable in the air. When heated by itself it suffers but little decomposition at a low red heat; but when heated to redness in contact with organic substances it is completely decomposed. Carbonic acid, chlorhydric acid, carbon, as well as water and hydro-carbons in some cases, are the products of the reaction. The water, chlorhydric acid and carbonic acid are collected and weighed in the reagents above described. And in case hydrocarbons or chloride of carbon are generated they are burned (compare Carbon, Method 7), and their carbon weighed as carbonic acid.

When the porcelain tube has been heated throughout to redness, and no more bubbles of gas are seen in the nitrate of lead bulbs, the tube is swept clean by a slow current of nitrogen and the fire is extinguished. The object of the solution of stannous chloride is to absorb the excess of chlorine resulting from the decomposition of the chloroplatinate. After the absorption bulbs have been weighed, those which contain potash are reattached to the porcelain tube, the latter is reheated, and a stream of oxygen is passed through it to consume the free carbon left in the tube at the close of the previous reaction. In case any nonvolatile chloride of carbon condense in the tube some oxide of copper should be placed in the tube before the latter is reheated. At the end of the experiment pass a current of chlorine into the hot tube to revivify the chloroplatinate of potassium, so that the tube shall be left in readiness for the reception of matter for a new analysis. (A. Mitscherlich, *Zeitsch. analyt. Chem.*, 1868, **7**. 272.)

Chlorous Acid.

Principle. Oxidizing power.

Applications. Estimation of chlorous acid and chlorites.

Method. See Chlorine and Chlorates,—notably the paragraphs on action of a chlorate upon a ferrous salt, and upon nitrous acid.

Cholesterin.

Principle. Solubility in ether and inertness as regards potash.

Application. Separation of cholesterin from fats, and estimation of cholesterin in animal and vegetable substances.

Method. Treat the substance to be examined with ether, as long as anything dissolves; remove any water which may settle from the ethereal solution, and decant or filter the latter so that it may be perfectly clear. Distil and evaporate the ethereal solution to dryness upon a water bath, and weigh the residue. Boil the residue, or a weighed portion of it, with an excess of a clear concentrated alcoholic solution of caustic potash upon a water bath for several hours, then distil off the alcohol and dissolve the residue in water to a thin solution, shake this solution with ether, allow the mixture to settle, and decant the ether, taking care to repeat the process several times. The ethereal solution thus obtained contains the cholesterin in a condition of almost absolute purity. The weight of the cholesterin is found by evaporating the solution to dryness. In case the cholesterin solution were not pure the residue obtained by saponification might be shaken, while still warm, with dilute potash lye, and again treated with ether after the mixture had become cold. None of the soap will go into solution in the ether unless there is a lack of water or alkali in the mixture. By treating the soap solution with chlorhydric acid after the removal of the cholesterin, and washing the acid liquor with ether, the fatty acids may be dissolved out and weighed after the ether has been evaporated, or after their conversion into sodium salts. (Hoppe-Seyler, *Zeitsch. analyt. Chem.*, 1866, **5**. 422.)

Chromates.

For a method of analyzing those metallic chromates which are decomposable by heat, see under Chromic Acid, decomposition of by heat.

Chromic Acid.

[Compare biChromate of Potassium.]

Chromic acid may be estimated as Chromate of Barium, Chromate of Lead, and Chromate of Mercury, or better, by determining the amount of efficient oxygen contained in it. It may also be reduced to the condition of sesquioxide and thrown down as Hydrate of Chromium.

To separate chromic acid from sesquioxide of chromium, determine in one portion of the material the chromic acid, according to one of the methods described below, then reduce another portion as below, and estimate the whole of the chromium as Hydrate of Chromium, or oxidize this second portion, as described under Chromate of Sodium, and estimate the whole of the chromium as Chromic Acid.

Principle I. Oxidizing power.

Applications. Reduction of CrO_3 as a preliminary in the estimation of Cr, or to the separation of H_2SO_4 from CrO_3 (Methods 1 and 9). Valuation of chromates, such as bichromate of potassium and chrome yellow, separation of Cr from Na, K, NH_4, Ba, Sr, Ca, Mg, Co, Ni, Mn, Zn (Method 1, B), Cd, Pb, Ag, Hg, Cu, Bi, Sb, As, Sn, Au, and Pt. Separation of Pb from CrO_3. Estimation of Fe in compounds and ores of iron. Estimation of Arsenious Acid, ferrous salts, and other compounds oxidizable by nascent chlorine (Method 1, B). Estimation of nitrous acid in presence of nitric acid. Valuation of nitrite of sodium (Method 5). Valuation of crude ferrocyanide of potassium (Method 6).

Methods.

1. *Reduction of Chromic Acid by Chlorhydric Acid.*

A. *Gravimetric.* Chromic acid or a chromate may be readily reduced to the condition of sesquichloride of chromium by boiling the dry substance with an excess of concentrated chlorhydric acid. The process is preferable to those in which alcohol, sulphuretted hydrogen, or sulphurous acid are employed to effect the reduction. The solution of sesquichloride of chromium obtained should be largely diluted with water before any reagent is added to it. To separate chromium from the alkali-metals in such a solution, add ammonia water at once to throw down Hydrate of Chromium; but to separate chromium from sulphuric acid, or from barium, precipitate these substances as Sulphate of Barium before attempting to throw down the chromium. — To separate Cr from Co, Ni, Mn and Zn, add Carbonate of Barium to throw down Hydrate of Chromium (or better, adopt the volumetric method described in § B).

B. *Volumetric.* See Arsenious Acid, oxidation of by nascent chlorine; and Chlorine, action of on As_2O_3 and on KI, for descriptions of apparatus proper for effecting the decomposition and absorbing the chlorine which is set free. See also biChromate of Potassium.

2. *Reduction of Chromic Acid by a Ferrous Salt.*

A. To estimate chromic acid, acidify with sulphuric acid the not too dilute solution of the chromate to be examined, and add to it the solution of a definite quantity of iron, prepared as explained under Chlorine (action of on ferrous salts). The mixed solution should be strongly acid, and should of course contain an excess of the ferrous salt. The reduction of the chromic acid by the iron solution is instantaneous. By means of a standard solution of Permanganate of Potassium, or of biChromate of Potassium, proceed to estimate how much of the iron in the mixed solution has been left in the state of a ferrous salt. The difference between the amount of iron thus found and that originally taken, will show how much iron has been changed from a ferrous salt to a ferric salt by the chromic acid in the substance; — the reaction may be supposed to be as follows: —

$$6FeO + 2CrO_3 = 3Fe_2O_3 + Cr_2O_3.$$

In case permanganate of potassium is to be employed, take care to dissolve the standard iron in diluted sulphuric acid, and to dilute largely the solution into which the permanganate is to be poured. Instead of dissolving a fresh piece of metallic iron for each experiment, a weighed quantity of pure sulphate of iron, prepared as described under Chlorine, may be taken; or, in case a considerable number of determinations are to be made all at once, measured portions of a standard solution of ferrous chloride may be used. The method yields excellent results (Schwarz). To determine the chromic acid in dry chromate of lead, place a weighed quantity of this substance in a mortar, pour upon it the standard solution of iron and a quantity of chlorhydric acid, and grind the mixture thoroughly. Add water, and proceed to estimate the residual ferrous salt (Mohr, *Titrirmethode*, 1855, **1**. 240). See further Chromate of Lead.

B. Another way of estimating chromic acid depends upon the titration of the ferric salt formed. For the details of this process see Chlorates, action of upon ferrous salts. To avoid the effects of any interference with the final reaction which might be occasioned by the presence of chloride of chromium in the solution to be titrated, put a corresponding quantity of chloride of chromium into the normal iron solution before standardizing it, so that similar conditions may obtain in all cases. (Braun, *Zeitsch. analyt. Chem.*, 1867, **6**. pp. 63, 54).

C. For the method of estimating iron see biChromate of Potassium.

3. *Reduction of chromic acid by Oxalic Acid.*

A. *Gravimetric.* Treat a mixture of the chromate to be analyzed and oxalate of sodium with sulphuric acid, and collect and weigh the Carbonic Acid formed, or estimate the Carbonic Acid by the loss.

$$2CrO_3 + 3C_2Na_2O_4 + 6H_2SO_4 = 6CO_2 + Cr_2O_3, 3SO_3 + 3(Na_2O, SO_3) + 6H_2O.$$

For each grm. of chromic acid present in the substance 2.25 grms. of oxalate of sodium will be required. Oxalate of ammonium, or oxalate of barium may be used instead of oxalate of sodium in case either of the fixed alkalies are to be estimated in the residue. For the details of the process see binOxide of Manganese (Vohl, *Annalen Chem. und Pharm.*, **63**. 398).

The process may be employed in connection with the so-called method of "Limited Oxidation" (see Carbon, p. 75), in which certain organic substances are oxidized by chromic acid, while others are unacted upon. It is claimed for the method of limited oxidation, as thus applied, that it will be found useful in studying the true composition of many organic products, such as quinine, thein, morphine, the essential oils, etc., and that by means of it the nature and quantity of impurity in an organic substance may be determined. — As has been already stated under Carbon, the acids of the acetic or fatty series, when once produced by the action of dilute chromic acid on organic substances, withstand all further oxidation by that reagent. But the oxidation of any organic substance to the state of fatty acid takes place with almost mathematical precision, as is readily proved by the close agreement between the amount of oxygen actually consumed in experiments for effecting the oxidation, and the amount required by theory. The amount of oxygen consumed might be determined by estimating the sesquioxide of chromium which has resulted from the reduction of the chromic acid, but there are objections to that method, inasmuch as ammonia cannot be used for precipitating the sesquioxide and the washing out of the soda salts or potassium salts, resulting from the use of fixed alkalies, is difficult. Moreover, the atomic weight of chromium is still so uncertain that the calculation ought not to be based upon it in cases where great accuracy is required. A better way is to determine the chromic acid which has escaped reduction, and subtract this amount from the known quantity of chromic acid taken. — The estimation of the chromic acid may be effected by means of oxalic acid, as described above. Determine in the first place the strength of the chromic solution by treating a weighed quantity of it with an excess of oxalic and sulphuric acids, and absorbing and weighing the carbonic acid formed. Then digest a weighed quantity of the organic substance to be analyzed with a weighed excess of the chromic solution, and after the reaction is complete treat the mixture with oxalic and sulphuric acids as before, in order to determine how much chromic acid remained unreduced. The difference between the two amounts of carbonic acid indicates how much oxygen has been consumed.

The following apparatus may be employed:—For the generating vessel provide a flask with an outlet tube fused into its neck. Close the flask with a perforated cork carrying a stop-cock tube, the lower extremity of which does not project far into the flask. To the end of the tube which projects from the flask, connect the top of a bulb pipette, the other end of which passes to the bottom of a small two-necked Woulfe's bottle, containing not quite enough sulphuric acid to fill the bulb of the pipette. This Woulfe bottle serves to dry the gas, which is, however, further dried by passing it through a U-tube containing pumice stone moistened with sulphuric acid. To this U-tube is attached the absorption apparatus, which consists of a pipette and Woulfe bottle, as before, but containing a solution of caustic potash; the second neck of the Woulfe bottle is connected with a set of Liebig's potash bulbs, followed by a small drying tube charged with lumps of caustic potash. The absorption apparatus is of such size that the whole of it when charged only weighs about 190 grammes. To use the apparatus, the absorption apparatus is detached and weighed, after it has been allowed to stand for some time in cold water, and to hang for half an hour in the balance case; or better, in a room of constant temperature. It is then reconnected with the apparatus, and the cold chromic solution is poured into the generating flask. Before charging the flask some fragments of tobacco pipe should be placed in it in order to prevent the liquid from bumping when it subsequently comes to be heated. After the introduction of the chromic solution pour in the oxalic solution, together with some dilute acid, both cold, by means of an ordinary funnel. No loss of carbonic acid need be apprehended at this stage, since some minutes elapse after the mixture is made before any of that gas is evolved. Thrust the cork into the flask and close the stop-cock which it carries. In the course of a few minutes the evolution of carbonic acid commences, and large bubbles of air are forced out through the drying and absorption vessels. After the comparatively large amount of air which the apparatus contains has been expelled, the absorption of carbonic acid is confined almost entirely to the Woulfe bottle. After a short time air will begin to be sucked back into the apparatus through the Liebig bulbs. When this happens heat the generating flask gently, but not enough to force gas beyond the Woulfe bottle. When no more carbonic acid goes forward heat the contents of the generating flask to boiling, so that the steam may drive out the carbonic acid. After the steam has fairly passed into the sulphuric acid for a few seconds remove the lamp, and the moment the sulphuric acid begins to rise in the pipette open the stop cock above the generating flask, so that air may enter and prevent any regurgitation. During the process of boiling out some air will have begun to enter the potash-apparatus. Suck now a little air out of the potash-apparatus, plug its outer end and leave it for a short time to absorb the carbonic acid out of the rest of the apparatus. To remove the last traces of carbonic acid suck a little air through the apparatus. Then detach the absorption apparatus, place it in cold water, and allow it to stand in the balance case, as before, before weighing. The increase of weight represents the carbonic acid. Instead of absorbing and weighing the carbonic acid, it might be estimated from the loss, though less conveniently. (Chapman, *Journal London Chem. Soc.*, 1867, **20**. 227.)

As regards the quantity of material to be taken, it may be said that Chapman operated upon 0.5 to 0.8 grm. of matters like alcohol and butyric ether. The chromic solution contained about 7 per cent of its weight of bichromate of potassium. The mixture of organic matter and chromic solution was allowed to digest for several hours at the ordinary temperature, in a sealed tube if need were, and was afterwards heated for a few minutes to 100° before being transferred to the generating flask. In the case of substances heated in sealed tubes, air must be drawn through the liquid after it has been placed in the flask, in order to remove dissolved carbonic acid.

B. *Volumetric.* Hempel has proposed to titrate chromates by means of a standard solution of oxalic acid. But according to Mohr (*Titrirmethode*, 1855, **1.** 240), the reaction between the two substances is slow and incomplete, unless the solutions employed are concentrated, and made strongly acid with sulphuric acid. The value of the solution of oxalic acid employed has to be determined by means of permanganate of potassium.

4. *Reduction of chromic acid by Stannous Chloride.* Weigh out a small quantity of the chromate to be analyzed (0.3 to 0.5 grm. of bichromate of potassium, for example), place it in a flask, pour upon it a measured quantity of a standard solution of protochloride of tin, and add a drop or two of a solution of iodide of potassium and a little starch paste. The chloride of tin solution is standardized as explained below, and enough is taken that it may be in slight excess as regards the chromic acid. Shake the flask vigorously, and pour into it from a burette a one-tenth normal solution of biChromate of Potassium until a persistent tint of blue is visible throughout the green liquid. — According to Mohr (*Titrirmethode*, 1855, **1.** 268) no good results can be obtained by attempting to work directly upon the substance with the standard solution of stannous chloride. When such a solution is poured from a burette into the solution of a chromate which has been acidulated with chlorhydric acid and treated with iodide of potassium and starch, the blue color of the iodide of starch will disappear, it is true, but the reaction proceeds so slowly that it is hard to determine the precise moment at which it is complete. The results of analyses made in this way not only differ from those obtained by the process above recommended, but they fall below the truth. For the influence of the oxygen dissolved in the water of the solutions see protoChloride of Tin.

The method may be applied to the valuation of chrome yellow (chromate of lead) as follows: — Weigh out one or two grms. of the chrome yellow, place it in a mortar together with some strong chlorhydric acid, rub the mixture with the pestle, and allow a standard solution of stannous chloride to flow upon it from a burette, until the yellow color of the chromate has been completely changed to the green of chloride of chromium. As soon as most of the chloride of lead has settled, decant the supernatant liquid into a flask or beaker, rub the residue again with chlorhydric acid and a little more of the chloride of tin, and again decant. Repeat these operations until the residue in the mortar is perfectly white. Rinse the mortar with chlorhydric acid, add some iodide of potassium and starch paste to the solution in the flask, and titrate with one-tenth normal bichromate of potassium as above. On the addition of the iodide of potassium some yellow iodide of lead will be precipitated at first, but it will soon dissolve again. Instead of a standard solution of stannous chloride, 1.5 to 2 grms. of the double salt of chloride of tin and chloride of ammonium may be weighed out and rubbed in the mortar with the chrome yellow and some chlorhydric acid (Mohr, *loc. cit.*, p. 270).

To standardize the chloride of tin, in the first place, proceed as follows: — Fill a couple of Mohr's burettes, one with the chloride of tin solution, and the other with a one-tenth normal solution of bichromate of potassium. Let 1 c.c. of tin solution flow into a flask, add to it some iodide of potassium and starch paste, and pour in the bichromate of potassium solution until the liquid is blue. The object of this first trial is to enable the operator to dilute the tin solution understandingly. It is well to dilute it to such an extent that each c. c. of it shall be of approximately the same value as a c. c. of the solution of the bichromate. After this dilution has been effected the true standardizing begins. To this end run out 10 or 20 c. c. of the tin solution, add the iodide of potassium and starch, and then the bichromate to blue coloration. Repeat the operation a second time, with a quantity of the tin solution different from that taken for the first trial, or as many times as may be necessary in order to obtain results which agree closely. It is to be remembered that the standard of a tin solution is destroyed by diluting with water which contains air. (Mohr, *Titrirmethode*, 1855, **1.** 257).

5. *Reduction of chromic acid by Nitrous Acid.* To estimate a nitrite in presence of a nitrate Tichborne (*Chemical News*, 1865, **12.** 147) proceeds as follows: — In case the substance to be examined is commercial nitrite of sodium, which is usually contaminated with carbonate and nitrate of sodium, dissolve a weighed quantity of the substance in a tolerably large quantity of water, and estimate the carbonate of sodium with standard sulphuric acid (see Alkalimetry), taking care not to add an excess of the acid. To hit the precise point of neutrality drops of the liquid may be placed upon litmus paper, and the latter allowed to become dry before each new addition of the acid; or a mixture of iodide of potassium and starch paste may be employed as the indicator, since a drop of the nitrite will not color the mixture blue until all the carbonate of sodium has been neutralized. In the next place weigh out 3 grms of pure bichromate of potassium for every 2 grms. of

the nitrite taken, dissolve it in some water in a well stoppered bottle, add an excess of sulphuric acid to the mixture, and cool the flask in a freezing mixture of Glauber's salt and muriatic acid. Cool the nitrite solution in a similar way, and pour it carefully into the chromic acid bottle in such a way that it shall float above the chromic acid without mixing with it to any great extent. Then close the bottle, remove it from the freezing mixture, twirl it around, and let it stand so as to acquire the temperature of the room. In the course of from half an hour to an hour the reaction will be complete. The nitrous acid will have reduced its equivalent of chromic acid, and by collecting and weighing the sesquioxide of chromium which has been formed, the amount of nitrite of sodium in the sample can be estimated. See Hydrate of Chromium for the precautions to be observed in presence of chromic acid.

6. *Reduction of chromic acid by Ferrocyanide of Potassium.*

A. To estimate chromic acid, dissolve from 0.3 to 0.7 grm. of the chromate to be analyzed in water, acidulate with chlorhydric acid, dilute with water to about 150 c. c., and pour into the solution from a burette a standard solution of ferrocyanide of potassium, until the whole of the chromic acid has been reduced, and the ferrocyanide just shows itself to be in excess. The standard solution may be prepared by dissolving 40 grms. of ferrocyanide of potassium in water to the volume of a litre; and the completion of the reaction is indicated when a drop of the chrome solution touched to a drop of a decidedly acid solution of ferric chloride upon a white porcelain plate, colors it greenish. A green point is first seen at the edge of the drop, and quickly increases. Since some experience is required in order to appreciate readily this final reaction, it is well to weigh out two equal quantities of material for each analysis, and to employ one of the quantities for obtaining an approximative result. Almost enough of the ferrocyanide solution to reduce the whole of the chromic acid may then be added at once to the second sample, and the test then applied methodically after the addition of each c. c. of the standard liquor until the proper green tinge appears in the ferric chloride. The reaction between chromic acid and ferrocyanide of potassium may be represented by the following equation: —

$$CrO_3 + 3K_4FeCy_6 + 6HCl = 3K_3FeCy_6 + CrCl_3 + 3KCl + 3H_2O.$$

(Rube, *Journ. prakt. Chem.*, **95**. 53).

B. *For estimating the value of crude ferrocyanide of potassium*, and the fused mass from which it is made, grind up a sample of the fused cake, weigh out ten grms. of the fine, sifted powder, place it in a dish and boil it in about, but no less than, 150 c. c. of water, together with some recently precipitated ferrous carbonate. Wash the contents of the dish into a flask of 250 c. c. capacity, and heat the liquid for half an hour, or an hour, upon a sand bath. Then add as much carbonate of lead to the solution as may be needed to free it from sulphur, cool the contents of the flask by placing it in water, fill it with water to the mark, and shake it vigorously. Throw the mixture into a filter and collect the filtrate in a dry glass. There will be obtained 230 or 240 c. c. of a liquid, every 50 c. c. of which correspond nearly enough to 2 grms. of the dry substance taken. To 50 c. c. of this filtrate add 300 c. c. of cold water decidedly acidulated with sulphuric acid, but not with chlorhydric acid, and proceed to estimate the ferrocyanide of potassium by means of a standard solution of chromic acid, the value of which has been previously determined against pure ferrocyanide of potassium. — The chromic acid solution should be of such strength that 100 c. c. of it represent 2 grms. of ferrocyanide of potassium. The oxidation of the ferrocyanide proceeds rapidly, and is known to be complete when a drop of the liquid touched to a drop of ferric chloride on a white plate no longer gives a green or blue, but a reddish-brown coloration. It is essential to success that oxidizable matters, such as sulphide of potassium, etc., be removed from the solution before adding the chromic acid. But by operating in very dilute solutions and at a low temperature, the action of cyanides and sulphocyanides upon the chromic solution may be avoided. In concentrated and strongly acid solutions the sulphocyanides present in the crude ferrocyanide of potassium would reduce some of the chromic acid. In alkaline solutions the oxidation is found to be irregular. The process is rapid, and yields sufficiently accurate results (E. Meyer, *Zeitsch. analyt. Chem.*, 1869, **8**. 508).

7. *Reduction of chromic acid by Iodide of Potassium.* A description of the attempts of Zulkowsky to estimate chromic acid by adding iodide of potassium to the acid solution, letting the mixture stand during half an hour or an hour, and then titrating the iodine which has been set free with a standard solution of hyposulphite of sodium, will be found in *Journ. prakt. Chem.*, **103**. 351, and *Zeitsch. analyt. Chem.*, 1869, **8**. 74. The process is not yet perfected.

8. *Reduction of chromic acid by Chloride of Ammonium.* See Chromate of Ammonium, instability of.

9. *Reduction of chromic acid by Sulphurous Acid, Sulphureted Hydrogen and Alcohol.* Instead of employing chlorhydric acid alone, as in Method 1, for reducing chromic acid as a preliminary to the precipitation of hydrate of chromium, the reduction may be effected either by heating the solution of the chromate with a mixture of chlorhydric acid and alcohol, or by acidulating the solution of the chromate with chlorhydric acid, and passing sulphuretted

hydrogen through the mixture, or by adding a strong solution of sulphurous acid and heating the mixture gently. According to Genth (*Chemical News*, 1862, p. 32), it is best to employ the sulphurous acid in alkaline solution. Instead, therefore, of acidulating with chlorhydric acid before adding the sulphurous acid, an excess of sulphurous acid should first be added to the solution of the chromate. The mixture should then be carefully heated to boiling, a slight excess of ammonia added to it, and the boiling continued for several minutes. The reduction will be far more rapid and complete than if the solution had been left acid. — For a dry chromate, or for the highly concentrated solution of a chromate, chlorhydric acid is the best reducing agent, as in Method 1. For tolerably strong solutions a mixture of chlorhydric acid and alcohol may be used, or, perhaps better, the solution may be evaporated somewhat, and then treated with chlorhydric acid alone. To decompose chromate of lead, a mixture of strong alcohol and concentrated chlorhydric acid may be employed, and the Chloride of Lead collected and weighed in case the substance to be analyzed contain no other insoluble matter. Native chromate of lead must be very finely powdered or elutriated in order that it may be completely reduced. For dilute solutions, the use of sulphuretted hydrogen or sulphurous acid is to be recommended. The method by sulphuretted hydrogen often comes into play in the separation of chromium from the heavy metals which form insoluble sulphides; sulphuretted hydrogen is passed into the acidulated mixed solution, either at once or after the chromic acid has been reduced with sulphurous acid. When sulphuretted hydrogen alone is employed as the reducing agent, the acid liquid supersaturated with the sulphuretted hydrogen must be allowed to stand in a moderately warm place until the sulphur which is set free has completely subsided. According to H. Rose, the reduction by means of sulphuretted hydrogen, though less rapid than that by sulphurous acid, is to be preferred on the whole, since in presence of sulphurous acid hydrate of chromium is somewhat difficultly precipitable by ammonia. When alcohol is used, the excess of it, and the products of its decomposition, must be expelled from the liquid before the hydrate of chromium can be precipitated with ammonia. The use of sulphurous acid, as above, is common in the analysis of chrome-iron ore. It is noteworthy that chromic acid is not reduced by hyposulphite of sodium.

A volumetric process of estimating chromic acid by means of a standard solution of sulphite of sodium has been described by O'Neill (*Chemical News*, 1862, p. 199). The value of the sulphite solution is determined before each series of analyses by titrating a weighed quantity of pure bichromate of potassium dissolved in water and acidulated with sulphuric acid. The completion of the process is indicated either by the change of color, or by the appearance and disappearance of the blue reaction on the addition of a mixture of iodide of potassium and starch paste. But the results obtained in this way by O'Neill have been shown to be erroneous by Genth (*Chemical News*, 1862, p. 32) and by Oudesluys (*Ibid*, p. 254); the amount of chromium found being much below the truth. — Fresenius (*Zeitsch. analyt. Chem.*, 1862, **1.** 500) has remarked that the process of O'Neill is manifestly bad in view of the observations of Lenssen & Lœwenthal. These chemists have shown that it is true of sulphurous acid as it is of protoChloride of Tin, that when mixed with even a small quantity of chromic acid (or with permanganate of potassium) in presence of water, the oxygen dissolved in the latter suddenly becomes active, and converts an equivalent quantity of the sulphurous acid into sulphuric acid; and this, in spite of the fact that chromic acid can only be completely decomposed by sulphurous acid when the latter is largely in excess. Even when there is much iodhydric acid in the mixture of chromic and sulphurous acids, the dissolved oxygen still becomes active, for chromic acid is not decomposed by iodhydric acid when sulphurous acid is present, except after long standing. But by completely decomposing the chromic acid solution with strong iodhydric acid in the first place, it can be afterwards correctly titrated with sulphurous acid. (Lenssen & Lœwenthal, *Journ. prakt. Chem.*, 1862, **86.** pp. 194, 209).

10. *Reduction of chromic acid by organic Coloring Matters.* See biChromate of Potassium.

Principle II. Decomposition of by heat.

Application. Analysis of compounds of chromic acid and sesquioxide of chromium, and of other hydrated metallic chromates.

Method. Place a weighed quantity of the hydrated chromate to be analyzed in a weighed bulb tube of hard glass, connect a weighed chloride of calcium tube with the bulb tube, pass a current of dry air through the apparatus, heat the bulb tube cautiously till all the water has been expelled from the chromate and absorbed by the weighed chloride of calcium tube, and finally heat the bulb tube to dull redness to decompose the last traces of chromic acid. The loss of weight of the bulb tube represents the sum of the weights of water and oxygen which have been expelled; the gain in weight of the chloride of calcium tube gives the water, and the difference between the weight of water and oxygen and the weight of the water will give the oxygen. Every three equivalents of oxygen found correspond to two equivalents of chromic acid. The sesquioxide of chromium left in the bulb tube may be collected and weighed in case of need. — The process is well adapted for the analysis of hydrated precipitates and basic salts, which from being partially soluble in water cannot be washed, but have to be prepared for analysis by pressure between folds of filter paper. The saline matters which result from mother liquor left adhering to the precipitate, are finally washed away from the residue left in the bulb tube after ignition, and their weight is thus determined and allowed for. (Vogel, *Journ. prakt. Chem.*, 1859, **77.** 484; Storer & Eliot, *Proceedings American Academy*, 1861, **5.** 197.)

Chromate of Ammonium.

Principle. Instability of the salt when heated.

Applications. Separation of Na, K, Ba, Sr, Ca, Mg and Cr, from chromic acid; analysis of chromate of ammonium.

Method. Mix the finely powdered, dry chromate with 4 or 5 times its weight of dry, powdered chloride of ammonium in a porcelain crucible, and heat the mixture cautiously until the whole of the chloride of ammonium is expelled. The residue will contain insoluble sesquioxide of chromium and soluble chlorides of the metals enumerated (Bahr, *Journ. prakt. Chem.*, **60.** 60). A single ignition with the chloride of ammonium is usually sufficient to decompose the chromate completely, but it is well, after having weighed the residue, to reignite it with a fresh quantity of chloride of ammonium, and observe whether the second residue has the same weight as the first.

For the analysis of chromate of ammonium it will be sufficient to ignite a weighed quantity of the salt, and weigh the residual sesquioxide of chromium. To separate sesquioxide of chromium from chromic acid, ignite one portion of the substance with chloride of ammonium, and determine the total chromium, and in another portion estimate the Chromic Acid by some one of the methods given under that head.

Chromate of Barium.

Principle 1. Insolubility in water and in saline solutions.

Applications. Separation of Cr from Mg, Al, Fe, Co, Ni, Mn and Zn (Methods A and B). Separation of Ca, Ba and Sr (Method C). Volumetric estimation of SO_3 (Method D).

Methods. In case the chromium is not already in the form of chromic acid, convert it to that state by some one of the methods of oxidation described under Oxide of Chromium. Then proceed as follows: —

A. Neutralize the solution exactly, add a solution of nitrate of barium and let the precipitate settle, then collect, wash, dry and weigh it after gentle ignition. The common impression that it is better to throw down chromic acid in the form of chromate of lead rather than in that of chromate of barium is incorrect (H. Rose).

According to Richards (*American Journ. Sci.*, 1869, **48.** 200), chromate of barium, though soluble to an appreciable extent in pure water, is insoluble in tolerably strong saline solutions, even in presence of acetic acid. On the other hand chromate of barium, like the sulphate, is liable to drag down chloride of barium and nitrate of barium, and consequently requires to be thoroughly washed. The washing should be effected by means of some saline solution competent to dissolve out the contaminating salts without acting upon the chromate.

According to A. H. Pearson (*American Journ. Sci.*, 1869, **48.** 198), a solution of acetate of ammonium may be employed with advantage in the sense indicated by Richards. Pearson proceeds as follows: — The mixed solution of chromic acid and magnesium, for example, which may contain much saline matter resulting from chlorate of potassium and nitric acid used for oxidizing the chromium, is diluted with water, a little over neutralized with ammonia, and then treated with enough acetic acid to make it slightly acid. After the acidulated solution has become cold, a solution of chloride of barium is added to it in slight excess, and the mixture left at rest for 10 or 12 hours. The clear supernatant liquor is then decanted into a filter, the precipitate is washed by decantation with a cold solution of acetate of ammonium, — prepared just previous to use, if need be, by adding acetic acid to diluted ammonia water, — and finally transferred to the filter, rinsed with water, dried, and heated in a crucible to expel the last traces of water and of the ammonium salt. The mixture of precipitate and liquid must be allowed to stand several hours before filtering, lest some of the chromate pass through the pores of the filter and render the filtrate cloudy.

B. Gibbs (*American Journal of Science*, 1861, **39.** 58) directs that the alkaline solution, in which chromic acid is to be estimated, be neutralized with acetic acid and then mixed with a slight excess of acetate of barium. In case sulphate and chromate of barium be thrown down together, the chromic acid may be reduced by boiling with chlorhydric acid and alcohol, and the sulphuric acid and oxide of chromium subsequently separated in the usual way.

C. *For separating Ba, Ca and Sr*, Fleischer (*Chemical News*, 1869, **19.** 290) precipitates the three metals together as carbonates (see Carbonate of Calcium), and weighs the dry mixed precipitate. He then dissolves the precipitate in an excess of standard chlorhydric acid, dilutes with water, heats to expel carbonic acid, as explained under Acidimetry, and titrates the excess of acid with one-half normal ammonia water to determine how much of the acid has been neutralized by the mixed carbonates. The neutralized solution is then mixed with bichromate of potassium solution and an excess of ammonia water, the precipitate collected and washed and the amount of chromate of barium contained in it is determined volumetrically by means of ferrous sulphate, as described under Chromic Acid. The amounts of calcium and strontium in the mixture may then be calculated from the weight of the mixed carbonates and the amount of standard acid expended, after subtracting that which was neutralized by the carbonate of barium.

D. *For estimating Sulphuric Acid*, Wildenstein (*Zeitsch. analyt. Chem.*, 1862, 1. 323) proceeds as follows:—Put the substance to be analyzed in a short-necked flask of about 200 c. c. capacity, dissolve it in from 45 to 55 c. c. of water, heat the liquid to boiling and allow a standard solution of chloride of barium (1 c. c. equal about 0.015 grm. SO_3) to flow into the flask from a burette, until all the sulphuric acid has been precipitated; then boil for a minute or a minute and a half. Take care always not to add any great excess of the chloride of barium over and above what is actually needed to precipitate the sulphuric acid. Next add, in case the solution is acid, enough dilute ammonia water to a little more than neutralize the liquid, and to the still boiling mixture, no matter whether it be clear or cloudy, pour in from a burette a standard solution of neutral chromate of potassium until the excess of chloride of barium is all thrown down. — The solution of chromate of potassium should be of such strength that 1 c. c. of it is equal to about 0.01 grm. of sulphuric acid; it should be added by half c. c. to the solution under examination until the supernatant liquid in the flask shows a distinct yellow color. No special indicator is needed to show when the titration is finished, since the prevalence of the yellow color of the neutral chromate of potassium is of itself sufficient to mark the completion of the process. Even after the addition of the first half c. c. of the chromate of potassium the liquid will become clear enough, after a little shaking, to show whether it is colorless, provided no great excess of chloride of barium is present, and as successive portions of the chromate of potassium are added the liquid clears itself more and more readily, so that there is no difficulty in determining when the process is finished, and but a few minutes are needed for the titration. — To make assurance doubly sure single drops of the standard solution of chloride of barium may be dropped into the now clear liquid, which has been made yellow by addition of chromate of potassium, until it becomes colorless. Usually a very few drops will accomplish this result, and at the most no more than 0.3 c. c. will be required. Beginners not yet acquainted with the process may filter the yellow liquid before proceeding to make it colorless by adding the drops of chloride of barium. The process is easy and rapid and yields satisfactory results for technical purposes.

Principle II. Decomposition of by alkaline carbonates.

Applications. Separation of Ba from CrO_3.

Method. Boil the finely divided chromate with the solution of an alkaline carbonate, pour off the liquid and boil the residue a second time with a fresh quantity of the alkaline solution. Filter off the insoluble Carbonate of Barium, neutralize the filtrate with nitric acid and throw down the chromic acid as Chromate of Mercury. Two consecutive boilings are sufficient to completely decompose chromate of barium, and change it to Carbonate of Barium and chromate of the alkali. But with a single boiling the decomposition is incomplete, for when mixed with any considerable quantity of a chromate of an alkali the solution of the alkaline carbonate ceases to act upon chromate of barium. — The decomposition of the chromate of barium might be effected at the ordinary temperature, but the operation would require a long time and repeated pouring away and renewal of the alkaline liquid. It is to be observed that carbonate of barium may be completely changed to chromate of barium, even at the ordinary temperature, by means of a solution of normal chromate of potassium. — By fusion with an alkaline carbonate (see Chromate of Sodium) the decomposition of chromate of barium is far less complete than by the wet method above described.

Properties. Precipitated chromate of barium is of a light lemon-yellow color, slightly soluble in water, but as good as insoluble in various saline solutions, as has been stated above. It dissolves readily in the mineral acids, but not in acetic acid. It is not precipitated from solutions containing citrate of sodium. It is somewhat more readily decomposed than sulphate of barium by boiling solutions of the alkaline carbonates. The composition of chromate of barium is:—

BaO	153	60.35
CrO_3	100.5	39.65
	253.5	100.00

Chromate of Bismuth.

Principle. Insolubility in water.

Applications. Estimation of Bi in all compounds of that metal which dissolve in nitric acid to form nitrate of bismuth, provided no substance which, like citric acid, prevents the precipitation, be present. Separation of Bi from Cd.

Method. The solution to be analyzed must be as nearly neutral as possible. Hence in case it contains much free nitric acid evaporate it upon a water bath until the excess of acid is expelled. Then pour it into a warm solution of bichromate of potassium contained in a porcelain dish, taking care to stir the mixture and to have the chromate of potassium slightly in excess. Use water acidulated with nitric acid to wash out the dish that contained the bismuth solution. Boil the contents of the dish for ten minutes, with stirring; then decant the supernatant liquid into a weighed filter and wash the precipitate in the dish by boiling it repeatedly with fresh portions of water, and decanting the liquid into the filter. After awhile transfer the precipitate to the filter, wash it thoroughly with boiling water, dry at about 112°, and weigh. — In the case of separating bismuth from cadmium the filtrate from the chromate of bismuth is concentrated by evaporation and the cadmium precipitated as Carbonate of Cadmium. — The precipitate which forms when the bismuth solution is added to the bichromate should be orange-yellow colored and dense throughout. If it is flocculent and has the color of yolk of egg that is a sign that too little chromate of potassium has been used. In that event add some more of the bichromate, and boil the mixture until the precipitate has the proper appearance. Care must always be taken, however, to avoid adding too much of the bichromate. The process yields very satisfactory results. (Lœwe, *Journ. prakt. Chem.*, **67**. 464.)

According to Fresenius the volumetric method of Pearson (*Phil. Mag.* (4.), , 204), depending on the principle in question, is not to be commended, since it is based upon the mistaken assumption that chromate of bismuth is insoluble in dilute nitric acid, while, in fact, it is only insoluble in that liquid when mixed with a sufficient excess of chromate of potassium.

Properties. The precipitate produced by adding bichromate of potassium in slight excess to a neutral solution of nitrate of bismuth is a dense orange-yellow powder which settles readily. It is insoluble in water, even in presence of some fr e chromic acid; but is soluble in chlorhydric and nitric acids. It may be tried at 110°–113° without alteration (Lœwe). It is not precipitated from solutions which contain citrate of sodium (Spiller). Contrary to the statement of Pearson, the formula of the precipitate is Bi_2O_3, 2 CrO_3 [Not Bi_2O_3, CrO_3]. (Lœwe and Fresenius). The composition of the dry precipitate is:—

Bi_2O_3	468	69.95
$2CrO_3$	201	30.05
	669	100.00

Chromate of Calcium.

Principle. Decomposition of by alkaline carbonates.

Application. Separation of Ca from CrO_3.

Method. See Chromate of Barium. One single boiling with the alkaline carbonate will be sufficient to completely decompose chromate of calcium (H. Rose).

Chromate of Chromium.

Principle. Decomposition of by heat. See Chromic Acid, decomposition of by heat.

Chromate of Copper.

Principle. Oxidizing power of.

Application. Estimation of sulphur in organic compounds.

Method. The substance to be analyzed and the chromate of copper are mixed in the ordinary way in a combustion tube of hard glass, and the ignition is conducted in the usual manner, as described under Carbon. The precautions to be observed are that an abundance of the chromate of copper is mixed with the substance to be analyzed; that the combustion is effected in a large and capacious tube, so that a wide channel may be left above the mixture; that the combustion be proceeded with very slowly, and that the anterior part of the tube is kept at a temperature lower than that at which sulphuric acid can be expelled from sulphate of copper. The reaction between the chromate and the organic substance may easily be violent enough to project matter from the tube, unless care be taken to moderate it. — When the combustion has ceased treat the contents of the tube, which consist of a mixture of the oxides of copper, and of chromium, and chromate and sulphate of copper, with strong chlorhydric acid, add some alcohol and heat the mixture in order to reduce the chromic acid to oxide of chromium. When the liquid exhibits a pure green color, filter it, and in the hot filtrate throw down the sulphuric acid as Sulphate of Barium (*Otto, Zeitsch. analyt. Chem.*, 1868, **7**. 117).

According to Otto, the merits of the process consist in the large proportion of active oxygen contained in the chromate, whence a comparatively small quantity of it will effect a quicker and more complete oxidation than can be obtained with a mixture of saltpetre and carbonate of sodium; that the glass tube is not acted upon; that the contents of the tube can consequently be easily discharged, and that there is no nitrate of barium, or, at the most, a faint trace, to go down with the sulphate. The entire process can be finished in 3 or 4 hours. — To prepare the chromate of copper precipitate pure nitrate of copper with pure chromate of potassium, and wash the precipitate 3 or 4 times with water, to remove most of the nitrate of potassium. Longer washing would remove some of the chromic acid, and make the salt more and more basic.

Chromate of Lead.

Principle 1. Insolubility in water acidulated with acetic acid, and fixity at 100°.

Applications. Estimation of Pb in those compounds of that metal which are soluble in water and nitric acid, excepting such as contain substances like citric acid, which prevent the precipitation. Estimation of chromic acid. Separation of chromic acid from Ca, Sr, Mg and SO_3. Estimation of sulphuric acid in sulphates, and of phosphoric acid in phosphates,—also in manures, urine, etc.

Method 1. To *estimate Lead* proceed as follows:—

A. *Gravimetric.* If the solution to be examined is not already distinctly acid make it so with acetic acid; then add bichromate of potassium in excess, and if free nitric acid has been present add enough acetate of sodium to remove the free nitric acid, and replace it with free acetic acid. Let the mixture stand in a warm place to settle; collect the precipitate on a weighed filter; wash with cold water, dry at 100° and weigh. Or the precipitate may be collected on an unweighed filter, and ignited at a low heat in a porcelain crucible, if care be taken to remove the powder completely from the paper, and to burn the latter by itself. The process yields accurate results, but, according to H. Rose, in view of better processes, there are comparatively few cases in which it can be used with advantage.

Instead of drying and weighing the washed precipitate as above, Schwarz (*Annalen Chem. und Pharm.*, **84.** 92) treats it with chlorhydric acid and a measured quantity (an excess) of a solution of ferrous chloride of known strength. Sesquichloride of chromium, chloride of lead and ferric chloride are formed. The mixture is filtered to remove the chloride of lead, and the excess of ferrous chloride is determined in the filtrate by means of permanganate of potassium, as explained under Chromic Acid (reduction of by a ferrous salt), or in some other appropriate way. — The difference between the amount of iron taken and that found in the filtrate gives the quantity which has been oxidized by the chromate of lead, and every three equivalents of iron thus oxidized correspond to one equivalent of lead. The process, though complicated and rarely applicable, yields tolerably accutate results. Mohr (*Titrirmethode*, 1855, **1.** 199 and **2.** 107) has made some slight, and, as it seems, ill-founded objections to the process, but later experiments of Fresenius have shown it to be exact. Compare Chromic Acid, reduction of by ferrous salts.

B. *Volumetric.* Add ammonia or carbonate of sodium to the nitric acid solution of the lead as long as the precipitate redissolves on shaking. Add a solution of acetate of sodium in not too small quantity, and pour into the solution from a burette a one-tenth normal solution of bichromate of potassium (containing 14.73 grms. to the litre) till the precipitate begins to settle rapidly. Then place a number of drops of a neutral solution of nitrate of silver on a porcelain plate, and proceed more cautiously with the addition of the bichromate, adding only 2 or 3 drops at a time, and stirring thoroughly after each addition. A few seconds after each new addition of the bichromate, as soon as the precipitate has settled, so as to leave a tolerably clear liquid, take up a drop of the liquid and touch it to one of the drops of nitrate of silver on the white plate. A distinct red coloration, due to chromate of silver, will appear as soon as a small excess of the bichromate has been added to the lead solution. Any particles of chromate of lead which may be brought into the drop of nitrate of silver remain suspended there without reacting upon the silver. 0.1 c. c. of the bichromate solution used should be deducted as an allowance for the excess. Unless a sufficient quantity of acetate of sodium be present the liquid will become yellow colored, through excess of bichromate, before the reaction with silver is obtained. In that event add some more acetate of sodium and 1 c. c. of a standard solution of lead containing 0.0207 grm. of lead in 1 c. c. Complete the process in the usual way, and deduct another c. c. from the amount of bichromate used, as an allowance for the lead added. All metals whose chromates are insoluble must be removed before the method can be employed. If iron be present it must be in the form of a ferric salt (Schwarz, *Dingler's polytech. Journ.*, **169.** 284).

C. *To estimate Sulphuric Acid.* Besides the one-tenth normal solution of bichromate of potassium described in B, prepare a two-tenths normal solution of nitrate of lead (33.1 grm. to the litre). Mix the solution of the sulphate to be tested with a measured quantity of the standard lead solution, taking care that a small excess of lead shall remain unprecipitated. Collect the sulphate of lead upon a filter and wash it. Mix the filtrate with acetate of sodium, and titrate the lead with the standard solution of bichromate, as in B. Each c. c. of the lead solution which was expended in precipitating sulphuric acid represents 0.008 grms. of sulphuric acid. The process yields satisfactory results, though they are naturally a trifle too low in view of the solubility of sulphate of lead. The presence of salts like acetate and nitrate of ammonium, which increase the solubility of sulphate of lead, must of course be avoided as far as may be possible. — Instead of filtering the sulphate of lead, the mixture of liquid and precipitate may be allowed to settle in a graduated cylinder, and a portion of the clear liquid taken out with a pipette for analysis, without any great sacrifice of accuracy. Calcium and several other metals which would interfere with the process may be separated with carbonate of ammonium. Chlorhydric acid must be got rid of by evaporating with an excess of nitric acid on a water bath, for in a concentrated solution some chloride of lead would go down with the sulphate. — If phosphoric or arsenic acids are present the process becomes less simple, but it is still possible to employ it by first precipitating the sulphuric acid with the lead solution, from a nitric acid solution; then, after filtration and addition of acetate of sodium, the phosphoric or arsenic acids as phosphate or arseniate of lead. The excess of lead is then determined in the last filtrate; the phosphate or arseniate of lead is dissolved in nitric acid; the solution is mixed with a great excess of bichromate of potassium, and the precipitated chromate of lead collected and titrated with ferrous chloride, as in A.

To separate sulphuric from chromic acid the latter may either be reduced with sulphuretted hydrogen, at the risk of converting a little of the latter into sulphuric acid, or the two acids may be precipitated together with the lead solution, and the chromic acid estimated in the precipitate, or in another portion of the original substance (Schwarz, *Dingler's polytech. Journ.*, **169**. 289).

D. *To estimate Phosphoric Acid.* The same standard solutions are needed as in C, and the operations are similar in principle to those described in C. An aqueous solution of the phosphate of an alkali metal, or a nitric acid solution of the phosphate of an alkaline earth is mixed with an excess of the lead solution, and then with an excess of acetate of sodium. Triphosphate of lead (3 PbO, P_2O_5) goes down as a flocculent precipitate, which, after having been allowed to settle, is collected upon porous filter paper and washed. The excess of lead is then determined in the filtrate, as in C. For each c. c. of the two-tenths normal lead solution 0.004733 grms. of phosphoric acid may be allowed (*i. e.*, two-thirds of 0.0071). The processes of filtering and washing are not so easy as they are with the sulphate of lead of Method C. But by warming the mixture it is possible to make the precipitated phosphate somewhat more compact, after which operation it is well to bring the mixture to a definite volume, and to take out a portion of it with a pipette for analysis, either before or after filtering through a dry filter. — According to Mohr, the liquid must always be absolutely free from suspended particles of phosphate of lead, for this salt would be decomposed by the solution of bichromate.

The presence of calcium does no harm. If iron or aluminum be present they will be thrown down as phosphates when acetate of sodium is added to the solution, and must be separated by filtration. If the precipitate contains no other metal than iron the amount of the latter may be determined, in most cases and 1 equivalent of P_2O_5 allowed for each equivalent of Fe_2O_3 found. Phosphate of iron cannot be decomposed by nitrate of lead, but the phosphoric acid can easily be separated from it by means of magnesia. — In case sulphate of calcium, or magnesium, or a large proportion of chlorides are present in the mixture to be analyzed, the solution may be treated at the temperature of boiling with carbonate of sodium, and the mixed precipitate of carbonate and phosphate of magnesium and calcium may be dissolved in nitric or acetic acid for the analysis, after the sulphate and chloride of sodium have been washed out of it. Some phosphoric acid is lost in the operation, since a small proportion of the precipitate remains dissolved in the excess of carbonate of sodium, and the proportion thus dissolved increases with the amount of the carbonate of sodium. For estimating phosphoric acid in urine the titration should be preceded with a precipitation by chloride of calcium and carbonate of sodium. The process is said to yield very satisfactory results (Schwarz, *Dingler's polytech. Journ.*, **169**. 294).

Instead of the two-tenths normal solution of lead employed by Schwarz, Mohr (*Zeitsch. analyt. Chem.*, 1863, **2**. 253) recommends a three-tenths normal solution (49.671 grms. to the litre), and enjoins that care be taken not to add any too great excess of it to the solution of the phosphates. — According to Mohr, the process is almost absolutely correct when neutral solutions are operated upon, but is less exact, though still exceedingly valuable for technical purposes, when the liquid is acid, as when the phosphate of an alkaline earth is dissolved in acetic acid. The trouble lies in the fact that the indicator (see Chromate of Silver) is less delicate in the presence of acetic acid than it is in aqueous solutions. It is of the first importance, therefore, to have as little free acetic acid as possible in the liquid to be titrated, and to keep the volume of the liquid as small as possible. — For the estimation of phosphoric acid in urine Mohr adds magnesia mixture to the liquid, collects and washes the Phosphate of Magnesium and Ammonium and dissolves it in acetic acid before proceeding with the titration.

Method 2. To estimate Chromic Acid mix the solution with acetate of sodium in excess, add acetic acid to strong acid reaction and finally a solution of acetate of lead as long as any precipitate falls. Collect and weigh the precipitate as in Method 1. The results are accurate.

Properties. Chromate of lead is a bright yellow precipitate, insoluble in water and acetic acid, and but little soluble in other dilute acids. It is as good as insoluble in a dilute solution of acetate and nitrate of ammonium, slightly acidulated with acetic acid. For Brown's experiments on its solubility in nitric acid see Dictionary of Solubilities. It is readily decomposed by hot chlorhydric acid, and dissolves easily in hot oil of vitriol and in caustic alkalies. — According to Mohr (*Titrirmethode*, 1855, **1**. 199), chromate of lead may be completely precipitated from a solution of nitrate of lead, even when the latter is rather strongly acidulated with nitric or chlorhydric acid, provided a distinct excess of bichromate of potassium be added to the acid liquid. It appears that, although chromate of lead by itself is decomposed and dissolved to a certain extent by strong nitric and chlorhydric acids, this action cannot occur in presence of an excess of chromate of potassium. It would seem as if the excess of acid went to combine with the potassium of the bichromate, and as if chromate of lead were insoluble in chromic acid. Chromate of lead dries thoroughly at 100°; it is permanent in the air, and suffers no change when heated to temperatures lower than its melting point, excepting that when hot it exhibits a reddish brown color. It fuses at a red heat, and at higher temperatures it gives off oxygen and is reduced to a basic chromate of lead, mixed with sesquioxide of chromium. When heated in contact with organic matters it readily gives up oxygen to the carbon and hydrogen which they contain. Its composition is:—

PbO	223	68.93
CrO_3	100.5	31.07
	323.5	100.00

Principle II. Oxidizing power of, and fact of its fusing at a red heat.

Applications. Estimation of carbon in organic substances.

Methods. See Carbon, pp. 73, 74, 76.

For use in organic analysis, chromate of lead may be prepared as follows:—Add a solution of bichromate of potassium in slight excess to a clear filtered solution of acetate of lead, slightly acidulated with acetic acid. Wash the precipitate repeatedly by decantation with water, then collect it upon a piece of cotton or linen cloth, and again wash it thoroughly. Dry the powder, put it in a Hessian crucible and heat the latter to bright redness, until its contents are fairly fused. Then pour the melted chromate upon a slab of stone or iron, pulverize the sheet of solid chromate, sift the powder through a tolerably fine metallic sieve, and keep the powder in stoppered bottles. When properly prepared the chromate of lead will appear as a heavy, dirty, yellowish-brown powder. It should be free from substances soluble in water, and from any contamination with organic matter, such as dust. To test whether it be free from organic matter, heat some of it to redness in a tube of hard glass, and conduct into lime water any gas which may be evolved. The formation of any precipitate of carbonate of calcium would indicate impurity. — Chromate of lead may be used over and over again indefinitely, if it be powdered, moistened with nitric acid, dried, and fused after each combustion. For the second time of using the nitric acid may be omitted; it will then be sufficient to roast, fuse and powder the chromate. (Vohl, *Annalen Chem. und Pharm.*, **106**. 127.) When used for the combustion of sulphur compounds in long tubes, the chromate of lead may be used three or four times without refusion, and may afterwards be treated by Vohl's process, above described, just as if no sulphur were present. (Carius, *Annalen Chem. und Pharm.*, **116**. 28).

Chromate of Mercury. (Mercurous chromate).

Principle. Insolubility in water.

Applications. Estimation of chromic acid. Separation of chromic acid from Na, K, Ca, Sr, Mg and Cr.

Method. Make the cold solution very slightly acid with nitric acid, so that it may redden litmus paper only very feebly, and add a solution of mercurous nitrate as long as any precipitate falls. Then carefully add a few drops of ammonia water until the precipitate turns slightly brown. The precipitate is somewhat voluminous at first, but on standing it becomes heavy and settles well. After the mixture has been allowed to settle, collect the precipitate upon a filter and wash it with cold water which has been mixed with a small quantity of mercurous nitrate. If pure water were used, cloudy washings would run through the filter. Ignite the dry precipitate in a platinum crucible, and weigh the sesquioxide of chromium that is left. Even when the precipitate is contaminated with small quantities of insoluble chloride or sulphate of mercury, nothing but oxide of chromium will be left after strong ignition. But the method should nevertheless be avoided when the solution to be analyzed contains any very large quantity of a sulphate or chloride. — The process is specially commended by H. Rose, who finds it decidedly preferable to the processes depending on the insolubility of chromate of lead and chromate of barium. It is somewhat less exact when applied to the separation of chromic acid from oxide of chromium, since a small quantity of the latter is liable to go down with the chromate of mercury.

Properties. The chromate formed by precipitating mercurous nitrate with bichromate of potassium is a basic compound ($4Hg_2O, 3CrO_3$). It is a bright red powder, which turns black on exposure to light. Ammonia water converts it to a black powder, about half the chromium in which is in the state of sesquioxide. It dissolves very sparingly in cold water, but more freely in boiling water, partly in the form of mercuric salt. It is slightly soluble in ammonium salts. Chlorhydric acid decomposes it with formation of mercurous chloride and chromic acid. Nitrous acid or a nitrite reduces it. At a red heat it is resolved into oxygen gas, mercury vapor and sesquioxide of chromium.

***Mono*Chromate of Potassium.**

See Chromate of Sodium. For the use of monochromate of potassium as an indicator in the volumetric estimation of silver and chlorine, see Chromate of Silver.

***Bi*Chromate of Potassium.**

Principle. Oxidizing power.

Its Applications are numerous in volumetric analysis. As a material for preparing standard solutions it has the merits of being cheap, easily kept, easily weighed and readily obtained in a state of purity. Valuation of bichromate of potassium (Methods 1, 4). Preparation of a standard solution of iodine (Method 1). Estimation of iron (Method 2); of oxalic acid (Method 3); of tin (Method 4); of antimonious and arsenious acids (Method 6); and of coloring matters (Method 7).

Methods.

1. *Reduction by Chlorhydric Acid.* See Arsenious Acid and Chromic Acid. To determine the value of a sample of bichromate of potassium weigh out 0.3 or 0.4 grm. of it, place it in the flask of an apparatus such as is described under Chlorine (action of upon KI, and upon As_2O_3), pour upon it a considerable excess of pure fuming chlorhydric acid, free from any contamination of chlorine or sulphurous acid, and heat the flask. After 2 or 3 minutes boiling the whole of the chlorine will have passed out of the flask, and have liberated a corresponding amount of iodine in the absorbing vessel in case iodide of potassium were placed in it, or have oxidized an equivalent amount of arsenious acid in case arsenite of sodium is used as the absorbent. No loss of chlorine need be apprehended at the moment of adding the chlorhydric acid to the bichromate, for the evolution of that gas does not begin until the mixture is heated. When the absorption liquor has become quite cold, transfer it to a beaker and estimate the Iodine or the Arsenious Acid. Three atoms of chlorine are liberated for every molecule of chromic acid. The decomposition of the bichromate by the chlorhydric acid is rapid and complete, and although it may sometimes happen that a little of the volatile compound chlorochromic acid ($CrCl_3$, CrO_3) goes forward no harm is done, inasmuch as this compound liberates just as much iodine as would have been liberated by its components. (Bunsen, *Annalen Chem. und Pharm.*, **86**. 279).

To prepare a solution of iodine of known strength, Bunsen conducts a known quantity of chlorine into an excess of iodide of potassium, in aqueous solution. The known quantity of chlorine is obtained by heating pure bichromate of potassium with chlorhydric acid, as above described. To this end weigh out about 0.35 grm. of the pure bichromate, which has been dried by heating it just to fusion, and proceed as above. Enough of the solution of iodide of potassium must be taken to hold dissolved all the iodine which is set free. The cold, perfectly clear brown liquid is rinsed into a beaker and is known to contain a definite quantity of iodine, for each molecule of the bichromate taken corresponds to 6 atoms of iodine. With proper care this method answers very well, but it requires skilful manipulation, and is less convenient than the ordinary method of weighing out pure Iodine and dissolving it in iodide of potassium. It has the disadvantage, moreover, of depending on the atomic weight of chromium, which is still somewhat in doubt.

2. *Reduction of the bichromate by Ferrous Salts.* See the corresponding heading under Chromic Acid. See also Ferrous Salts. To prepare a normal solution of bichromate of potassium, dissolve 14.759 grm. (1 equivalent) of the pure salt in water, and dilute the solution to the volume of 1 litre. 50 c. c. of such a solution correspond to 0.84 grm. of metallic iron, or the whole litre to 0.6 equivalent (= 16.8 grms.) of iron, which may be converted from the state of protoxide to that of sesquioxide. Care must be taken to use perfectly pure and dry bichromate. To dry it, heat it in a porcelain crucible until it begins to fuse, allow it to cool in a dessicator, and weigh out the required quantity when the salt is cold. Besides the ordinary normal solution another solution should be prepared ten times more dilute (see Alkalimetry, p. 18), containing only 0.01 equivalent of the bichromate in the litre. It is well to test the correctness of the standard solution thus prepared, by using it to oxidize a known weight of pure iron — The details of an actual estimation of iron are as follows:—Mix the rather dilute solution of the Ferrous Salt with enough chlorhydric acid, or dilute sulphuric acid, to make it decidedly acid, and add a standard solution of biChromate of Potassium, until a drop of the liquid ceases to yield a blue color when tested with ferricyanide of potassium. The bichromate solution must be poured slowly from a burette into the iron solution, and the latter should be constantly stirred with a thin glass rod. The liquid, which is at first almost colorless, soon acquires a pale green tint, which gradually changes to a darker chrome green. From time to time a small drop of the liquid should be allowed to fall from the stirring rod upon a drop of a solution of ferricyanide of potassium on a white plate. — To save time, a number of drops of the ferricyanide solution should be put upon the plate beforehand, and the iron liquor should be added to these drops in regular order. As soon as the test ceases to show a strong blue reaction, the bichromate must be added slowly and carefully, and towards the close the test should be applied after each new addition of two drops, or even one drop of the bichromate solution. Towards the very last a couple of drops of the iron solution should be taken for the ferricyanide test, and an appreciable time should be allowed for the appearance of the reaction. The oxidation is complete when no further blue coloration of the ferricyanide is manifested. The reaction being exceeding sensitive it is easy to hit the exact moment when the oxidation is complete. In order to make the results as accurate as possible, it is best to employ two solutions of the bichromate, the one comparatively strong and the other weak. A normal solution may be used at first, but just at the end of the process a one-tenth normal solution should be employed. The iron

solution must be kept decidedly acid in order to prevent the deposition of any brown chromate of chromium, upon which the ferrous salt would exert scarcely any deoxidizing action (Penny. Schabus). — The process has special merit in that the standard solution of the bichromate can be easily made and kept. For the methods of changing a ferric salt to the condition of ferrous salt, to prepare it for analysis see Ferric Salts.

An indirect gravimetric method of estimating iron, proposed by Vohl (*Annalen. Chem. und Pharm.*, **94**. 218), depends in part upon this principle. The solution of ferrous salt, which must be highly concentrated, is mixed with a slight excess of bichromate of potassium, and the excess of the bichromate is determined by heating it with an oxalate and collecting and weighing the carbonic acid which is evolved (see under Chromic Acid). The quantity of iron is calculated from the weight of the bichromate that was consumed in oxidizing it. The process is complicated and difficult.

3. *Reduction of the bichromate by Oxalic Acid.* Mix the oxalate to be analyzed with an excess of bichromate of potassium, add sulphuric acid to the mixture, and collect and weigh the carbonic acid that is evolved. 2 equivalents of carbonic acid will be obtained for each equivalent of oxalic acid in the substance. In case free oxalic acid is to be examined, supersaturate it slightly with ammonia water after weighing before proceeding with the analysis. For forms of apparatus proper for effecting the decomposition see Carbonic Acid, volatility of (Vohl). Compare Chromic Acid. The same principle is involved in one of the methods of estimation of iron, described under the head reduction of the bichromate by a ferrous salt.

4. *Reduction of the bichromate by Stannous Chloride.*

A. For the method of estimating chromic acid in this way see the analogous heading under Chromic Acid.

B. To estimate tin bring the metal into solution in chlorhydric acid, in a flask, or if it be already in solution acidulate the liquor strongly with chlorhydric acid, add a small quantity of a solution of iodide of potassium and some thin starch paste, shake the flask continually and pour into it from a burette the standard solution (one-tenth normal) of bichromate of potassium, prepared as above described, until the entire liquid suddenly becomes blue. From the amount of bichromate of potassium consumed reckon the quantity of tin in the solution. In order to do this correctly several chemists have attempted to determine once for all how much bichromate of potassium is needed to oxidize 100 parts of pure tin dissolved in chlorhydric acid. Penny and Streng, for example, found that 83.2 pts. of the bichromate were needed. The reason of the need of this empirical calculation is to be sought in the incomplete decomposition of the bichromate by the tin solution rather than in any error in the accepted atomic weight of tin. — In case the tin solution to be examined contains stannic chloride as well as stannous chloride, determine the amount of the latter in one portion of the solution, and in another portion throw down the whole of the Tin by means of zinc, dissolve in chlorhydric acid and by means of the bichromate solution determine the whole of the tin (Streng, *Poggendorff's Annalen*, **92**. 57; Mohr, *Titrirmethode*, 1855, **1**. 264). The process is but little esteemed. — According to Mohr (*loc. cit.*, pp. 260, 262, 267), the results are liable to vary slightly from one another in a manner not readily explicable. The results always come out a little too low when the iodide of potassium and starch are added directly to the tin solution, as above described. — Better results are obtained in operating as follows:— Take a measured quantity of the bichromate solution, acidulate it with chlorhydric acid, and pour the tin solution into it from a burette until the pure green color of chloride of chromium appears. Add yet another c. c. of the tin solution, then mix with the liquid the iodide of potassium and starch and pour in the bichromate of potassium from the same burette whence the measured quantity was taken, until the appearance of the blue color. Compare the similar heading under Chromic Acid.

5. *Reduction of the bichromate by Ferrocyanide of Potassium.* See Chromic Acid. The reaction between bichromate of potassium and ferrocyanide of potassium is as follows:—

$$6K_4FeCy_6 + K_2O, 2CrO_3 + 14HCl = 6K_3FeCy_6 + 8KCl + 2CrCl_3 + 7H_2O.$$

Hence 21.122 grms. of crystallized ferrocyanide of potassium are equivalent to 14.759 grms. of bichromate of potassium (E. Davy, *Phil. Mag.*, (4), **21**. 214).

6. *Reduction of the bichromate by Arsenious and by Antimonious Acids.* The details of the method for estimating Antimonious Acid (see p. 31) and Arsenious Acid (p. 46) are as follows:— Prepare a standard solution of bichromate of potassium which shall contain about 2.5 grms. to the litre. Weigh out exactly 5 grms. of pure arsenious acid, dissolve it in a solution of carbonate of sodium, slightly acidulate the solution with chlorhydric acid, then add 100 c. c. more of chlorhydric acid of 1.12 sp. gr., and dilute to a litre. Each c. c. of this solution will contain 0.005 grm. of arsenious acid, and will correspond to 0.007374 grm. of antimonious acid. Dissolve about 1.1 grm. of iron wire in 2 c. c. of dilute sulphuric acid, prepared by mixing 1 volume of concentrated acid with 4 volumes of water, and dilute the solution to the volume of a litre. — To establish the relation between the bichromate solution and the iron solution pour 10 c. c. of the bichromate solution into a beaker from a

burette, add 5 c. c. of chlorhydric acid, and 50 c. c. of water, and pour in the iron solution rapidly from a burette, until the liquid is green. Continue to add the iron solution by portions of 1 c. c. each, and test the liquid after each such addition, by touching a drop of it to a drop of a tolerably dilute, freshly prepared solution of ferricyanide of potassium upon a white porcelain plate, until a distinct blue reaction of the ferrous salt is obtained. Then add 0.5 c. c. of the solution of bichromate, and afterwards the iron solution, two drops at a time, till the blue reaction, with the ferricyanide, just occurs. Read off both the burettes and calculate how much of bichromate solution corresponds to 10 c. c. of the iron solution. Since the iron solution cannot be kept for any length of time without oxidizing, this experiment must be repeated before each new series of analyses. — To determine the relation between the bichromate solution and the solution of arsenious acid, transfer 10 c. c. of the latter to a beaker, add 20 c. c. of chlorhydric acid of 1.12 sp. gr., and from 80 to 100 c. c. of water, and pour in the solution of bichromate from a burette till the yellow color of the liquid shows that the bichromate is present in excess. Then wait a few minutes and pour in from the other burette the iron solution to slight excess, as above, then again 0.5 c. c. of the bichromate, and finally the iron solution, two drops at a time, until the blue reaction with the ferricyanide appears. Deduct from the total amount of bichromate solution employed, the quantity which corresponds to the iron used, and proceed to calculate how much antimonious acid corresponds to 100 c. c. of the chromate solution, that is to say, how much Sb_2O_3 will be converted into Sb_2O_5 by this quantity of the bichromate. See further under Antimonious Acid. The water used to dilute the solution of arsenious acid, above, must be measured, since the action of bichromate of potassium upon arsenious and antimonious acids is normal only when the liquid contains at least one-sixth of its volume of chlorhydric acid of 1.12 sp. gr. (Kessler). Fresenius (*Zeitsch. analyt. Chem.*, 1869, **8.** pp. 154, 155, 159) found the process to yield excellent results when applied to the estimation of exceedingly small quantities of antimony.

7. *Reduction of the bichromate by organic Coloring Matters.* As applied to the valuation of indigo the process is as follows: — Prepare a standard solution of bichromate of potassium containing 7.66 grms. to the litre. 1 c.c. of this solution corresponds to 0.01 grm. of pure indigotine, or to 10 c. c. of a solution which contains 1 grm. of indigotine to the litre. Reduce to fine powder a quantity of the Indigo to be tested, dry it at 100°, weigh out precisely 1 grm. and mix it in a capsule with 12 grms. of fuming sulphuric acid. Cover the dish with a glass plate and leave it at rest for 24 hours, at a temperature of 20° to 22°. By the action of the fuming acid the indigotine is dissolved, while the foreign matters are carbonized or destroyed. Dilute the sulphuric solution with water, filter and bring the filtrate to the volume of 1 litre. Measure out 100 c.c. of the solution into a porcelain dish by means of a pipette, boil it gently, add 10 c. c. of chlorhydric acid, and pour the standard solution of bichromate into the boiling liquid from a burette, until the instant when the greenish tint of the liquid disappears, and is replaced by a yellowish orange. A little practice is necessary in order that the completion of the reaction may be hit with certainty. — The result of the first trial should be regarded as a mere approximation, and should always be controlled by a second experiment upon another 100 c. c. of the solution. The number of tenths of c. c. of the bichromate solution consumed gives the percentage of indigotine in the sample. The amount of bichromate required to decolorize a given weight of pure indigotine, taken in the form of sulphindigotic acid, has been found in practice to agree exactly with the quantity required by theory to change the indigotine to isatine, or rather to sulphisatic acid. As long as there is any indigotine present to be oxidized, the bichromate acts upon it alone. 3 molecules, or 393 parts of indigotine require 1 molecule, or 297.2 parts of the bichromate. In other words, 100 parts of indigotine require 75.6 parts of the bichromate. In case the indigo contained any ferric oxide, this would be reduced in the process of solution, and the ferrous salt formed would consume an equivalent quantity of the bichromate. Such indigo should be leached with dilute boiling chlorhydric acid to remove the iron before it is subjected to analysis. The same treatment is advisable in case the indigo contains much carbonate of calcium. (Penny's process, as used by Ernest Schlumberger, and reported by Schützenberger in his *Matières Colorantes*, 1867, **2.** 560). The process is applicable for the valuation of other coloring matters besides indigo.

Principle II. Power of precipitating lead from its solutions. See Chromate of Lead.

For use as a reagent, bichromate of potassium may often be found pure in commerce. The commercial salt, however, is liable to be contaminated with sulphate of potassium. To free it from this impurity dissolve some of the salt incompletely in warm water, and let it crystallize. Repeat this process of crystallization until sulphuric acid can no longer be detected in the product. In order to test for sulphuric acid, reduce some of the bichromate by boiling it with pure, strong, chlorhydric acid under a hood, dilute the product with water, add a drop of chloride of barium, and let the mixture stand for some time. The bichromate must be perfectly free from sulphur, in case it is to be used for estimating sulphur in organic compounds, and it should be as pure as possible when used for making standard solutions. But in case it is to serve for estimating carbon or a carbon compound by heating it with the substance and sulphuric acid, no special degree of purity is necessary.

Chromate of Silver.

Principle. Deep red color and insolubility in water.

Application. Use as an indication in the volumetric estimation of chlorine, silver and lead.

Method. See Chloride of Silver and Chromate of Lead. When a solution of nitrate of silver is added to a solution of neutral chromate of potassium chromate of silver is immediately precipitated as a bright, blood-red powder, but sparingly soluble in water. According to Mohr (*Titrirmethode*, 1856, **2**. 25), 1 part of it dissolves in 6700 parts of water at 17.5°, and in 3700 parts of boiling water. Its color is so deep that it can be readily seen, even when largely diluted by a white precipitate or a yellow liquid. It is immediately decomposed by soluble metallic chlorides, with formation of insoluble chloride of silver and a soluble chromate of the metal. It is decomposed also by solutions of the alkaline carbonates, and is readily soluble in free acids.

When used as an indicator in the volumetric process of estimating phosphoric acid by Chromate of Lead some annoyance is occasioned by the easy decomposition of the chromate of silver by the acid in the liquid. As the volume of the liquid increases, moreover, the indications of the chromate of silver become less sharp. It may be said, in short, that silver as an indicator of chromic acid is very delicate in neutral solutions, cannot be used at all with the mineral acids, and is so much the less delicate in presence of acetic acid in proportion as there is more of the acid in the solution. Acetic acid does not prevent the formation of the red color completely, as the smallest quantity of nitric acid would, but it delays its appearance remarkably. It should be used, therefore, only in the smallest possible quantity, and the volume of the liquid should be kept as small as may be practicable.

To test the limit of action of the indicator single drops of a one-tenth normal solution of bichromate of potassium were added to 50 c. c. of water, and after each such addition a little of the water was touched to a drop of nitrate of silver upon a white plate. The first and second drops gave no reaction, but the third gave a visible red color. A small quantity of acetic acid being then added to the water, its power of reacting with nitrate of silver disappeared, and was not recovered until 0.4 c. c. of the solution of bichromate had been added. Even then a little acetate of sodium had to be added to the water, for in its absence the bichromate in reacting upon the nitrate of silver set free enough nitric acid to hinder for a long time the separation of chromate of silver. With the acetate of sodium a reaction was obtained after the third drop of bichromate, but in its absence there was no reaction until the fifth drop. As a rule, in estimating phosphoric acid, when the liquids are not too strongly acid a correction of 0.4 c. c. may be applied to compensate for this solubility of the indicator (Mohr, *Zeitsch. analyt. Chem.*, 1863, **2**. 254).

Chromate of Sodium.

Principle. Fixity when heated and solubility in water.

Applications. Separation of chromic acid from Ba, Sr, Ca, Mg, Zn, Mn, Co, Ni, Fe.

Method. Weigh out some of the chromate to be analyzed, mix it in a platinum crucible with four parts of carbonate of sodium and fuse the mixture thoroughly. Treat the fused mass with hot water, and filter the solution in order to separate the insoluble carbonates or oxides from the soluble chromate of sodium. In case manganese be present the fusion must be made in a bulb tube, in a stream of carbonic acid. Since the insoluble carbonates or oxides thus obtained are liable to be contaminated with alkali they cannot be weighed directly. — Instead of fusing with carbonate of sodium alone a mixture of that substance and carbonate of potassium may be used. The mixture has the merit of melting at a comparatively low temperature. The Chromates of Barium, Calcium and Strontium may be more readily and completely decomposed by boiling with an excess of a solution of an alkaline carbonate than by the method of fusion just described (H. Rose, *Journ. prakt. Chem.*, **66**. 166, and Handbuch 6te Aufl., **2**. 384).

For the various methods of converting oxide of chromium into chromate of sodium see sesquiOxide of Chromium, action of oxidizing agents upon.

Chromate of Strontium.

Principle. Decomposition of by alkaline carbonates.

Application. Separation of chromic acid from strontium.

Method. See Chromate of Barium and Chromate of Sodium. A single boiling with a solution of carbonate of sodium is sufficient to decompose chromate of strontium completely (H. Rose).

Chromium is determined either as sesquiOxide or as a Chromate of Barium, Lead or Mercury, or indirectly by the reduction of Chromic Acid. It is noteworthy that although chromium cannot be precipitated by itself as a basic acetate, some of it will go down with iron and aluminum when these metals are thus precipitated (Gibbs, *American Journ. Sci.*, 1865. **39**. 61).

Cinchona Bark. [Compare Quinine and Cinchonin].

Principle. Insolubility of the cinchona alkaloids in aqueous alkaline solutions, as well as their solubility in acids and in ether, chloroform and alcohol.

Methods.

Method of Rabourdin (A). A quantity of the bark having been reduced to powder and sifted through a fine hair sieve, the powder (40 grammes of the gray bark or 20 grammes of the yellow, Calisaya, bark) is packed in a percolation cylinder and exhausted with water acidulated with chlorhydric acid. To prepare the acidulated water mix 20 grms. of strong chlorhydric acid with 1 kilog. of water. As soon as the percolate passes off colorless and tasteless add to it 5 or 6 grms. of caustic potash together with 15 grms. of chloroform. Shake the mixture during several minutes and then let it stand. In the course of half an hour a dense whitish deposit consisting of chloroform charged with the cinchona alkaloids will separate from the water. The water is then decanted, and the chloroformic solution washed repeatedly with water by decantation. Sometimes the separation of the chloroform from the water is instantaneous and complete, so that the red transparent liquid left floating above the chloroform may be immediately poured off. The washed chloroformic solution of the alkaloids is transferred to a porcelain capsule, and evaporated to dryness upon a water bath. In the case of red bark or Calisaya bark the residue, consisting of alkaloids in a condition of tolerable purity, is simply weighed. But for the pale or cinchonine barks the process must be carried further. — The residue left on evaporating the chloroform contains cinchonic red as well as cinchonine. To remove this impurity treat the residue with water acidulated with chlorhydric acid, which will dissolve the whole of the alkaloids and a portion of the cinchonic red. Filter the liquid and add to it, drop by drop, with constant stirring, ammonia water diluted with 15 or 20 parts of water, as long as the cinchonic red continues to be precipitated in reddish brown flakes, and until white curdled flakes of cinchonine, not removed by agitation, begin to appear. In this way the cinchonic red is first precipitated before any of the alkaloid goes down. Filter, wash the precipitate with a small quantity of water and precipitate the alkaloids in the filtrate by adding an excess of ammonia. The precipitate is collected upon a filter, washed, dried and weighed (Rabourdin, *U. S. Dispens.*, 1867, p. 295, note; *Handw.*, **1.** 470).

2. *Method of Rabourdin* (B). Exhaust 40 grms. of gray bark or 10 grms. of red or Calisaya bark, at the ordinary temperature, with water acidulated with 3 or 4 per cent of chlorhydric acid, until the filtrate tastes only slightly bitter. Add to the acid liquor a quantity of soda lye, collect the precipitate and wash it with a little water. The soda throws down the alkaloids, while it holds in solution the tannic acid, cinchonic red and other coloring matters, and the resin with which the acid solution of the alkaloids was contaminated.

To purify the precipitate it may be treated with a quantity of chlorhydric acid insufficient to dissolve the whole of it, and the solution mixed with ammonia, whereby the quinine is thrown down white and pure. Or the precipitate may be dissolved in a slight excess of chlorhydric acid, the solution filtered if need be, and then treated with dilute ammonia until a brown precipitate appears, and a colorless liquor can be obtained by filtration. On adding ammonia to this colorless filtrate quinine is precipitated together with slight traces of cinchonine. This precipitate is washed, dried in the air and weighed. In case too much ammonia is added in the first place, so that white flocks are mixed with the brown precipitate, a little of the acidulated water must be added to redissolve them (Rabourdin, *Journ. Pharm. et Chim.*, 1861, p. 408, cited by Van der Burg, *Zeitsch. analyt. Chem.*, 1865, **4.** 288).

With regard to the foregoing process Van der Burg (*loc. cit.*, pp. 292, 289) reports that it is exceedingly difficult, if not impossible, to extract all the alkaloids from cinchona bark by means of four per cent. chlorhydric acid at the ordinary temperature. Even with the utmost care a relatively large proportion of the alkaloids remains in the residue of the percolation. As a means of extracting the alkaloids from bark the dilute chlorhydric acid is distinctly inferior to the lime and alcohol employed by de Vrij. The process was formerly held in considerable esteem on account of its cheapness and supposed accuracy.

According to Van der Burg, the largest amount of alkaloids is obtained when the solution is concentrated to a small bulk before the precipitation, and finally treated with a large excess of ammonia. No absolutely colorless solutions can be obtained by the method of partial precipitation with ammonia, even when so much ammonia is used that a part of the alkaloids is precipitated together with the coloring matters; at the best the solution will still be of a light yellow color. It is none the less true, however, that a considerable quantity of coloring matter free from alkaloids can be thus thrown down by ammonia before the alkaloids themselves are precipitated. This method of fractional precipitation is on the whole to be preferred to the other process of purification by partial solution. It is to be observed that the cinchona alkaloids are less soluble in strong than in dilute ammonia water. — The alkaloids finally obtained always contain traces of alkaline earths and of coloring matters. They should be dried over sulphuric acid before weighing, since the air-dried alkaloids always retain considerable and varying quantities of water. The statement of Rabourdin that an assay of cinchona can be finished in an hour is erroneous. More than an hour is required for the process of percolation alone, and the process is incomplete even then, as has been already stated. A much larger quantity of the acidulated water should be used than is directed by R. More than twice as much as would appear from his statement that from 100 to 120 grms. of filtrate are obtained from 10 grms. of bark. Not only quinine but all the other cinchona bases are precipitated, so that the method gives only the total amount of alkaloids.

3. *Method of de Vrij.* Mix a weighed quantity of the powdered bark, dried at 100° with 0.25 of its weight of hydrate of calcium, and boil the mixture for 5 minutes with ten times its weight of alcohol of 0.852 sp. gr. Throw the mixture upon a warm filter and wash with small portions of boiling alcohol, until the filtrate is equal to 20 times the weight of the bark taken. Add dilute acetic acid to the alcoholic filtrate until a slight acid reaction persists, then evaporate the liquid to dryness upon a water bath, treat the residue with water, throw the aqueous mixture upon a filter and wash with water, until a portion of the clear filtrate ceases to become cloudy on the addition of an alkali. The aqueous solution thus obtained contains all the alkaloids of the bark, while the Kinic Acid and fatty and resinous matters remain upon the filter.

Evaporate the aqueous solution of the alkaloids to a small bulk upon a water bath, and mix the concentrated liquor with a slight excess of strong milk of lime to precipitate the alkaloids. Throw the precipitate upon a small filter, and wash it with the smallest possible quantity of cold water. If the operation be properly conducted, the quantity of water necessary to remove the coloring matter will be so small that the trace of alkaloids dissolved by it may be safely neglected. — After washing with water, dry the filter with its contents, and boil the whole repeatedly with alcohol of 0.819 sp. gr. to dissolve the alkaloids. Filter the alcoholic solution, collect the filtrate in a weighed platinum capsule, dry upon a water bath and weigh. The total amount of alkaloids is thus obtained; for the methods of separating them: See Quinine; Iodhydrate of Quinidin; Cinchonin. (De Vrij, *Zeitsch. analyt. Chem.*, 1865, **4.** pp. 202, 274.)

According to Van der Burg (*Zeitsch. analyt. Chem.*, 1865, **4.** pp. 287, 292) the method of de Vrij is faulty, inasmuch as it is very difficult to extract the whole of the alkaloids from cinchona bark by means of lime and alcohol, and that a certain quantity of the lime goes into solution to be afterwards reckoned as cinchonin. But the lime and alcohol are nevertheless more efficient agents than dilute chlorhydric acid (see Rabourdin's method, above) for extracting the alkaloids. Van der Burg consequently recommends de Vrij's method with this modification, that the operation of boiling the bark with lime and alcohol be several times repeated, so that the final residue shall retain only an insignificant quantity of the alkaloids Moreover, after the evaporation of the alcoholic solution acidulated with acetic acid, and the subsequent treatment with water and filtration, he prefers to precipitate the alkaloids with caustic soda, and to separate the coloring matter by redissolving the precipitate in dilute chlorhydric acid and partial precipitation with ammonia, as directed above, in the second description of Rabourdin's process. The purified alkaloids are finally to be dissolved in alcohol, evaporated in a tared platinum dish, dried and weighed. Since quinidin is somewhat soluble in pure water, and quinine still more so, it is well to collect by themselves the washings of the soda precipitate, as soon as they cease to exhibit a strong alkaline reaction, and begin to taste slightly bitter; to shake the liquor with ether, to collect and evaporate the ethereal solution, and add the residue to the precipitate, before proceeding to treat the latter with alcohol.

4. *Process of Winckler.* Digest 1000 grains of the finely powdered bark with 6 ounces of alcohol of 80 per cent upon a water bath, until it is completely exhausted. When cold, strain the tincture through thin, close linen. Digest the residue anew with 3 ounces of alcohol, and strain the mixture as before. Yet again treat the residue in like manner with alcohol. Unite the several alcoholic solutions, filter and treat the filtrate at the ordinary temperature with a mixture of equal parts of freshly slaked lime and crude well burnt animal charcoal, of which about 500 grains will be required. Shake the mixture frequently, and let it macerate until the supernatant liquid is observed to be colorless. In most of the genuine barks the liquid is soon decolorized, but in those containing kinonic acid the process is imperfect. Decant and filter the decolorized liquid; shake the residue repeatedly with small quantities of alcohol, and afterwards throw it upon a filter and wash it with alcohol. By distilling off the alcohol from the filtrate there is obtained a mixture of alkaloids, fatty matter, cinchonic red, and any kinonic acid which may have existed in the bark. — To remove the impurities transfer the matter to a small evaporating dish, and wash out the distilling vessel with water acidulated with sulphuric acid. Add some more sulphuric acid to the dish, until it is in slight excess, heat the mixture, then allow it to cool, and filter off the kinonic acid and other impurities which have been precipitated. Add a slight excess of ammonia to the filtrate, and evaporate to dryness at a gentle heat the mixture of precipitate and liquid. Remove the sulphate of ammonium from the residue by means of a small quantity of very cold water, and dry and weigh the residual alkaloids. Though the alkaloids thus obtained are not absolutely pure, their amount affords a tolerably

good indication of the value of a bark. Winckler states that the barks yield to the manufacturer quite as much as is obtained in this way, and generally from 0.12 to 0.25 per cent more, in consequence of the loss in working being less on the large scale (Winckler, *American Journ. Pharm.*, **25**. 343, through *U. S. Dispens.*, 1867, p. 295). — The best yellow Calisaya bark, the finest red bark, and the finest fibrous Carthagena bark (soft Pitaya), each yield about 3 or 4 per cent of alkaloids; while between these and the barks of lowest value there is every grade of productiveness, down to a mere trace of alkaloid matter (*U. S. Dispens.*).

5. *Process of Hager.* See Picrate of Cinchonin, etc.

6. The British Pharmacopœia gives the following methods of testing the various colored barks.

A. *Test of Yellow Cinchona.* Boil 100 grains of the bark reduced to very fine powder, for quarter of an hour, in a fluid ounce of distilled water acidulated with ten minims of chlorhydric acid, and allow it to macerate for 24 hours. Transfer the whole to a small displacement tube, and after the fluid has ceased to percolate add at intervals about an ounce and a half of similarly acidulated water, or add until the fluid which passes through is free from color. Add to the percolated fluid a solution of subacetate of lead, until the whole of the coloring matter has been removed, taking care that the fluid remains acid in reaction. Filter and wash with a little distilled water. To the filtrate add about 35 grains of caustic potash, or so much as will cause the precipitate which is at first formed to be nearly redissolved, and afterwards 6 fluid drachms of pure ether. Then shake briskly, and having removed the ether repeat the process twice with three fluid drachms of ether, or until a drop of the ether employed leaves on evaporation scarcely any perceptible residue. Lastly, evaporate the mixed ethereal solution in a capsule. The residue, which consists of nearly pure quinine, when dry, should weigh not less than 2 grains, and should be readily soluble in dilute sulphuric acid.

B. *Test for Pale Cinchona.* 200 grains of the bark, treated in the manner directed in the test for yellow cinchona, with the substitution of chloroform for ether, should yield not less than two grains of alkaloids.

C. *Test for Red Cinchona.* 100 grains of the bark, treated in the manner directed in the test for yellow cinchona, with the substitution of chloroform for ether, should yield not less than 2 grains of alkaloids. (From the *Dispensatory of the U. S.*, 1867, p. 253.)

Cinchonin.

Principle. Insolubility in aqueous alkaline solutions.

Application. Estimation of cinchonin in extracts of cinchona bark from which quinine and quinidin have been separated.

Method. Evaporate the solution upon a water bath to expel alcohol, if any be present, dissolve the residue in water acidulated with acetic acid and add soda lye to the solution. Collect the precipitate upon a filter, wash it with water and dry and weigh it. The precipitate consists, according to circumstances, of cinchonin or of a mixture of cinchonin and cinchonidin. In case Cinchona Bark is extracted with lime and alcohol the cinchonin precipitated as above is apt to be contaminated with lime (van der Burg). It may consequently be well to treat the dry precipitate with alcohol, to collect the alcoholic filtrate in a tared platinum dish, and evaporate it to dryness for a second weighing (De Vrij, *Zeitsch. analyt. Chem.*, 1865, **4**. 203; van der Burg, *ibid*, pp. 275, 287). According to De Vrij, small quantities of cinchonin can only be determined with certainty by means of a polarization apparatus.

Citric Acid.

Principle I. Power of neutralizing alkalies.

Application. Valuation of solutions of the acid.

Method. Color the solution with solution of logwood and titrate with standard caustic soda (see Acidimetry) (Mohr).

For estimating the value of lemon juice the English practice is to neutralize a measured sample of the juice either with chalk or with a standard solution of carbonate of sodium, and to calculate from the amount of alkali expended an equivalent quantity of citric acid. According to Ogston, these methods invariably give too high an estimate of the value of the juice. Any foreign acids which may be present, as well as any aluminum salt which may be precipitated, go to increase the amount of alkali used and the weight of citric acid calculated.

Principle II. Inability of the acid to influence the plane of vibration of a ray of polarized light.

Application. According to Buignet, it is easy to detect on this principle any adulteration of the acid with Tartaric Acid, by means of a polarization apparatus (*Journ. Pharm. et Chim.*, **40**. 252, and *Zeitsch. analyt. Chem.*, 1862, **1**. 234).

Cobalt is usually weighed as metallic Cobalt, protosesquiOxide of Cobalt, Sulphate of Cobalt, or Nitrite of Cobalt and Potassium. It is often thrown down as Sulphide and sometimes as Hydrate of Cobalt.

Principle. Fixity when heated.

Applications. Estimation of cobalt in the precipitated oxide and in the chloride, nitrate, carbonate and other salts of cobalt, when they contain no fixed impurities (Method A).

Separation of Co from Ba, Sr, Ca and Mn (Method B).

Method A. Place the dry substance in a weighed porcelain crucible provided with a perforated platinum cover. Or, in case the substance to be analyzed is a liquid, evaporate it to dryness in the crucible. By means of a bent tube fitted to the hole in the cover pass a stream of hydrogen gas into the crucible. The crucible must be heated strongly, since metallic cobalt obtained by reducing the oxide with hydrogen at a low heat is so finely divided that it takes fire on coming in contact with the air. This tendency to take fire is specially marked in cases where the oxide of cobalt being impure the reduced metal is mixed with infusible substances. — After the crucible has been allowed to cool in the atmosphere of hydrogen, and has been weighed, it is well to ignite it again in the same way, or again and again, until the weight becomes constant. Finally, in case it is the precipitated oxide which is under examination, a quantity of water should be poured into the crucible to dissolve any alkali which may have been retained by the precipitate. Decant this water after a while, and treat the metallic cobalt with fresh doses of water as long as the liquid continues to turn red litmus paper blue and to leave a residue on evaporation. After the washing has been completed dry the metal and again ignite it in a stream of hydrogen. The amount of matter dissolved out by the water is usually less than 0.2 per cent of the weight of the impure cobalt (H. Rose). — Unless the oxide of cobalt has been precipitated in platinum vessels by means of alkali absolutely free from silica, the metallic cobalt should be dissolved in acid after it has been weighed and the solution evaporated to dryness on a water bath, in order that the silica which contaminates it may be separated (Johnson). According to Fresenius, the process yields strictly accurate results only when the compound to be analyzed is free from sulphuric acid and alkali. He finds it impracticable to obtain an absolutely pure product by boiling the impure metallic residue with water. The impure metallic powder, after having been repeatedly boiled with water, still continues to give an alkaline reaction with turmeric paper when the latter is left in contact with it for some time.

Method B. In case cobalt is to be separated from another metal the process remains essentially the same as in A, only that the mixed chlorides of cobalt and the other metal are operated upon, and that a chloride of the other metal is left in the crucible, and must be dissolved out with water before the metallic cobalt can be weighed. If the metal to be separated be an alkali metal, care must be taken not to heat the crucible above low redness, so as to avoid volatilizing any of the alkaline chloride. It is perhaps best in this case to effect the reduction in a bulb tube instead of a crucible. — In the case of manganese the reduction may be effected in a crucible provided the substance to be analyzed is a chloride. But in case the mixture of cobalt (or of cobalt and nickel) and manganese is not in the form of chlorides it is well to precipitate the whole as Hydrates, by means of soda lye; to place the dried precipitate, or a portion of it, in a bulb tube; to heat the bulb to moderate redness and to pass in a current of dry chlorhydric acid gas until the metals are wholly converted to chlorides, and no more water is given off. A long time is required to complete this operation. As soon as it is finished heat the bulb strongly and pass in dry hydrogen gas until only a slight cloud of chloride of ammonium is formed, when a glass rod moistened with ammonia water is held at the end of the tube. The chlorides of cobalt and of nickel are reduced to the metallic state, while the chloride of manganese remains unaltered.

After the bulb tube has been allowed to cool in the current of hydrogen, it is placed in a cylinder with cold water—which has been boiled to expel air—so that the undecomposed chloride may dissolve and leave the Cobalt (and Nickel) free. A portion of the chloride of manganese usually refuses to dissolve, and remains suspended in the liquor in the form of brown flakes; it may, however, be readily separated by decantation from the far heavier metallic cobalt. The latter must finally be dried, again ignited in an atmosphere of hydrogen and weighed. Care must be taken not to heat strongly enough to volatilize chloride of manganese. The process yields accurate results (H. Rose).

Properties. As obtained by reducing the chloride or nitrate in a stream of hydrogen, metallic cobalt is a grayish black powder, which is attracted by the magnet and is more difficultly fusible than gold. In case the reduction is made at a low heat the finely divided metal will take fire in the air and burn to protosesquioxide, but nothing of the kind occurs when the reduction is effected at an intense red heat. The powder does not decompose water, even at the temperature of boiling, unless an acid be present. It dissolves readily in nitric acid and in hot concentrated sulphuric acid. When obtained by reducing oxide of cobalt which has been precipitated by caustic alkalies, the metal exhibits an alkaline reaction due to alkali retained by the oxide. The amount of this impurity is rarely more than 0.2 or 0.3 per cent, and it may be removed by means of hot water.

Cobalticyanide of Copper.

Principle. Insolubility in water.

Application. Estimation of Co in a solution of cobalticyanide of potassium. Separation of Co from Ni.

Method. Supersaturate the solution of cobalticyanide of potassium with acetic acid, boil the acid liquor and pour into it a solution of sulphate of copper as long as any precipitate falls. Boil the mixture for some time longer, then collect the precipitate of cobalticyanide of copper upon a filter. Finally decompose the precipitate by boiling it with a solution of caustic potash and weigh the Oxide of Copper obtained. Calculate the amount of cobalt from that of the oxide of copper found. (Liebig, *Annalen Chem. und Pharm.*, **65.** 244.) The process is less simple and convenient than that which depends upon the insolubility of cobalticyanide of mercury.

Cobalticyanide of Mercury.

Principle. Insolubility in water.

Application. Estimation of Co in a solution of cobalticyanide of potassium. Separation of Co from Ni.

Method. Neutralize the solution of cobalticyanide of potassium almost, but not quite, completely with nitric acid. There is no harm in leaving the liquid slightly alkaline. Add a solution of mercurous nitrate, made as nearly neutral as possible, as long as any precipitate falls. Collect the precipitate upon a filter, wash it with water, ignite it intensely in the air, then reduce it to metallic Cobalt in a stream of hydrogen and weigh. Or, less accurately, instead of reducing the precipitate, ignite it in a crucible in free air and weigh as protosesquiOxide of Cobalt. The precipitated cobalticyanide is white, or gray from admixture of mercurous oxide, and heavy; it settles readily and is easily washed. The process is preferable to that which depends on the insolubility of cobalticyanide of copper (Wœhler, *Annalen Chem. und Pharm.*, **70.** 256). According to Gauhe (*Zeitsch. analyt. Chem.*, 1866, **5.** pp. 79, 83), this process is applicable only in the absence of chlorine compounds. In case chlorides be present some dichloride of mercury will go down with the cobalticyanide, and on igniting the mixture some of the cobalt will be lost through volatilization of chloride of cobalt.

Cobalticyanide of Potassium.

Principle. Solubility in water and power of resisting decomposition.

Application. Separation of Co from Ni and Mn.

Method. Add cyanhydric acid to the mixed solution of cobalt and nickel, which must be free from other metals excepting manganese, then a solution of caustic potash and heat the mixture until every thing has dissolved. Or, instead of cyanhydric acid and potash, use a solution of cyanide of potassium free from cyanate. Heat the reddish yellow solution to boiling in order to expel the free cyanhydric acid. By this process the double cyanide of cobalt and potassium in the solution is changed to cobalticyanide of potassium, while hydrogen is evolved. But the original double cyanide of nickel and potassium remains unaltered. The solution of cobalticyanide of potassium is neither decomposed by chlorhydric, sulphuric nor nitric acids, nor by potash lye, either at the ordinary temperature or on boiling. From the mixed solution the nickel may be precipitated by means of oxide of mercury as sesquiOxide of Nickel, or by chlorine as black perOxide of Nickel. Neither the chlorine nor the mercuric oxide has any action upon the cobalticyanide of potassium. The cobalt may then be precipitated from the filtrate as Cobalticyanide of Mercury (Liebig, *Annalen Chem. und Pharm.*, **65.** 244; **87.** 128). See Cyanide of Manganese, for the separation of manganese from cobalt.

This process has been subjected to a critical examination by Gauhe (*Zeitsch. analyt. Chem.*, 1866, **5.** pp. 75–83), who finds that though cobalt may be well nigh completely separated from nickel by means of oxide of mercury in a mixture of the double cyanide of nickel and potassium and cobalticyanide of potassium, the process is nevertheless ill suited for the analysis of mixtures which contain cobalt in any other form. But since the conversion of the double cyanide of cobalt and potassium into cobalticyanide of potassium can never be made absolutely complete, even by long continued boiling with an excess of cyanide of potassium, there will occur only comparatively few cases in which the use of oxide of mercury as above will be found advantageous. — But on the other hand it is easy to convert the double cyanide of cobalt and potassium completely into cobalticyanide of potassium by leading chlorine gas into the alkaline liquid. Hence that modification of Liebig's method in which chlorine is employed to precipitate the nickel is to be recommended. Excellent results were, in fact, obtained by means of it. It is to be observed, however, that the peroxide of nickel thrown down by the chlorine must not be ignited and weighed directly, since it retains very persistently no inconsiderable quantity of alkali. — It is possible indeed, as Gauhe shows, to change the double cyanide of cobalt and nickel into cobalticyanide of potassium by means of chlorine alone, in the same way that ferrocyanide of potassium is changed to ferricyanide by the use of chlorine, but it is nevertheless better in practice to change the larger part of the double cyanide into cobalticyanide by boiling with cyanide of potassium, as Liebig has directed, and to effect the completion of the process by the chlorine employed to precipitate the nickel.

With regard to the use of oxide of merçury as a precipitant of the nickel, Gauhe shows that though a satisfactory separation of cobalt from nickel can be thus made when the conditions are favorable, a little cobalt does invariably go down with the nickel. This source of error would be hardly worth mentioning, however, if there were no other objections to the process in which oxide of mercury is used. As Rose has previously remarked, the filtrate from the precipitate produced by oxide of mercury is apt to become turbid immediately after it has passed through the paper, and to deposit a yellowish white precipitate on cooling. This precipitate also contains a minute quantity of cobalt, whence it would appear as if the small amount of cobalt which goes down with the nickel when oxide of mercury is used is in the form of a difficultly soluble compound of cobalt and cyanogen, probably in part cobalticyanide of mercury, some of which remains dissolved at first, but afterwards separates in the filtrate and makes it turbid. The conception of H. Rose, who ascribed the turbidity to the presence of a difficultly soluble basic cyanide of mercury, is manifestly incorrect.

Cobalticyanide of Zinc.

Principle. Decomposability by acids.

Application. Separation of Co from Zn.

Method. Acidulate the solution of the two metals with chlorhydric acid, add as much of a solution of good commercial cyanide of potassium as is required to completely redissolve the precipitate which forms at first, taking care to use a distinct excess of the cyanide. Boil the solution for some time with occasional addition of a drop or two of chlorhydric acid, but not in sufficient quantity to make the solution acid. Finally mix the solution with an excess of chlorhydric acid, set the flask in an oblique position and boil until the cobalticyanide of zinc which is precipitated at first has all redissolved, and the whole of the cyanhydric acid not combined with the cobalt is expelled. Add an excess of soda or potash lye, and boil until the liquid is clear. The whole of the cobalt is now in the form of Cobalticyanide of Potassium, while the zinc is in the form of zincate of sodium. Precipitate the zinc as sulphide, and the cobalt as Cobalticyanide of Mercury. The separation of the two metals is said to be complete, and the process simple.

Cobaltite of Cobalt.

See protosesquiOxide of Cobalt.

Cobaltous Salts.

Principle. Oxidability of, with precipitation of the sesquioxide. See sesquiOxide of Cobalt. See also Chlorine.

Cochineal.

Principle I. Power of the coloring matter (carminic acid) to reduce oxygen compounds.

Applications. Valuation of cochineal and analagous substances, such as lac-dye, lac-lake and kermes.

Methods.

A. *Oxidation by Ferricyanide of Potassium in alkaline solution.* Dissolve one grm. of the powdered cochineal in 36 grms. of a weak solution of caustic potash, mix the solution with 24 grms. of water and pour in a standard solution of ferricyanide of potassium, drop by drop, from a burette until the reddish purple color of the liquid has changed to brownish yellow. To prepare the standard solution dissolve 5 grms. of the ferricyanide in water, and dilute to the volume of 1 litre. The process has merit. (Penny, cited by Schützenberger in his *Matières Colorantes*, 1867, **2.** 359).

B. *Oxidation by Chlorine* (applied in the form of bleaching powder). An old method, devised by Robiquet, consisted in determining how much of a solution of bleaching powder was required to decolorize the decoction of a known weight of cochineal; the value of the bleaching powder solution being determined at the moment of use by testing it against a special sample of cochineal kept as a type or standard. According to Schützenberger (*loc cit*, p. 359), the process was faulty, since the chlorine acted upon other organic matters besides the coloring substance (carminic acid).

Principle II. Power of the coloring matter to form insoluble compounds with aluminum and with lead.

Applications. Valuation of cochineal, etc.

Method A. Determine how many volumes of a standard solution of alum are required to precipitate completely the coloring matter in any given sample of cochineal, and compare the result with that obtained by operating upon a standard sample (Anthon).

Method B. Instead of alum, as in A, Bloch employs a standard solution of neutral acetate of lead, which precipitates carminic acid readily. An objection to the process is found in the fact that the lead salt is liable to precipitate other organic substances besides the coloring matter.

Principle III. Solubility in ammonia water.

Application. Assay of carmine lake.

Method. Heat the lake with ammonia water until the coloring matter has dissolved. collect, wash and dry at 100° the insoluble residue and weigh it.

Principle IV. Colorific power.

Applications. Valuation of cochineal, etc. (Methods A and B). Use as an "indicator" in processes of akalimetry (Method C).

Method A. The method of testing cochineal ordinarily employed in dye houses and print works consists either in dyeing a definite quantity of mordanted cloth with 1 gramme of cochineal, as in testing Madder, or in printing upon woollen a known weight of the cochineal, ground and mixed as it would be for application in actual practice. In either case a similar sample of cloth is dyed or printed with an equal weight of good cochineal kept as a standard sample. This method is simple and easily executed, and is esteemed. It does not give accurate results, however, in case the cochineal has been mixed with the coloring matter of Brazil wood. To detect the presence of the latter pour some lime water into a dilute decoction of the cochineal. If the cochineal is pure the liquid will become completely colorless, but it will retain a tolerably intense violet tint in case any brazilin is present. (Persoz.)

Method B. Weigh out 1 grm. of the cochineal to be tested, and 1 grm. of a good sample of cochineal kept as a standard, and boil each sample by itself in a litre of water during the same space of time. Filter into litre jars and dilute each filtrate to the volume of 1 litre; shake the solutions, pour an equal quantity of each into the tubes of a Colorimeter, and determine how much water must be added to the deeper solution to bring it to the same hue as the other.

Method C. The use of cochineal as an indicator has already been sufficiently explained under Alkalimetry, p. 19. Besides the merit of distinctness and extreme delicacy which permits the use of much more dilute solutions than can be employed with litmus, the indifference of cochineal to carbonic acid is a great advantage. Moreover the solution, when prepared as described under Alkalimetry, may be kept for any length of time in closed vessels without decolorization or alteration. According to Luckow, cochineal is quite indifferent to carbonic or sulphydric acids, since carminic acid is stronger than either of them. This is practically true for solutions of considerable strength. Hence a standard alkali for technical analysis may be made by simply dissolving a known weight of pure carbonate of sodium in a definite volume of water, and from the standard solution thus obtained a standard acid may easily be prepared. — In effecting the neutralization it is not necessary to expel carbonic acid by boiling, as it is when litmus is used. Still the presence of carbonic acid does tend to obscure the sharpness of the final reaction, and it is best to employ caustic alkali for the standard liquor, as is explained under Alkalimetry, especially in nice investigations. The influence of carbonic acid is seen at once when a caustic and carbonated alkali are operated upon side by side. With the caustic alkali the point of neutralization, or, rather of supersaturation, is shown by a prompt and decisive change from a tint in which orange predominates to one in which the orange disappears and violet is most marked. But in presence of carbonic acid the change is somewhat gradual, and though a red color is produced it is modified by an orange tint even in presence of a large excess of alkali, so that there is sometimes a disagreeable uncertainty as to the point of neutralization. When caustic alkali is used a trifle less of it will be found needful to neutralize a given volume of acid than is required of a carbonated solution, and no doubt will exist as to the point of saturation. — The indifference of cochineal to carbonic acid is a great advantage in nice analyses, in that the time consumed for effecting neutralization is without influence on the result. When litmus is used and the point of neutralization is reached, a short exposure to the air suffices to redden the liquid again. If the operator is obliged to proceed slowly, he will require somewhat more alkali than when he operates rapidly, for a portion of it is neutralized by carbonic acid from the atmosphere. With cochineal the result is independent of the small amount of carbonic acid that can come from the air. Moreover, the permanence of the color permits the operator to compare the products of several titrations one with the other. (Johnson, *American Journ. Sci.*, 1863, **35**. 282.) Cochineal cannot be used in the presence, even of minute quantities, of salts of iron.

Colorimetry.

A term applied to processes of analysis, in which the amount of a colored substance in a solution is determined by observing the intensity of the color. An instrument known as a *Colorimeter* is employed for measuring the depth of color, and the process consists either in comparing the solution to be tested with a standard solution of the substance, and diluting the stronger colored liquid until the two solutions are alike; or in comparing the solution to be tested with a number of different standards until one is found with which it agrees; or in varying the depth of the solution of the liquid to be tested until it exhibits the same intensity of color as the normal liquid, and then measuring the depth of the stratum; or in neutralizing the color of the solution to be tested by means of a plate of glass or a sheet of liquid of complementary color. In this last case the instrument employed is called a Complementary Colorimeter, and the depth of the stratum of liquid needed to obliterate the complementary color is measured, in order to determine the amount of the substance sought for.

Colorimetric methods are employed for the valuation of coloring matters, such as Indigo and Cochineal; for the estimation of Bromine and Iodine, cobalt, nickel, chromium, iron (see Acetate, Chloride and Nitrate of Iron), and the carbon in cast iron, and particularly for the estimation of Copper. Since the descriptions of the processes have been made with reference to the estimation of copper or some other special substance, it will be best to state them here in the same narrow sense. The methods evidently admit of application in the analysis of very many substances which form colored solutions.

A. *Jacquelain's copper test.* To prepare the standard solution of copper dissolve 0.5 grm. of pure metallic copper in nitric acid, supersaturate the liquid with ammonia water and dilute to the volume of 1 litre. Measure off 5 c. c. of the liquid into a glass tube, and seal the latter at the lamp. Dissolve in nitric acid a quantity of the substance to be analyzed, supersaturate with ammonia as before and dilute with water in a graduated cylinder to the volume of 200 c. c., or to 150, 100 or 50 c. c., if need be. Measure off 5 c. c. of the diluted liquid into a long tube, which has been graduated to tenths of centimetres and which has the same diameter as the tube that contains the standard solution. Set both the tubes in front of a sheet of white paper, and dilute the contents of the graduated tube with water until the same color is exhibited by the liquids in both the tubes. — Since the quantity of copper, 0.0025 grm., in the standard solution is known, and since the liquid under examination has been diluted to a determined volume, the amount of copper in the weighed sample can readily be calculated from the amount of water used in the process of dilution. If, for example, 50 c. c. of water had to be added to the 5 c. c. of liquid in the graduated tube, every 5 c. c. of the diluted liquid will contain 0.0025 grm. of copper, like the 5 c. c. of standard liquid in the sealed tube. So that the 50 c. c. will contain 0.025 grm. of copper. And if it happened that two grms. of ore were taken and that the ammoniacal solution was diluted to the volume of 200 c. c., it would appear that the sample contained 0.1 grm. of copper. — The only object in sealing the tube which contains the standard solution is to keep the latter from evaporating. It may sometimes be more convenient to have the standard liquid in an open tube. Thus in case the liquid to be tested is lighter colored than the standard liquid, proceed as follows:— pour 5 c. c. of the standard liquid into the graduated tube and 5 c. c of the liquid to be tested into a short ungraduated tube of the same diameter, and dilute the contents of the graduated tube until both the liquids have the same color. Note the quantity of water used and calculate how much copper is contained in each 5 c. c. of the diluted liquid. The further calculation will then be as above. (Jacquelain, *Comptes Rendus* 8me Juin 1846, through Kerl and v. Hubert.) The process is said to be exact and easy of execution. It may be employed for the analysis of regulus, and alloys such as brass, bronze, pack-fong, and gold and silver coin and ware. It can hardly be applied to the valuation of slags and ores which contain iron unless special pains be taken to remove the latter. (See below.)

B. *Von Hubert's modification of Jacquelain's test.* According to v. Hubert, several important considerations have been neglected in Jacquelain's description of his process. The internal diameter and length of the tube are not matters of indifference. Any given blue liquid would necessarily appear darker in a wide than in a narrow tube, since a thicker layer of it would be looked at. The tube ought to be so narrow that the liquid shall appear so light colored that the comparison of the blue tints can be readily made. An internal diameter of 1 c. m. is well suited for the purpose. The tube should be graduated to 35 c. c.; it is well to have it 46 c. m. long, in order that there may be plenty of room to shake its contents after the addition of the water. — A single standard solution is hardly sufficient, for certain differences in the tone of the blue color are noticeable when the standard solution contains much more or much less copper than the liquid to be examined. It is not easy, for example, to arrive at a satisfactory result when liquids containing 0.1 or 0.8 grm. of copper are compared with Jacquelain's standard liquid, which contains 0.5 grm. If quantities of 0.1, 0.2, 0.3, 0.4, 0.5, 0.6, 0.7, 0.8, 0.9 and 1 grm. of pure copper are dissolved in nitric acid, supersaturated with ammonia and diluted with distilled water to 200, 400, 600, 800, 1000, 1200, 1400, 1600, 1800 and 2000 c. c., it will be found that 5 c. c. of the solution prepared from 1 grm. of copper when compared with 5 c. c. of the solution prepared with 0.1 grm. will exhibit a different blue tone. The color of the former will be more "fiery" than that of the latter, and this peculiarity will diminish through the entire series of liquids above named, until in the solution which contains 0.1 grm. there will be seen a tinge of greenish blue. If only a single standard liquid were used some difficulty would be met, especially by unpractised eyes in attempting to compare with it liquids which contain more or less copper. The liquids showing the greenish blue tinge would be held to be deeper colored than they really are.

To avoid this difficulty v. Hubert prepares two standard liquids, the one from 0.1 grm. of copper and the other from 1 grm., and dilutes the former to 200 c. c., and the latter to 2000, so that the relation of 5 c. c. to 0.0025 grm. may remain unchanged. In case then a different tone of color appears on diluting the liquid in the graduated tube, and especially if this occurs at the very commencement of the process of dilution, the two standard liquids may be mixed in the short tube in such proportions that the color of the mixture shall appear to the eye identical with that of the liquid in the graduated tube. If the liquid in the graduated tube shows a brighter blue than the other, put more of the solution of 1 grm. of copper into the short tube, but if its tone tend toward greenish blue add more of the solution of 0.1 grm. of copper.

In preparing ores and products of the smelt works for analysis, take care to reduce them to the finest possible powder. Slags, in particular, must be finely powdered in order that strong chlorhydric acid may decompose them completely. Let the mixture of powder and acid stand for some hours at a moderate heat, or in the case of ores, until the sulphur which separates is no longer brown or green colored, but yellow. — The oxide of iron which falls when the acid solution of an ore is supersaturated with ammonia water carries down a certain amount of copper in insoluble combination, especially when the proportion of copper in the liquid is large. The copper thus dragged down may amount to as much as 0.9 per cent in a 20 per cent ore; it cannot be removed from the oxide of iron by long continued washing with hot water. To avoid this loss treat the moist, thoroughly washed, precipitate upon the filter with a very small quantity of concentrated chlorhydric acid, again add an excess of ammonia water to the solution thus obtained, collect the precipitate upon a filter, wash it with a little hot water, and add the filtrate and washings to the first ammoniacal solution.

The quantity of the substance to be weighed out varies with the amount of copper contained in it,— 4 or 5 grms. will be enough of an ore of 0.1 to 2 per cent; 2 grms. of an ore of 2 to 20 per cent, and 1 grm. of an ore of more than 20 per cent. — The first dilution of the solution should depend upon the depth of its color. In general, the solution of a 0.1 to 2 per cent ore may be diluted to 150 or 100 c. c. and an ore of more than 2 per cent to 200 c. c.,— in a graduated cylinder. Care must of course be taken to prepare the standard solutions and the solutions to be tested at some one common temperature as nearly as may be. To this end it is well to cool the liquids by leaving the vessels which contain them for half an hour or so in running water, or in freshly drawn well water. Since the number of c c. of water used is directly proportional to the per cent of copper in the weighed sample, it follows that if 2 grms. of a 60 per cent ore were taken, if the solution were diluted to 200 c. c. and 5 c. c. of the diluted solution measured out for the test, there would be obtained a column of liquid equal to 60 c. c. But it would be well nigh impossible to manipulate with such a tube or to mix the last drops of water thoroughly with the rest of the liquid. Hence, for all ores that contain over 40 per cent of copper, no more than 2 c. c. of the liquid to be tested should be measured out with a pipette into the graduated tube; but for ores of less than 40 per cent copper 5 c. c. of the liquid should be taken for the test.

In making the comparison between the standard liquid and the liquid to be tested, it is best to hold the two tubes parallel with one another in the right hand before a sheet of white paper, while with the left hand the paper is pressed firmly against the tubes. But in order to avoid disturbance from white stripes, due to the glass, which would be seen if the tubes were held vertically, it is well to hold the tubes in a slanting position, at an angle of about 45°, though still parallel with one another. In case the liquid to be tested is much darker than the standard solution, it should first be diluted to such an extent that it shall be nearly of the color of the standard, though still manifestly darker than the latter. Then observe whether the tone of color in the two liquids is the same, and if it be not, proceed to adjust the standard solution as above directed. When the standard has been brought to the required tone proceed to dilute further the liquid to be tested, adding the water at last, drop by drop, until the color of the two liquids is as nearly alike as it can be, and the liquid to be tested still left a shade darker than the standard. Leave the tube at rest for a minute, note the number of c. c. and tenths of c. c. of water which have been used, and again add water drop by drop, until the liquid to be tested appears a trifle lighter colored than the standard. Again note the amount of water used and take the mean of the two observations as the correct reading. The calculation has been explained in A.

In case the ore contains cobalt or nickel add an excess of powdered white marble to the acid solution to precipitate the copper, while the cobalt and nickel are left in solution. Collect the precipitate, dissolve it in chlorhydric acid and proceed as above. To remove manganese, place the ammoniacal solution in a porcelain dish, add carbonate of potassium and heat the mixture for several minutes. The manganese will go down in the form of Carbonate while the copper remains dissolved; filter the blue solution and proceed as above.

The process has the merit of cheapness and accuracy. (Von Hubert, *Jahrbuch k. k. geologisch. Reichsanstalt*, Wien, 1850, **1**. pp. 415, 562.)

C. *Bishof's modification*. A cheap and simple apparatus devised by G. Bischof, Jr., for facilitating the comparison of copper solutions and other colored solutions, is figured in *Dingler's Polytech. Journ.*, **184**. 433, and *Zeitsch. analyt. Chem.*, 1867, **6**. 459.

D. *Heine's test for copper slags*. Heine was the first who published a description of a colorimetric assay of copper. (See *Bergwerksfreund*, 1839, **1**. 33 and **17**. 405.) His purpose was to have a ready method of determining the small proportion of copper in the furnace slag of the works at Mansfeld, but the process has been found useful for testing slags in many other localities. The details of it are as follows :— A number of standard solutions are prepared by dissolving a known weight of pure copper in nitric acid, supersaturating the solution with ammonia water and diluting with measured quantities of water to any desired extent. These solutions are kept in a row of glass stoppered bottles of colorless glass, of precisely the same capacity and of quadrangular shape. In the first bottle there may be, say 0.004 grm. copper to 25 c. c. of water, in the second 0.003, in the third 0.002 and in the fourth 0.001 grm. The intensity of the blue color will be proportionate to the amount of copper in the solutions, and by comparing any other solution with this series of standards the amount of copper contained in it may be inferred from its color, and the amount of water that has been added to it. It is essential that the liquids should be very dilute, otherwise the color will be so intense that no just comparisons can be made.

A weighed quantity of the slag was treated with concentrated nitric acid, the solution was supersaturated with ammonia water, the precipitated hydrates of iron and aluminum were allowed to settle or were removed by filtration, the filtrate was diluted with water until it had the same volume as the standard liquids; it was then poured into a quadrangular bottle similar to those which contained the standard liquids, and its color compared with that of the several standards. The value of the liquid under examination was deduced from that of the standard solution, which it most nearly resembled. If need were, the solution under examination was diluted with water until it resembled some one of the standard liquids. In that event only the ordinary volume was placed in the bottle for comparison, and the whole of the solution was finally measured in a graduated cylinder. During the examination the bottles were placed in a good light between a window and the observer.

Exceedingly poor slags, no matter whether they have been decomposed by fusion with carbonated alkali, or by acids should be filtered, after the addition of acid in order to remove silicic acid, and treated with sulphydric acid to precipitate the copper. The precipitated sulphide may then be roasted, the residue dissolved in nitric acid, and the solution supersaturated with ammonia. It will not do to add ammonia immediately to the original acid solution, since much copper would be carried down with the slimy precipitate; when much iron is present the iron and copper may be precipitated together with sulphide of sodium, and the sulphide of iron afterwards dissolved out with dilute sulphuric acid. Or the precipitated hydrate of iron may be redissolved and again precipitated, as in Hubert's process above. — Or the slag may be decomposed by Mohr's (*Zeitsch. analyt. Chem.*, **1**. 143) process, as follows: Place the powder in a porcelain dish, together with some sulphuric acid, water and nitric acid, cover the dish, evaporate to dryness and heat the residue until no more fumes of sulphuric acid are visible; then treat the residue with boiling water. Sulphate of copper together with a little iron will go into solution, while insoluble basic sulphate of iron, etc., will remain in the residue.

The process furnishes useful approximations, and is still somewhat extensively used for testing slags and shales which contain but little copper; it is less well adapted, however, for testing richer materials (v. Hubert; Kerl). Heine (*Bergwerksfreund*, **17**. 405) claims that for substances poor in copper, such as slags, his process is preferable to v. Hubert's; with such materials he can determine the copper accurately to 0.03 per cent, while v. Hubert can only be sure to 0.3 per cent. According to Heine, the oblong or quadrangular form of the bottles is a point of considerable importance. According to Müller (*Bergwerksfreund*, **18**. 118), the color of an ammoniacal copper solution depends very much upon the amount of ammonia present. For exact experiments a solution of ammonia of known strength should be prepared (see Acidimetry), and an account should be kept of the quantity added after the acid liquor has been neutralized. Nitrate of copper yields a more intense blue with ammonia than sulphate of copper. The blue of the solution appears to be more intense when held before a gray wall than before a white one. After a while the solution becomes greenish, and loses its intensity.

The processes of Heine and Jacquelain have been applied by Eggertz (*Berg und Hutten Zeitung*, 1862, p. 218,) to the estimation of copper in iron and iron ores.

E. *Earlier copper test of Dehms.* Place eleven. not too narrow test tubes, of similar diameter, in a rack and fill them with solutions of ammonio-sulphate of copper, of strengths varying from 5. 6, 7—15 equivalents of copper, to the litre, in terms of milligrammes, ($H = 1.$) Pour the solution to be tested into a test tube similar to the rest, and place the tube between each pair of the standard solutions until its proper place is determined. In this way there will always be found two limits between which the real value of the solution must lie. Blue glasses can be found in commerce, of a tint so nearly like that of the ammonio-copper solution, that they may be employed with advantage as standards of comparison instead of solutions of known strength. (Dehms, *Zeitsch. analyt. Chem.*, 1864, **3**. 218.) For the later methods of Dehms, and for A. Mueller's copper test, see below.

F. *Herapath's method of estimating Iron, Cyanogen and Sulphocyanogen.* For the analysis of waters containing minute quantities of iron Herapath employs a process founded on the colorific power of ferric sulphocyanide. A standard solution of perChloride of Iron, containing a little less than one one hundredth of a grain of metallic iron per cent, is prepared by dissolving 1 grain of iron in chlorhydric acid, with the addition of a little nitric acid, evaporating nearly to dryness and diluting to 10,000 gr. measures with distilled water at 15.5°; from this solution other standard solutions of different strengths are prepared as required, and colored by the addition of a few drops of sulphocyanide of potassium. A convenient quantity of the water to be examined, generally half a gallon, is evaporated to dryness, and the saline residue dissolved in chlorhydric acid. The solution is boiled with a few drops of nitric acid, filtered to separate silica, etc., and the iron precipitated as a hydrate, by means of ammonia. The hydrate of iron is collected upon a filter and well washed with water, after which it is dissolved in the smallest possible quantity of chlorhydric acid, and the solution placed in a phial or tube of known capacity. A few drops of a solution of sulphocyanide of potassium are then added to the iron solution, and the mixture is diluted with water until it reaches a mark upon the tube corresponding to 1000 water-grain measures. The depth of tint of the diluted solution is then compared with that of several standard solutions prepared as above described, contained in tubes or phials of similar diameter, in which certain known quantities of iron, ranging from the one-thousandth to the one-fourth of a grain, are contained in the same bulk of solution. At the moment of comparison the tubes are placed against a sheet of white paper and held between the eye and diffused daylight. By operating in this way the one-thousandth of a grain of iron per gallon may be estimated with the greatest readiness. For small quantities of material the process commends itself for convenience and accuracy. — In some cases it may be found preferable to employ but one standard solution; in that event the proportion of iron is determined by measuring the quantity of water which it required to lighten the tint of the liquid tested so as to render it identical with that of the standard solution, or *vice versa.* (T. J. Herapath, *Journ. London Chem. Soc.*, 1853, **5.** 27.)

The same principle has been employed by Herapath, (*Chemical Gazette*, 1853, **11.** 294,) for the estimation of iron in the ashes of plants and animal matter, in guano and other manures, etc. The ash or other matter is treated with boiling chlorhydric acid and the mixture evaporated to dryness. The residue is moistened with strong chlorhydric acid and warmed until everything soluble in the acid has dissolved; water is then poured upon the mass and the silica separated by filtration. The acid filtrate is heated to boiling after addition of a few drops of nitric acid, and supersaturated with ammonia water, which precipitates hydrate of iron, either pure or combined with phosphoric acid, and usually more or less contaminated with the phosphates of calcium and magnesium. Collect the precipitate on a filter and wash, dry and weigh it. Then redissolve it in boiling chlorhydric acid and transfer the solution to a burette. Pour 5, 10 or 20 measures of the solution into the colorimeter tube, add an excess of sulphocyanide of potassium, and dilute the liquid to the mark upon the tube, which may correspond to 500, 1000, 10,000 or 20,000 water grain measures. The tint is then compared with that produced by the gradual addition of a standard solution of ferric chloride to water impregnated with sulphocyanide of potassium, and contained in another tube of similar dimensions placed beside the first against a sheet of white paper. From the number of measures of the standard iron solution that are expended in bringing the fluids to a uniform tint, the amount of iron in the solution to be examined is ascertained.

The same principle has been applied also by Herapath (*Chemical Gazette*, 1853, **11.** pp. 295, 296) to the estimation of *cyanogen, cyanhydric acid and sulphocyanhydric acid.* When sulphide of ammonium and ammonia water are added to an aqueous solution of cyanhydric acid, and the mixture is heated with the addition of flowers of sulphur; or when a mixture of cyanhydric acid, ammonia water and quinquisulphide of ammonium is gently heated, the cyanhydric acid is soon converted into sulphocyanide of ammonium. The substance to be tested is treated either by distillation with dilute sulphuric acid, or in any other appropriate way, so that an aqueous solution of cyanhydric acid shall be obtained, and the latter is converted to sulphocyanide of ammonium, as has been explained. After such conversion, the liquid is carefully evaporated to dryness, the residue is dissolved in boiling water, the solution filtered and placed in a colorimeter tube, where it is treated with a slight excess of ferric chloride. The tint produced is then compared from time to time with that of a liquid in another colorimeter tube, containing a weak solution of ferric chloride, to which a standard solution of sulphocyanide of ammonium, or sulphocyanide of potassium, is gradually added, until the two correspond. From the amount of sulphocyanide solution thus expended, that of the cyanhydric acid in the substance analyzed is calculated. According to Herapath the process yields excellent results, and is specially valuable where minute quantities of cyanogen are to be determined. By using small sized tubes and operating with care, the three-thousandth of a grain of cyanhydric acid may be estimated with the greatest exactness. — When the process is employed for estimating sulphocyanide of potassium in saliva, the latter should be evaporated to dryness on a water bath, and the residue treated with dilute chlorhydric acid. The filtered solution is placed in a colorimeter tube, a drop or two of a solution of ferric chloride is added to it, and the mixture is diluted with water to the mark. The tint is then compared with that of an iron solution, to which a standard solution of sulphocyanide of ammonium is added, as above described. (Herapath, *loc. cit.*)

G. *Houton's Indigo test.* The colorimeter of Houton-Labilladiére (*Description d'un colorimétre*, Rouen, 1827) consisted of two cylindrical tubes of the same glass and of the same diameter and thickness. The tubes were of about half an inch bore and 13 or 14 inches long. The tubes were closed at one end, and at a distance of about five-sixths of the length from the closed end they were divided into two parts of equal capacity, the second half being graduated into 100 divisions. To receive these tubes a small wooden box was prepared, and blackened upon the inside. This box had two contiguous holes in its upper part for the reception of the tubes, and at one end, immediately behind the tubes, there were two rectangular slits corresponding to the diameters of the tubes. At the other end of the box, facing the tubes, was an eye-hole, so that by holding the box to the light, and looking at the tubes through this hole, any difference of hue in the colored liquids placed in the tubes could be readily appreciated.

The process of testing was as follows: Draw a fair sample of the indigo to be tested, reduce it to fine powder by trituration and sifting. Weigh out 20 grains of the powder, place it in a dry flask, pour upon it 400 grains of concentrated sulphuric acid of 1.845 sp. gr., and heat the mixture to 38° to 43° for about an hour, with occasional shaking. In the same way and at the same time treat 20 grains of a standard sample of indigo, which is kept for the purpose of comparison; when the indigo has dissolved let the liquids cool, and pour each solution into a separate two-quart bottle. Wash the matter which adheres to the flasks into the bottles, and add enough water to fill the bottles. Shake the bottles thoroughly, measure out 10,000 fluid grains of each solution into tall beakers or cylinders, and let the liquids settle. Pour a portion of the clear solutions into the colorimeter tubes to the 0° mark, place the tubes in the holes in the box and look at them through the eye-hole. If a difference in the depth of color is observed add water to the deeper colored liquid until the contents of both the tubes exhibit the same tint. After each addition of water close the tube with the finger, and shake the liquid thoroughly. The amount of water added is finally read off from the graduated portion of the tube.

H. *Mueller's Complementary Colorimeter.* A principle somewhat different from that of the methods thus far described has been employed by A. Mueller (*Journn. prakt. Chem.*, 1852, **60**. 474; 1855, **66**. 193; **99**. 337, 366, also in a special pamphlet entitled *Das Complementär-Colorimeter*, Chemnitz, 1854. Further, *Zeitsch. analyt. Chem.*, 1863, **2**. 143; 1864, **3**. 407). Instead of comparing similar quantities of two liquids, as above, Mueller determines the degree of intensity of the color of a liquid by finding how deep a layer of it is required to neutralize the color of a glass plate—made of glass so colored that it is complementary to the color of the solution to be tested,—and then measuring the thickness of the layer. — The process of Müller is evidently one of great merit. It requires a special, though simple apparatus, which may be obtained of the dealers in German chemical wares. Figures and descriptions of the apparatus will be found in the original papers of Mueller, cited above. Without the aid of a figure it would not be easy to describe the process clearly.

I. *The later processes of Dehm's*, depend upon the same principle as the process of A. Müller. Descriptions of them, and figures of the apparatus required, will be found in Fresenius's *Zeitsch. analyt. Chem.*, 1864, **3**. pp. 219, 494. This apparatus also can readily be obtained from Germany.

Copper may be determined as metallic copper, or as Oxide, Carbonate, Sulphide, or Sulphocyanide; by various volumetric processes depending upon the deoxidation of Copper Salts by reducing agents; and by processes depending upon the color of its solutions (See Colorimetry). Descriptions of the various methods of assaying copper ores in the dry way will be found in the treatises of Kerl, Percy, and Mitchell. Compare Watt's *Dict. Chem.*, **2**. 63. None of these dry assays come fairly within the scope of the present work, since the results obtained by means of them are, at the best, mere approximations to the truth.

Principle I. Insolubility in dilute acids when in presence of metallic zinc, iron or cadmium, and when subjected to the influence of a galvanic current. [Compare Copper Salts.]

Applications. Assay of copper salts and ores. Separation of Cu from other metals not precipitated by zinc, cadmium or iron. Notably from Fe, Zn, Mn, Ni, Co and the metals of the alkalies, alkaline earths and earths. Also from Bi and Pb (Method 1, B).

Method 1. *Precipitation of the copper by Zinc.*

A. The copper must be either in sulphuric or chlorhydric acid solution; in case nitric acid be present it must be expelled by evaporating the solution with sulphuric or chlorhydric acid. Place the solution, which should be tolerably dilute, in a weighed platinum dish, throw in a lump of pure zinc and acidulate the liquid, if need be, with chlorhydric acid, so that there may be a moderate evolution of hydrogen. But if the solution is already so acid that the evolution of hydrogen is violent, dilute with water until the effervescence moderates. Take care to cover the dish with a watch glass, in order that no particles of the liquid may be lost. The separation of copper begins at once, a large proportion of it being deposited upon the platinum in the form of a solid coating, while another portion separates in red, spongy masses. If the solution be concentrated the proportion of the loose, spongy precipitate will be larger. There is no need of heating the contents of the dish, though the deposition of the copper would be hastened by the application of heat; but there must always be enough free acid present to keep up the evolution of hydrogen.

After the lapse of an hour or two the whole of the copper will have separated. To make sure of this point take out a few drops of the supernatant liquid from time to time, and test them with sulphuretted hydrogen water; not even a brown tint should be imparted to the liquid. Prove that the last particles of zinc have been dissolved, by feeling for hard lumps with a glass rod, and observing that no further evolution of hydrogen occurs on adding some chlorhydric acid. Then press the spongy copper together with the glass rod, decant the liquid and immediately fill the dish with boiling water, again decant, add more water and repeat these operations until the washings are wholly free from acid. Rinse the copper once or twice with strong alcohol, to remove the water, dry at 100°, cool and weigh. — Instead of a platinum capsule the precipitation may be effected almost as well in a porcelain crucible, or dish, or in a beaker. A little more time will then be required, however, owing to the absence of galvanic action between the platinum and zinc,—unless indeed a weighed strip of clean platinum foil be placed in contact with the zinc,— and the whole of the copper will be deposited in the form of loose masses, not adherent to the dish.

Great care must be exercised in washing this spongy precipitate that no minute particles of the copper are carried away by the water and lost. The dry, spongy precipitate had better be ignited in a current of hydrogen before weighing it. I have observed that an appreciable loss of weight usually occurs, — varying in my experiments from 0.5 to 2 per cent of the dry copper,—when the copper precipitated by zinc from sulphuric acid solutions is ignited in hydr gen. (Storer, *Memoirs American Acad.*, 1860, **8.** 47.) In a single experiment, reported by Fresenius, no such loss was observed. When carefully conducted the process yields very accurate results. It is an excellent process, provided only that pure zinc can be procured. But it is essential that the zinc employed shall dissolve in chlorhydric acid without leaving any residue. Such zinc can rarely be found in commerce, and it cannot readily be prepared anywhere, excepting from ores of exceptional purity. The common statement of the books that zinc can be freed from lead by redistillation is an error. (Fresenius; Compare Wollaston's method of precipitating Cadmium, p. 56.) It is necessary to avoid using a piece of zinc much larger than is needed, as in that case after all the copper is precipitated, a galvanic current may be formed between the copper and the remaining zinc, which may cause a small portion of the dissolved zinc to be deposited upon the copper in the metallic state and to mix itself so intimately with the copper that its subsequent separation by chlorhydric acid is difficult. (H. Rose.)

The method now in question is a very old one. It was used, for example, by Vauquelin as long ago as 1798 (*Annales de Chimie*, **28.** 50). Pfaff in 1825 (*Pfaff's Handbuch analyt. Chem.*, **2.** 294) speaks of it in the following terms: Zinc is to be preferred to iron as the precipitant, since the precipitated copper may be removed from zinc more readily than it can be from iron; almost the whole of the copper will fall off in fine scales on the slightest agitation. In order to be sure that all the copper is precipitated test the liquid with sulphuretted hydrogen. Immediately after the precipitation collect the copper on a filter, wash it rapidly and dry it on blotting paper at not too high a heat, lest it oxidize. — For Mohr's description of the process see *Annalen Chem. und Pharm.*, **96.** 215. It differs in no essential particular from that given above, excepting that no platinum capsule is employed.

B. Instead of weighing the metallic copper, this metal may be dissolved in the solution of a ferric salt, and the amount of ferrous salt thus formed may be estimated by titration. See below, Principle III. If the solution contains nitric acid or metals reducible by zinc, such as bismuth or lead, the copper may be precipitated from an ammoniacal solution. To this end mix the acid solution with an excess of ammonia water, filter if need be, add a quantity of zinc filings or powder to the clear solution, and heat the latter moderately until its blue color has disappeared and all the copper is precipitated. Wash the copper thoroughly with hot water in the first place, then digest it with dilute sulphuric acid to remove the excess of zinc, and finally wash with water and add the ferric salt.

C. An important modification of the process has recently been introduced by Steinbeck, who uses the impure zinc of commerce to precipitate the copper, and subsequently estimates the latter by titration with cyanide of potassium. (See Copper Salts.) In this case the small amount of lead which ordinary zinc contains does no harm. — As applied to the poor ores of Mansfeld, the details of Steinbeck's process are as follows:—Weigh out 5 grms. of the powdered ore, roast it, if it contain much sulphur or bitumen, put it in a flask and pour upon it 40 or 50 c. c. of crude chlorhydric acid of 1.16 sp. gr. then add either 1 c. c. or 6 c. c, according to the character of the ore, of diluted nitric acid, prepared by mixing equal volumes of water and nitric acid of 1.2 sp. gr. Let the mixture digest at a gentle heat on a sand bath for half an hour and finally heat the liquid to strong boiling for 10 or 15 minutes. In this way not only the copper, but the iron, lead, zinc, nickel, cobalt and manganese which accompany it are dissolved as chlorides. No more than 0.01 to 0.03 per cent of copper is left in the insoluble residue. By operating in this methodical way no trace of nitric acid or of any oxide of nitrogen will be left in the solution. — Filter the solution into a beaker of about 400 c. c. capacity, in which a strip of platinum foil and a bit of zinc rod have previously been placed. The zinc may contain as much as 0.1 to 0.3 per cent of lead without detriment to the process. Copper begins to be precipitated during the filtration and the precipitation is speedily finished, thanks to the galvanic action between the zinc and platinum and to the entire absence of nitric acid from the concentrated hot solution. At the end of from one-half to three-quarters of an hour after the beginning of the filtration no trace of copper can be detected in the solution. The precipitated copper is in part attached to the platinum foil, while another part floats about in spongy masses. After the excess of zinc has been removed the copper is washed by decantation repeatedly with clear well water, and is then dissolved in nitric acid and titrated with cyanide of potassium (see Copper Salts, Steinbeck's process). In washing the precipitated copper care should be taken to decant the wash water into a large dish, and to collect any particles of copper which may be deposited there. (Steinbeck, *Zeitsch. analyt. Chem.*, 1869, **8**. 9.)

Method 2. Precipitation of the copper by Iron. (The so-called *Swedish Assay.*) Acidulate the copper solution with sulphuric acid, and in case any nitric acid be present expel it completely by evaporating the solution. Then dilute with water, heat the solution to boiling, and as soon as it boils place in it a plate of iron which has been made clean and bright by filing, or, better, a rather broad strip of sheet iron which has been cleaned by immersion in dilute sulphuric acid. Continue to heat the solution until all the copper has been precipitated, as is determined by testing with sulphuretted hydrogen, then remove the copper from the iron, wash it with boiling water and dry it. In order to remove any carbon which may have been derived from the iron heat the copper in a glass tube, first in a current of air and then in a current of hydrogen, in order to reduce the oxide of copper which has been formed. Allow the copper to cool in the current of hydrogen and weigh it, after air has been admitted to the cold tube. In case the copper is collected on a filter, and the latter is burned by itself after the copper has been removed from it as completely as possible, there will be formed from the filter ash a little silicate of copper which cannot be completely reduced by hydrogen. It may be reckoned as oxide of copper after subtracting the weight of the filter ash. Iron is less convenient than zinc as a precipitant of copper, as has been already set forth in Method 1. (Berzelius, and Dumenil. *Schw.*, (N. R.) **3**. 445, as cited by Pfaff in his *Handbuch analyt. Chem.*, 1825, **2**. 295.) But little can be added to-day to Pfaff's concise account of the Swedish assay, though it is often better to use chlorhydric than sulphuric acid. The use of hot air, as a preliminary to the ignition in hydrogen, is no longer regarded as necessary, it being customary to heat the copper directly in hydrogen as soon as it is sufficiently dry.

The ignition in hydrogen is essential, since the copper precipitated by iron is rarely so pure that it will not lose something on being thus ignited. This ignition is necessitated less by the oxide of copper formed during the process of drying the precipitate, than by the presence of an impurity consisting of organic matter, derived from the iron, which adheres to the spongy copper,—probably some one of the hydro-carbons which are generated by the action of acids upon iron. Mere oxidation of the copper could readily be avoided by washing away the water from it with strong alcohol before drying it, and indeed when no alcohol is used the bright surfaces of compact specimens of copper precipitated by iron will remain untarnished when rapidly dried in the air, at temperatures even as high as 110° or 115°. It is easy to convince one's self of the presence of impurity in the precipitated copper by heating a little of it in a small bulb tube with narrow outlet, such as is used for detecting arsenic. A quantity of water will be seen to collect as a sublimate in the narrow, cold part of the tube, while the copper in the bulb becomes bright and lustrous. A distinct empyreumatic odor is at the same time perceptible. Carbonic acid is given off, as may be proved by testing with lime water, and ammonia also is sometimes evolved. The amount of impurity thus retained by porous spongy copper may amount to 2 or 2.5 per cent of the weight of the precipitate. In a series of 50 or 60 experiments, in which the copper was precipitated from sulphuric acid solutions, I found the loss of weight on ignition in hydrogen to vary from 1.5 to 4, or even 5 per cent. When precipitated from chlorhydric acid solutions the copper is usually crystalline and compact, and doubtless contains less impurity.

It has often been proposed to calcine the metallic copper to oxide before weighing it, or to convert it to oxide by treating it with nitric acid and subsequently igniting, but the operation is far less convenient and certain than the reduction by hydrogen. According to Reischauer (*Zeitsch. analyt. Chem.*, 1864, **3**. 139) it is as good as impossible to convert small quantities of finely divided metallic copper completely into cupric oxide by long continued ignition in a current of oxygen gas.

Selected pieces of the best Russian sheet iron, about 3.5 inches long by 2.5 inches broad, are well suited for the precipitation of copper. Only such pieces as have a perfectly smooth and even surface should be used, for those which are corrugated or uneven will be acted upon unequally by the acid copper solution, and some copper might be lost in the cavities thus formed upon the sheet. The corners of the bits of sheet iron had better be rounded off with a file, lest by their corrosion particles of metallic iron drop off and contaminate the copper. Before being used the sheet iron must be soaked in dilute sulphuric acid until the glazed coating of silicate of iron upon it is loosened to such an extent that it can be washed off. One of these bits of iron is made to lean against the side of the beaker in which the precipitation is effected, so that the largest possible surface of the iron shall be exposed, and that the sheet may be equally corroded upon both sides by the acid liquid.

At the close of the precipitation most of the copper will be found detached from the iron in one or more loose spongy masses. Any small particles of copper which remain adhering to the iron may be rubbed off with the finger, taking care not to disturb the black coating of impurities from the iron, which forms upon the surface of the latter beneath the copper. Care must be taken, moreover, to avoid the precipitation of an insoluble subsalt of iron when the copper comes to be washed.

If after the removal of the iron the clear acid solution were largely diluted with hot water, an abundant precipitate of a basic salt of iron would immediately be formed, especially if the precipitation has been effected in a sulphuric acid solution, and the same thing would happen if hot water were poured upon the copper from which the clear acid solution had just been decanted. To avoid this trouble, wash the copper once or twice with cold water before any hot water is poured upon it, or, in case of need, rinse the precipitate once or twice with very dilute chlorhydric acid before washing it with water. — In conducting the assay it is of importance that the solution should be dilute, since, in that event, the precipitation of the copper goes on more regularly and is sooner completed; it should be warm, not only that the copper may be precipitated more rapidly, but in order to avoid the formation of a basic iron salt which is very apt to contaminate the copper when it is precipitated from a cold solution. This insoluble basic iron salt is liable to form during the precipitation even in warm solutions, if they are not sufficiently acidulated. If a solution is at the same time rather concentrated and but feebly acid, a portion of the copper may adhere to the iron so firmly that it cannot be rubbed off.

The chief difficulty in the assay is to distinguish between minute particles of metallic copper, which are liable to float away in the wash water and be lost, and equally minute particles of impurities from the iron which are apt to remain with and contaminate the precipitated copper. After some practice, however, the operator can overcome this difficulty and can obtain, constantly, tolerably satisfactory results. As has already been said, it is usually easier to obtain good results with chlorhydric acid solutions than in those acidulated with sulphuric acid (Storer, *Memoirs American Acad.*, 1859, **8.** pp. 43-48).

The following judgment of the process has recently been published by the Directors of the Mining and Smelting works at Mansfeld, (*Zeitsch. analyt. Chem.*, 1869, **8.** 2). The Swedish assay is inconvenient on account of the multiplicity of operations and the difficulty of conducting it. It is true that the results leave little to be desired when all the requirements are fulfilled, but the operator must have had long experience in order that he may know the assay thoroughly, and must continually exercise the greatest possible care. The chief difficulty lies in the operation of precipitating the copper. The precipitation must occur at a certain temperature, the solution must neither be too hot nor too cold, lest the copper fasten itself too firmly to the iron. It is necessary also to have the right quantity of acid, so that there may not be a great excess of it to occasion too violent action and to dissolve too much of the iron, while there is still enough to prevent the formation of basic iron salts which would contaminate the copper. The solution should, moreover, have neither too large nor too small a volume. It is specially important that the operator be present at the moment when all the copper has been precipitated, in order that the iron may be immediately removed and the acid liquors decanted from the copper. All these operations and precautions require much practice, experience, patience and care, in order that useful results may be obtained. It is true that from 4 to 6 assays may be carried on simultaneously, but the operator must watch them carefully, especially towards the close. The process is distinctly inferior to that of Steinbeck (see under Method 1) or to that of Luckow, by galvanic precipitation (see Copper Salts).

According to Johnson (in his New York edition of Fresenius), the precipitation by iron succeeds well when iron can be obtained, which dissolves in dilute acid without the separation of weighable quantities of black particles or flakes. If the copper solution be cold, dilute and nearly neutral when the iron is first placed in it, the copper has little adhesion to the iron, and may be readily detached from it for the purpose of weighing. If, as soon as the iron is coated with copper, some 20 c. c. of chlorhydric acid are added to the solution, and the liquid is heated nearly to boiling, and maintained at that temperature without actual ebullition, the rest of the copper is deposited as a spongy, coherent mass, which, with due care, may be removed from the iron and washed without falling to pieces. In case the copper cannot readily be washed by decantation, it may be gathered on a small filter and the latter subsequently burned.

It is a matter of great importance that the last traces of nitric and nitrous acids be removed from the copper solution before the introduction of the iron, for if either of these acids be present, even in minute proportion, it is well nigh impossible to precipitate the last portions of the copper. The nitric acid may be removed by evaporating the solution as nearly to dryness as can be done upon a water bath, with two successive portions of pure sulphuric acid, or until no vestige of blue crystals can be seen in the residue; or, better, the nitric acid solution may be twice evaporated to dryness with an excess of ordinary strong muriatic acid.

In order to bring a copper ore or alloy into solution with the use of the least possible quantity of nitric acid, Mohr (*Titrirmethode*, 1855, p. 363) heats the substance with chlorhydric acid, and adds nitric acid drop by drop, so long as any thing dissolves, taking care to wait after each addition until no more nitric oxide gas is given off. The chlorhydric acid solution is finally heated to boiling to destroy the last portions of nitric acid. Or, better, in case iron is not to be determined in the solution, some ferrous sulphate may be added and the mixture boiled until all the nitric acid is destroyed.

F. P. Pearson (*American Journ. Sci*,. 1869, 4 . 194) proceeds as follows:—Place a weighed quantity of the powdered ore in a porcelain dish, together with some Chlorate of Potassium, 5 grms. of an 18 per cent ore will be enough for an assay, and a small teaspoonful of the chlorate may be added to it. Invert a glass funnel, having a bent stem, in the dish, and pour into the dish rather more ordinary strong nitric acid than would be sufficient to cover the powder. Place the dish on a water bath, and after some time throw in small quantities of chlorate of potassium at frequent intervals, until free sulphur can no longer be seen in the dish. As a general rule, it is safer and more convenient to heat the mixture on a water bath than upon sand, though the oxidation of the sulphur is more rapid when the mixture of nitric acid and chlorate is heated to actual boiling. When the last particles of sulphur have been destroyed remove the funnel from the dish, rinse it with water, and collect the washings in a beaker by themselves. Allow the liquid in the dish to become cold, pour upon it a quantity of ordinary strong chlorhydric acid, rather larger than the quantity of nitric acid taken at first, evaporate the mixed solution to dryness and heat the dry residue, to render silica insoluble. Treat the residue with water and without filtering, wash the contents of the dish into the beaker which contains the rinsings of the funnel. Heat the liquid nearly to boiling, add to it some 25 c. c. of a strong solution of ferrous sulphate, and boil the mixture for 4 or 5 minutes, in order to destroy the small quantity of nitric acid which has previously escaped decomposition. The ferrous salt seldom or never acts instantaneously, but the reducing action proceeds rapidly when once begun. If need be, add more of the ferrous solution, little by little, until the entire contents of the beaker become dark colored or almost black, and no more gas is disengaged. To be sure that all the nitric acid has been reduced place a drop of the boiled liquid upon porcelain, and test it with ferricyanide of potassium. Filter the boiled liquid into a wide beaker and precipitate the copper on a sheet of iron, as above. It is easy to oxidize the whole of the sulphur in the ore by means of the mixture of nitric acid and chlorate; and by means of the ferrous sulphate the last traces of nitric acid may be got rid of far more quickly, easily and certainly, than by the old method of evaporating with several successive portions of chlorhydric acid.

According to Gibbs (*American Journ. Sci.*, 1867, **44**, 212), the finely pulverized ore may be mixed in a porcelain crucible with 3 or 4 times its weight of a mixture of one molecule of bisulphate of potassium and one of nitrate of potassium. Heat the mixture slowly to low redness,—best in a muffle. The sulphides are oxidized completely without the least frothing of the hot mixture. Add enough strong sulphuric acid to convert all the sulphate of potassium into bisulphate, and again carefully heat the crucible until its contents fuse to a clear mass. The crucible is not attacked by the flux, and the cold mass may usually be readily separated from the crucible. On dissolving the mass in water the whole of the iron and copper are found in the state of sulphates, and the insoluble silica may be separated by filtration. The whole operation requires about an hour. In the case of ores containing much bisulphide of iron it is best to heat the powdered ore as long as sulphur is given of before treating it with the oxidizing mixture.

Another method, proposed by Mohr (*Zeitsch. analyt. Chem.*, 1862, **1**. 143) is as follows.—Place 5 grms. of the powdered ore in a porcelain dish of 10 c. m. diameter, together with some sulphuric acid, water, and nitric acid. Cover the dish with a large watch glass and heat the mixture gently. Much spirting will occur, while a quantity of free sulphur separates and envelopes some of the ore. Dry the liquid by applying a stronger heat, remove the watch glass and increase the heat until the sulphur burns and the free acid is volatilized. Allow the dish to cool and add some more nitric acid and a little sulphuric acid; if red fumes appear it is a sign that some of the ore is still undecomposed. In that event evaporate to dryness and burn off the sulphur for the second time. In the case of rich ores this operation may have to be repeated a third time. The process has the merit of removing almost completely any lead, antimony or tin which might be contained in the ore.

For other methods of decomposing sulphuretted ores see above, under Method 1, Steinbeck's modification; and below, under Copper Salts, Parkes's copper assay.

Method 3. Precipitation of the copper by Cadmium. Instead of zinc, cadmium may be used with advantage for precipitating copper. Its merits are that it can be obtained pure, and that it dissolves with less violence than zinc in acid solutions. It should be employed in the form of a rod rather than in that of foil, since portions of the latter are apt to break off in small pieces, especially in a strongly acid solution, which are difficult to remove. (Classen, *Journ. prakt. Chem.*, **96**. 259.)

Method 4. *Precipitation of the copper by the galvanic current.* The process of Luckow, (*Zeitsch. analyt. Chem.*, 1869, **8**. 26. *Chemical News*, 1869, **19**. 221), for which a premium was recently awarded by the Directors of the Mansfeld copper mines, is as follows:—Weigh out several portions, each of 2 or 3 grms., of the finely-powdered, well sampled ore. Place each of the powders on a piece of sheet iron and roast them over the flame of a Bunsen lamp, taking care to stir them occasionally with a stiff platinum wire. With the ores of the Mansfeld district the whole of the bituminous matter may be thus burned off in about 7 minutes. When the pieces of sheet iron have become cold, brush off the powders with a camel's-hair brush, upon sheets of glazed paper, such as book-binder's use, and transfer them to small beakers having perfectly flat bottoms both within and without. The beakers should be 2 inches high, and about 1.5 inch in diameter. — Pour upon each of the powders 2 or 3 c. c. of nitric acid of 1.2 sp. gr., together with 10 or 15 drops of pure concentrated sulphuric acid. Place the beakers upon a sand bath, cover them with perforated watch glasses, or with funnels from which the stems have been cut, and heat them moderately at first. When the contents of the beakers have become almost dry increase the heat so as to evaporate and expel the whole of the free sulphuric acid. This operation will require from three-quarters of an hour to an hour. The purpose of the sulphuric acid is to increase the oxidizing power of the nitric acid and to combine with any calcium which may be present. It is well to add some 10 or 20 drops of chlorhydric acid to the original mixture of nitric and sulphuric acids, since the rapidity of the evaporation is thereby promoted, and the mixture of acid and ore is less liable to bump and spirt than it is when no chlorhydric acid has been added. Or instead of proceeding as above, measure out for each assay, from burettes kept charged for the purpose, 4 c. c. of a mixture of equal bulks of strong sulphuric acid and water, 6 c. c. of nitric acid, and about 25 drops of chlorhydric acid.

When all the sulphuric acid has been expelled, the beaker is removed from the sand bath and allowed to cool. The crust of solid matter at the bottom of the beaker is barely moistened with a mixture of 1 part of nitric acid of 1.2 sp. gr., and 6 parts of water, and then carefully pierced in several places with the sharpened point of a glass rod. The rod, as well as the cover and sides of the beaker, are then washed with the nitric acid, diluted as above, and the beaker is about half filled with this liquid. A few drops of a concentrated solution of tartaric acid are finally added to the solution, and a coil of platinum wire is carefully immersed in it. This wire may be about one-twentieth of an inch thick and 7.5 inches long; for two-thirds of its length it is twisted into the form of a flat coil, the outermost convolution of which is so large that it touches the sides of the beaker. The rest of the wire is left straight, and is made to project upwards, as if it were the continuation of an axis at the centre of the coil. By means of binding screws the upper end of the platinum wire is connected with the copper pole of a galvanic battery. The coil is immersed in the beaker so far that it shall rest upon the undissolved residue of the evaporation; its convolutions serve to keep the wire in place at the middle of the beaker. If the operation has been carefully conducted, the liquid above the residue in the beaker will be nearly clear. If it be very cloudy, as is likely to be the case when the liquid has spirted during the evaporation, add to it 1 or 2 c. c. of a concentrated solution of nitrate of barium, and mix this reagent with the rest of the liquid by carefully moving the platinum coil up and down in the beaker; then leave the liquid at rest for a few minutes before subjecting it to the action of the galvanic current. It is not necessary, however, to wait until the liquid is perfectly clear before beginning to precipitate the copper, nor is it essential that all the copper in the residue of the evaporation shall be brought into solution at the start. The last of it will gradually dissolve in the dilute acid during the process of precipitation. A cylinder-shaped piece of platinum foil is next placed in the beaker above the coil of wire. — To make this cylinder take a flat piece of foil about 2.5 inches long by 1.25 inch broad, and lap one end of it over a piece of platinum wire so that it shall be attached firmly to the wire, then bend the foil into the shape of a cylinder, or, rather, of a half cylinder, for there must be some open space left in order that the foil may be placed in the beaker so as to enclose the axis of the coil. This cylinder is immersed in the liquid to such a depth that its lower edge shall be about one-tenth of an inch (not more than one-seventh of an inch) above the flat coil, while the wire which projects upwards from the axis of the coil is at the centre of the cylinder. If the beaker is half full of acid about three-quarters of the height of the cylinder will be immersed. — The wire attached to the cylinder is fastened to the zinc pole of the battery by means of binding screws, properly fitted to the arms of an iron stand.

In a very short time after the admission of the current the cylinder of platinum foil which forms the negative pole of the battery is covered with a layer of metallic copper, which gradually spreads from below upwards, while bubbles of gas escape from the surface of the coil of platinum wire and materially facilitate the action of the dilute acid upon the copper in the residue of the evaporation. The simplest way of determining when all the copper has been precipitated is to add some dilute nitric acid to the liquid in the beaker, and to observe whether any copper is deposited upon the fresh surface of platinum which has been thus brought into the field of action. If no film of copper becomes visible in the course of 5 or 10 minutes, the process is finished. Or, a bit of thin platinum wire may be immersed in the liquid in contact with the cylinder, and watched to see if it remains bright. Or, test a drop of the liquid with sulphuretted hydrogen.

When all the copper has been deposited, lift the platinum cylinder out of the liquid, without disconnecting it from the battery, and immerse it several times in a beaker of pure hot water. Then loosen the binding screw, remove the platinum cylinder from the battery, wash off the water from the copper by means of alcohol thrown from a wash bottle, dry the copper at 90°—100° and weigh it. — The precipitation may be conveniently effected by means of Krüger's battery (figured in *Zeitsch. analyt. Chem.*, 1869, **8.** 31).

The time required for the precipitation varies according to the strength of the galvanic current. The whole of the copper will often be precipitated in the course of 3 or 4 hours, while at other times 5 or 6 hours will hardly be sufficient. In no case which has occurred in the Mansfeld experience has more than 8 hours been required when the battery was in order. It is consequently well enough, in most cases, to let the galvanic current act for 8 hours, for the sake of certainty. A single assay may sometimes be made in 5 or 6 hours, and never requires more than 10 hours. 18 assays can be finished with 18 small batteries in 12 working hours; or, if the deposition of copper be allowed to go on by night, 18 assays can be made with 9 batteries in 24 hours, or 24 assays with 12 batteries, without trouble.

Of the poor Mansfeld ores at least 2 grms. must be taken for each assay. It is well, for that matter, to take always one and the same quantity of any given ore. Instead of heating the ore upon iron it may just as well be roasted in the porcelain crucible in which it has been weighed; the operation thus becomes a trifle simpler and cleaner. By careful manipulation it is easy to avoid a turbid liquid when the residue of the evaporation is treated with dilute nitric acid. The addition of nitrate of barium should be carefully avoided, except in cases of marked turbidity, for the sulphate of barium formed is liable to attach itself to the precipitated copper, and either vitiate the analysis, or necessitate a double weighing of the product.

It is important not to disconnect the battery before the precipitated copper has been removed from the acid liquid in the small beaker, lest some of it be dissolved by the free nitric acid there. The battery should be disconnected at the moment when the copper is immersed in the beaker of hot water.

In order that the process may succeed with rich copper ores, some slight modification of the foregoing process is required. The following details are adapted for a 60 per cent sulphuretted copper ore, which contains also 0.3 to 0.4 per cent of silver. Digest 2 grms. of the powdered ore with nitric acid, until the sulphur separates in small globules. Pour into the solution, from a burette, dilute standard chlorhydric acid, until all the silver is precipitated, and note how much of the acid is required. This standard acid is of such strength that 1 c. c. of it is equivalent to 1 milligrm. of silver (see Chloride of Silver). Neutralize the solution with ammonia water, then add 15 or 20 drops of nitric acid, and filter. Dilute the filtrate to the volume of 200 c. c., and proceed to precipitate the copper by means of the galvanic current, as explained above. The cylinder of platinum foil must present a larger surface than before, in order that it may receive the comparatively large quantity of copper, and that the time of precipitation may be shortened. Special care must be taken that the copper is deposited in a compact state, and not as a loose sponge. A battery of 5 small, or 4 somewhat larger, lead-zinc couples will be needed,—one that can generate from 50 to 75 c. c. of mixed oxygen and hydrogen in 30 minutes, by the decomposition of water, will be sufficient. A current weaker than this would precipitate the copper very slowly or incompletely, while with a stronger current the copper is liable to be deposited in too loose a form. The presence of sulphates in the solution is to be avoided as far as possible. So, too, as regards chlorhydric acid, every trace of an excess of it hinders the compact precipitation of the copper.

It is of the first importance that the precipitated copper be removed from contact with the acid liquid before the battery is disconnected, since owing to the large surface of the platinum cylinder the nitric acid would be able to act upon the copper to a considerable extent even in a very short time. The easiest way is to wash out the acid liquid from the beaker, without disturbing the platinum cylinder or interrupting the current. To this end set a glass tube, attached to a funnel, in the beaker, so that the bottom of the tube reaches to the bottom of the beaker, and let rain water run through the funnel out of a vessel placed above it so that the acid liquid may be pressed upwards, and flow away over the edge of the beaker, or through a hole pierced in its side near the top. When all the acid has thus been washed away disconnect the battery, and wash the copper with water and alcohol in the usual way. Copper may thus be precipitated completely, as a hard, firm, glistening coating upon the platinum, even from solutions which contain comparatively large quantities of it. (*Zeitsch. analyt. Chem.*, 1869, **8.** 37.)

Several other proposals to estimate copper by galvanic precipitation have been made of late years—to say nothing of the old experiments of Fischer. Thus Gibbs (*American*

Journ. Sci., 1865, **39.** 64) has shown that copper may be accurately estimated in this way in solutions of the sulphate:—The deposition being made upon a platinum capsule which forms the negative electrode of a Bunsen's battery of one or two cells in rather feeble action, and the current being so regulated that the copper may be precipitated as a compact, bright coating. The positive electrode, consisted of a stout platinum wire plunged into the surface of the copper solution at its centre. The time required for the precipitation varied from 1 to 3 hours.

So, too, Luckow, in 1865, (*Dingler's polytech. Journ.*, **177.** 296), insisted on the merit of the process as applied to sulphuric acid solutions of copper ores, salts and alloys. At a later date Luckow (see *Zeitsch. analyt. Chem.*, 1869, **8.** 23) extended his process to the analysis of nitric acid solutions of copper. He found that even a comparatively feeble galvanic current is competent to precipitate the whole of the copper in excellent condition, from solutions which contain no more free nitric acid than 0.1 grm. to the c. c. The precipitation goes on regularly and easily, and is far less influenced by the strength of the current than by the presence of free sulphuric acid in the solution. — The process has great merit, inasmuch as it admits of the previous easy separation of tin and antimony, as Oxides, and of silver as chloride, as has been explained in a previous paragraph. Even the presence of lead does no harm, since in a solution which contains free nitric acid the action of the galvanic current will cause all the lead to be precipitated in the form of peroxide, at the positive pole of the battery, while the copper is attaching itself to the negative pole. Of the various metals likely to occur with copper it may be said that those of the alkalies, alkaline earths and earths, as well as chromium, iron, cobalt, nickel and zinc, will not be thrown down by the galvanic current from acid solutions; that mercury, silver, and bismuth will be precipitated like copper, and that lead and manganese will be completely precipitated at the positive pole, as peroxides. Silver also, in part, will go down as a peroxide. In case a mixture of copper and mercury is operated upon, the mercury will be precipitated before the copper, in the form of liquid drops, and an amalgam of copper and mercury will form as soon as any copper is precipitated. Silver and copper are precipitated well nigh simultaneously, but bismuth only after most of the copper has been deposited. Arsenic and antimony are precipitated from a nitric acid solution, only long after all the copper has gone down.

According to Lecoq de Boishaudran (*Soc. Chim. de Paris*, 1867, p. 468), the process of precipitating copper by means of the galvanic current, is rendered less exact by the presence of sulphate of iron, for the ferric salt into which the latter is converted at the positive pole of the battery, rapidly acts upon copper in presence of a free acid. To avoid this difficulty, Lecoq (*Ibid.*, 1869, p. 35) directs that the apparatus be arranged in such manner that the copper may be freed from the liquid as rapidly as possible, without interruption of the galvanic current. At the close of the precipitation he draws off the liquid through a syphon, taking care to push down the positive electrode, meanwhile, so that it may almost touch the bottom of the platinum crucible which forms the negative electrode, in order that the current may not be interrupted. Successive quantities of dilute sulphuric acid are then poured upon the metal, without interrupting the current, until the mother liquor has been completely removed. The washing is then finished with hot water. By operating in this way accurate results may be obtained, unless a very large excess of ferric salt is present. But since the loss of copper is proportional to the amount of ferric salt, when the other conditions of the experiment remain unchanged, it is possible to determine this proportion once for all, and to apply a correction for it. Lecoq uses 3 of Bunsen's elements, of medium size and weakly charged, as the source of the galvanic current, and allows the latter to act during 4 or 5 hours. In presence of nickel, cadmium, and zinc (but particularly of nickel), only a weak current should be maintained, lest traces of the foreign metal be precipitated with the copper. But the nickel or other metal thus precipitated can always be readily removed by washing with acid in the manner just described, the action of the syphon being merely made intermittent.

A simple method of precipitating copper by galvanic action was devised by Ullgren, in 1866 (see *Journ. prakt. Chem.*, **102.** 477, and more fully in *Zeitsch. analyt. Chem.*, 1868, **7.** 442). He proceeds as follows:—Bind a piece of moist bladder tightly over one end of a glass tube about 10 c. m. long and 14 m. m. wide; fill the tube five-sixths full of a saturated solution of chloride of sodium, and close its open end loosely with a cork through which a strip of zinc, which has been rolled into the form of a cylinder about 0.5 m. m. in diameter, is made to pass in such wise that it can be slipped up and down at will. At the upper end of the cylinder the strip of zinc is left flat, and some 4 or 5 m. m. broad; this flat strip, which serves as a conductor, is attached to a strip of platinum foil of similar width. — The copper solution to be analyzed is placed in a platinum capsule of from 35 to 40 c. c. capacity. This capsule is placed on a wet pane of glass upon the strip of platinum with which the zinc is connected, the bladder end of the tube is sunk a little below the surface of the liquid in the capsule, and the tube is clamped firmly in that position. The zinc cylinder is then immersed to a depth of 1 c. m. in the solution of chloride of sodium. The purpose of the wet glass beneath the capsule is to ensure a connection between the platinum of the capsule and the strip. The copper is deposited upon the inside of the capsule, and the only care to be taken is to avoid a current strong enough to occasion any considerable development of hydrogen, for in that event the copper would be but loosely attached to the platinum. No trouble of this sort will be experienced if the zinc is merely touched to the solution of salt in the beginning, and afterwards gradually immersed more deeply, from time to time, as occasion may require.

The platinum capsule should be clean and have been recently ignited, and the solution to be analyzed should be so dilute that it contains no more than from 0.1 to 0.15 grm. of copper for a capsule of the dimensions above given. The copper should be in the form of a sulphate, and the solution should be acidulated to such an extent that it may contain about 1 c. c. of free sulphuric acid for every 30 c. c. of solution. In case the proportion of copper in the substance to be analyzed is totally unknown, it is best to dilute the solution to a determined volume, to measure out a small quantity of the diluted solution and to add water to this portion until it has a volume of 30 c. c.

With an apparatus of the size described the copper will usually all be precipitated in the course of 4 or 5 hours. The copper is carefully washed, first with water, then with alcohol, to remove the water, dried and weighed. If arsenic be present, it will be deposited in black strips after a part of the copper has been precipitated. In case zinc (or cobalt, nickel, iron, or manganese,) is to be determined, as well as copper, a cylinder of cadmium or of aluminum may be placed in the chloride of sodium instead of the cylinder of zinc. When cadmium is thus used it is not easy to employ sulphuretted hydrogen to test whether all the copper has been precipitated; but by leaving the apparatus in action for a couple of hours after copper has ceased to be deposited on a fine needle placed in the solution, there will be no risk of leaving any copper unprecipitated.

Principle II. Fixity of when heated in hydrogen gas.

Applications. Purification of precipitated copper. Estimation of copper in mixtures of oxide of copper and metallic copper.

Method. Place the substance in a porcelain crucible, provided with a perforated cover, conduct a stream of dry hydrogen gas into the crucible and heat the latter with a lamp. After a while remove the lamp, and let the contents of the crucible cool in the atmosphere of hydrogen. Or, instead of the crucible, put the substance to be reduced in a porcelain boat, and place the latter in a tube through which hydrogen may be conducted. In some few cases a bulb tube of hard glass may be better than either the crucible or the boat. Instead of a perforated cover for the crucible, the bowl of an ordinary tobacco pipe may be inverted in or upon the porcelain crucible, and the hydrogen passed through the stem of the pipe (Allen).

Where a small quantity of oxide of copper has been collected upon a large filter, it is best to reduce the oxide with hydrogen, as above, after burning the filter. There will always be a considerable loss of material if the oxide be moistened with nitric acid, and then ignited for the purpose of re-oxidizing the copper which has been reduced by the combustion of the filter. — In the analysis of certain cyanides of copper it is not sufficient to heat the compound in hydrogen, since a quantity of carbon or nitrogen, amounting to about 3 per cent of the weight of the residue, remains mixed or combined with the copper. To remove this impurity the copper may be heated alternately in oxygen and hydrogen, until the weight of the metal remains constant. (Reischauer, *Zeitsch. analyt. Chem.*, 1864, **3**. 139.)

Principle III. Power of reducing ferric salts. (Compare din Oxide of Copper.)

Application. Estimation of metallic copper by titration, instead of by directly weighing it.

Method. In order that copper may be accurately estimated in this way, it must have been precipitated by absolutely pure zinc, in the first place, and afterwards digested with dilute sulphuric acid, until the last traces of the precipitant have been dissolved. Pour upon the washed but still moist spongy copper a quantity of a solution of ferric chloride, free from any contamination of ferrous salt, and acidulate the mixture with chlorhydric acid. The copper will dissolve speedily, while a corresponding quantity of the ferric chloride is reduced to the condition of ferrous chloride:—

$$Cu + 2FeCl_3 = CuCl_2 + 2FeCl.$$

When the copper has all dissolved, dilute the liquid, and determine how much of the ferrous salt has been formed, by titrating with permanganate of potassium. Every 56 parts, by weight of iron found, correspond to 31.7 parts of copper. (Fleitmann, *Annalen Chem. und Pharm.*, **98**. 141).

If it were only possible to obtain pure zinc readily and cheaply, this process would have a certain value in cases where a large number of analyses have to be made simultaneously. It has of course no significance in the case where only a single determination is to be made. Were it not for the sake of saving time it would always be best to weigh the copper directly.

Properties. Copper fuses only at a white heat. Solid copper suffers no change in dry air, or in moist air free from carbonic acid, but gradually tarnishes in moist air which contains carbonic acid. Finely divided copper, such as is obtained by precipitation, oxidizes rather quickly in moist air, especially when the air is warm. When ignited in the air a layer of black oxide forms upon the surface of the metal. Copper dissolves easily in nitric acid, and in hot, strong sulphuric acid. But in chlorhydric acid and in dilute cold sulphuric acid it dissolves so slowly that it has often been said to be insoluble in these liquids. According to Mohr (*Titrirmethode*, 1855, **1**. 204), precipitated copper dissolves slowly in dilute chlorhydric and sulphuric acids, with reduction of the latter to sulphurous acid. The action is sufficiently rapid that the solution will give a distinct reaction for copper with sulphuretted hydrogen, ten minutes after the addition of the acid. For an elaborate set of experiments on the action of chlorhydric acid upon solid metallic copper, out of contact with the air see Lœwe, *Zeitsch. analyt. Chem.*, 1865, **4**. 361. Lœwe finds that while solid metallic copper is undoubtedly soluble to a certain extent, the action of the chlorhydric acid, out of contact with the air, is well nigh insignificant, unless it is concentrated. In proportion as the acid is more dilute, so much the less does it act upon the copper. Copper dissolves slowly in ammonia water if air be freely admitted to it, but out of contact with the air ammonia has no solvent action upon it. When brought in contact with an ammoniacal solution of a copper salt, or with a solution of chloride of copper in chlorhydric acid, out of contact with the air, metallic copper reduces the cupric salts in these solutions to the state of cupreous salts, which remain dissolved in the ammonia, or in the chlorhydric acid, as the case may be. Copper dissolves with considerable facility in solutions of ferric salts, especially when a free acid is present.

For the use of copper as a reagent in organic analysis, see under Carbon, p. 64. When pure copper is required for other purposes, as for the indirect assay of copper, described under Copper Salts, or that of iron, described under Ferric Salts, it is best to precipitate a quantity of the metal from a solution of the sulphate upon a clean iron plate. The precipitated copper is then boiled with chlorhydric acid, to remove the last traces of iron, washed dried, fused, cast into bars and rolled into thin sheets (Fuchs). Japanese copper, as found in commerce, is often pure enough for analytical purposes. — The following tests will show whether the copper is pure enough for use:—It must dissolve completely in nitric acid, and no trace of a precipitate of the hydrates of iron, lead, etc., must be left, even after long-continued standing, when the solution is saturated with ammonia water. Neither should any precipitate of chloride of silver appear when chlorhydric acid is added to the solution in dilute nitric acid. After all the copper has been thrown down with sulphuretted hydrogen, the filtrate should leave no residue on evaporation.

Copper Salts.

Principle. Oxidizing power of.

Applications. Estimation of copper in salts and alloys. Assay of copper ores. Separation of copper from Fe, Mn, Zn, Co, Ni and various other metals.

Methods.

1. *Reduction of the Copper Salt by Cyanide of Potassium.* The process depends upon the fact that when a solution of cyanide of potassium is slowly added to the ammoniacal solution of a copper salt the azure blue color of the latter gradually disappears, so that the liquid becomes light violet red towards the close, and at last completely colorless. The completion of the reaction is well defined.

According to Liebig (*Annalen Chem. und Pharm.*, **95**. 118), dicyanide of copper, cyanide of ammonium, caustic potash and free cyanogen are formed, and the latter, acting upon the free ammonia, gives urea, cyanide of ammonium and formiate of ammonium.

A. *Parkes's copper test.* To prepare the standard solution of cyanide of potassium proceed as follows:—Dissolve 2000 grains of photographer's cyanide of potassium, or about 2880 grains of the ordinary cyanide, in 4 pints of water. On the other hand clean some pure electrotype copper, by means of chlorhydric, or dilute nitric acid, and wash and dry it. Weigh out three pieces of the copper, each of from 5 to 10 grains. Dissolve them in separate pint flasks, in dilute nitric acid and boil the liquids to expel nitrous fumes. Dilute each of the solutions with water to the volume of about half a pint, supersaturate them with ammonia water, and allow them to stand until cold. Fill a Mohr's burette with the solution of cyanide of potassium, and allow the latter to flow, little by little, into one of the flasks until the blue color of the copper solution is almost discharged and is replaced by a faint lilac tint. Note the number of c. c. of the cyanide solution which have been expended, again fill the burette and titrate the liquids in the second and in the third flasks in the same way. Take the mean of the three results as the true value of the cyanide solution. Or if the results of the first and second trials agree closely it may be unnecessary to proceed with the third. — *The actual assay* is as follows:—Place a known weight of the powdered ore in a flask or beaker provided with a glass cover, and moisten it with strong sulphuric acid. Then add strong nitric acid and digest the mixture at a gentle heat until nitrous fumes cease to be given off. Add small quantities of nitric acid, from time to time, as occasion may require. When the evolution of nitrous fumes has wholly ceased, dilute the solution with water, reheat the solution, transfer it without filtering to a pint flask and dilute it with water to the volume of one-half or three-quarters of a pint. When the ore is a sulphide, it may be, with ordinary care, completely oxidized by the sulphuric and nitric acids; but if any globules of sulphur remain, they should be picked out after the dilution and burned, and the residue, if any there be, should be dissolved in nitric acid, and added to the main solution. If any difficulty is experienced in dissolving the basic sulphates formed, some chlorhydric acid may be added with advantage. — To the diluted solution of the ore add an excess of ammonia water, and let the solution stand until it is as cold as the air of the apartment. Then, without heeding the precipitated hydrate of iron which may be present, pour in the standard solution of the cyanide from a burette, gradually and cautiously, by small portions, until the blue color is nearly discharged, and there remains only a faint lilac tint. Take care, meanwhile, to shake the copper solution frequently. Read off the number of c. c. of cyanide expended, and calculate, from the known value of that solution, how much copper was contained in the sample. — From one-half to three-quarters of an hour will be required for the complete decolorization of the solution, according as more or less copper is present. The cyanide being added very slowly, especially towards the end of the process, the last tint should remain permanent, or nearly so, for about 10 minutes. To aid in recognizing the tint of the solution, a white glazed tile, or piece of white paper, should be placed beneath and behind the flask during the titration.

When sesqui-oxide of iron is present, it imparts a greenish appearance to the ammoniacal solution, and the proper tint of the solution is then best observed by placing the eye on a level with the top of the liquid; after a little practice the alteration in the tint of the oxide of iron, which occurs will afford a sufficient indication until near completion, the reddish brown color becoming more distinct as the assay is proceeded with. It will not do to remove the hydrate of iron by filtration before adding any of the cyanide, for the iron precipitate would then retain a portion of the copper solution with such tenacity that it could not all be removed, either by ammonia or by several hours' washing with boiling water, but if any difficulty is experienced in observing the tint towards the completion of the titration, the iron precipitate may then be removed by filtration without risk of losing much copper. When the assay is finished, the hydrate of iron is completely free from copper, since that which was retained at first gradually goes into solution as the decolorization proceeds. The iron thus freed from copper may be estimated by means of biChromate or Permanganate of Potassium (see Ferric Salts).

Certain metals, such as silver, nickel, cobalt and zinc, interfere with the titration, more of the cyanide of potassium being then required to decolorize the copper solution. Silver may be precipitated with chlorhydric acid, and removed by filtration from the original acid liquor; but when cobalt, nickel or zinc are present, the copper must first be precipitated, either as Copper, or as Sulphide of Copper, and redissolved in acid before the titration can be proceeded with. — According to Field (*Chemical News*, **19**. 253), manganese interferes in the contrary sense; less than the normal quantity of cyanide of potassium being sufficient to effect the decoloration when it is present. The manganese can be got rid of by adding carbonate of ammonium and a few drops of Bromine, and heating the solution. Arsenic does not interfere, except when present with iron. In that event arseniate of iron dissolves in the ammonia, and, in the presence of copper, forms a brownish green solution. This difficulty can be easily remedied by adding some solution of sulphate of magnesium; arseniate of magnesium is formed, and the solution, after the lapse of a few minutes, acquires the proper color. The assay can then be made, without filtration, in the usual way.

The standard solution of cyanide of potassium should be kept in bottles of green glass, since lead glass is rather easily acted upon by the solution. Even the green glass is slowly acted upon, a thin scaly deposit being formed. The cyanide solution, moreover, slowly decomposes on keeping. This circumstance is not of any material consequence where many assays are to be made, for considerable time is required to make any decided alteration in the strength of the solution. For practical purposes the standard solution will not require to be checked oftener than once a week, if it has been prepared from the best cyanide of potassium. A solution prepared from common cyanide of potassium can only be kept for a few days without becoming discolored and muddy, but when the solution is rapidly consumed after having been standardized, it answers very well for assay purposes. (Parkes, *Mining Journal*, 1851; through Percy's *Metallurgy*.) — The process has been extensively used, particularly for assaying ores which consist chiefly of oxides, carbonates, silicates and oxychlorides, and for assaying slags; in general, it has proved to be well adapted for assaying copper ores. Like all other methods, it requires experience to be skilfully conducted; but that it is capable of yielding correct results, within 0.1 or 0.2 per cent, has been fully proved by some thousands of assays made by various persons. Some difficulties will always be met with in taking up any new process, but no process should be condemned by an operator, because he fails to succeed in his first trials (Percy, *Metallurgy*, 1861, p. 479).

The process is improperly credited to C. Mohr in some German treatises. (See *Fresenius' Quant. Anal.; Mohr's Titrirmethode*, 1856, **2**. 91, *and Annalen Chem. and Pharm.*, **94**. 198.) Flajolot (*Annales des Mines*, 1862, **2**. 313, and *Dingler's polytech. Journ.*, **168**. 217) also describes the process as if he (F.) were its discoverer.

Renewed attention has recently been called to the process by Steinbeck, to whom a premium has been awarded by the directors of the Mansfeld copper mines.

Steinbeck's method of procedure is as follows:—5 grms. of the ore are treated as described under Copper, and the precipitated copper is washed. The spongy metallic copper, some of which may still be adhering to the platinum foil, is then dissolved in moderately warm diluted nitric acid, prepared by mixing equal bulks of water and pure nitric acid of 1.2 sp. gr. For copper ores which contain no more than 6 per cent of copper, 8 c. c. of the diluted acid will be sufficient, but for ores richer than 6 per cent 16 c. c. of the acid should be taken. After a little practice the operator can judge from the bulk of the spongy metal how much of the acid to take. The solution of nitrate of copper is allowed to cool, is then mixed with 10 c. c. of a solution of ammonia, prepared by mixing one volume of ammonia water of 0.93 sp. gr. with 2 volumes of water, and is finally titrated with a standard solution of cyanide of potassium, of such strength that 1 c. c. of it corresponds to 0.005 grm. of copper, until the blue color disappears. — In the case of ores of more than 6 per cent copper where 16 c. c. of the diluted nitric acid are used, dilute the solution to the volume of 100 c. c. divide it into two portions each of 50 c. c., add to each portion 10 c. c. of the ammonia water, above described, and proceed with the titration. The small amount of lead with which the copper is contaminated, derived from the impure zinc employed as the precipitant, does no harm. It is precipitated as a hydrate on the addition of the ammonia, and imparts a slight milkiness to the liquid. The presence of small quantities of zinc (less than 5 per cent of the copper present) does not interfere with the visibility of the point of decoloration, but any larger proportion of zinc than 5 per cent should be excluded by carefully washing the precipitated copper. The solutions to be titrated must never be warm; they should always be allowed to acquire the temperature of the laboratory.

The standard solution of cyanide of potassium is made of such strength that 1 c. c. shall correspond to 0.005 grm. of copper. Since 5 grms. of substance are taken for an analysis, 1 c. c. will consequently represent 0.1 per cent of copper in accordance with the proportion.

$$5 : 0{,}005 :: 100 : x \; (= 0.1)$$

The number of c. c. of the cyanide solution expended in destroying the blue color of the copper solution multiplied by 0.1 will consequently give the percentage of copper in the sample. In works where many assays have to be made daily, new quantities of the cyanide solution must be standardized so often that no trouble is experienced from alteration of the liquid. It will be enough if the standard be checked each week. Starting with the powdered ore 6 assays can be made in this way within 4 hours, or 20 assays in a working day of 7 1-2 hours. (Stienbeck, *Zeitsch. analyt. Chem.*, 1869, **8.** 13; *Chemical News*, 1869, **19.** 207).

The chief merit of Steinbeck's process seems to consist in the methodical and systematic way in which the nitric acid and ammonia are employed. It had been shown long ago by Liebig (*Annalen Chem. und Pharm.*, **95.** 118), that the quantity and degree of concentration of the ammonia employed has a marked influence upon the character of the reaction between the cyanide of potassium and the copper salt. Experiments of Fresenius had also shown that neutral salts of ammonium could interfere to a certain extent with the reaction. But by operating always with similar quantities of ammoniacal compounds, for any one kind of ore, these sources of inaccuracy are practically eliminated. The idea of precipitating the metallic copper by means of commercial zinc is an excellent one; it follows naturally from the observation of Field that the presence of metallic lead does not interfere with the conduct of Parkes's process.

Fleck's modification. Instead of neutralizing the acid solution of copper with ammonia water, as directed by Parkes, Fleck (*Polytech. Centralblatt*, 1859, p. 1313) uses a solution of sesquicarbonate of ammonium (1 to 10), and heats the mixture to about 60°; and in order to make the end reaction plainer, he adds a couple of drops of a solution of ferrocyanide of potassium (1 to 20). Neither the blue color nor the clearness of the copper solution is altered by this addition. The cyanide solution is standardized against a copper solution of known strength. On dropping the cyanide solution into the blue copper solution warmed to 60°, the odor of cyanogen is perceptible, and the color gradually disappears. When the ammonio-copper compound is destroyed, the solution becomes red through the formation of ferrocyanide of copper, though no precipitate falls, and with the addition of the final drop of cyanide of potassium this red color vanishes in its turn, and the liquid is left completely colorless. — The process is convenient, and appears to be an improvement upon that of Parkes. According to Fresenius, the presence of ammonium salts may exert a certain disturbing influence, as in Parkes's method. Hence the standardizing of the cyanide and the actual assay, should be performed under circumstances as nearly similar as they can be made.

C. Mohr's method of estimating Cyanogen is the converse of Parkes' copper assay, the ammoniacal solution of a cyanide being titrated with a standard solution of a copper salt, until a persistent blue color pervades the liquid. The cyanide solution is placed in a porcelain dish, and is stirred continuously during the titration. Each drop of the copper solution gives a deep blue color at the point where it touches the ammoniacal cyanide, but this color disappears on stirring so long as there is any of the cyanide left undecomposed; the disappearance of the color is immediate at first, but becomes more gradual towards the close. (C. Mohr, *Annalen Chem. und Pharm.* **94**. 198, and **95**. 110; Mohr's *Titrirmethode*, 1856, **2**. 8). For certain degrees of concentration and quantities of ammonia the process gives correct results; but under different circumstances, as regards the amount and strength of the ammonia, a different quantity of the copper salt will be required for the same quantity of cyanhydric acid (Liebig, *Annalen Chem. und Pharm.* (**95**. 118).

2. *Reduction of the copper salt by Grape-sugar.* When the solution of a copper salt is mixed with a sufficient quantity of tartrate of potassium or tartrate of sodium, and an excess of caustic soda, a deep blue clear solution is produced. If this solution be then warmed and mixed with a sufficient quantity of grape-sugar, the whole of the copper will be thrown down after a short time in the form of dinoxide. The principle is applied both to the estimation of copper and of sugar, as will appear below.

A. *Schwarz's copper test.* Dissolve the ore or alloy to be tested in nitric acid, or if the substance to be analyzed is a salt, dissolve it in water. Mix the cold solution with a solution of normal tartrate of potassium in a capacious porcelain dish, and add an excess of caustic soda or caustic potash. Mix the dark blue liquid with a sufficient quantity of a solution of grape-sugar or milk-sugar, and heat the mixture on a water bath until the liquid shows a brown color on the border, which is a sign that the whole of the copper is precipitated, and that the alkali begins to act upon the sugar with formation of brown-colored products. When the precipitate has subsided, filter. The filtrate is, in most cases, of a deep brown color, and shows a muddy yellowish layer at the point of contact with the wash water. But this turbidity disappears on stirring the liquid, and is not due to the presence of copper. Wash the dinoxide of copper with hot water, until the washings are perfectly colorless; but leave in the dish any particles of the precipitate which may firmly adhere to it. Finally estimate the copper by means of a ferric salt, as explained under dinOxide of Copper.

B. *Fehling's method of estimating Sugar.* Two general methods for estimating sugar depend upon the principle now in question. Either there may be added to a solution of copper of known strength the exact quantity of grape-sugar required to reduce the whole of the copper salt to the state of dinoxide; or the copper solution may be added in excess, and the amount of dinoxide which separates determined. The former method is usually employed when practicable; the latter being resorted to whenever the liquid to be tested is so dark-colored that it would be difficult to determine the point at which the reduction of the copper salt and precipitation of the dinoxide is accomplished.

The quantity of copper salt reduced is proportional to that of the grape-sugar added: — 1 molecule of grape-sugar ($C_6H_{12}O_6$) = 180 reduces 5 molecules of oxide of copper (CuO) = 397. Or 100 parts by weight of anhydrous grape-sugar correspond to 220.5 parts of cupric oxide. Hence if the quantity of copper salt reduced be known, the quantity of grape-sugar required to reduce it may be readily calculated. Milk-sugar reduces the solution of the copper salt directly, but in a different proportion from grape-sugar; —where one equivalent of the latter reduces 10 equivalents of oxide of copper, 1 equivalent of milk-sugar will reduce only 7 or 8 equivalents. It is best therefore to convert milk-sugar to grape-sugar by boiling the solution for an hour or so with a little sulphuric acid, before subjecting it to analysis.

First method (applicable to clear solutions). To prepare the standard solution of copper, grind some fresh crystals of pure sulphate of copper to powder, dry the powder by pressing it between folds of filter paper until the particles of the powder no longer cling to the paper, weigh out exactly 34.639 grms. of the powder, and dissolve it in about 200 c. c. of water. Dissolve in another vessel 173 grms. of pure crystallized Rochelle salt in 480 c. c. of a pure solution of caustic soda of 1.14 sp. gr. Add the first solution gradually to the second, taking care to wash out the last portions of it, and dilute the blue mixture to the volume of 1 litre. 10 c. c. of this solution contain 0.34639 grm. of sulphate of copper, and correspond to 0.05 grm. of anhydrous grape-sugar. The solution should be kept in a cool, dark place in tightly stoppered bottles filled to the top; for by the action of light, or the absorption of carbonic acid from the air, the solution might be changed to such an extent that it would deposit dinoxide of copper on being heated.

When properly prepared and in good condition, the copper solution will remain unaltered, even when strongly boiled; it is only on the addition of grape-sugar that any dinoxide of copper is precipitated. Before using the solution it should be tested as follows: — Mix 10 c. c. of the solution with 40 c. c. of water, or with a dilute solution of caustic soda, if there is reason to believe that the liquid has absorbed carbonic acid, and boil the mixture for some minutes; in case even the smallest quantity of dinoxide of copper separates out in this experiment, the solution is unfit for use. — The solution of sugar to be tested must be highly dilute; it should contain no more than 0.5 per cent of sugar. In case the first experiment shows that the sugar solution is too strong, dilute it with a definite quantity of water, and repeat the trial.

For the actual analysis measure 10 c. c. of the copper solution, from a pipette or burette, into a small flask or porcelain dish, add 40 c.c. of water, or of a very dilute solution of soda, in case water should make the solution turbid, heat the mixture until it boils gently, and let the sugar solution flow in slowly, and by small portions, from a burette, divided to tenths of c. c. The liquid will exhibit a greenish-brown tint, after the addition of the first few drops of the sugar solution, due to the suspension of particles of the red dinoxide and yellow hydrated dinoxide in the blue liquor. In proportion as more sugar is added, the precipitate becomes larger, acquires a redder tint, and subsides more speedily. When the precipitate presents a deep red color, remove the lamp, allow the mixture to settle a little, and place the flask on white paper; or, in case a porcelain dish has been employed, set it in an inclined position, so that the faintest tint of bluish green may be detected. To make quite sure that the precipitation is complete, pour a small portion of the clear supernatant liquid into a test tube, add a drop of the sugar solution, and heat the tube. A yellowish red precipitate will form if the least trace of the copper salt has been left undecomposed. In case any precipitate does form, pour the contents of the test tube back into the dish or flask, and continue to add the solution of sugar until the reaction is complete. The volume of the solution of sugar which has been expended contains 0.05 grm. of anhydrous grape-sugar.

When the precipitation is finished it is well to try whether the operation has been a thoroughly successful one, that is to say, whether the solution is really free from copper, sugar, and brown products of the decomposition of sugar. To this end filter off a portion of the liquid while it is still hot, and heat some of the filtrate with a drop of the copper solution, and some of it with a little of the sugar solution; or, instead of testing with sugar, acidify the liquid and add sulphuretted hydrogen. The filtrate itself must be colorless, and free from the least tinge of brown. If any excess of copper or sugar is found by these trials, the experiment should be repeated. Except in practised hands, the first experiment will usually yield only an approximate result. In the second experiment take care to add almost the whole of the sugar solution required, and then proceed cautiously to the end, adding only two drops at a time. Bear in mind that the copper solution must be strongly alkaline from first to last. In case the sugar solution is acid add enough soda to the copper solution to counteract the acidity. The process yields very satisfactory results.

The second method (applicable to colored liquors) requires the same solutions as the first. For the analysis, transfer 20 c. c. of the copper solution to a porcelain dish, and add to it 80 c. c. of water, or of a highly dilute solution of caustic soda. Or perhaps better, take a still larger volume of the copper solution, and dilute it in the same proportion. Add a measured quantity of the dilute solution of sugar to the contents of the dish, but not enough to reduce the whole of the copper salt, and heat the mixture for about ten minutes on a water bath. When the reduction is complete, wash the dinOxide of Copper by decantation with boiling water, pass the decanted liquors through a weighed filter, then transfer the precipitate to the filter, dry at 100° and weigh. Or determine the copper by one of the volumetric methods, depending upon the reduction of ferric chloride, as explained under dinOxide of Copper. — It is important to wash the dinoxide in the manner above described, for if the liquid were allowed to cool in contact with the precipitate, the latter would gradually redissolve in proportion as it absorbed oxygen from the atmosphere and changed to cupric oxide (Fehling, *Annalen Chem. und Pharm.*, **72**. 106, and **106**. 75; Neubauer, *Archiv. Pharm.*, (2) **72**. 278).

The foregoing methods may be applied immediately to the estimation of sugar in the juice of grapes, apples, and other fruits, after the liquid has been properly diluted. So also to brewer's wort, the filtrate from distiller's mash and to diabetic urine. The other substances besides sugar contained in these liquids generally exert no perceptible reducing action upon the copper salt. In case there is reason to apprehend the presence of any reducing agent, beside sugar, mix the liquid in a measuring flask with a solution of acetate of lead, until the foreign matters are precipitated, then dilute with water to the mark, let the mixture settle, filter through a dry filter, and estimate the sugar in the filtrate (Fehling). In case the amount of grape-sugar in fermented liquids is to be determined with the greatest possible accuracy, the foregoing process of purification should be resorted to in order to remove a reducing substance known as glucic acid (Graham & others, *Journ. London Chem. Soc.*, **5**. 235), which accompanies the sugar in such liquids. — To clarify dark colored vegetable juices, heat a measured quantity of the liquid just to boiling; add a few drops of milk of lime, which will usually produce a copious precipitate of albumen, coloring matters, salts of calcium, etc., filter the liquid through bone-black, wash the precipitate thoroughly, and add the washings to the filtrate. Then dilute the filtrate to 10, 15 or 20 times its original volume (Neubauer). — To determine the milk-sugar in milk, remove the Casein by adding acetic acid at the temperature of boiling, clarify the whey with a little white of egg and filter. Boil the filtrate for 1 hour with a little sulphuric acid, dilute the liquid to 10 times the volume of the original milk, and proceed as above.

The processes in question apply only to grape-sugar, fruit-sugar, milk-sugar and glucose. In case cane-sugar is to be estimated in this way, as in the analysis of many vegetable juices, such as the juice of the sugar cane, beet root, maple, etc., it must first be changed to grape-sugar by boiling with an acid, as will be described under Sugar. Starch and Dextrin, and substances which contain these bodies must be treated like cane-sugar.

3. *Reduction of the copper salt by Iodide of Potassium.* When the solution of a cupric salt is mixed with an excess of a solution of iodide of potassium, diniodide of copper and free iodine are formed in accordance with the following equation:—

$$CuSO_4 + 2KI = CuI + I + K_2SO_4.$$

The insoluble diniodide of copper is precipitated, but the free iodine remains dissolved in the excess of iodide of potassium, and its amount may readily be determined by titration with hyposulphite of sodium, or in any other suitable way, as explained under Iodine. Each atom of iodine (= 127) found corresponds to one atom of copper, (= 63.4). — The actual analysis may be made as follows:—dissolve the copper compound in sulphuric acid, best to a neutral solution, though the presence of a moderate excess of free sulphuric acid does no special harm. Dilute the solution to some definite volume in a measuring flask, to such an extent that each 100 c.c. of liquid may contain from 1 to 2 grms. of cupric oxide. Put 10 c.c. of a solution of iodide of potassium (1 to 10) into a large beaker, add 10 c. c. of the copper solution, mix the two solutions and proceed without delay to estimate the iodine which is set free. The copper solution must not contain any free nitric or free chlorhydric acid, ferric salts or nitrous acid; and all other substances which would decompose iodide of potassium, must also be carefully excluded. If these rules be strictly attended to, tolerably accurate results may be obtained by the process, though it can hardly be regarded as a method of wide applicability. It is commended by Fresenius, and by F. Mohr, (*Zeitsch. analyt. Chem.*, 1864, **3**. 139) as well suited for estimating small quatities of copper. (De Haen, *Annalen Chem. und Pharm.*, 1854, **91**. 237.) It is important not to leave the mixture of cupric salt and iodide of potassium standing. The titration of the free iodine must be proceeded with at once.

De Haen's method has been recommended by E. O. Brown, (*Journ. London Chem., Soc.*, 1857, **10**. 65), for estimating copper in alloys, such as gun metal and bronze, and in those varieties of commercial copper which contain no very large quantities of lead or iron. According to Percy it has been extensively used at the Royal Arsenal, Woolwich, to this end. As described by Percy (Metallurgy), the process is as follows:— Dissolve 8 or 10 grains of the alloy in nitric acid, and boil, to expel nitrous acid. Dilute the acid solution with a little water, and add carbonate of sodium, until a portion of the copper remains precipitated. Add an excess of pure acetic acid to the mixture, transfer the solution to a pint flask, and dilute it further with water. Drop about 60 grains of iodide of potassium in the form of crystals, free from iodate, into the flask, and allow the salt to dissolve. Then proceed to titrate the free iodine with a solution of hyposulphite of sodium. In case iron is present, the copper must be precipitated as Sulphide, and the latter redissolved, before the process now in question can be employed. — According to Mohr, (*Titrirmethode*, 1855, **1**. 388) impurities may be got rid of by supersaturating the nitric acid solution with ammonia water, filtering and acidulating the filtrate with chlorhydric acid. The nitrous acid is thus changed to nitric acid, which does no harm, and the iron precipitated as a hydrate; though if much iron be present, some copper will be dragged down by the ferric hydrate, and so lost. Fresenius, on the contrary, does not approve of Mohr's suggestion. He could not obtain satisfactory results by operating in this way, since a solution of nitrate of ammonium mixed with some chlorhydric acid, will in a very short time begin to liberate iodine from solutions of iodide of potassium. — It will be observed that Brown's method of working, does away with this objection of Fresenius.

For a description of the process by Rümpler, see *Journal prakt. Chem.*, **105**. 193.

The process can be used in conjunction with Chloride of Tin, as follows:—Place a weighed quantity of the copper salt in a flask, dissolve it in water acidulated with sulphuric acid, add an excess of a solution of iodide of potassium, and a little thin starch paste. Shake the flask continually, and pour into the blue liquid from a burette a standard solution of protochloride of tin, until the blue color of the iodide of starch has disappeared. The liquid will then be white and cloudy from separation of subiodide of copper. Note how much of the chloride of tin solution has been used, and from another burette pour a one-tenth normal solution of biChromate of Potassium into the flask, until the blue color of the iodide of starch reappears. As only half the iodine is set free, 2 equivalents of copper must be reckoned for each equivalent of iodine. The method is of but narrow applicability, since it cannot be employed in presence of free nitric, chlorhydric or acetic acids, ferric salts, or other substances which decompose iodide of potassium. The solution must not be too dilute, nor should the mixture of copper solution and iodide of potassium be left for any length of time before the tin solution is added to the mixture. (Mohr, *Titrirmethode* 1855, **1**. 277.) Another method of estimating copper by means of iodide of potassium and chloride of tin, is referred to in *Zeitsch analyt. Chem.*, 1869, **8**. 14. Two standard solutions are employed, one of iodide of potassium, and the other of chloride of tin.

4. *Reduction of copper salt by Sulphite of Sodium.* Supersaturate the acid solution of copper with ammonia water, add a solution of sulphite of sodium to the mixture, and boil

until the cupric salt is reduced. Expel the excess of sulphurous acid by boiling with chlorhydric acid, and estimate the Copper by means of ferric chloride and permanganate of potassium. The method is said to be easier of execution than that depending on the reduction of the copper salt with grape-sugar, since it is difficult to wash out the whole of the sugar from the precipitated dinoxide (Terreil, *Comptes Rendus*, 1858, **46**. 230).

5. *Reduction of the copper salt by Hypophosphorous Acid.* Add to the cold, slightly acid, and not too dilute solution of sulphate of copper, an excess of a soluble hypophosphite, such as hypophosphite of magnesium, for example, and heat the mixture gradually, until after standing for some minutes between 80° and 90°, the hydride of copper has entirely separated in coherent masses. It is best not to allow the liquid to boil, in order that the sudden evolution of hydrogen gas which would result from decomposition of the hydride of copper, if the latter were heated to 100° may be avoided, and that the precipitate may be obtained in a spongy, coherent form. It is easy to determine when the precipitation is complete, by testing a drop of the solution with sulphuretted hydrogen water on a white plate.

No filter need be used, if the precipitation be effected in an assay flask; for the precipitate may easily be washed by decantation in the flask, and afterwards transferred to a porcelain crucible by the method of inversion, described under Chloride of Silver. After having been dried, the precipitate is gently ignited in a current of hydrogen, and is weighed as Copper. It is essential that the substance to be analyzed should be in the form of a sulphate, and that the solution should contain a little free sulphuric acid. If nitric acid be present, the precipitation will be incomplete; and when chlorhydric acid or a chloride is present, the process fails entirely, the copper being then reduced to dichloride which remains in solution.

The presence of iron in the form of sulphate does no harm, and copper can be perfectly well estimated also in presence of manganese, nickel and zinc. But the process is inapplicable in presence of arsenic and antimony, even when tartrates are added to the solution. As has been already stated, the solution must not be too dilute. But the precipitation of the copper will be complete if a saturated solution of the sulphate is diluted with not more than 10 times its bulk of water before the addition of the hypophosphite. The process yields excellent results. Gibbs and others, (*American Journ. Sci.*, 1867, **44**. 210).

As A. Wurtz (*Bulletin Soc. Chim. de Paris*, 1866, p. 194) has shown, hydride of copper is precipitated when a mixed solution of copper and hypophosphorous acid is heated to 70° and the hydride when exposed to the temperature of boiling, is reduced to metallic copper with evolution of hydrogen.

For Gibbs's method of bringing the metal in an ore to the form of a sulphate, see under Copper, precipitation of by iron.

6. *Reduction of the Copper salt by Hyposulphite of Sodium.* See Sulphide of Copper.

7. *Reduction of the copper salt by hot Hydrogen.* See Copper, fixity of.

8. *Reduction of the copper salt by metallic Iron.* (Compare the processes described under the head of Copper.) A process described by C. Mohr (*Annalen Chem. und Pharm.*, **92**. 97, and Mohr's *Titrirmethode*, 1855, **1**. 203), depends upon the reduction of a slightly acid solution of the copper salt by iron and titration of the ferrous salt produced.

$$CuSO_4 + Fe = Cu + FeSO_4.$$

Place the copper solution in a glass stoppered bottle, acidulate it very slightly with a few drops of chlorhydric acid, add about 0.25 part of pure chloride of sodium, and a number of pieces of soft iron wire. Close the bottle and let it stand at a temperature of 30° or 40°. The reduction proceeds rapidly, so that the whole of the copper will be precipitated in the course of 1 or 2 hours. It is essential that the solution must not be too strongly acid, lest some iron dissolve directly without taking part in the reduction. Neither must the solution be heated too strongly, lest a basic ferric salt be formed which has no action upon the permanganate of potassium. — When the reduction is complete, as may be seen by the absence of color in the liquid, and by the fact that a drop of it tested with sulphuretted hydrogen no longer gives a black precipitate, the solution is diluted with water to the volume of 300 or 500 c. c., and portions of 50 or 100 c. c. are taken out with a pipette and titrated with permanganate of potassium. (See Ferrous Salts.) The purpose of the chloride of sodium, as well as the small quantity of free acid, is to facilitate the reduction of the copper salt by increasing the conducting power of the liquid. According to Fresenius, the process gives unreliable results, sometimes as much as 12 per cent too high. It would hardly be applicable in any event to the analysis of ores or alloys containing iron, in spite of Mohr's proposal to determine the iron by itself in a separate portion of the substance, after reducing the ferric salt with zinc, and making an allowance for it.

9. *Reduction of the copper salt by metallic Copper.*

A. *Levol's indirect copper test.* Supersaturate the copper solution with ammonia water, and pour the clear solution into a wide mouthed bottle, which can be closed air tight with a glass stopper. Warm the bottle and dilute the solution with boiling water, until the bottle is almost completely filled with it. Place a weighed sheet of bright pure copper foil (see Copper as a reagent) in the liquid, close the bottle, and set it aside until the blue liquid has become completely colorless. Then take out the strip of copper, wash it clean and dry and weigh it. The loss of weight represents the copper which was contained in the original solution: —

$$CuCl_2 + Cu = 2CuCl.$$

The process, which is an adaptation of Fuchs's indirect iron assay (see Ferric Salts), has merit in so far as it is applicable to the analysis of copper solutions in almost any of the acids, even nitric acid, no matter how strongly acid the original solution may be. There must of course always be used such an excess of ammonia that the whole of the subsalt of copper formed shall remain dissolved in it. But on the other hand, the process has the great disadvantage of requiring a long time for its completion. With a liquid containing about 1 grm. of cupric oxide, and using a strip of copper weighing 4 or 5 grms., the experiment takes about 4 days for its completion. (H. Rose.) — According to Phillips (*Annalen Chem. und Pharm.*, **81**: 208) and Erdmann (*Journ. prakt. Chem.*, **75**. 211), the process is apt to give incorrect results, and to indicate more copper than the solution really contained. But Plessy and Moreau (*Berg und hütten. Zeitung*, 1860, p. 10, and 1861, p. 168) have sought to improve the process. See also the recent improvements in the corresponding iron assay, under Ferric Salts.

B. *Runge's indirect copper test* consists in boiling the solution of copper, acidulated with chlorhydric acid, but free from nitric acid, ferric salts and other oxidizing agents, with a weighed strip of metallic copper. After the fluid has been decolorized, the strip of copper is washed and weighed, as in A. The process has fallen into disrepute. For the action of acids and other agents upon the strip of copper, see Copper, properties of.

Creatin.

Principle I. Solubility in water and difficult solubility in alcohol.

Applications. Estimation of creatin in flesh

Methods.

A. The old method of Liebig (*Annalen Chem. und Pharm.*, 1847. **62**. 287, 293) of extracting creatin from flesh was as follows:— The finely minced flesh was well kneaded with water, and the fluid expressed by wringing the mixture, by small portions, in coarse cloths, care being taken to stir the residue of the first expression with water, and to again express the liquid. The extract was gradually heated to boiling to coagulate the albumin and coloring matters, and from the filtrate the phosphoric and sulphuric acids were precipitated by baryta water. The filtrate from the barium precipitate was carefully concentrated upon a water bath to about one-twentieth of its volume. The syrupy liquid was then set aside in a moderately warm flask to crystallize. The crystals were separated from the mother liquor by filtration [and washed first with spirit and then with strong alcohol?]. In evaporating the solution of creatin care must be taken that the sides of the dish above the liquid are no hotter than the liquid, lest a part of the liquid be changed, and a brown ring formed, which would color the liquid in case it dissolved in it. — The process of Liebig yields less creatin than the improved methods described below. It is seldom used now-a-days. (Compare, Nawrocki, *Zeitsch. analyt. Chem.*, 1865, **4**. 339.) A modification of Liebig's process recently employed by Valentiner (see *Zeitsch. analyt. Chem.*, 1863, **2**. 29) consisted in extracting the flesh with spirit, according to Stædeler (see below), and afterwards precipitating with baryta water, as above.

B. Stædeler (*Journ. prakt. Chem.*, 1857, **72**. 256) pours upon the minced flesh, or upon that which has been rubbed in a mortar with moderately coarse glass powder, from 1 to 1.5 times its bulk of spirit, heats the mixture gently upon a water bath with constant stirring, and presses out the liquid. After evaporating the alcohol, acetate of lead is added to the solution, the mixture is filtered and the excess of lead removed from the filtrate by sulphuretted hydrogen. The filtrate from the sulphide of lead is evaporated to a syrup, and the creatin allowed to crystallize from the latter. According to Nawrocki (*Zeitsch. analyt. Chem.*, 1865, **4**. pp. 338, 339), it is not well to use strong hot alcohol for extracting creatin from flesh, since some of the fatty matters which go into solution in the alcohol cannot be completely separated from the crystallized creatin without occasioning the loss of some of the latter. — In remarking on Stædeler's process, Neubauer (*Zeitsch. analyt. Chem.*, 1863, **2**. 25) admits that the extraction with spirit is easy, and that the creatin obtained is remarkably pure. He was repelled from the process, however, on finding that the mother liquor from which the creatin had crystallized contained a substance other than creatinin, precipitable by chloride of zinc, which would of course interfere with the determination of creatinin. But since Neubauer has himself shown by subsequent researches that there is really no creatinin in the mother liquor in question, this objection to Stædeler's process would seem to fall to the ground.

The difficulty just mentioned may be avoided by extracting the flesh with water instead of with spirit, and in this sense Neubauer (*Zeitsch. analyt. Chem.*, 1863, **2**. 26) has devised the following process: —

C. Thoroughly mix some 200 to 250 grms of the finely minced flesh with an equal weight of water, and heat the mixture during 10 or 15 minutes to 55° or 60°, upon a water bath, with constant stirring, so that the albumin may begin to coagulate. Decant the liquid and press out the residue by twisting it, by small portions, in a piece of cotton cloth. Stir up the residue with some 60 to 80 c. c. of water, and press it thoroughly a second time.

Heat the combined filtrates over a lamp, with constant stirring, until the albumin is completely coagulated, and filter the liquid after it has become cold. — In order to avoid the formation of much glue, it is important not to heat the moistened flesh any longer than is absolutely necessary. 10 or 15 minutes heating at 55° or 60° is amply sufficient to bring all the soluble ingredients of the flesh into solution, and to contract the flesh so that it may be easily pressed with the hand in a porcelain strainer. — To the thoroughly cold filtrate from the albumin add a solution of acetate of lead, as long as any precipitate is produced, but avoid adding any appreciable excess of the acetate. Let the mixture stand for an hour; then throw it upon a plaited filter, wash it twice and precipitate the excess of lead in the filtrate with sulphuretted hydrogen. After filtering the sulphide of lead, there will be obtained a liquid clear as water, which may be concentrated over a small gas flame at first, if care be taken that the liquid does not boil; it must afterwards be evaporated upon a water bath to the consistence of a thin syrup of bright yellow color. This syrup is set aside for 2 or 3 days in a cool place, in order that the creatin may crystallize. — The evaporation had better be made at first in a large dish, but when the liquid has been reduced to a volume of 40 or 50 c. c., it may be decanted and washed into a small dish, and there further evaporated to a volume of about 5 c. c. Care must be taken, however, not to push the evaporation too far, nor to expose the residue too long to the action of heat. By the continued action of heat the residue gradually acquires a brownish or brown color, while some of the creatin is destroyed. To avoid this decomposition of creatin it is well to frequently remove the dish from the water bath during the evaporation, and to swing the liquid about in the dish, so that no dry rings of residue may form upon the sides of the dish. In case the syrup obtained is brown, no confidence can be placed in the results obtained from it; but if the operation has been carefully conducted there will finally be found an abundant crop of crystals of colorless creatin beneath a light yellowish mother liquor.

The crystals of creatin are collected upon a weighed filter as follows: — Decant the mother liquor into a very small beaker, and let the crystals drain. The decanted liquid, though almost absolutely clear, contains some small isolated crystals of creatin; it may be mixed with 2 or 3 times its volume of alcohol of 88 per cent in order to make it fit for filtration, and then poured into a small plain filter which has previously been moistened with alcohol. After the mother liquor has been filtered, wash out the crystals from the dish by means of the filtrate, or in case of need, with alcohol of 88 per cent, wash the dish and crystals once or twice with alcohol of 88 per cent, and afterwards with absolute alcohol, dry at 100° and weigh. The anhydrous creatin thus obtained is completely colorless, leaves no trace of residue when burned, and is absolutely free from creatinin. — According to Neubauer, and Nawrocki also, there is no creatinin in normal flesh; that which has been obtained by other observers has resulted from the decomposition of creatin. See below, under properties of creatin. When the operation of extracting creatin, as above described, is carefully conducted, no creatinin can be detected in the mother liquor, from which the crystals of creatin have separated. (Neubauer, *loc. cit.*, pp. 31, 32.)

D. Nawrocki (*Zeitsch. analyt. Chem.*, 1865, **4**, 332) employs a combination of Stædeler's and Neubauer's methods, as follows: — From 15 to 30 grms. of muscle is carefully freed from lymph and blood by means of a cloth, and placed in a weighed flask full of 95 per cent alcohol. The flask is carefully closed, and again weighed; it is then warmed slightly, and left at rest during 12 or 24 hours, in order that the muscle may harden and be made fit for comminution. The alcohol is then decanted into a clean dish, and the muscle is cut into small pieces by means of scissors, and rubbed fine with some pure sand. Pour some almost boiling water upon the powder, and heat the mixture from 3 to 5 minutes (but no longer) upon a water bath, with constant stirring. — It is better to treat the muscle with hot water several times, and to press the residue between whiles, than to digest it for half or three quarters of an hour upon the water bath, for in the latter case it is easy to form a quantity of glue which disturbs the subsequent operations. After each digestion in hot water the muscle is pressed with the hand in a bit of linen cloth upon a funnel. The pressed residue should be taken from the cloth, and broken down with a glass rod before pouring the hot water upon it. From 3 to 5 doses of hot water are sufficient to extract the whole of the creatin. The aqueous extract and washings are now mixed with the alcohol in which the muscle was hardened, and the liquid is left upon a slightly warm water bath until the alcohol has evaporated. It is then filtered through very porous paper — such as that made from wool, which is used in the arts for filtering coffee. The perfectly cold filtrate is treated with the smallest possible quantity of acetate of lead which will suffice to completely precipitate the phosphoric and sulphuric acids and the albumin which has remained in solution. But, as Neubauer has urged, an excess of the acetate must carefully be avoided if correct results are to be obtained. When a sufficient quantity of the acetate has been added, the precipitate settles rapidly, and the liquid, which was opalescent before, becomes clear and transparent. Since the flocculent, adhesive precipitate easily clogs the pores of the filter, as much of the supernatant liquid as possible must be decanted from the precipitate. It is well also to place a filter of the coarse coffee paper within an ordinary filter in order to prevent the clogging of the latter. The precipitate is washed with water as quickly as possible. In case it sticks closely to the filter loosen it with a glass rod. The reason for haste in washing is that some of the albuminate of lead is liable to be decomposed by carbonic acid from the air, so that the filtrate might become cloudy from the presence of albumin. — The clear filtrate is freed from lead by means of sulphuretted hydrogen, the sulphide of lead is separated by filtration, and the filtrate is evaporated at a very moderate temperature. The temperature of the water bath must be closely attended to; a broad, flat dish should be employed, and the liquid within it should be frequently agitated to prevent the formation of dry rings. It is well to have the opening at the middle of the water bath so small that only a very small portion of the bottom of the dish can come into immediate contact with the aqueous vapor. By the exercise of due care and patience the liquid may be reduced to a volume of 2 c.c. without becoming yellow to any noticeable extent. Sometimes it remains absolutely colorless. After the syrup has been left for 2 or 3 days to crystallize, the crystals are carefully separated from the mother liquor and collected on a weighed filter. In case the mother liquor is too thick, which is seldom the case, or when it has almost completely evaporated on standing, it may be diluted with a few drops of 50 per cent spirit. The mother liquor should be allowed to drain away completely from the filter before beginning to wash. The washing is effected by means of 88 per cent alcohol followed by absolute alcohol. The crystals are dried at 100° and weighed between two watch glasses. It is best to weigh twice, or until the results of two weighings agree, in order to be sure that the creatin is anhydrous.

The process gives satisfactory results, and is doubtless the most convenient and exact method of estimating creatin hitherto published. It is to be commended in default of better methods, though it can lay no claim to the precision of the ordinary processes of inorganic analysis. — In Nawrocki's control experiments the weight of the creatin found varies, at the most, from 0.01 to 0.02 per cent of the weight of the muscle taken, and from 2 to 7 per cent of the weight of the creatin found. But since the quantities of creatin range from 0.01 to 0.02 per cent, and the limits of error from 0.01 to 0.02 per cent, the error of observation in an extreme case might amount to one-tenth of the entire quantity of creatin. Hence the utmost care must be exercised in the experiments, and in the drawing of conclusions from their results.

Principle II. Decomposition of by dilute, boiling acids, with formation of creatinin.

Application. Estimation of creatin in flesh.

Method. Prepare an aqueous extract of flesh, as described above, under Principle I, Nawrocki's method, and, after the alcohol has been evaporated, add to the solution a few drops of very dilute sulphuric acid to remove the albumin. To the filtrate from the albumin add from 3 to 5 c. c. of dilute sulphuric acid, prepared by mixing 1 volume of the monohydrated acid with 19 volumes of water, and boil the acidulated liquid 6 or 8 hours. Then decompose the sulphate of creatinin by heating the liquid with carbonate of barium, filter, acidulate the filtrate with acetic acid, concentrate it to a small bulk, mix with several volumes of alcohol, and after several hours filter and add an alcoholic solution of chloride of zinc, as explained under Creatinin. Or instead of using sulphuric acid, acidulate the extract of flesh with acetic acid, heat the mixture to coagulate albumin, and boil the filtrate for 6 hours with addition of about 2 per cent of acetic acid of 25 per cent. (Sarokow; Nawrocki, *Zeitsch. analyt. Chem.*, 1865, **4.** 340.) The process yields considerably less creatin than the direct method as employed by Neubauer and by Nawrocki, and described under Principle I.

Instead of operating as above, Valentiner (*Zeitsch. analyt. Chem.*, 1863, **2.** 29) boils the filtrate from albumin with chlorhydric acid, adds acetate of sodium to the boiled liquor, and precipitates the creatinin with chloride of zinc.

Properties. Creatin crystallizes in small rectangular prisms, which are transparent, brilliant and pearly, and permanent in the air. The crystals contain 2 molecules of water of crystallization, which are expelled at the temperature of 100°, 12·17 per cent of the weight of the crystals being lost. Anhydrous creatin, in which form the substance is usually weighed, must consequently be multiplied by 1.1374, in order to obtain the weight of crystallized creatin to which it is equivalent. 1 part of crystallized creatin dissolves in 75 parts of water at 18° in a very small quantity of boiling water, and in 9400 parts of cold, strong alcohol. It is much more easily soluble in weak alcohol than in strong, and is nearly or quite insoluble in ether. It does not act on vegetable colors. It dissolves in cold baryta water without change, but is decomposed when boiled with baryta water. It dissolves in dilute acids with but little, if any, change, but when boiled for a long time with dilute sulphuric, chlorhydric or acetic acids, or heated for some time with strong acids, it is completely converted to creatinin. It is changed to creatinin also, when its aqueous solution is heated, and this fact explains not only the small percentage of creatinin found in flesh by the earlier observers, but the false impression that creatinin is a normal constituent of flesh (Neubauer, *Zeitsch. analyt. Chem.*, 1863, **2.** 33; Nawrocki, *Ibid*,

1865, **4.** 336). Creatin dissolves in tolerably large quantity in a cold concentrated solution of chloride of zinc, and wart-like crystals of a compound of the two substances form readily, especially on the addition of alcohol (Neubauer, *loc. cit.*, p. 31).

Creatinin and Chloride of Zinc.

Principle. Sparing solubility of the compound in strong alcohol.

Applications. Estimation of creatinin and of creatin after it has been changed to creatinin, as above described.

Method. To estimate creatinin in urine measure out 300 c. c. of the liquid, add to it milk of lime until an alkaline reaction persists, and afterwards a solution of chloride of calcium as long as any precipitate falls. Let the mixture stand for an hour or two, then filter and evaporate the filtrate, and wash-water on a water-bath almost to dryness, as rapidly as possible. Stir into the concentrated warm liquor 30 or 40 c. c. of alcohol of 95 per cent, pour the mixture into a beaker, wash out the dish with a small quantity of alcohol, and let the mixture stand 4 or 5 hours in the cold, in order that all insoluble matters may separate. Filter the liquid through the smallest possible filter, and wash the precipitate with small portions of alcohol after it has been allowed to drain thoroughly. In case the volume of the filtrate is much larger than 50 c. c. set the beaker on a warm iron plate, let the liquid evaporate to the bulk of 40 or 50 c. c., and afterwards set it aside to cool. To the thoroughly cooled liquid add 0.5 c. c. of an alcoholic solution of chloride of zinc, absolutely free from acid, and of specific gravity 1.2. Stir the mixture persistently, since such agitation greatly facilitates the separation of the precipitate, then cover the beaker with a glass plate and set it aside in a cellar for 3 or 4 days. After the lapse of this time collect the crystals on a weighed filter, taking care to use the mother liquor, or filtrate for washing the last portions of crystals out of the beaker. When all of the crystals are upon the filter let them drain thoroughly, and finally wash them with small portions of alcohol until the washings are colorless, and no longer give any reaction for chlorine. The washing should be thorough, but not too long. Dry the filter and crystals at 100°, and weigh (Neubauer, *Annalen Chem. und Pharm.* 1861, **119.** 35).

According to Winogradoff (*Zeitsch. analyt. Chem.*, 1869, **8.** 100), some inconvenience will be experienced if the foregoing process is applied directly to the estimation of creatinin in diabetic urine. The sugar in such urine is reduced by evaporation to the condition of a viscous mass, which is very difficulty miscible with alcohol. Moreover, some of the creatinin is converted into creatin during the process of evaporation, which is necessarily tedious, so that less creatinin is obtained than was really present in the urine. According to Gaehtgens, this difficulty can be in good part overcome by mixing the syrupy residue with fine sand before treating it with alcohol, but in that event some of the sugar is dissolved by the alcohol, and is liable to interfere with the subsequent deposition of the compound of creatinin and chloride of zinc. — The best way of proceeding is to destroy the sugar by fermentation at the start. To this end mix 500 c. c. of the urine with a quantity of pure yeast, and set it aside in a tolerably warm place. After the completion of the fermentation add a mixture of milk of lime and chloride of calcium, and filter the mixture after it has stood for 2 hours. Evaporate the filtrate on a water bath to the consistence of a thick syrup, mix the residue with 100 c. c. of alcohol of 95 per cent, transfer the mixture to a flask, and let it stand for several hours in the cold with occasional shaking. Then filter, evaporate the filtrate to about 50 c. c., add the solution of chloride of zinc, and let the mixture stand in the flask for a week. Collect the deposit on a small weighed filter, and wash it with hot alcohol until the washings cease to give any reaction for chlorine, then weigh. In case the microscopic examination of the deposit reveals the presence of any foreign substance, confirm the foregoing result by determining the amount of zinc in the deposit. To this end ignite the contents of the filter, expel the chlorine with nitric acid, boil and wash the residue with water and dry, ignite and weigh the Oxide of Zinc. (Gaehtgens, *Zeitsch. analyt. Chem.*, 1869, **8** 101.)

The process is simpler when creatinin is to be estimated in pure aqueous solutions. The following details apply to experiments made by Neubauer (*loc. cit.*, p. 35) to test the accuracy of his process. A standard solution of pure creatinin was prepared by dissolving 0.8938 grm. of the substance in 2 or 3 c. c. of water, and diluting with absolute alcohol to the volume of 160 c. c. Portions of 50 c. c. each of this solution were measured out, and to each portion there was added 0.5 c. c. of an alcoholic solution of chloride of zinc of 1.195 sp. gr. After the mixture had been allowed to stand 48 hours in a cellar, the crystals were collected and weighed, as above described. From 99 to 99.2 per cent of the creatinin taken was found again in each experiment.

Nawrocki (*Zeitsch. analyt. Chem.*, 1865, **4.** 335) estimates creatinin in the mother liquor from creatin as follows: — Mix the solution with twice its volume of alcohol of 95 per cent. Let the milky mixture stand at rest during 18 or 24 hours, and filter to separate an albuminous substance which collects upon the bottom and sides of the beaker. Add to the clear filtrate 5 or 6 drops of a concentrated alcoholic solution of chloride of zinc. After 3 or 4 days collect the crystals of the compound of creatinin and chloride of zinc on a weighed filter, wash them first with alcohol of 70 or 75 per cent, and afterwards with absolute alcohol, until chlorine can no longer be detected in the filtrate. Dry at 100° and weigh. In applying the process for the estimation of the creatinin obtained by decomposing Creatin. as above described, the first drops of chloride of zinc produce a flocculent precipitate which must be removed by filtration before the solution is set aside to crystallize.

A process employed by Schottin (cited in *Annalen Chem. und Pharm.*, **119.** 40) and others for estimating creatinin in urine, consisted in evaporating the urine. extracting the residue with spirit, and treating the extract with baryta water, or with acetate of lead, — after the alcohol had been expelled from it by evaporation; the mixture was then filtered, the excess of barium or of lead was removed from the filtrate, and the latter evaporated to the consistence of a syrup from which the compound of creatinin and chloride of zinc was then obtained. This process, however, yields very much less creatinin than that of Neubauer, above described, and is objectionable, inasmuch as some of the creatinin appears to be changed to creatin during the evaporations, and so lost.

The original crude process of Liebig was to treat the urine with lime water and chloride of calcium, to separate the precipitate by filtration, and evaporate the filtrate until the greater part of the salts had crystallized. The mother liquor poured off from the crystals in question was treated with one-twenty-fourth of its weight of a syrupy solution of chloride of zinc, and the mixture set aside for some days to crystallize.

Properties. As precipitated from alcoholic solutions, the compound of creatinin and chloride of zinc is a crystalline powder of light yellowish color. Under the microscope this powder appears to be composed of well defined balls of various sizes, transparent and yellowish. When magnified 400 diameters, concentric stripes are plainly visible. When precipitated from a solution which contains sugar and chloride of sodium, the presence of transparent octahedral crystals may be detected with the microscope. In case there are amorphous white particles commingled with the compound of chloride of zinc and creatinin, a compound containing phosphoric acid is present, and this phosphoric acid would go to contaminate the zinc if that substance were determined for the sake of control. — 1 part of the compound is soluble in about 9200 parts of alcohol of 98 per cent, and in about 5700 parts of alcohol of 87 per cent, at a temperature of 15° to 20°. It is readily soluble in water, and slightly soluble in ether; yeast has no action upon it. Its composition may be represented by the formula,

$$C_4H_7N_3O,\ ZnCl_2.$$

It may here be mentioned that creatinin, when in aqueous solution, is slowly changed to creatin, especially when in presence of lime or ammonia.

Cyanhydric Acid.

[Compare Cyanogen, below, and in the finding list in the Appendix. See also Cyanogen Compounds.]

Principle. Volatility of.

Application. Estimation of cyanhydric acid in all cyanides from which it may be completely expelled by heating with chlorhydric acid. Separation of cyanogen from many metals.

Method. Fit a cork, or caoutchouc stopper, with two perforations to a small flask. Through one hole of the cork thrust the point of a Mohr's burette, and through the other hole a bent glass tube, the outer end of which is fitted air tight to a tubulated receiver, in which a small quantity of a solution of caustic soda has been placed. Attach to the receiver a U-tube, and place in it also a small quantity of a solution of caustic soda. Fill the burette with chlorhydric acid. Place the cyanide to be analyzed in the flask, together with some water, warm the mixture and, by opening the clip of the burette, let the chlorhydric acid flow into the flask by small portions, until an excess of it has been added. Heat the flask until the whole of the cyanhydric acid has been expelled, and estimate the acid in the distillate as explained under Cyanide of Silver, Cyanide of Silver and of Potassium, or Cyanide of Potassium. The reason why the chlorhydric acid is added by instalments is that free cyanhydric acid is easily changed to formic acid when heated with the mineral acids. By operating as above, however, there will never be any free chlorhydric acid in the flask until the close of the operation, after all the cyanhydric acid has been expelled.

Cyanide of Cadmium.

Principle. Solubility in an aqueous solution of cyanide of potassium.

Applications. Separation of Cd from Bi and Pb.

Method. See Carbonate of Bismuth. Compare Cyanide of Copper.

Cyanide of Cobalt.

Principle. Solubility in an aqueous solution of cyanide of potassium.

Applications. Separation of Co from Ni, Mn, Zn, Ca, Ba, Sr, Mg and Al.

Methods. See the Cobalticyanides of Potassium and of Zinc; Cyanide of Manganese, and Carbonate of Cobalt. For separating Co from Al it is only necessary to add an excess of a solution of cyanide of potassium, and avoid warming the mixture. Hydrate of Aluminum will remain undissolved. For separating Co from Mg, precipitate with carbonate of potassium, add an excess of cyanide of potassium to dissolve the carbonate of cobalt; add some more carbonate of potassium, and boil the mixture down to dryness. On treating the residue with water all the Carbonate of Magnesium will be left undissolved. (Haidlen &

Fresenius, *Annalen Chem. und Pharm.*, 1842, **43**. 141.)

Cyanide of Copper.

Principle. Solubility in an aqueous solution of cyanide of potassium; — or solubility in water of the double cyanide of copper and potassium.

Applications. Separation of Cu from Bi and Pb.

Method. See Carbonate of Bismuth, insolubility of in cyanide of potassium. — It is to be remembered that commercial cyanide of potassium contains carbonate of potassium, and that when a solution of it is added to a solution of bismuth or lead, carbonates of these metals are thrown down. — To prepare the filtrate for the estimation of the copper, boil it with some chlorhydric acid, to which a little nitric acid has been added. (Haidlen & Fresenius, *Annalen Chem. und Pharm.*, 1842, **43**. pp. 134, 135, 143.)

Cyanide of Gold.

Principle. Solubility in an aqueous solution of cyanide of potassium;—or solubility in water of the double cyanide of gold and potassium.

Applications. Separation of Au from Bi and Pb.

Method. See Carbonate of Bismuth. To estimate the gold in the solution, boil the latter with chlorhydric acid to expel the cyanhydric acid, and precipitate the Gold with a reducing agent.

Cyanide of Manganese.

Principle. Difficult solubility of the cyanide in an aqueous solution of cyanide of potassium.

Application. Separation of Mn from Co.

Method. Add a considerable quantity of cyanhydric acid to the solution of the two metals, then some soda or potash lye, and warm the mixture. The cyanide of cobalt which is precipitated at first, will dissolve completely, while most of the cyanide of manganese remains undissolved. Filter and boil the filtrate with finely pulverized oxide of mercury to precipitate the rest of the Oxide of Manganese. Ignite the two manganese precipitates together, and weigh as Manganite of Manganese. Estimate the cobalt in the filtrate, as explained under Cobalticyanide of Potassium. See also under that head the process to be followed for separating cobalt from a mixture of nickel and manganese.

Cyanide of Mercury.

Principle I. Solubility in water.

Applications. Separation of cyanogen from metals, as a step preliminary to its estimation.

Method. Many metallic cyanides may be completely decomposed by boiling them, either as solutions, or in the form of fine powder, with an excess of oxide of mercury, all the cyanogen being obtained in solution as cyanide of mercury, while the metals are changed to oxides. This fact is true, not only of the simple cyanides, but of many double cyanides, such as Prussian blue, ferro- and ferri-cyanides, and the double cyanide of nickel and potassium; but not of the cobalticyanides. In general terms it may be said that the process is applicable to those compounds of cyanogen which are not easily soluble in nitric acid, and precipitable by nitrate of silver. — The details of the process are as follows: — Boil the compound in water, to which a decided excess of oxide of mercury has been added, for a few minutes, or until complete decomposition has been effected. In case iron is present add nitric acid, drop by drop, to the hot liquor until the alkaline reaction has almost, but not quite disappeared, in order that the liquid, which would be muddy if not thus treated with acid, may be made fit for filtration. Collect the sesquiOxide of Iron on a filter, and determine the cyanogen in the filtrate by precipitation, as Cyanide of Silver using the requisite precautions (H. Rose' *Zeitsch. analyt. Chem.*, 1862, **1**. 297). Attempts to substitute oxide of silver for oxide of mercury in this process resulted in failure. Not only does some of the cyanogen escape in the form of ammonia on long continued boiling of the mixture, but the cyanide of silver finally obtained is impure, and there is always much less of it than there should be (H. Rose, *loc. cit.*, p. 304).

Principle II. Power of resisting nitric acid.

Application. Separation of Hg from Cd.

Method. Add a solution of cyanide of potassium to the mixture until the precipitate which forms at first has redissolved, then add an excess of very dilute nitric acid, and boil. The cyanide of mercury is not decomposed, but the cyanides of cadmium and potassium are changed to nitrates. When all the free cyanhydric acid has been expelled, throw down the cadmium as a Carbonate, and estimate the mercury in the filtrate therefrom. (Haidlen & Fresenius, *Annalen Chem. und Pharm.*, 1842, **43**. 145.)

Principle III. Decomposability of by various agents.

Application. Separation of cyanogen from mercury, as a step preliminary to its estimation.

Methods. Cyanide of silver is not precipitated when a soluble silver salt is added to a solution of cyanide of mercury. On the contrary, a somewhat soluble crystalline compound of the two cyanides is formed. Hence the necessity of resorting to indirect methods in order to convert the cyanogen of the cyanide of mercury into Cyanide of Silver, or into some other form in which it may be estimated.

A. One way is to distil the cyanide with chlorhydric acid, as described under Cyanhydric Acid, volatility of.

B. Another way is to precipitate the mercury as Sulphide of Mercury, though after that has been done it is still difficult to estimate the cyanogen in the filtrate.

C. Another way is to precipitate the Mercury as such, by means of phosphorous acid.

D. A far better method recently devised by H. Rose (*Zeitsch. analyt. Chem.*, 1862, **1.** pp. 294–296), is as follows:—Dissolve the cyanide of mercury in 300 parts of water, and add to the solution about 2 parts of nitrate of zinc dissolved in ammonia water. Add more than enough pure potash lye to the mixture to combine with the whole of the cyanhydric and nitric acids, and afterwards as much ammonia water as may be needed to make the solution clear. Add sulphuretted hydrogen water by instalments, until the last portion added gives only a perfectly white precipitate of sulphide of zinc. Let the mixture stand for quarter of an hour and then filter. Wash the mixed precipitate of sulphide of mercury and sulphide of zinc with very dilute ammonia water, which seems to take up the cyanide of zinc better than water would. The filtrate contains cyanide of zinc, nitrate of zinc and ammonia. It does not smell in the least of cyanhydric acid. To determine the cyanogen add nitrate of silver to the filtrate, acidulate with dilute sulphuric acid, wash the precipitate somewhat by decantation, and then heat it with a dilute solution of nitrate of silver in order to remove any cyanide of silver which may have been precipitated with it, before throwing it upon the filter. Wash, dry and weigh the Cyanide of Silver in the usual way. Good results may be obtained also by using only ammonia water and omitting the potash lye after the addition of the nitrate of zinc; and the point is one of importance in cases where potassium is to be determined, as well as when no potash lye free from contamination of chlorides is to be had.

E. Still another method devised by Rose (*Zeitsch. analyt. Chem.*, 1862, **1.** 296) depends on the reduction of the mercuric cyanide by means of metallic cadmium. The cyanide of mercury dissolved in from 25 to 30 parts of water is digested in a stoppered bottle for 36 hours with an equal quantity of cadmium filings. The liquid is then poured out and treated with nitrate of silver and nitric acid in the usual way. A small quantity of cyanide of cadmium which adheres to the bottle as a white deposit, must be dissolved in acetic acid and added to the rest. The process yields approximative results when applied as above, but is worthless when the solution of cyanide of mercury is highly dilute. Neither zinc nor iron can be used in place of the cadmium.

Cyanide of Nickel.

Principle. Solubility in an aqueous solution of cyanide of potassium.

Application. Separation of Ni from Co, Ca, Ba, Sr, Mg and Al.

Method. See Cobalticyanide of Potassium and Cyanide of Cobalt.

Cyanide of Palladium.

Principle. Insolubility in water.

Application. Estimation of Palladium.

Method. Add carbonate of sodium to the solution of protochloride of palladium until the latter is almost neutralized, then add a solution of cyanide of mercury, and leave the mixture in a warm place. If the solution is dilute some little time will be required in order that the yellowish white precipitate of protocyanide of palladium may subside completely. Collect the precipitate upon a filter, wash, dry and ignite it, and weigh the Palladium which is left in the crucible. In case the original solution contains any nitric acid it must be evaporated to dryness with an excess of chlorhydric acid, before proceeding with the precipitation, in order that the precipitate obtained may not deflagrate when ignited (Wollaston).

Cyanide of Potassium.

Principle I. Decomposition of by free iodine.

Applications. Volumetric estimation of cyanhydric acid and of cyanide of potassium.

Method. The reaction upon which the process depends may be expressed by the equation:

$$KCN + 2I = KI + ICN.$$

The standard solution of iodine may be prepared by dissolving 40 grms. of iodine in a litre of alcohol, and testing with hyposulphite of sodium to determine its real strength; or, perhaps better, by dissolving the iodine in a solution of iodide of potassium, as explained under Iodine. 5 grms. of the cyanide of potassium to be examined are weighed out, and dissolved in water to the volume of 500 c. c. 50 c. c. of this solution corresponding to 0.5 grm. of the salt are measured out into a flask of 2 litres capacity, and there mixed with 1 litre of water and 100 c. c. of Seltzer water; the purpose of the latter being to change the alkali present to the condition of a bicarbonate, which will not absorb iodine. The flask is placed on white paper, and the standard iodine solution poured into it, until a permanent yellow color is produced. The cyanide tested must be free from sulphide of potassium. The average strength of the commercial cyanide is said to be about 55 per cent. — If the substance to be examined is an aqueous solution of free cyanhydric acid, mix it carefully with a solution of soda to alkaline reaction, add some carbonic acid water, either Seltzer water or soda water, until the fluid no longer turns turmeric paper brown, and add the iodine solution as before, until the liquid becomes permanently yellowish. 2 atoms of iodine correspond to 1 equivalent of cyanogen, cyanhydric acid or a cyanide. The method gives satisfactory results, but is not applicable to the valuation of bitter almond water (Fordos & Gelis, *Journ. prakt. Chem.*, **59.** 255). According to Souchay (*Zeitsch. analyt. Chem.*, 1863, **2.** pp. 176, 180) the close of the reaction is indicated plainly enough. — The process may be used for separating CN from Cl, Br and I, by determining the cyanogen in one portion of the mixture as above, and throwing down and weighing everything in the form of insoluble silver salts (see Chloride of Silver) in another portion.

Principle II. Power of reducing copper salts.

Applications. Estimation of Cu and of CN.

Method. See Copper Salts, reduction of by cyanide of potassium.

For use as a reagent, cyanide of potassium may be obtained pure enough in commerce. It should be of a pure milk white color, free from specks of iron or charcoal, and should dissolve in water to a perfectly clear liquid. It should neither contain any sulphide of potassium nor silicic acid. In other words it should give a white precipitate when tested with acetate of lead, and should leave no residue insoluble in water when evaporated to dryness, after addition of chlorhydric acid. The uses of the cyanide are to precipitate certain metallic cyanides, as above described, and to dissolve certain other cyanides, oxides, sulphides and carbonates (see, for example, Carbonate of Cobalt). It is used also as a reducing agent, as explained under the heading Copper Salts. The commercial cyanide, prepared by Liebig's method, always contains some carbonate and cyanate of potassium. It precipitates many metals as carbonates, when added to their solutions.

Cyanide of Silver.

Principle I. Insolubility in water acidulated with nitric acid.

Applications. Estimation of cyanhydric acid in pure aqueous solutions, as well as in bitter almond water, cherry laurel water, etc. Separation of CN from Na, K, Cu, Sr, Ba, Mg, Zn, Cd, Mn, Co, Ni, Pb, Cu and Bi; but not from Hg. Separation of H_2CN from As_2O_3, As_2O_5, CrO_3, SO_3, P_2O_5, B_2O_3 and HFl. Precipitation of CN as a preliminary step to its separation from Cl, Br and I. Estimation of silver. Separation of Ag from Hg, Cu and Cd.

Methods. The method now in question is esteemed to be the best, and is indeed almost the only direct method of estimating cyanogen. But there are comparatively few compounds of cyanogen from which cyanide of silver can be precipitated directly. Hence the necessity of several modifications of the process, as will be described.

A. *Gravimetric.*

If the substance to be analyzed is a *solid cyanide of an alkali metal* pour upon it a solution of nitrate of silver, and afterwards, water, and acidulate slightly with nitric acid. Allow the precipitate to settle without warming the mixture, then collect it upon a weighed filter, wash with hot water and dry at 100°, or collect the precipitate upon an unweighed filter, and instead of drying at 100°, ignite it in a porcelain crucible until it ceases to lose weight. The ignition may be completed in a quarter of an hour at the temperature of redness. It is neither necessary nor admissable to heat the silver to the melting point, nor is any thing gained by igniting the cyanide of silver in a current of hydrogen. The metals in the substance analyzed are determined in the filtrate from the cyanide of silver after the excess of silver has been separated therefrom. The object in adding the nitrate of silver to the solid cyanide rather than to its solution is to avoid the loss of those traces of cyanhydric acid which exhale from aqueous solutions. It is for a similar reason that the nitric acid is added after, and not before, the silver salt. Care must always be taken to avoid an unnecessary excess of nitric acid, for cyanide of silver is somewhat soluble in that acid, especially if it be warm (H. Rose, *Zeitsch. analyt. Chem.*, 1862, **1.** 199).

In the case of a solution of *free cyanhydric acid in pure water* mix the rather dilute solution with a solution of nitrate of silver in excess; then add a little nitric acid and proceed as before. — But for estimating cyanhydric acid in *bitter almond water or cherry laurel water*, add ammonia water in strong excess to the mixture, after the addition of the nitrate of silver, until the precipitate has all dissolved, and then immediately add nitric acid to slight acid reaction. The precipitate thus obtained will contain the whole of the cyanhydric acid which was present in the original solution, but any precipitate thrown down directly from the original solution, or from that solution after acidulation with nitric acid, would contain only a part of the cyanogen. The addition of the ammonia water is an indispensible necessity to success, but, on the other hand, the ammonia must not be allowed to act for any length of time, lest a portion of the cyanogen be lost (Souchay, *Zeitsch. analyt. Chem.*, 1863, **2.** pp. 177, 180).

According to Feldhaus (*Zeitsch. analyt. Chem.*, 1864, **3.** 36), bitter almond water contains, when fresh, cyanhydrate of benzaldehyde, free benzaldehyde, free cyanhydric acid and cyanide of ammonium. But only the last two of these solutions are acted upon by a solution of nitrate of silver at the ordinary temperature. In order that the cyanogen of the cyanhydrate of benzaldehyde may be combined with silver it is necessary to decompose this compound by means of potash or ammonia. But care must be exercised in doing this, for an excess of the alkali must be used of necessity, since it is a matter of experience that one equivalent of the alkali—or, indeed, several equivalents—are insufficient to effect this decomposition; while, on the other hand, any long continued action of the alkali would induce other decompositions, and tend to diminish the amount of precipitable cyanogen; hence it has long been customary to add ni-

☞ **This signature is not to be bound in the completed volume.**

APPENDIX TO PART I.

EXAMPLES FOR PRACTICE.

Examples are here given more for the sake of roughly indicating the method which will be followed in the appendix to the finished work, than on account of any special merit which may attach to the examples themselves.

Though none of these examples can be regarded as specially difficult, the beginner, before undertaking either of them, will of course perform several simpler experiments,—such as are described in almost any treatise on quantitative analysis. See, also, a few simple examples on the next page, at the close of the list.

The following examples are taken from Part I. (pages 1 to 112). — The completed volume will contain five such parts.

Analysis of Marble.

Grind a quantity of pure white marble, or better, Iceland Spar, to powder, dry the powder at 100° and weigh out three portions of it, each of about one gramme.

Place one portion in a beaker covered with a watch glass, dissolve it in dilute chlorhydric acid and determine the calcium by precipitation, as Carbonate of Calcium.

In another portion estimate the Carbonic Acid by Johnson's method, see p. 84; and in the third portion estimate the carbonic acid by the method of absorption in alkali, see p. 91.

To control the amount of calcium, weigh out a fourth small portion (about 0.3 grm.) of the original marble, heat it intensely over a blast lamp to expel the carbonic acid, and weigh the oxide of calcium which is left. Again heat and weigh, and repeat the operations until the results of two successive weighings are the same. The amount of calcium in the residue is found by the proportion:—

$$56\left(=\begin{matrix}\text{Molec.}\\\text{wt. of}\\\text{CaO}\end{matrix}\right) : 40\left(=\begin{matrix}\text{Wt. of an}\\\text{atom of}\\\text{Ca}\end{matrix}\right) :: \begin{matrix}\text{Wt. of}\\\text{the}\\\text{residue}\end{matrix} : x\left(=\begin{matrix}\text{Wt. of Ca}\\\text{in the}\\\text{residue.}\end{matrix}\right)$$

Valuation of Arsenious Acid.

Select a sample of arsenious acid, pure enough to volatilize completely when heated on platinum foil. Weigh out two portions of the substance, each of about 0.5 grm., and estimate the arsenic contained in them, by the methods described under Arseniate of Lead, and Arseniate of Magnesium and Ammonium.

Estimation of Lead.

Weigh out about one gramme of pure, dry nitrate of lead, dissolve it in water and determine the lead by the method described under Carbonate of Lead.

Analysis of Sulphide of Antimony.

Treat about 0.3 grm. of the pure, crystallized native sulphide, as directed under Antimony Compounds.

Estimation of Nitric Acid.

Weigh out about 0.5 grm. of pure, dry nitrate of potassium, and proceed as directed under Arsenious Acid (Method F).

Estimation of Ammonia.

Weigh out from 0.6 to 1 grm. of pure chloride of ammonium, and proceed as directed under Ammonia, volatility of (Method A).

Separation of Iron from Magnesium.

Dissolve about 0.2 grm. of the finest iron wire in chlorhydric acid mixed with nitric acid. Add to the solution about 0.8 grm. of pure, recently crystallized Epsom salt, dilute the mixture and proceed to precipitate the iron as Acetate of Iron. Determine magnesium in the filtrate as Phosphate of Magnesium and Ammonium.

Separation of Antimony from Tin.

See Antimony, Method B.

Valuation of Carbonate of Sodium.

1. By Alkalimetry, through neutralization.
2. By fusion with bichromate of potassium, see p. 80.
3. By Rumpf's method, see p. 88.

Valuation of Acetic Acid.

By several of the methods described under that head.

Valuation of Bleaching Powder.

By Penot's method (the chlorine acting upon arsenious acid in an alkaline solution). See Arsenious Acid.

Valuation of Spirit.

See Alcohol, Principle III.

Separation of Magnesium from Sodium.

Weigh out from 0.3 to 0.4 grm. of pure, dry oxide of magnesium, prepared by igniting the oxalate; and about 0.5 grm. of pure, dry chloride of sodium. Dissolve the mixture in dilute chlorhydric acid, and proceed as described under Carbonate of Magnesium and of Ammonium. Determine the Sodium in the filtrate as Chloride of Sodium.

Analysis of Organic Compounds.

A. Prepare a quantity of oxalate of lead, by precipitation; wash it carefully, dry at 100° and weigh out three portions, of about 0.6 grm. each, for analysis. Burn one sample with oxide of copper, after Liebig (see Carbon, Principle II, Method 1), another by Bunsen's modification (p. 61) of that process, and a third with oxide of copper and oxygen, as explained under Method 2.

Though oxalate of lead contains no hydrogen, the student should nevertheless weigh the chloride of calcium tube before and after each combustion, in order to gain some idea of the errors which may be introduced through the absorption of atmospheric moisture by oxide of copper.

B. Weigh out about 0.4 grm. of pure, crystallized tartaric acid, which has been previously powdered and dried at 100°. Estimate the carbon and hydrogen by Method 2, above cited. The amount of oxygen in the tartaric acid is obtained from the difference. The composition of pure tartaric acid is: —

C_4	=	48	=	32
H_6	=	6	=	4
O_6	=	96	=	64
		150		100

Estimation of Carbon in Cast Iron.

See Carbon, Principle I, Methods B and C.

Separation of Bromine from Chlorine.

Dissolve about 0.8 grm. of pure bromide of potassium, and from 2 to 3 grms. of pure chloride of sodium in two or three hundred c. c. of water; precipitate with nitrate of silver (see Bromide of Silver), and treat portions of the dry, weighed precipitate as described under Bromide of Silver, Principles II and III.

Estimation of Carbonic Acid in Atmospheric Air.

By Pettenkofer's process, see page 97.

Estimation of Carbonic Acid in "Soda-water."

By the method of Fresenius, described on page 101.

Indirect separation of Barium from Strontium.

See page 80.

As examples of simpler experiments, fit for absolute beginners, may be mentioned: —

A. The precipitation of iron, by ammonia, as hydrated sesquioxide, and its estimation as ferric oxide. As the starting point, weigh out from 0.3 to 0.4 grm. of fine, tough, iron wire and dissolve it in pure, dilute chlorhydric acid mixed with a small quantity of nitric acid.

B. The estimation of sodium in chloride of sodium; by adding an excess of pure, diluted sulphuric acid to a weighed quantity (from 0.5 to 1 grm.) of pure fused chloride of sodium, evaporating to expel chlorhydric acid and the excess of sulphuric acid, and weighing the sulphate of sodium.

C. Estimation of chlorine in chloride of sodium, by precipitating, with nitrate of silver, as chloride of silver.

D. Estimation of silver in chloride of silver by decomposing the latter with pure metallic zinc, in presence of dilute sulphuric acid, and weighing the metallic silver.

FINDING LISTS,

FOR METHODS OF SEPARATING ELEMENTS.

ALUMINUM : Separation of from

As : See MgO, $(NH_4)_2$ O. As_2O_5, — and Arseniomolybdate of Ammonium.

Ba : See Al_2O_3, Acetate of.

Ca : See Al_2O_3, Acet.; — BaO, CO_2, — and MgO, CO_2.

Co : See Al_2O_3, Acet.; — Na_2O, Al_2O_3 ; — BaO, CO_2, — and CoO, CO_2.

Fe : See Al_2O_3, Acet. ; — Na_2O, Al_2O_3, — and BaO, CO_2.

Li : See Al_2O_3, Acetate of.

Mg : See Al_2O_3, Acet.; — Na_2O, $2B_2O_3$; — BaO, CO_2, — and MgO, CO_2.

Mn : See Al_2O_3, Acet.; — Na_2O, Al_2O_3 ; — BaO, CO_2, — and MnO, CO_2.

Ni: See Al_2O_3, Acet.; — Na_2O, Al_2O_3; — BaO, CO_2, — and NiO, CO_2.

P: See BaO, CO_2.

K: See Al_2O_3, Acetate of.

Na: See Al_2O_3, Acetate of.

Sr: See Al_2O_3, Acetate of.

Ur: See Al_2O_3, Acetate of.

Zn: See Al_2O_3, Acet., — and BaO, CO_2.

[For the decomposition of refractory aluminates see Na_2O, Al_2O_3; — BaO, CO_2, — and Na_2O, $2B_2O_3$].

AMMONIUM: Separation of from

All the elements : See Ammonia (Principle III) and Ammonium Salts.

ANTIMONY : Separation of from

Sb: See Sb_2O_3.

As: See Na_2O, Sb_2O_5; — MgO, $(NH_4)_2O$, As_2O_5; — Sb, — and As.

Bi: See Sb_2O_5.

Cd: See Sb_2O_5.

Co: See Sb_2O_5; — Sb_2O_3, — and Sb.

Cu: See Sb_2O_5; — Sb_2O_3, — and Sb.

Au: See Sb_2O_3, — and Sb.

Fe: See Sb_2O_5; — Sb_2O_3, — and Sb.

Pb: See Sb_2O_5.

Mg: See Sb_2O_5.

Mn: See Sb_2O_5.

Hg: See Sb_2O_5.

Ni: See Sb_2O_5; — Sb_2O_3, — and Sb.

K: See Hg_2O, Sb_2O_5.

Ag: See Sb_2O_5; — Sb_2O_3, — and Sb.

Na: See Hg_2O, Sb_2O_5.

S: See Sb_2O_3.

Sn: See Na_2O, Sb_2O_5, — and Sb.

Zn: See Sb_2O_5.

ARSENIC : Separation of from

Al: See MgO, $(NH_4)_2O$, As_2O_5, — and Arseniomolybdate of Ammonium.

Sb: See Na_2O, Sb_2O_5; — MgO, $(NH_4)_2O$, As_2O_5; — Sb_2O_5; — Sb; — As.

As: See MgO, $(NH_4)_2O$, As_2O_5, — and As_2O_3.

Ba: See Hg_2O, As_2O_5; — Fe_2O_3, As_2O_5; — K_2O, As_2O_5; — Ur_2O_3, As_2O_5; — As; — As_2O_3; — As_2O_5; — BaO, CO_2, — and Arseniomolybdate of Ammonium.

Bi: See As; — As_2O_3, — and Arseniomolybdate of Ammonium.

Br: See AgBr.

Cd: See Ag_2O, As_2O_5; — MgO, $(NH_4)_2O$, As_2O_5; — As; — As_2O_3, — and Arseniomolybdate of Ammonium.

Ca: See Hg_2O, As_2O_5; — Fe_2O_3, As_2O_5; — K_2O, As_2O_5; — Ur_2O_3, As_2O_5; — As; — As_2O_3; — BaO, CO_2, — and Arseniomolybdate of Ammonium.

Cr: See Arseniomolybdate of Ammonium.

Co: See MgO, $(NH_4)_2O$, As_2O_5; — Hg_2O, As_2O_5; — Fe_2O_3, As_2O_5; — As; — As_2O_3; —BaO, CO_2, — and Arseniomolybdate of Ammonium.

Cu: See Hg_2O, As_2O_5; — As_2O_5; — MgO, $(NH_4)_2O$, As_2O_5; — K_2O, As_2O_5; — As; — As_2O_3, — and Arseniomolybdate of Ammonium.

Fe: See MgO, $(NH_4)_2O$, As_2O_5; — As_2O_5; — K_2O, As_2O_5; — As; — As_2O_3, — and Arseniomolybdate of Ammonium.

Pb: See Hg_2O, As_2O_5; — As_2O_5; — As; — As_2O_3, — and Arseniomolybdate of Ammonium.

Mg: See Hg_2O, As_2O_5; — As_2O_5; — Ur_2O_3, As_2O_5; — As; — As_2O_3, — and Arseniomolybdate of Ammonium.

Mn: See MgO, $(NH_4)_2O$, As_2O_5; — As_2O_5; — Fe_2O_3, As_2O_5; — K_2O, As_2O_5; — As; — As_2O_3; — BaO, CO_2, — and Arseniomolybdate of Ammonium.

Hg: See As; — As_2O_3, — and Arseniomolybdate of Ammonium.

Ni: See MgO, $(NH_4)_2O$, As_2O_5; — Hg_2O, As_2O_5; — Fe_2O_3, As_2O_5; — As; — As_2O_3; — BaO, CO_2, — and Arseniomolybdate of Ammonium.

K: See MgO, $(NH_4)_2O$, As_2O_5; — Hg_2O, As_2O_5; — As_2O_5; — Fe_2O_3, As_2O_5; — Ur_2O_3, As_2O_5, — and Arseniomolybdate of Ammonium.

Ag: See As; — As_2O_3, — and Arseniomolybdate of Ammonium.

Na: See MgO, $(NH_4)_2O$, As_2O_5; — Hg_2O, As_2O_5; —As_2O_5; — Fe_2O, As_2O_5; — SnO_2, As_2O_5; — Ur_2O_3, As_2O_5, — and Arseniomolybdate of Ammonium.

Sr: See Hg_2O, As_2O_5; — Fe_2O_3, As_2O_5; — K_2O, As_2O_5; — Ur_2O_3, As_2O_5; — As; — As_2O_3; — BaO, CO_2, — and Arseniomolybdate of Ammonium.

S: See As_2O_3.

Sn: See SnO_2, As_2O_5, — and As.

Ur: See MgO, $(NH_4)_2O$, As_2O_5.

Zn: See MgO, $(NH_4)_2O$, As_2O_5; — Hg_2O, As_2O_5; — As_2O_5; — Fe_2O_3, As_2O_5; — K_2O, As_2O_5; —Ur_2O_3, As_2O_5; — As; — As_2O_3; — BaO, CO_2, — and Arseniomolybdate of Ammonium.

BARIUM: Separation of from

Al: See Al_2O_3, Acetate of.

As: See Hg_2O, As_2O_5; — As_2O_5; — Fe_2O_3, As_2O_5; — K_2O, As_2O_5; — Ur_2O_3, As_2O_5; — As; — As_2O_3; — BaO, CO_2, — and Arseniomolybdate of Ammonium.

Ca: See CO_2, — and CaO,CO_2.

C: See C, pp. 69, 79, 80; — and BaO, CO_2.

Co: See BaO, CO_2, — and CoO, CO_2.

H: See B_2O_3.

Fe: See Fe_2O_3, Acet., — and BaO, CO_2.

Mg: See BaO, CO_2.

Mn: See BaO, CO_2, — and MnO, CO_2.

Ni: See BaO, CO_2, — and NiO, CO_2.

P: See BaO, CO_2.

K: See BaO, CO_2, — and MgO, CO_2.

Na: See BaO, CO_2.

Sr: See CO_2, — and BaO, CO_2.

Zn: See BaO, CO_2.

BISMUTH: Separation of from

Sb: See Sb_2O_5.

As: See As; — As_2O_3, — and Arseniomolybdate of Ammonium.

Cd: See Bi_2O_3, CO_2, — and CdO, CO_2.

Cl: See Bi.

Cu: See Bi_2O_3, CO_2, — and CuO, CO_2.

Au: See Bi_2O_3, CO_2.

Pb: See Bi.

Mn: See Bi_2O_3, CO_2, — and MnO, CO_2.

Hg: See Bi_2O_3, CO_2.

Ag: See Bi_2O_3, CO_2.

S: See Bi.

BORON: Separation of from

Most metals other than alkali-metals: See Na_2O, $2B_2O_3$, — and CO_2.

Br: See AgBr.

K: See MgO, B_2O_3.

Na: See MgO, B_2O_3.

BROMINE: Separation of from

Most metals: See AgBr.

As: See AgBr.

B: See AgBr.

C: See HgBr; — AgBr; — Br, — and C, pp. 64, 68, 71, 73, 74.

Cl: See AgBr; — KBr; — NaBr, — and Br.

Cr: See AgBr.

Fl: See AgBr.
I: See KBr, — and Br.
P: See AgBr.
Si: See AgBr.
S: See AgBr.

CADMIUM: Separation of from

Sb: See Sb_2O_5.
As: See Hg_2O, As_2O_5; — MgO, $(NH_4)_2O$, As_2O_5; — As; — As_2O_3, — and Arseniomolybdate of Ammonium.
Bi: See Bi_2O_3, CO_2, — and CdO, CO_2.
C: See CO_2.
Cu: See CdO, CO_2, — and CuO, CO_2.
Pb: See CdO, CO_2, — and PbO, CO_2.
Mn: See CdO, CO_2, — and MnO, CO_2.

CALCIUM: Separation of from

Al: See Al_2O_3, Acet.; — BaO, CO_2, — and MgO, CO_2.
As: See Hg_2O, As_2O_5; — Fe_2O_3, As_2O_5; — K_2O, As_2O_5; — Ur_2O_3, As_2O_5; — As; As_2O_3; — BaO, CO_2, — and Arseniomolybdate of Ammonium.
Ba: See CO_2, — and CaO, CO_2.
C: — See C, pp. 69, 79, 80; — CO_2, — and CaO, CO_2.
Co: See CaO, CO_2, — and CoO, CO_2.
Fe: See Fe_2O_3, Acet., — and BaO, CO_2.
Mg: See CaO, CO_2.
Mn: See CaO, CO_2, — and MnO, CO_2.
Ni: See CaO, CO_2, — and NiO, CO_2.
P: See BaO, CO_2.
K: See BaO, CO_2; — CaO, CO_2, — and MgO, CO_2.
Na: See CaO, CO_2.
Sr: See CO_2, — and CaO, CO_2.
Zn: See CaO, CO_2.

CARBON: Separation of from

All bases: See CO_2, — and C, p. 64.
Ba: See BaO, CO_2, — and C, pp. 69, 79, 80.
B: See CO_2.
Br: See HgBr; — AgBr; — Br, — and C, pp. 64, 68, 71, 73, 74.
Cd: See CO_2.
Ca: See CO_2; — CaO, CO_2, — and C, pp. 69, 79, 80.
Cl: See C, pp. 64, 68, 71, 73, 74.
Cu: See CO_2.
H: See C.
I: See C, pp. 64, 68, 71, 73, 74.
Fe: See C, pp. 57, 58, 71, 75, 76.
Pb: See CO_2.
Mg: See CO_2, — and MgO, CO_2.
Ni: See CO_2.
N: See C, pp. 63, 68, 71, 73.
K: See CO_2; — K_2O, CO_2, — and C, pp. 69, 79, 80.
Na: See C, pp. 69, 79, 80.
Sr: See CO_2, — and C, pp. 69, 79, 80.
S: See Ag_2O, CO_2; and C, pp. 64, 71, 73, 74.

CHLORINE: Separation of from

Bi: See Bi.
Br: See AgBr; — KBr; — NaBr, — and Br.
C: See C, pp. 64, 68, 71, 73, 74.
I: See KBr, — and Br.

[For the estimation of chlorine "(chlorimetry)" see As_2O_3].

CHROMIUM: Separation of from

As: See Arseniomolybdate of Ammonium.
Br: See AgBr.
Co: See BaO, CO_2.
Fe: See BaO, CO_2.
Mn: See BaO, CO_2.
Ni: See BaO, CO_2.
Zn: See BaO, CO_2.

[For the decomposition of refractory Chromites see BaO, CO_2, — and Na_2O, B_2O_3].

COBALT: Separation of from

Al: See Al_2O_3, Acet.; — Na_2O, Al_2O_3; — BaO, CO_2, — and CoO, CO_2.
Sb: See Sb_2O_5; — Sb_2O_3, — and Sb.
As: See MgO, $(NH_4)_2O$, As_2O_5; — Hg_2O, As_2O_5; — Fe_2O_3, As_2O_5; — As; — As_2O_3; — BaO, CO_2, — and Arseniomolybdate of Ammonium.
Ba: See BaO, CO_2, — and CoO, CO_2.
Ca: See CaO, CO_2, — and CoO, CO_2.
Cr: See BaO, CO_2.
Fe: See Fe_2O_3, Acet., — and BaO, CO_2.
Ni: See (CaO, CO_2; CoO, CO_2).
Sr: See CoO, CO_2.

COPPER: Separation of from

Sb: See Sb_2O_5; — Sb_2O_3, — and Sb.
As: See Hg_2O, As_2O_5; — As_2O_5; — MgO, $(NH_4)_2O$, As_2O_5; — K_2O, As_2O_5; — As; — As_2O_3, — and Arseniomolybdate of Ammonium.
Bi: See Bi_2O_3, CO_2, — and CuO, CO_2.

Cd : See CdO, CO_2, — and CuO, CO_2.
C : See CO_2.
Pb : See CuO, CO_2, — and PbO, CO_2.
Mn : See CuO, CO_2, — and MnO, CO_2.

FLUORINE : Separation of from

Br : See AgBr.

GOLD : Separation of from

Sb : See Sb_2O_3, — and Sb.
Bi : See Bi_2O_3, CO_2.
Pb : See PbO, CO_2.

HYDROGEN (and Water) : Separation of from

Ba : See B_2O_3.
C : See C.
K : See B_2O_3.
Na : See B_2O_3.
Sr : See B_2O_3.

IODINE : Separation of from

Br : See KBr, — and Br.
C : See C, pp. 64, 68, 71, 73, 74.
Cl : See KBr, — and Br.
[For the estimation of iodine, see As_2O_3].

IRON : Separation of from

Al : See Al_2O_3, Acet.; — Na_2O, Al_2O_3, — and BaO, CO_2.
Sb : See Sb_2O_5; — Sb_2O_3, — and Sb.
As : See MgO, $(NH_4)_2O$, As_2O_5; — As_2O_5; — K_2O, As_2O_5; — As; — As_2O_3, — and Arseniomolybdate of Ammonium.
Ba : See Fe_2O_3, Acet., — and BaO, CO_2.
Ca : See Fe_2O_3, Acet., — and BaO, CO_2.
C : See C, pp. 57, 58, 71, 75, 76.
Cr : See BaO, CO_2.
Co : See Fe_2O_3, Acet., — and BaO, CO_2.
Fe : See Fe_2O_3, Acet.; — BaO, CO_2, — and MgO, CO_2.
Li : See Fe_2O_3, Acetate of.
Mg : See Fe_2O_3, Acet., — and BaO, CO_2.
Mn : See Fe_2O_3, Acet.; — Fe_2O_3, As_2O_5; — Fe_2O_3, Benz.; — BaO, CO_2, — and MnO, CO_2.
Ni : See Fe_2O_3, Acet.; — Fe_2O_3, As_2O_5; — Fe_2O_3, Benz., — and BaO, CO_2.
P : See BaO, CO_2.
K : See Fe_2O_3, Acetate of.
Si : See Na_2O, $2B_2O_3$.
Na : See Fe_2O_3, Acetate of.
Sr : See Fe_2O_3, Acet., — and BaO, CO_2.
Ur : See Fe_2O_3, Acetate of.
Zn : See Fe_2O_3, Acet.; — BaO, CO_2, — and Fe_2O_3, Benz.

LEAD : Separation of from

Sb : See Sb_2O_5.
As : See Hg_2O, As_2O_5; — As_2O_5; — As; — As_2O_3, — and Arseniomolybdate of Ammonium.
Bi : See Bi.
Cd : See CdO, CO_2, — and PbO, CO_2.
C : See CO_2.
Cu : See CuO, CO_2, — and PbO, CO_2.
Au : See PbO, CO_2.
Mn : See PbO, CO_2, — and MnO, CO_2.
Hg : See PbO, CO_2.
Ag : See PbO, CO_2.

LITHIUM : Separation of from

Al : See Al_2O_3, Acetate of.
As : See Fe_2O_3, As_2O_5, — and Arseniomolybdate of Ammonium.
Fe : See Fe_2O_3, Acetate of.
Si : See CaO, CO_2, — and BaO, CO_2.

MAGNESIUM : Separation of from

Al : See Al_2O_3, Acet.; — Na_2O, $2B_2O_3$; — BaO, CO_2, — and MgO, CO_2.
Sb : See Sb_2O_5.
As : See Hg_2O, As_2O_5; — As_2O_5; — Ur_2O_3, As_2O_5; — As; — As_2O_3, — and Arseniomolybdate of Ammonium.
Ba : See BaO, CO_2.
Ca : See CaO, CO_2.
C : See CO_2, — and MgO, CO_2.
Fe : See Fe_2O_3, Acet., — and BaO, CO_2.
Mn : See MgO, CO_2, — and MnO, CO_2.
K : See MgO, $(NH_4)_2O$, As_2O_5; — BaO, CO_2; — MgO, CO_2, — and [MgO, CO_2; $(NH_4)_2O$, CO_2].
Na : See MgO, $(NH_4)_2O$, As_2O_5, — and [MgO, CO_2; $(NH_4)_2O$, CO_2].

MANGANESE : Separation of from

Al : See Al_2O_3, Acet.; — Na_2O, Al_2O_3; — BaO, CO_2, — and MnO, CO_2.
Sb : See Sb_2O_5.
As : See MgO, $(NH_4)_2O$, As_2O_5; — As_2O_5; — Fe_2O_3, As_2O_5; — K_2O, As_2O_5; — As; — As_2O_3; — BaO, CO_2, — and Arseniomolybdate of Ammonium.
Ba : See BaO, CO_2, — and MnO, CO_2.
Bi : See Bi_2O_3, CO_2, — and MnO, CO_2.

Cd : See CdO, CO_2, — and MnO, CO_2.
Ca : See CaO, CO_2, — and MnO, CO_2.
Cr : See BaO, CO_2.
Cu : See CuO, CO_2, — and MnO, CO_2.
Fe : See Fe_2O_3, Acet. ; — Fe_2O_3, As_2O_5 ; — Fe_2O_3, Benz. ; — BaO, CO_2, — and MnO, CO_2.
Pb : See PbO, CO_2, — and MnO, CO_2.
Mg : See MgO, CO_2, — and MnO, CO_2.
P : See MnO, CO_2.
Sr : See MnO, CO_2.
Zn : See MnO, CO_2.

MERCURY : Separation of from

Sb : See Sb_2O_5.
As : See As; — As_2O_3, — and Arseniomolybdate of Ammonium.
Bi : See Bi_2O_3, CO_2.
Pb : See PbO, CO_2.

NICKEL : Separation of from

Al : See Al_2O_3, Acet. ; — Na_2O, Al_2O_3 ; — BaO, CO_2, — and NiO, CO_2.
Sb : See Sb_2O_5; — Sb_2O_3, — and Sb.
As : See MgO, $(NH_4)_2O$, As_2O_5 ; — Hg_2O, As_2O_5; — Fe_2O_3, As_2O_5; — As; — As_2O_3; — BaO, CO_2, — and Arseniomolybdate of Ammonium.
Ba : See BaO, CO_2,—and NiO, CO_2.
Ca : See CaO, CO_2, — and NiO, CO_2.
C : See CO_2.
Cr : See BaO, CO_2.
Co : See [CaO, CO_2; CoO, CO_2].
Fe : See Fe_2O_3, Acet. ; — Fe_2O_3, As_2O_5 ; — Fe_2O_3, Benz., — and BaO, CO_2.
Sr : See NiO, CO_2.

NITROGEN : Separation of from

C : See C, pp. 63, 68, 71, 73.

[For a method of estimating nitric acid, see As_2O_3].

PHOSPHORUS (and Phosphoric Acid) : Separation of from

Al : See BaO, CO_2.
Ba : See BaO, CO_2, — and Ag_2O, CO_2.
Br : See AgBr.
Ca : See BaO, CO_2, — and Ag_2O, CO_2.
Fe : See BaO, CO_2.
Mn : See MnO, CO_2.
K : See Ag_2O, CO_2.
Na : See Ag_2O, CO_2.
Sr : See BaO, CO_2, — and Ag_2O, CO_2.

POTASSIUM : Separation of from

Al : See Al_2O_3, Acetate of.
Sb : See Hg_2O, Sb_2O_5.
As : See MgO, $(NH_4)_2O$, As_2O_5 ; — Hg_2O, As_2O_5 ; — As_2O_5 ; — Fe_2O_3, As_2O_5 ; — Ur_2O_3, As_2O_5, — and Arseniomolybdate of Ammonium.
Ba : See BaO, CO_2, — and MgO, CO_2.
B : See MgO, B_2O_3.
Ca : See BaO, CO_2 ; — CaO, CO_2, — and MgO, CO_2.
C : See CO_2; — K_2O, CO_2, — and C, pp. 69, 79, 80.
H : See B_2O_3.
Fe : See Fe_2O_3, Acetate of.
Mg : See MgO, $(NH_4)_2O$, As_2O_5 ; — BaO, CO_2 ; — MgO, CO_2, — and [MgO, CO_2; $(NH_4)_2O$, CO_2].
Si : See CaO, CO_2, — and BaO, CO_2.

[See also Alkalimetry, for the estimation of potassium, and the separation of KHO from K_2O, CO_2.

SILICON : Separation of from

Br : See AgBr.
Fe : See Na_2O, $2B_2O_3$.
Li : See CaO, CO_2, — and BaO, CO_2.
K : See CaO, CO_2, — and BaO, CO_2.
Na : See CaO, CO_2, — and BaO, CO_2.

[For the decomposition of refractory silicates see B_2O_3; — Na_2O, B_2O_3; — BaO, CO_2; —CaO, CO_2, — and K_2O, CO_2].

SILVER : Separation of from

Sb : See Sb_2O_5; — Sb_2O_3, — and Sb.
As : See As; — As_2O_3, — and Arseniomolybdate of Ammonium.
Bi : See Bi_2O_3, CO_2.
Pb : See PbO, CO_2.

SODIUM : Separation of from

Al : See Al_2O_3, Acetate of.
Sb : See Hg_2O, Sb_2O_5.
As : See MgO, $(NH_4)_2O$, As_2O_5 ; — Hg_2O, As_2O_5 ; — As_2O_5 ; — Fe_2O_3, As_2O_5 ; — SnO_2, As_2O_5; — Ur_2O_3, As_2O_5, — and Arseniomolybdate of Ammonium.
Ba : See BaO, CO_2.
B : See MgO, B_2O_3.
Ca : See CaO, CO_2.
C : See C, pp. 69, 79, 80.
H : See B_2O_3.

Fe : See Fe_2O_3, Acetate of.
Mg : See MgO, $(NH_4)_2O$, As_2O_5, — and [MgO, CO_2; $(NH_4)_2O$, CO_2].
Si : See CaO, CO_2,—and BaO, CO_2.
Sn : See SnO_2, As_2O_5.

[See also Alkalimetry, for the estimation of sodium and the separation of NaHO from Na_2O, CO_2].

STRONTIUM : Separation of from

Al : See Al_2O_3, Acetate of.
As : See Hg_2O, As_2O_5; — Fe_2O_3, As_2O_5; — K_2O, As_2O_5; — Ur_2O_3, As_2O_5; — As; — As_2O_3; — BaO, CO_2, — and Arseniomolybdate of Ammonium.
Ba : See CO_2.
Ca : See CO_2, — and CaO, CO_2.
C : See CO_2, — and C, pp. 69, 79, 80.
Co : See CoO, CO_2.
H : See B_2O_3.
Fe : See Fe_2O_3, Acet., — and BaO,CO_2.
Mn : MnO, CO_2.
Ni : See NiO, CO_2.
P : See BaO, CO_2.

SULPHUR : Separation of from

Sb : See Sb_2O_3.
As : See As_2O_3.
Bi : See Bi.
Br : See AgBr.
C : See Ag_2O, CO_2; and C, pp. 64, 71, 73, 74.

TIN : Separation of from

Sb : See Na_2O, Sb_2O_5, — and Sb.
As : See SnO_2, As_2O_5, — and As.
Na : See SnO_2, As_2O_5.

URANIUM : Separation of from

Al : See Al_2O_3, Acetate of.
As : See MgO, $(NH_4)_2O$, As_2O_5.
Fe : See Al_2O_3, Acetate of.

WATER : See Hydrogen, above.

ZINC : Separation of from

Al : See Al_2O_3, Acet., — and BaO, CO_2.
Sb : See Sb_2O_5.
As : See MgO, $(NH_4)_2O$, As_2O_5; — Hg_2O, As_2O_5; — As_2O_5; — Fe_2O_3, As_2O_5; — K_2O, As_2O_5; — Ur_2O_3, As_2O_5; — As; — As_2O_3; — BaO, CO_2, — and Arseniomolybdate of Ammonium.
Ba : See BaO, CO_2.
Ca : See CaO, CO_2.
C : See CO_2.
Cr : See BaO, CO_2.
Fe : See Fe_2O_3, Acet., — Fe_2O_3, Benz.,—and BaO, CO_2.
Mn : See MnO, CO_2.

☞ This signature is not to be bound in the completed volume.

FINDING LISTS,

[RELATING TO PART II.]

FOR METHODS OF SEPARATING ELEMENTS.

ALUMINUM: Separation of from

Ba: See Na_2O, CO_2.
Ca: See Na_2O, CO_2.
Cr: See BaO, CrO_3.
Cu: See Cu.
Fe: See $FeCl_2$.
Mg: See Na_2O, CO_2.
Mn: See Na_2O, CO_2; — Cl.
Hg: See HgCl.
K: See 2KCl, $PtCl_4$.
Ag: See AgCl.
Sr: See Na_2O, CO_2.
Zn: See ZnO, CO_2.

AMMONIUM: Separation of from

Ba: See $2NH_4Cl$, $PtCl_4$.
Ca: See $2NH_4Cl$, $PtCl_4$.
Cr: See CrO_3.
Li: See $2NH_4Cl$, $PtCl_4$.
Mg: See $2NH_4Cl$, $PtCl_4$.
Ag: See AgCl.
Na: See $2NH_4Cl$, $PtCl_4$.
Sr: See $2NH_4Cl$, $PtCl_4$.

[For the estimation of NH_3, see further NH_4Cl and $2NH_4Cl$, $PtCl_4$.]

ANTIMONY: Separation of from

Ba: See $SbCl_3$.
Ca: See $SbCl_3$; — $CaCl_2$.
Cr: See CrO_3.
Co: See $SbCl_5$; — $CoCl_2$.
Cu: See $SbCl_5$; — $CuCl_2$.
Au: See $SbCl_5$.
Pb: See $SbCl_5$; — $PbCl_2$.
Ni: See $SbCl_5$; — $NiCl_2$.
Pt: See $SbCl_5$.
K: See $SbCl_3$.
Ag: See $SbCl_5$; — AgCl.
Na: See $SbCl_3$.
Sr: See $SbCl_3$.
Zn: See $ZnCl_2$.

[For the estimation of Sb_2O_3 see further K_2O, $2CrO_3$.]

ARSENIC: Separation of from

Ba: See $AsCl_3$.
Ca: See $AsCl_3$; — $CaCl_2$.
Cl: See AgCl.
Cr: See CrO_3.
Co: See $AsCl_3$; — $CoCl_2$.
Cu: See $AsCl_3$ — $CuCl_2$.
Au: See $AsCl_3$.
Pb: See $AsCl_3$; — $PbCl_2$.
Ni: See $AsCl_3$ — $NiCl_2$.
Pt: See $AsCl_3$.
K: See $AsCl_3$; — 2KCl, $PtCl_4$.
Ag: See $AsCl_3$; — AgCl.
Na: See $AsCl_3$.
Sr: See $AsCl_3$.
Zn: See $ZnCl_2$.

[For the estimation of arsenic see further CrO_3; — K_2O, $2CrO_3$.]

BARIUM: Separation of from

Al: See Na_2O, CO_2.
NH_4: See $2NH_4Cl$, $PtCl_4$.
Sb: See $SbCl_3$.
As: See $AsCl_3$.
Bi: See $BiCl_3$ (basic).
Cl: See HCl.
Cr: See CrO_3; — $(NH_4)_2O$, CrO_3; — BaO, CrO_3; — Na_2O, CrO_3.

Co: See $CoCl_2$; — Co.
Cu: See Cu.
Pb: See Cl.
Mn: See Cl.
Hg: See HgCl; — $HgCl_2$; — $NiCl_2$.
K: See 2KCl, $PtCl_4$.
Ag: See AgCl.
Sr: See BaO, CrO_3.
Zn: See ZnO, CO_2.

BISMUTH: Separation of from

Ba: See $BiCl_3$ (basic).
Cl: See $BiCl_3$ (basic); — Bi_2O_3, $2CrO_3$.
Ca: See $BiCl_3$ (basic); — $CaCl_2$.
Cr: See CrO_3.
Co: See $BiCl_3$ (basic); — $CoCl_2$.
Cu: See $BiCl_3$ and $BiCl_3$ (basic); — $CuCl_2$; — Cu.
Pb: See $BiCl_3$; — $PbCl_2$.
Mn: See $BiCl_3$ (basic).
Hg: See $BiCl_3$ (basic).
Ni: See $BiCl_3$ (basic); — $NiCl_2$.
Pt: See $2NH_4$ Cl, $PtCl_4$; — 2KCl, $PtCl_4$.
K: See $BiCl_3$ (basic).
Ag: See $BiCl_3$; — AgCl.
Na: See $BiCl_3$ (basic).
Sr: See $BiCl_3$ (basic).
Zn: See $BiCl_3$ (basic).

BORON: Separation of from

Cl: See AgCl.
K: See 2KCl, $PtCl_4$.

[For the separation of B_2O_3 see further HCl.]

BROMINE: Separation of from

Cl: See AgCl; — NaCl; — Cl.
I: See Cl.

[For the estimation of bromine see further $SnCl_2$; — Cl.]

CADMIUM: Separation of from

Bi: See $BiCl_3$ (basic); — Bi_2O_3, $2CrO_3$.
Cr: See CrO_3.
Cu: See Cu.
Hg: See HgCl.
Pt: See $2NH_4Cl$, $PtCl_4$; — 2KCl, $PtCl_4$.
Ag: See AgCl.

CALCIUM: Separation of from

Al: See Na_2O, CO_2.
NH_4: See $2NH_4Cl$, $PtCl_4$.
Sb: See $SbCl_3$; — $CaCl_2$.
As: See $AsCl_3$; — $CaCl_2$.
Ba: See BaO, CrO_3.
Bi: See $BiCl_3$ (basic); — $CaCl_2$.
Cl: See HCl.
Cr: See CrO_3; — $(NH_4)_2O$, CrO_3; — CaO, CrO_3; — PbO, CrO_3; — Hg_2O, CrO_3; — Na_2O, CrO_3.
Co: See $CoCl_2$; — Co.
Cu: See Cu.
Pb: See Cl.
Mg: See $CaCl_2$; — $MgCl_2$.
Mn: See Cl.
Hg: See $CaCl_2$; — HgCl.
Ni: See $NiCl_2$.
K: See 2KCl, $PtCl_4$.
Ag: See AgCl.
Sn: See $CaCl_2$.
Zn: See ZnO, CO_2.

CARBON: Separation of from

Cl: See $CaCl_2$; — $CuCl_2$; — $PbCl_2$; — KCl; — $ZnCl_2$.
H: See $CaCl_2$; — 2KCl, $PtCl_4$.
Fe: See $FeCl_3$.
N: See 2KCl, $PtCl_4$.
K: See KCl.
Na: See Na_2O, CO_2; — NaCl.
Sr: See SrO, CO_2.
S: See CuO, CrO_3.

[For the oxidation of carbon see Chlorates; — CuO, CrO_3; — PbO, CrO_3.]

[For the estimation of C see further 2KCl, $PtCl_4$; — CrO_3; of CO_2 see further ZnO, CO_2.]

CHLORINE: Separation of from

As: See AgCl.
Ba: See HCl.
Bo: See AgCl.
Br: See AgCl; — NaCl; — Cl.
Ca: See HCl.
C: See $CaCl_2$; — $CuCl_2$ — $PbCl_2$ — $ZnCl_2$.
Cl: See AgCl; — Cl.
Cr: See AgCl.
H: See $CuCl_2$.
I: See NaCl; — Cl.
Pb: See $PbCl_2$.
Mg: See HCl; — $HgCl_2$.
Hg: See $HgCl_2$.
N: See AgCl.
P: See AgCl.

Pt: See HCl; — NaCl.
K: See HCl; — KCl.
Ag: See AgCl.
Si: See AgCl.
Na: See HCl; — NaCl.
Sr: See HCl.
S: See AgCl.

[For the estimation of chlorine see further Chlorates; — NH_4Cl; — $CuCl_2$; — HgCl; — KCl; — AgCl; — $SnCl_2$.]

CHROMIUM: Separation of from

Al: See BaO, CrO_3.
NH_4: See CrO_3.
Sb: See CrO_3.
As: See CrO_3.
Ba: See CrO_3; — $(NH_4)_2O$, CrO_3; — BaO, CrO_3; — Na_2O, CrO_3.
Bi: See CrO_3.
Cd: See CrO_3.
Ca: See CrO_3; — $(NH_4)_2O$, CrO_3; — CaO, CrO_3; — PbO, CrO_3; — Hg_2O, CrO_3; — Na_2O, CrO_3.
Cr: See CrO_3; — $(NH^4)_2O$, CrO_3; — Hg_2O, CrO_3.
Co: See CrO_3; — BaO, CrO_3; — Na_2O CrO_3.
Cu: See CrO_3; — Cu.
Au: See CrO_3.
Fe: See $FeCl_2$; — BaO, CrO_3; — Na_2O, CrO_3.
Pb: See $PbCl_2$; — CrO_3.
Mg: See CrO_3; — $(NH_4)_2O$, CrO_3; — BaO, CrO_3; — PbO, CrO_3; — Hg_2O, CrO_3; — Na_2O, CrO_3.
Mn: See CrO_3; — BaO, CrO_3; — Na_2O, CrO_3.
Hg: See CrO_3.
Ni: See CrO_3; — BaO, CrO_3; — Na_2O, CrO_3.
Pb: See CrO_3.
K: See KCl; — CrO_3; — $(NH_4)_2O$, CrO_3; — Hg_2O, CrO_3.
Ag: See AgCl; — CrO_3.
Na: See NaCl; — CrO_3; — $(NH_4)_2O$, CrO_3; — Hg_2O, CrO_3.
Sr: See CrO_3; — $(NH_4)_2O$, CrO_3; — PbO, CrO_3; — Hg_2O, CrO_3; — Na_2O, CrO_3; — SrO, CrO_3.
S: See CrO_3; — PbO, CrO_3.
Sn: See CrO_3.
Zn: See CrO_3; — BaO, CrO_3; — Na_2O, CrO_3.

[For the estimation of chromium see further. $CrCl_3$; — CrO_3: of CrO_3, see $SnCl_2$.]

COBALT: Separation of from

Sb: See $SbCl_5$; — $CoCl_2$.
As: See $AsCl_3$; — $CoCl_2$.
Ba: See $CoCl_2$; — Co.
Bi: See $BiCl_3$ (basic); — $CoCl_2$.
Ca: See $CoCl_2$; — Co.
Cr: See CrO_3; — BaO, CrO_3; — Na_2O, CrO_3.
Cu: See Cu.
Mn: See $CoCl_2$; — $MnCl_2$; — Co; — K_2O, cobalticyanide.
Hg: See $CoCl_2$.
Ni: See Cl; — CuO, cobalticyanide; — Hg_2O, cobalticyanide; — K_2O. cobalticyanide.
Pt: See $2NH_4Cl$, $PtCl_4$; — 2KCl, $PtCl_4$.
K: See 2KCl, $PtCl_4$; — CuO, cobalticyanide; — Hg_2O, cobalticyanide.
Ag: See AgCl.
Sr: See SrO, CO_2; — $CoCl_2$; — Co.
Sn: See $CoCl_2$; — $SnCl_4$.
Zn: See ZnO, cobalticyanide.

[For the estimation of cobalt see further $CrCl_3$; — $CoCl_2$; — $SnCl_2$.]

COPPER: Separation of from

Al: See Cu.
Sb: See $SbCl_5$; — $CuCl_2$.
As: See $AsCl_3$; — $CuCl_2$.
Ba: See Cu.
Bi: See $BiCl_3$ (normal and basic); — $CuCl_2$; — Cu.
Cd: See Cu.
Ca: See Cu.
Cr: See CrO_3; — Cu.
Co: See Cu.
Fe: See Cu.
Pb: See $PbCl_2$; — Cl; — Cu.
Mg: See Cu.
Mn: See Cu.
Hg: See $CuCl_2$; — HgCl; — $HgCl_2$.
Ni: See Cu.
Pt: See $2NH_4Cl$, $PtCl_4$; — 2KCl, $PtCl_4$.
K: See 2KCl, $PtCl_4$; — Cu.
Ag: See AgCl.
Na: See Cu.
Sr: See Cu.
S: See Na_2O, CO_2.
Sn: See $CuCl_2$; — $SnCl_4$.
Zn: See Cu.

[For the estimation of Cu see further $SnCl_2$.]

For the estimation of Ferricyanhydric acid

see $SnCl_2$; — of Ferrocyanhydric acid see CrO_3.

FLOURINE: Separation of from

Cl: See AgCl.

GOLD: Separation of from

Sb: See $SbCl_5$.
As: See $AsCl_3$.
Cr: See CrO_3.
Pt: See 2KCl, $PtCl_4$.
Ag: See AgCl.
Sn: See $SnCl_4$.

HYDROGEN: Separation of from

C: See $CaCl_2$; — 2KCl, $PtCl_4$.
Cl: See $CuCl_2$.
K: See KCl.
Na: See NaCl.

[For methods of estimating water see $CaCl_2$; — $CoCl_2$; — NaCl.]

IODINE: Separation of from

Br: See Cl.
Cl: See AgCl; — NaCl; — Cl.

[For the estimation of Iodine see further $SnCl_2$; — Cl; — K_2O, $2CrO_3$.

IRON: Separation of from

Al: See $FeCl_2$.
C: See $FeCl_3$.
Cr: See $FeCl_2$; — BaO, CrO_3; — Na_2O, CrO_3.
Cu: See Cu.
Gl: See $FeCl_2$.
Mn: See Cl.
P: See PCl_3.
Pt: See $2NH_4Cl$, $PtCl_4$; — 2KCl, $PtCl_4$.
K: See 2KCl, $PtCl_4$.
Ag: See AgCl.
S: See Na_2O, CO_2.

[For the estimation of iron see further CuCl; — $FeCl_3$; — $SnCl_2$; — CrO_3; — K_2O, $2CrO_3$.]

LEAD: Separation of from

Sb: See $SbCl_5$; — $PbCl_2$.
As: See $AsCl_3$; — $PbCl_2$.
Ba: See Cl.
Bi: See $BiCl_3$; — $PbCl_2$.
Br: See $PbCl_2$.
Ca: See Cl.
Cl: See $PbCl_2$.
Cr: See $PbCl_2$; — CrO_3
Cu: See $PbCl_2$; — Cl; — Cu.
I: See $PbCl_2$.
Mg: See Cl.
Hg: See $PbCl_2$; — HgCl; — $HgCl_2$.
Ni: See Cl.
N: See $PbCl_2$.
K: See Cl.
Ag: See $PbCl_2$; — AgCl.
Na: See Cl.
Sr: See Cl.
Sn: See $PbCl_2$; — $SnCl_4$.
Zn: See Cl.

[For the estimation of lead see further $PbCl_2$; — $SnCl_2$.]

LITHIUM: Separation of from

NH_4: See $2NH_4Cl$, $PtCl_4$.
K: See LiCl; — 2KCl, $PtCl_4$.
Na: See LiCl.

MAGNESIUM: Separation of from

Al: See Na_2O, CO_2.
NH_4: See $2NH_4Cl$, $PtCl_4$.
Ca: See $CaCl_2$; — $MgCl_2$.
Cl: See HCl; — $HgCl_2$.
Cr: See CrO_3; — $(NH_4)_2O$, CrO_3; — BaO, CrO_3; — PbO, CrO_3; — Hg_2O, CrO_3; — Na_2O, CrO_3.
Cu: See Cu.
Pb: See Cl.
Mn: See Cl.
Hg: See HgCl; — $HgCl_2$.
Pt: See 2KCl, $PtCl_4$.
K: See $MgCl_2$; — $HgCl_2$; — AgCl; — 2KCl, $PtCl_4$.
Ag: See AgCl.
Na: See $MgCl_2$; — $HgCl_2$; — AgCl.
Sr: See SrO, CO_2.
Zn: See ZnO, CO_2.

MANGANESE: Separation of from

Al: See Na_2O, CO_2; — Cl.
Ba: See Cl.
Bi: See $BiCl_3$ (basic).
Ca: See Cl.
Cr: See CrO_3; — BaO, CrO_3; — Na_2O, CrO_3.
Co: See $CoCl_2$; — $MnCl_2$; — Co; — K_2O, cobalticyanide.
Cu: See Cu.
Fe: See Cl.
Mg: See Cl.
Ni: See $MnCl_2$; — $NiCl_2$; — Cl.
Pt: See $2NH_4Cl$, $PtCl_4$; — 2KCl, $PtCl_4$.
K: See Cl.

Ag: See AgCl.
Na: See Cl.
Sr: See SrO, CO_2; — Cl.
Zn: See ZnO, CO_2; — Cl.

[For the valuation of MnO_2 see further HgCl; — $SnCl_2$; — Cl.]

MERCURY: Separation of from

Al: See HgCl.
Ba: See HgCl; — $HgCl_2$.
Bi: See $BiCl_3$ (basic).
Cd: See HgCl.
Ca: See $CaCl_2$; — HgCl.
Cl: See $HgCl_2$.
Cr: See $PbCl_2$; — CrO_3.
Co: See $CoCl_2$.
Cu: See $CuCl_2$; — HgCl; — $HgCl_2$.
Pb: See $PbCl_2$; — HgCl; — $HgCl_2$.
Mg: See HgCl; — $HgCl_2$.
Ni: See $NiCl_2$.
Pt: See $2NH_4Cl$, $PtCl_4$; — 2KCl, $PtCl_4$.
K: See HgCl; — $HgCl_2$.
Ag: See $HgCl_2$; — AgCl.
Na: See HgCl; — $HgCl_2$.
Sr: See Hg Cl; — $HgCl_2$.
Zn: See $ZnCl_2$.

[For the estimation of mercury see further HgCl; — $HgCl_2$; — $SnCl_2$.]

NICKEL: Separation of from

Sb: See $SbCl_5$; — $NiCl_2$.
As: See $AsCl_3$; — $NiCl_2$.
Ba: See NiCl.
Bi: See $BiCl_3$ (basic); — $NiCl_2$.
Ca: See $NiCl_2$.
Cr: See CrO_3; — BaO, CrO_3; — Na_2O, CrO_3.
Co: See Cl;—CuO, cobalticyanide; — Hg_2O, cobalticyanide; — K_2O, cobalticyanide.
Cu: See Cu.
Pb: See Cl.
Mn: See $MnCl_2$; — $NiCl_2$; — Cl.
Hg: See $NiCl_2$.
Pt: See $2NH_4Cl$, $PtCl_4$; — 2KCl, $PtCl_4$.
K: See 2KCl, $PtCl_4$.
Ag: See AgCl.
Sr: See SrO, CO_2; — $NiCl_2$.
Sn: See $NiCl_2$; — $SnCl_2$.

[For the estimation of nickel see further $NiCl_2$; — $SnCl_2$.]

NITROGEN: Separation of from

Cl: See AgCl.
C: See 2KCl, $PtCl_4$.

[For the estimation of nitrous acid se CrO_3.]

PHOSPHORUS: Separation of from

Cl: See AgCl.
Cu: See Cu.
Fe: See PCl_3.
K: See 2KCl, $PtCl_4$.
Ag: See AgCl.

P_2O_5: (estimation) See PbO, CrO_3.

PLATINUM: Separation of from

Sb: See $SbCl_5$.
As: See $AsCl_3$.
Bi: See $2NH_4Cl$, $PtCl_4$; — 2KCl, $PtCl_4$.
Cd: See $2NH_4Cl$, $PtCl_4$; — 2KCl, $PtCl_4$.
Cl: See HCl; — NaCl.
Cr: See CrO_3.
Co: See $2NH_4Cl$, $PtCl_4$; — 2KCl, $PtCl_4$.
Cu: See $2NH_4Cl$, $PtCl_4$; — 2KCl, $PtCl_4$.
Au: See 2KCl, $PtCl_4$.
Fe: See $2NH_4Cl$, $PtCl_4$; — 2KCl, $PtCl_4$.
Mg: See 2KCl, $PtCl_4$.
Mn: See $2NH_4Cl$, $PtCl_4$; — 2KCl, $PtCl_4$.
Hg: See $2NH_4Cl$, $PtCl_4$; — 2KCl, $PtCl_4$.
Ni: See $2NH_4Cl$, $PtCl_4$; — 2KCl, $PtCl_4$.
Ag: See AgCl.
Sn: See $SnCl_4$.
Ur: See $2NH_4Cl$, $PtCl_4$; — 2KCl, $PtCl_4$.
Zn: See $2NH_4Cl$, $PtCl_4$; — 2KCl, $PtCl_4$.

[For the estimation of Pt see further $PtCl_4$; $2NH_4Cl$, $PtCl_4$; — 2KCl, $PtCl_4$.]

POTASSIUM: Separation of from

Al: See 2KCl, $PtCl_4$.
Sb: See $SbCl_3$.
As: See $AsCl_3$; — 2KCl, $PtCl_4$.
Ba: See 2KCl, $PtCl_4$.
Bi: See $BiCl_3$ (basic).
Bo: See 2KCl, $PtCl_4$.
Ca: See 2KCl, $PtCl_4$.
C: See KCl.
Cl: See HCl; — KCl.
Cr: See KCl; — CrO_3; — $(NH_4)_2O$, CrO_3; — Hg_2O, CrO_3.
Co: See 2KCl, $PtCl_4$; — CuO, cobalticyanide; — Hg_2O, cobalticyanide.
Cu: See 2KCl, $PtCl_4$; — Cu.
H: See KCl.
Fe: See 2KCl, $PtCl_4$.
Pb: See Cl.
Li: See LiCl; — 2KCl, $PtCl_4$.

Mg: See $MgCl_2$; — $HgCl_2$; —AgCl; —2KCl, $PtCl_4$.
Mn: See Cl.
Hg: See HgCl; — $HgCl_2$.
Ni: See 2KCl, $PtCl_4$.
P: See Ag_2O, CO_2; — 2KCl, $PtCl_4$.
Si: See KCl.
Ag: See AgCl.
Na: See KCl; — AgCl; — 2KCl, $PtCl_4$.
Sr: See SrO, CO_2; — 2KCl, $PtCl_4$.
S: See KCl; — 2KCl, $PtCl_4$.
Zn: See ZnO, CO_2; — 2KCl, $PtCl_4$.

[For the estimation of K see further KCl; 2KCl, $PtCl_4$.]

SILICON: Separation of from

Cl: See AgCl.
K: See KCl.
Na: See NaCl.

[For the decomposition of refractory silicates see Na_2O, CO_2.]

SILVER: Separation of from

Al: See AgCl.
NH_4: See AgCl.
Sb: See $SbCl_5$; — AgCl.
As: See $AsCl_3$; — AgCl.
Ba: See AgCl.
Bi: See $BiCl_3$; — AgCl.
Cd: See AgCl.
Ca: See AgCl.
Cl: See AgCl.
Cr: See AgCl; CrO_3.
Co: See AgCl.
Cu: See AgCl.
Au: See AgCl.
Fe: See AgCl.
Pb: See $PbCl_2$ — AgCl.
Mg: See AgCl.
Mn: See AgCl.
Hg: See $HgCl_2$; — AgCl.
Ni: See AgCl.
P: See AgCl.
Pt: See AgCl.
K: See AgCl.
Se: See AgCl.
Na: See AgCl.
Sr: See AgCl.
S: See AgCl.
Te: See AgCl.
Sn: See AgCl; — $SnCl_4$.
Ur: See AgCl.
Zn: See AgCl.

SODIUM: Separation of from

NH_4: See $2NH_4Cl$, $PtCl_4$.
Sb: See $SbCl_3$.
As: See $AsCl_3$.
Bi: See $BiCl_3$ (basic).
C: See Na_2O, CO_2; — NaCl.
Cl: See HCl; — NaCl.
Cr: See NaCl; — CrO_3; — $(NH_4)_2O$, CrO_3; — Hg_2O. CrO_3.
Cu: See Cu.
H: See NaCl.
Pb: See Cl.
Li: See Li Cl.
Mg: See $MgCl_2$; — $HgCl_2$; — AgCl.
Mn: See Cl.
Hg: See HgCl; — $HgCl_2$.
P: See Ag_2O, CO_2.
K: See KCl; — AgCl; — NaCl; — 2KCl, $PtCl_4$.
Si: See NaCl.
Ag: See AgCl.
Sr: See SrO, CO_2.
S: See NaCl.
Zn: See ZnO, CO_2.

[See further Carbonate of Sodium for the estimation of Na.]

STRONTIUM: Separation of from

Al: See Na_2O, CO_2.
NH_4: See $2NH_4Cl$, $PtCl_4$.
Sb: See $SbCl_3$.
As: See $AsCl_3$.
Ba: See BaO, CrO_3.
Bi: See $BiCl_3$ (basic).
Ca: See Cl.
C: See SrO, CO_2.
Cl: See HCl.
Cr: See CrO_3; — $(NH_4)_2O$, CrO_3; — PbO, CrO_3; — Hg_2O, CrO_3; — Na^2O, CrO_3; — SrO, CrO_3.
Co: See SrO, CO_2; — $CoCl_2$; — Co.
Cu: See Cu.
Pb: See Cl.
Mg: See SrO, CO_2.
Mn: See SrO, CO_2; — Cl.
Hg: See HgCl; — $HgCl_2$.
Ni: See SrO, CO_2; — $NiCl_2$.
K: See SrO, CO_2; — 2KCl, $PtCl_4$.
Ag: See AgCl.
Na: See SrO, CO_2.

Zn: See SrO, CO_2; — ZnO, CO_2.

[For the estimation of Sr see further SrO, CO_2.]

SULPHUR: Separation of from

C: See CuO, CrO_3.
Cl: See AgCl.
Cr: See CrO_3; — PbO, CrO_3.
Cu: See Na_2O, CO_2; — Cu.
Fe: See Na_2O, CO_2.
K: See KCl; — 2KCl, $PtCl_4$.
Ag: See AgCl.
Na: See NaCl.
Zn: See $ZnCl_2$.

[For the separation of S from metals see SCl.]

[For the estimation of H_2SO_4 see further AgCl; — BaO, CrO_3; —PbO, CrO_3; —of SO_2 see $SnCl_2$; —Cl; — of $H_2S_2O_2$ see Cl.

TIN: Separation of from

Ca: See $CaCl_2$.
Cr: See CrO_3,
Co: See $CoCl_2$; — $SnCl_4$.
Cu: See $CuCl_2$; — $SnCl_4$.
Au: See $SnCl_4$.
Pb: See $PbCl_2$; — $SnCl_4$.
Ni: See $NiCl_2$; — $SnCl_4$.
Pt: See $SnCl_4$.
Ag: See AgCl; — $SnCl_4$.
Sn: See $SnCl_2$.
Zn: See $ZnCl_2$.

[For the estimation of tin see further $SnCl_2$; — K_2O, $2CrO_3$.]

WATER: Compare Hydrogen.

[For methods of estimating water see $CaCl_2$; $CoCl_2$; — NaCl.]

ZINC: Separation of from

Al: See ZnO, CO_2.
Sb: See $ZnCl_2$.
As: See $ZnCl_2$.
Ba: See ZnO, CO_2.
Bi: See $BiCl_3$ (basic).
Ca: See ZnO, CO_2.
Cr: See CrO_3;—BaO, CrO_3; — Na_2O, CrO_3.
Co: See ZnO, cobalticyanide.
Cu: See Cu.
Pb: See Cl.
Mg: See ZnO, CO_2.
Mn: See ZnO, CO_2; — Cl.
Hg: See $ZnCl_2$.
Pt: See $2NH_4Cl$, $PtCl_4$; — 2KCl, $PtCl_4$.
K: See 2KCl, $PtCl_4$.
Ag: See AgCl.
Na: See ZnO, CO_2.
Sr: See SrO, CO_2; — ZnO, CO_2.
S: See $ZnCl_2$.
Sn: See $ZnCl_2$.
Va: See ZnO, CO_2.

[For the estimation of Zn see further ZnO, CO_2.]

www.ingramcontent.com/pod-product-compliance
Lightning Source LLC
LaVergne TN
LVHW050517100826
845148LV00002B/366